SOIL ANALYSIS

BOOKS IN SOILS, PLANTS, AND THE ENVIRONMENT

Series Editor
G. STOTZKY
Department of Biology
New York University
New York, New York

Soil Biochemistry, Volume 1, edited by A. D. McLaren and G. H. Peterson

Soil Biochemistry, Volume 2, edited by A. D. McLaren and J. Skujiņš

Soil Biochemistry, Volume 3, edited by E. A. Paul and A. D. McLaren

Soil Biochemistry, Volume 4, edited by E. A. Paul and A. D. McLaren

Soil Biochemistry, Volume 5, edited by E. A. Paul and J. N. Ladd

Soil Biochemistry, Volume 6, edited by Jean-Marc Bollag and G. Stotzky

Organic Chemicals in the Soil Environment, Volume 1, edited by C. A. I. Goring and J. W. Hamaker

Organic Chemicals in the Soil Environment, Volume 2, edited by C. A. I. Goring and J. W. Hamaker

Humic Substances in the Environment, by M. Schnitzer and S. U. Khan

Microbial Life in the Soil: An Introduction, by T. Hattori

Principles of Soil Chemistry, by Kim H. Tan

Soil Analysis: Instrumental Techniques and Related Procedures, edited by Keith A. Smith

Soil Reclamation Processes: Microbiological Analyses and Applications, edited by Robert L. Tate III and Donald A. Klein

Symbiotic Nitrogen Fixation Technology, edited by Gerald H. Elkan

Soil-Water Interactions: Mechanisms and Applications, edited by Shingo Iwata, Toshio Tabuchi, and Benno P. Warkentin

Soil Analysis: Modern Instrumental Techniques, Second Edition, edited by Keith A. Smith

Soil Analysis: Physical Methods, edited by Keith A. Smith and Chris E. Mullins

Growth and Mineral Nutrition of Field Crops, edited by N. K. Fageria, V. C. Baligar, and Charles Allan Jones

Semiarid Lands and Deserts: Soil Resource and Reclamation, edited by J. Skujins

Plant Roots: The Hidden Half, edited by Yaov Waisel, Amram Eshel, and Uzi Kafkafi

Additional Volumes in Preparation

SOIL ANALYSIS

Physical Methods

edited by
KEITH A. SMITH
Edinburgh School of Agriculture
Edinburgh, Scotland

CHRIS E. MULLINS
University of Aberdeen
Aberdeen, Scotland

MARCEL DEKKER, INC.   NEW YORK · BASEL · HONG KONG

Library of Congress Cataloging-in-Publication Data

Soil analysis: physical methods / edited by Keith A. Smith, Chris E. Mullins.
    p.   cm. -- (Books in soils, plants, and the environment)
    Includes bibliographical references and index.
    ISBN 0-8247-8361-1
    1. Soil physics--Methodology. I. Smith, Keith A.
II. Mullins, Chris E.   III. Series.
S592.3.S66 1991
631.4'1'0287--dc20   90-47074
                                         CIP

This book is printed on acid-free paper.

MARCEL DEKKER, INC.
270 Madison Avenue, New York, New York 10016

Current printing (last digit):
10 9 8 7 6 5 4 3 2

PRINTED IN THE UNITED STATES OF AMERICA

Preface

In soil physics more than in most subjects there has been a strong symbiotic relationship between the development of theory and practical applications on the one hand and the development of experimental methods on the other. For example, acceptance and use of the concept of matric potential, originally outlined by Buckingham in 1907, were greatly accelerated following Richards' description in 1942 of simple methods for constructing reliable tensiometers. More recently—over the past 20 years—the development of the neutron probe, which allows rapid nondestructive field measurement of soil water content, has led to an enormous upsurge of interest in soil water measurements and their practical applications and to the accelerated development of the theory of soil water movement. In comparison with soil chemistry and mineralogy, the development of techniques for soil physical measurements has been held back by the necessity for many measurements to be carried out in the field. Other techniques that require reliable "undisturbed" cores or samples on which to make laboratory measurements have also required considerable time and experience to perfect. However, the detailed specification of a number of soil physical measurement techniques, which is currently being undertaken by the Soil Quality working group of the International Standards Organisation, is an indication of both the increasing importance of soil physical measurements and the coming of age of a number of physical measurement methods. These methods can be regarded as being relatively stable, in the sense that their limitations, reliability, and areas of application are now well established.

Despite these current attempts at standardization, the choice of appropriate soil physical measurement techniques is often still uncharted ground full of pitfalls. Abandoned and unused soil physical equipment in research laboratories provides a sorry testimony to the difficulties faced by researchers who have been unable to obtain adequate advice on choice of appropriate methods. There are certainly many books, reviews, and technical notes that provide listings and descriptions of soil physical methods. However, most users require a brief guide to any necessary theory and to the techniques most suitable to their own particular application before they can make a choice and are ready to consult one of the many available compilations of methods. This book is written with the research scientist, agricultural or environmental adviser, and postgraduate student in mind: people who have some general background in soils but may not have received training in soil physics. Our aim has been to provide all the information required to allow the user to choose the technique most suited to his or her desired application. Each chapter contains a critical review of relevant theory and of measurement methods. Particular emphasis has been given to the merits, limitations, and range of application of each method. As well as the consideration of accuracy, this includes measurement time, ease of use, and cost. Lists of suppliers and references to construction details and to papers that describe the detailed use of each method are also provided.

Because of the close relationship between theory, measurement techniques, and applications in many areas of soil physics, this book should also provide a useful source of reference for undergraduate soil scientists. No collection of soil physical methods will cover everyone's requirements, and we have deliberately chosen to place an emphasis on soil physical factors that affect plant growth. Since a study of the plant root system itself may provide one of the best indications of whether soil physical factors are influencing plant growth, we have also chosen to include a chapter on the measurement of root systems in the soil.

We thank the contributors to this volume for their efforts, and the members of our families for their patience and tolerance when our attention to them was supplanted by our efforts to complete this work.

Keith A. Smith
Chris E. Mullins

Contents

Contributors

David Atkinson Department of Land Resources, The School of Agriculture, University of Aberdeen, Aberdeen, Scotland

Bruce C. Ball Soil Engineering Department, Scottish Centre of Agricultural Engineering, Penicuik, Midlothian, Scotland

John P. Bell Institute of Hydrology, Natural Environment Research Council, Wallingford, Oxfordshire, England

A. Glyn Bengough* Department of Plant and Soil Science, University of Aberdeen, Aberdeen, Scotland

Graeme D. Buchan Soil Science Department, Lincoln University, Canterbury, New Zealand

Donald J. Campbell Soil Engineering Department, Scottish Centre of Agricultural Engineering, Penicuik, Midlothian, Scotland

Andrée D. Carter Soil Survey and Land Research Centre, Shardlow, Derby, England

J. David Cooper Institute of Hydrology, Natural Environment Research Council, Wallingford, Oxfordshire, England

Tom J. Dean Institute of Hydrology, Natural Environment Research Council, Wallingford, Oxfordshire, England

---

*Current affiliation: Soil-Plant Dynamics Group, Cellular and Environmental Physiology Department, Scottish Crop Research Institute, Dundee, Scotland

Christiaan Dirksen   Department of Hydrology, Soil Physics and Hydraulics, Wageningen Agricultural University, Wageningen, The Netherlands

Catriona M. K. Gardner   Institute of Hydrology, Natural Environment Research Council, Wallingford, Oxfordshire, England

Nigel Gardner   Social Studies Faculty Centre, University of Oxford, Oxford, England

J. Kenneth Henshall   Soil Engineering Department, Scottish Centre of Agricultural Engineering, Penicuik, Midlothian, Scotland

Martin G. Hodnett   Institute of Hydrology, Natural Environment Research Council, Wallingford, Oxfordshire, England

Peter J. Loveland   Soil Survey and Land Research Centre, Cranfield Rural Institute, Silsoe, Bedfordshire, England

Lorna A. Mackie-Dawson   Plants Division, The Macaulay Land Use Research Institute, Aberdeen, Scotland

Chris E. Mullins   Department of Plant and Soil Science, University of Aberdeen, Aberdeen, Scotland

Michael F. O'Sullivan   Soil Engineering Department, Scottish Centre of Agricultural Engineering, Penicuik, Midlothian, Scotland

Malcolm J. Reeve   Soil Survey and Land Research Centre, Shardlow, Derby, England

Keith A. Smith   Department of Soil Science, The Edinburgh School of Agriculture, Edinburgh, Scotland

W. Richard Whalley   Physics Division, AFRC Institute of Engineering Research, Silsoe, Bedfordshire, England

Edward G. Youngs*   Soils Division, Rothamsted Experimental Station, Harpenden, Hertfordshire, England

---

Current affiliation: Department of Agricultural Water Management, Silsoe College, Cranfield Institute of Technology, Silsoe, Bedfordshire, England

SOIL ANALYSIS

1

Soil Water Content

CATRIONA M. K. GARDNER, JOHN P. BELL, J. DAVID COOPER,
TOM J. DEAN, and MARTIN G. HODNETT *Institute of Hydrology,
Wallingford, Oxfordshire, England*

NIGEL GARDNER *Social Studies Faculty Centre, University of
Oxford, Oxford, England*

I. INTRODUCTION

Measurement of soil water content is fundamental to many agricultural, forestry, hydrological, and civil engineering investigations
of soils. Requirements vary from the need for occasional determinations in the laboratory to continuous field monitoring at several
depths and in different locations. Several techniques have been
devised in response to these needs, but two are preeminent: the
thermogravimetric procedure for conducting measurements on samples in a laboratory, and the neutron method for in situ measurement. The principles of both are described in detail. Brief
accounts of the alternatives that are commonly used are also provided, but greater attention is devoted to recent developments in
the electrical capacitance and time domain reflectometry approaches,
and to the use of remote sensing techniques. A short review of
the concept of soil water content and its usage, and the different
approaches to its measurement and to field sampling precedes the
discussion of measurement methods.

Several reviews of soil water content measurement have been
published in the past, notably those by W. H. Gardner [1] and
Cope and Trickett [2]. More recently, another review has been
published by Schmugge et al. [3], Morrison [4] has provided very
useful summaries of the theory and practice of many of the techniques available, and the contribution by W. H. Gardner has been
revised and updated [5].

A. Soil Water Content

1. Definition

In the following pages, the term "soil water content" is used in the widely accepted sense to refer to the water that may be evaporated from a soil by heating to between 100 and 110°C until there is no further weight loss; this is the basis of the thermogravimetric method. It is important to be aware of the arbitrary nature of this definition, which is the ultimate reference against which other techniques are normally calibrated. As W. H. Gardner stated, "the choice of this particular temperature range appears not to have been based upon scientific consideration of the drying characteristics of soil" [1]. Its origin probably has more to do with the notion of ensuring evaporation of liquid or "free" water, and it is likely that its adoption relates to the relative ease with which determinations can be made by oven-drying soil samples.

Water is present in soil as water vapor and liquid. In addition, water molecules are adsorbed on the surfaces of colloidal materials, particularly clays, and are incorporated with hydroxyl groups within clay lattice structures. (The distinctions between thin films of water retained by surface tension and water that is adsorbed, and between adsorbed and structural water are, however, less precise than suggested.) Water vapor and structural water are disregarded in the conventional definition of soil water content. Structural water is immobile and is released only upon mineral decomposition, which requires heating to temperatures between 400 and 800°C. There are some exceptions; for example, structural water is eliminated from gypsum upon heating to only 80°C.

Adsorbed water does have a degree of mobility which becomes important at low water contents and may be exploited by drought-resistant plants. Heating to 110°C is not normally sufficient to remove adsorbed water; most is eliminated from clay surfaces at temperatures between 110 and 160°C. Heating also promotes the oxidation and decomposition of soil organic matter, however, and whereas this occurs to some extent at temperatures below 110°C, it is enhanced by higher temperatures. When dealing with organic soils, some inaccuracy is probable in the traditional determination because part of the weight loss may be due to organic matter changes. Alternative definitions of soil water content should be considered, therefore, when working with organic soils and soils containing gypsum.

The thermogravimetric definition of soil water content is not a disadvantage in most work because it is usually relative quantities of water in soil that are of interest. In many studies it is changes with time of the water held in a soil profile that are important. In others it is the quantity of water retained between the liquid and

and plastic limits, or between the "field capacity" and "wilting point" conditions, that are required. Changes in the quantity of water adsorbed on soil colloids do not normally occur under field conditions and so can be ignored. Several methods of water content determination are sensitive to all the water molecules present in a soil, but this information is effectively lost as they are calibrated against thermogravimetric determinations; it is subsumed into the intercept term of the linear calibrations derived for these techniques.

2. Units

Soil water content may be expressed either on a gravimetric or a volumetric basis, that is, kilograms per kilogram or cubic meters per cubic meter, respectively. In either case the value derived is dimensionless and can be regarded as a fraction, or a percentage. One can be obtained from the other if the soil bulk density is known

$$\theta = w\rho \tag{1}$$

where θ is volumetric soil water content, w the soil water content by weight, and ρ the appropriate dry soil bulk density.

Although recording of water content on a weight basis is common when referring to samples that have been chemically analyzed, for many other purposes expression on a volumetric basis is more useful, as multiplying by the soil depth gives the "depth" of water in this depth of soil, a figure compatible with the units used to measure rainfall, evaporation, transpiration, drainage, and irrigation.

If water content is monitored at several depths in a soil profile the depth interval, z_i, to which a measurement θ_i refers, is normally taken as the vertical distance between the two midpoints interposed between the measurement depth and the depths of the measurements immediately above and below it. The water content of soil profile, P, to a depth z, is obtained by summation of the water contents of each depth interval

$$P = \sum_{0}^{z} \theta_i z_i \tag{2}$$

The effect of this essentially is integration of a step function of the water content and is equivalent to trapezoidal integration. It has been shown that use of Simpson's rule can reduce the errors arising in this calculation [6]. Some soil water content measurement techniques inherently measure volumetric soil water content (e.g., the neutron method). Yet, because of the use of the thermogravimetric standard, soil bulk density measurements are also required to convert from a dry weight to volumetric basis before the calibration can be performed. This procedure introduces an additional source of error into the calibration, a point that is returned to in Section II.A. A technique can be no more accurate than the procedure used to calibrate it; hence particular care is required in both the density and the water content determinations when undertaking calibrations.

B. Direct and Indirect Measurements

Direct measurements entail removal of the soil water by evaporation, leaching, or a chemical process, and subsequent determination of the amount removed; the thermogravimetric method is the principal example. Direct measurements are beset with problems, principally due to the necessity for destructive sampling. Measurements cannot be repeated on the same sample of soil; hence replicate samples must be taken from a plot at any one time to determine the variance of the measurements at that time, and so to permit the analyst to ascertain whether they differ significantly from determinations on other occasions. This need for replication can result in the handling of very large numbers of samples. Practical difficulties are compounded if determinations deep in the profile are required. Furthermore, repeated sampling within the same area may well cause unacceptable damage to a crop or the soil. Provision must also be made for bulk density determinations if volumetric water content data are required.

Indirect methods depend on monitoring of some soil property that is a function of water content; for example, the neutron method is based on the detection of hydrogen nuclei in the soil, most of which are present in water molecules. Indirect methods may utilize either instrumentation placed in or on the soil or remote sensing techniques that involve sensors mounted on a platform over the soil, or on aircraft or satellites. Although indirect methods require calibration to yield water content data, some have the considerable advantage that measurements in situ are possible, and these can be repeated in the same position at other times. Other indirect methods require samples taken from the field but enable determinations to be conducted more rapidly than by the thermogravimetric procedure.

The principal advantage of in situ indirect measurements is that they give measures of change in soil water content directly. The variance in a change of water content obtained from repeated measurements on the same sample is simply

$$\text{est}[\text{var}(\Delta\theta)] = \text{var}(\Delta\theta) + \text{errors} \qquad (3)$$

whereas the variance associated with a water content change measured by making direct measurements on two sets of samples depends on the combination of the variances attached to each set of samples

$$\text{est[var}(\Delta\theta)] = \frac{2 \ \text{var}(\theta)}{\sqrt{n_1 n_2}} + \text{errors} \tag{4}$$

where n_1 and n_2 represent the number of samples within each set. If water content changes are the focus of interest, in situ measurements are therefore preferable.

The advent of a reliable, in situ field method, the neutron method, was consequently welcomed and widely adopted. The neutron method is not without its disadvantages, however; the method itself has some limitations in addition to its capital cost and the always-present radiation risk. These features have discouraged its use for continuous, automatic, in situ monitoring but have encouraged a continued exploration for alternative approaches. The methods being introduced today are based, in the main, on principles that have been known for years; the considerable recent improvements and innovations in electronics have permitted their development as reliable tools.

C. Sampling

Soil water content varies both with depth and areally on a variety of superimposed scales, but conventional methods provide point or approximately "point" measurements. Consideration should be given to the selection of measurement depths and to a strategy for sampling the experimental plot, field, or catchment spatially, to give the required accuracy in relation to the observed variance in water content. The sampling strategy devised will reflect the aim of the study and practical constraints such as staff and equipment availability, as well as the precision of the instrument.

Remote sensing techniques are suited only to estimates of near-surface moisture levels over large areas, but they avoid the problems associated with spatial extrapolation from point to areal estimates.

II. THERMOGRAVIMETRIC AND OTHER DIRECT METHODS

A. Thermogravimetric Determinations

The thermogravimetric method is very straightforward. A soil sample is placed in a heat-proof dish of known weight, weighed, dried in an oven set at a temperature of 100-110°C, removed and allowed to cool in a dessicator, then reweighed. This procedure is repeated until the sample attains a constant weight. The water content w of the sample is the mass of water per unit mass of dry soil

$$w = \frac{\text{mass of wet soil} - \text{mass of dry soil}}{\text{mass of dry soil}} \tag{5}$$

A good oven, which is ventilated and distributes the heat evenly, is required. It should not be overloaded. The efficiency of an
oven can be checked by loading it with subsamples of one mixed soil
sample and then checking the variation in the water contents measured [7]. The time required to dry a soil sample to constant weight
is influenced by the oven's efficiency and the size, condition, and
number of samples in it. If large sample numbers are involved, it
is easier to use the same drying time for all samples. An appropriate time can be determined by experimentation with a few samples
of the type being treated. In fact, an oven temperature of 105°C
and a 24-hour drying period are widely adopted standards used by,
for example, the Soil Survey of England and Wales [8]. However,
Reynolds [9], in a study of the thermogravimetric procedure found
that although 24 hours was adequate for most soils, a few took nearly 48 hours to reach a constant weight at 105°C. If the requirement
is to determine the water content of soil samples that have been air-
dried in preparation for chemical analyses, the drying time can
usually be reduced to between 6 and 12 hours, with an oven temperature of 105°C [8].

A constant weight can be assumed to have been achieved when
the changes between repeat weighings are less than 0.1% of the
original sample weight [10]. The interval between weighings should
be at least half an hour [10]; the British Standard for this method
of soil water content measurement states that a period of 4 hours
should elapse between successive weighings [11].

W. H. Gardner has provided an analysis of the random errors
accompanying gravimetric water content determinations associated
with varying degrees of weighing precision and accuracy [5]. Different authors have proposed the use of various sample sizes for
water content determinations ranging from 10 to 100 g and greater
for stony soils [12]. Reynolds [13] concluded that samples of 50-
100 g were appropriate, having experimented with samples of up to
500 g, and found that these took longer to dry but gave no improvement in results. Use of larger samples from stony soils is advisable;
the American Society for Testing and Materials has recommended minimum sample sizes according to the dimensions of the aggregates and
stones in the moist soil [14].

Reinhart [15] addressed the problem of water content measurement of stony soils caused by variation of the proportion of stone
in different samples, and his procedure is often used [16]. Separate determinations of the water content of the < 2 mm fraction and
of the volumetric stone content S (> 2 mm) are made; S may be estimated either by sieving large samples or by inspection of the sides
of a large soil pit. The volumetric water content of the fraction
smaller than 2 mm ($\theta_{<2}$) is then corrected by S to give a water
content estimate for the field soil

$$\theta = \theta_{<2}(1 - S) \tag{6}$$

The water content of the stone fraction θ_s is often considered to be negligible but will not be if the stones are porous [17]. It is then necessary to include θ_s in the water content estimate

$$\theta = \theta_{<2}(1 - S) + \theta_s S \tag{7}$$

To express soil water content on a volumetric basis, the dry bulk density of the soil must be known. Normally this is determined on the same sample by taking a sample of known volume. The error associated with volumetric water content determination is derived from the combined variances of the measurements of gravimetric water content and dry bulk density ρ.

$$\text{var}(\theta) = w^2 \, \text{var}(\rho) + \rho^2 \, \text{var}(w) \tag{8}$$

If these measurements are not made on the same sample, further errors may be introduced [18].

Because of its simplicity, the oven-drying method is easily abused. In particular, oven temperatures may not be regularly checked, and neither they, nor drying times, are usually reported. Yet because of the frequent use of thermogravimetric determinations as a reference against which to calibrate and investigate the accuracy of indirect methods of water content measurement, care should be taken in using this method and when comparing other results with it [5].

The advantages of the oven-drying gravimetric method are its simplicity, reliability, and low cost in terms of equipment requirements. Its major disadvantage is the necessity for destructive sampling. However, the time that the procedure takes is also a considerable disadvantage, and several attempts to reduce the time required to dry soil samples have been made.

In Bouyoucos's alcohol-burning method [19], the soil sample is saturated with an alcohol, which is then burnt, and the heat generated evaporates the soil water. This procedure is repeated until the sample reaches a constant weight. A complete determination can be made in about 20 minutes—more rapidly if several determinations are run concurrently [20].

Microwave ovens may be used and reduce drying times dramatically. However there are two problems:

1. Drying time increases with initial moisture content.
2. If a dry sample is left in a microwave oven, its temperature will continue to rise, resulting in weight changes other than those due to the evaporation of water.

Consequently, drying times have to be estimated for different types of sample. This method provides rapid results but is less accurate than use of a conventional oven [21].

B. Other Direct Methods

The principles of the alternative direct methods were summarized by W. H. Gardner [5]. They require one of three techniques:

1. Removal of the water from the sample, either by distillation or absorption in a desiccant and measurement of the quantity removed
2. Displacement of the water by another liquid and measurement of some property of the liquid
3. Measurement of the reaction, or reaction products displaced from a sample when a reagent is added

The techniques have recently been reviewed by Morrison [4], but most appear to be little used. The calcium carbide method, whereby the pressure of the acetylene gas produced when calcium carbide reacts with soil water is correlated with water content, is used for civil engineering purposes and is occasionally recommended for agricultural use [22]. Gas pressure chambers for conducting this method in the field are available commercially. The sulfuric acid procedure proposed by Baver [23], in which concentrated sulfuric acid is mixed with wet soil and the maximum temperature attained is recorded, has recently been evaluated [24]. It was found to give results that correlated well with the thermogravimetric method but were influenced by the presence of carbonate, or much organic matter or chloride. Both these methods are rapid, taking about 5 minutes per sample; they can be conducted in the field and require little equipment. They are suitable when a quick answer is more important than high accuracy.

III. NEUTRON METHOD

The neutron method for water content measurement makes use of the ability of hydrogen to slow down fast neutrons, emitted from a radioactive source, much more efficiently than other substances. The source and a slow neutron detector are normally housed in one unit, a "probe," which is lowered down a tube known as an "access tube," inserted into the soil. Some neutron meters are designed for use on the soil surface and are constructed to a different design. Fast neutrons are moderated (i.e., they are slowed down to "thermal" energies) upon collision with atomic nuclei, in particular hydrogen nuclei. Most of the hydrogen in soil is present in water molecules; changes in hydrogen concentration occur therefore, mainly as a result of changes in water content. Thus when a probe is used in soil, the number of fast neutrons that are slowed close to the source and the detector, and so detected, increases with an increase in water content. The number of slow neutrons detected is recorded as a "count rate," which is read by the user from a display or may be stored in a memory, which is interrogated later with a microcomputer. The count rate is converted to a water content by use of a soil-specific calibration relationship. For measurements at depth, access tubes are usually installed semipermanently. Within each, readings are taken at successive depths by lowering the probe down the access tube. Although measurements are most often made to about 1.5 m depth, they can easily be conducted at much greater depths in many soils, once the effort of access tube installation has been completed.

The neutron method was first proposed in the 1940s [25,26]. The first field tests of a neutron probe were published in 1950 [27], and this experience was soon built upon. By the mid-1950s portable instruments for field use had been developed in laboratories in North America [28,29] and Australia [30]. Commercial equipment soon became available. However, the instruments available today represent considerable refinements of the early ones, being more rugged, reliable, and stable, yet lighter and safer in terms of radioactivity, and recently full advantage has been taken of the progress in microprocessor technology to store and process the data.

In 1970 the International Atomic Energy Agency (IAEA) published a technical report on neutron moisture probes [31], which remains one of the most useful publications on the more theoretical aspects of this subject. Bell's contribution [32] on neutron probe practice provides a complementary guide to field usage of probes. Elements of both reports were more recently incorporated into a publication edited by Greacen [33], which reviews much of the pre-1981 literature and provides a lot of practical detail based on the experience of the authors. The present account aims to give an overview of the neutron method of soil water measurement; the aforementioned publications are frequently referred to as sources of further information and should be consulted if greater detail is required. The emphasis here is on the use of depth probes, which enable measurements to be made throughout the soil profile. Surface meters and their use are considered briefly in Section III.F. The theory of neutron moderation in soils is pertinent to both types of instrument. Dual-purpose probes, which measure soil bulk density by gamma-ray attenuation, as well as water content by neutron moderation, are available. The principles of the water content measurement are the same as detailed here; Chapter 7 describes the theory and use of gamma probes.

Neutron probe measurements are used in a wide variety of applications, not all of them relating to soils. Probably the most common use of the probe is to measure changes in soil profile water content at one or several places within a plot, field, or stream catchment, over one or more seasonal cycles. These data may then be used alone, or in combination with meteorological, tensiometer, and stream-flow measurements, to obtain an overall water balance for the relevant area [34-38]. They may be used to evaluate or calibrate simple models of soil water or catchment behavior [39,40]. The neutron probe does, however, provide a measure of both the water content distribution and the distribution of water content changes within a soil profile. There is therefore considerably more information available than can be had from a single figure for profile water content. This information can be used to build and evaluate more sophisticated models of water movement and plant water use, and to measure hydraulic properties of the soil in situ [41].

A. Neutrons and Neutron Moderation

A neutron is an uncharged particle of mass very slightly greater than a proton. Neutrons are arbitrarily classified according to their kinetic energy, measured in units of electron volts, as follows.

Fast: >1 keV
Intermediate: 2 eV to 1 keV
Epithermal: 0.5-2 eV
Thermal: 0.025-0.5 eV

The energy range, or energy spectrum, of a cloud of neutrons is described in probabilistic terms; there is only a probability that an individual neutron will possess a given amount of energy. Thermal neutrons are in equilibrium with the molecules of the surrounding medium, and their movement through the medium is controlled by the same diffusion laws that apply to the diffusion of gas molecules.

1. Neutron-Nuclei Interactions

Because they have no electrical charge, neutrons are not influenced by electrical fields. They undergo a variety of interactions with atomic nuclei when they come close to, or "collide" with, a nucleus. The nature of the interaction between a neutron and a nucleus depends on the energy of the neutron and the characteristics of the nucleus. The probability that collisions resulting in one type of interaction will occur when an isotope is irradiated with neutrons of given energy is defined by the interaction cross-section of the isotope. Interaction cross-sections are measured in units of area called barns; 1 barn is 10^{-28} m^2. The larger the cross-section, the greater is the probability that interactions will occur between the neutrons and nuclei.

The interaction cross-section of an individual nucleus is known as its microscopic cross-section. Values can be determined for all modes of neutron-nuclei interaction and neutron energy levels. The macroscopic cross-section of a unit volume (1 m^3) of soil for a particular type of interaction with neutrons of a given energy is calculated as the weighted sum of the microscopic cross-sections of the elements present.

Neutron-nuclei interactions take several forms but can be grouped under two headings: neutron scattering and neutron capture. Scattering takes place when a fast neutron collides with a nucleus, causing the direction of travel of the neutron to be changed and its velocity, hence its energy, to be reduced. Such collisions may be elastic—that is, kinetic energy and momentum are conserved—or inelastic, as when some of the neutron's energy is transferred to the interior of the nucleus, leaving it in an excited state and resulting in the emission of a photon of gamma radiation. Inelastic scattering is an insignificant process relative to elastic scattering, in the present context, and so is not discussed further. The elastic scattering microscopic cross-sections of most isotopes are small, less than 5 barns, and relatively constant at neutron energies between 2 eV and 2 MeV. Neutrons may also be "captured" in a collision; that is, they are absorbed by the nucleus as discussed in Section III.A below.

2. Elastic Scattering

The transfer of energy between fast neutrons and nuclei in the course of elastic scattering can be represented for a "head-on" collision by

$$E_1 - E_2 = \frac{4A}{(A + 1)^2} E_1 \tag{9}$$

where E_1 and E_2 represent the energy of the neutron before and after collision, respectively, and A is the mass of the nucleus in atomic mass units.

The loss of energy from the neutron is thus inversely related to the mass of the nucleus with which it collides. When such a collision takes place with a hydrogen nucleus, the neutron will lose all its energy. Less energy is lost in collisions with larger nuclei, as shown in Table 1. Heavier nuclei are most likely to cause a neutron to be deflected at a greater angle from its original path. This tends to reduce the distance neutrons move from a source during the slowing down process.

In reality, collisions occur at all angles, and many collisions are required to slow a fast neutron. Table 1 shows the average number

TABLE 1   The Effect of Elastic Collisions on Neutrons

Nucleus	Percentage energy lost in "head-on" collision	Average number of collisions required to slow 2 MeV neutron to < 0.5 eV
H	100	18
P	22.1	152
Si	13.8	252
Al	13.3	279
Fe	6.8	519

Source: From Ref. 42.

of collisions required to slow 2 MeV neutrons to thermal energy levels.   The table shows that hydrogen is the most efficient neutron moderator in soils.

3.   Neutron Capture

Two modes of neutron capture, or absorption, are important in soil water measurement.   Both involve epithermal and lower energy neutrons because the microscopic capture cross-sections of the majority of elements are insignificant at neutron energies above 1 eV.   One type of interaction is used in the detection of thermal neutrons; the other, known as radiative capture, causes the removal of thermal neutrons from the neutron cloud about the detector.   In slow-neutron detectors, neutrons are counted by monitoring the emission either of alpha particles or of protons, which takes place when neutrons are absorbed by ^{10}B, ^{6}Li, or ^{3}He, as described in the following section.

In radiative capture, a neutron is captured and absorbed by a nucleus, which then emits a photon of gamma radiation.   The radiative capture cross-sections of most soil constituents are between 0.1 and 1 barn, but a few have very large cross-sections, which contribute considerably to the macroscopic cross-section of the soil. These include gadolinium (46,000 barns), cadmium (2450 barns), boron (755 barns), chlorine (33 barns), manganese (33 barns), and iron (2.6 barns).   These elements occur in sufficient concentrations in some soils to have significant effects on the calibration.

B. Neutron Sources, Detectors, and
    Probe Geometry

1. Neutron Sources

Fast-neutron sources contain two elements: one emits alpha particles in the course of its radioactive decay; the other absorbs the
alpha particles and in the process emits neutrons. The second
"target" element is always beryllium. Its reaction with alpha particles is:

$$_4^9Be + {_2^4}He \rightarrow {_0^1}n + {_6^{12}}C + 5.74 \text{ MeV} \tag{10}$$

The neutron emitted gains some of the reaction energy plus some
of the alpha particle's energy.

Several alpha emitters are available, and the early workers experimented with many of them [43,44]. Today, the majority of
probes use an isotope of americium, ^{241}Am, as the alpha emitter.
It has a half-life of 458 years. Investigations continue into the
possibility of using curium and californium sources with less gamma
activity, such as ^{244}Cm with beryllium, and ^{252}Cf, which is a spon-
taneous neutron emitter but has a half-life of only 2.6 years [45].

With safety in mind, the activities of sources have been reduced
over the past three decades. Most probes now use 1.85 GBq (50 mCi)
or smaller sources. They are constructed to strict safety standards
and are known as "special form" sources. Beryllium, in the form of
a fine powder, and sintered americium oxide are contained within a
double-walled capsule of stainless steel, which is cylindrical or annular in shape. The working life of americium-beryllium sources is at
least 20 years, but regular tests for leakage should be conducted
[46].

2. Slow-Neutron Detectors

Detection efficiency is the prime requirement for thermal neutron
detectors used in probes. Greater efficiency enables use of lower
activity fast neutron sources. The isotopes ^{10}B, ^{3}He, and
^{6}Li all have very high probabilities of undergoing capture reactions
with neutrons of epithermal and lower energy levels, and are also
insensitive to high energy neutrons. Boron trifluoride and helium-
filled metal tube detectors are most common. Their capture reactions with slow neutrons are

$$_5^{10}B + {_0^1}n \rightarrow {_3^7}Li + {_2^4}He + 2.8 \text{ MeV} \tag{11}$$

$$_2^3He + {_0^1}n \rightarrow {_1^1}p + {_1^3}H + 760 \text{ KeV} \tag{12}$$

The detection efficiency of a detector falls very slowly with time; the useful life of a detector is at least 15 years. Helium-filled detectors are 4-5 times more efficient than similar boron trifluoride detectors [47] but more complicated to use electronically [32]. Boron trifluoride detectors are therefore often preferred. Both types require a stable voltage supply of 1-2 kV to operate. Lithium-enriched glass scintillation counters can give 100% efficiency of detection but require more complex electronics and are more delicate than gas counters. Their sensitivity to gamma radiation can be monitored independently of that to thermal neutrons by pulse height discrimination, so they can be used in a dual-purpose probe requiring only one detector, although much greater electrical stability is then necessary.

3. Probe Geometry

The arrangement of the source and detector within the probe contributes to its sensitivity to change in water content. Moreover, certain geometries result in a linear calibration over the range of soil water contents commonly encountered. Ideally, both source and detector would be placed at the same point, but the linearity of the calibration is acceptable if all parts of the detector lie within 60 mm of the source [32]. In some probes an annular source that fits around the midpoint of the detector tube is used, giving a symmetrical thermal neutron distribution about the detector. In other designs cylindrical sources are placed either adjacent to the midpoint of the detector, or below it. However there are two disadvantages associated with the latter arrangement. The detector is remote from the center of the neutron cloud to an extent that varies with soil water content, thus creating a nonlinear calibration. In addition, the influence on the readings of the soil-air interface and interfaces within the soil (described in Section III.C.2) is increased.

C. Instrument Design and Neutron Counting

1. Instrument Design

Nicolls et al. [48] have provided a very useful account of instrument design in relation to sensitivity, accuracy, precision and convenience of use. Probes designed for subsurface measurements comprise six parts, the probe, which is connected by a cable to the counting unit; the cable; the counting unit, which often contains the power supply; the power supply; the probe carrier; and finally a system for lowering the probe out of the carrier into an access tube and locating it at known depths (Fig. 1).

The probe comprises the neutron source, the detector, and some electronic circuitry. These are contained within a waterproof cylindrical metal casing, usually of aluminum or stainless steel. Probe

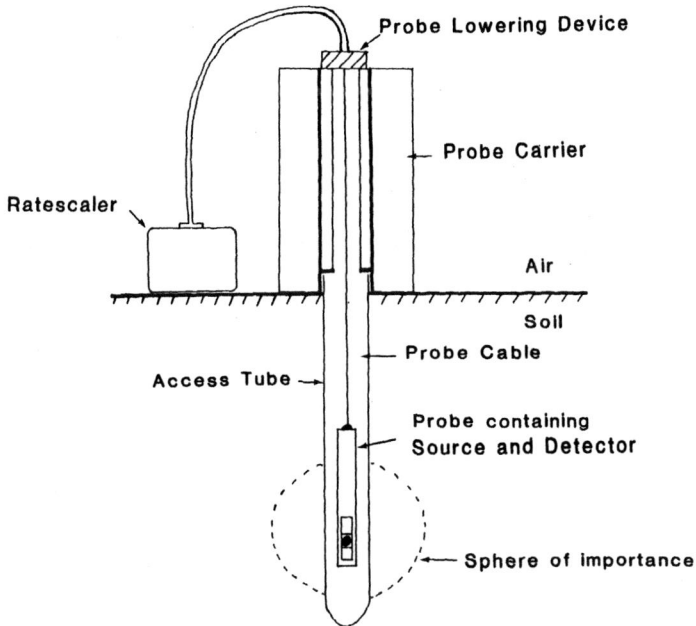

FIG. 1 The neutron probe: principal components. "Sphere of importance" designates the volume of soil that principally contributes to soil water measurement.

length varies, depending particularly on whether the pulse-discrimination circuitry and high-voltage generator are incorporated within it or in the counting unit. The most common size of probe outside diameter is 38 mm, designed for use in access tubes of outside diameter 44.5 mm (1.75 in.).

A strong waterproof cable is attached by waterproof connectors to the probe and the counter unit to relay power to the detector and pulses from it. Often the cable is also used to suspend the probe when it is lowered into access tubes. Designs that rely on the cable to carry the high voltage required by the detector are more susceptible to faults caused by moisture in the connections or by wear of the cable itself. It is preferable to have a cable that can be readily disconnected at both ends, to permit replacement if problems occur.

A counter unit measures the electronic pulses transmitted from the detector and displays the result. Ratemeters and scaler-timers have largely been superseded by ratescalers, which operate by counting for a preset time, typically between 8 and 64 seconds, and then calculating the count rate. Longer count times can be selected on some instruments for high-precision measurements. The power supply is normally housed within the counter unit, with the electronics necessary to provide the high voltage required by the detector, if these are not contained within the probe. Generally, rechargeable packs of nickel-cadmium or lead-acid cells are used.

The probe carrier stores the probe when it is not in use and limits radiation from it. Generally, the probe is locked into the carrier so that the source is surrounded by a shield of a hydrogen-rich material, usually polypropylene or paraffin wax. Designs vary; because of their shape and weight, some are easier than others to use in the field. The probe carrier also serves to locate the probe onto an access tube. A socket in the bottom of the carrier allows it to fit onto the top of the access tube, which protrudes from the ground. The probe passes through this and into the access tube.

The mechanism for lowering the probe into access tubes is fixed to the carrier. The probe may be suspended by the electrical connection cable, a fine chain, or a fine steel cable. Various methods are used to lower it to the required depth in the access tube. Some cables are marked or have stops at intervals which can be used to hold them. Friction devices, which actually record the amount of cable lowered, are more convenient to use and may be preferable to cable markings because errors resulting from cable stretching are avoided. However, greater precision of depth placement can be achieved by using stops on a separate wire that does not stretch.

2. Sphere of Importance

A neutron, emitted from the source of a probe, travels outward into the soil until it collides with an atomic nucleus. Some energy is lost in this collision, and the direction of travel is altered. The neutron then carries on in this new direction until it undergoes another collision. On average, neutrons migrate away from a source, but a small proportion return, after being slowed to thermal energies. Clearly, the further a neutron gets from the source, the smaller its chance of returning; this is particularly so once thermal energies have been attained, as the probability of absorption is then greatly increased. The soil closest to the probe therefore has the greatest influence on the count rate, with that at greater distances contributing progressively less. For working purposes, a "sphere of importance" can be defined; its center is at the source unless the detector is bottom placed, in which case the center of sensitivity is between the source and the midpoint of the detector; its radius is that which, if the soil and water surrounding the sphere were removed, would give a thermal neutron count that was a stated fraction of the count if the soil medium were infinite in extent [31].

A fraction of 0.95 is commonly used. Within the sphere, the soil nearest to the source contributes most to the count rate.

The size of the sphere of importance depends on three factors.

1. The soil water content: its radius decreases as water content increases because the greater hydrogen content causes more neutron scattering close to the probe, restricting the movement of neutrons away from it.
2. The macroscopic neutron scattering and capture cross-sections of the soil.
3. The energy spectrum of the neutrons emitted from the source (i.e., the *type* of radionuclide in the source *but not its strength*).

The radius of the sphere of importance of most subsurface probes with americium-beryllium sources is about 0.15 m in wet soil, increasing to about 0.5 m in very dry soil. There is therefore very little advantage to be gained from making readings at depth intervals of less than 0.1 m.

The fact that a water content measurement is made on a sizable volume of soil has important consequences when making measurements close to interfaces within a soil, or close to the soil surface, and when there is a marked water content gradient. If measurements are made through an interface between a wet and a dry soil, the measurements in the wet soil will indicate that the soil within about 0.15 m of the interface is drier than is actually the case. Conversely, the water content of the dry soil near the interface is overestimated (but to a lesser degree than the underestimation for the wet soil) [42]. The effect increases with the difference in water content between the layers. The shape of the water content profile is thus smoothed, and consequently neutron probes are not suitable for detailed measurements of water content profiles. The total profile water content is not greatly affected, and the slight underestimation due to interfaces is usually disregarded. However, Van Vuuren [49] has shown that the bias so introduced can be significant and advocates use of field calibrations, which allow (at least to some extent) for site-specific properties such as the presence of a water table.

For soils with layers that differ in terms of texture, chemistry, or density, it is important to use separate calibrations for measurements within each layer. However, the water content profile will still be smoothed across the boundaries.

The contrast at the soil-air interface is even more extreme and greatly affects measurements in the upper 0.15-0.30 m of the profile, depending on whether the soil is wet or dry. Neutrons of all energy levels are "lost" when the sphere of importance intersects the surface and much lower neutron counts are recorded. Various ways of overcoming this are described in Section III.D.4.

3. Random Counting Errors

Radioactive decay and thermal neutron counting are both random processes. This means that when repeated thermal neutron counts are made using the same time interval, the number of counts recorded varies. The error associated with this variation is usually the dominant uncertainty in the neutron method. Other errors may arise from errors in depth of placement, inaccurate calibration, thermal effects on the electronics, and warm-up. Because counting is a random process, repeated counts fit a Poisson distribution. If the number of counts that are integrated to obtain a mean count rate is increased, its standard deviation σ_R decreases according to

$$\sigma_R = \left(\frac{R}{t}\right)^{1/2} \tag{13}$$

where R is the count rate (in s^{-1}) and t the time in seconds. Alternatively, the standard deviation σ of the total number of counts N made in time interval t is given by

$$\sigma = (N)^{1/2} \tag{14}$$

Furthermore, the absolute error accompanying higher count rates, as in wet soils, is always greater than in dry soils because if counts are made over a fixed time interval, R is greater (Eq. 13), whereas if N is fixed, t is reduced.

The standard deviation σ_θ of a water content determination is given by

$$\sigma_\theta = a \frac{R}{R_s}\left(\frac{1}{Rt} + \frac{1}{R_s t_s}\right) \tag{15}$$

where a is the slope of the calibration curve, R_s is the standard count rate (in s^{-1}), and t_s the standard count time in seconds (Section III.C.4). Since the standard count itself introduces a small error, long standard count times of an hour or so should be used if possible to minimize that source of error.

For field measurement purposes, the advantages of the greater precision associated with, for example, counts of 64 seconds duration relative to 16 seconds, should be assessed. Figure 2 shows the relationship between water content error at varying water contents when count times of 16 and 64 seconds are used. Because of the spatial variability of soil water content, it is usually preferable to conduct measurements in many tubes using a short count time.

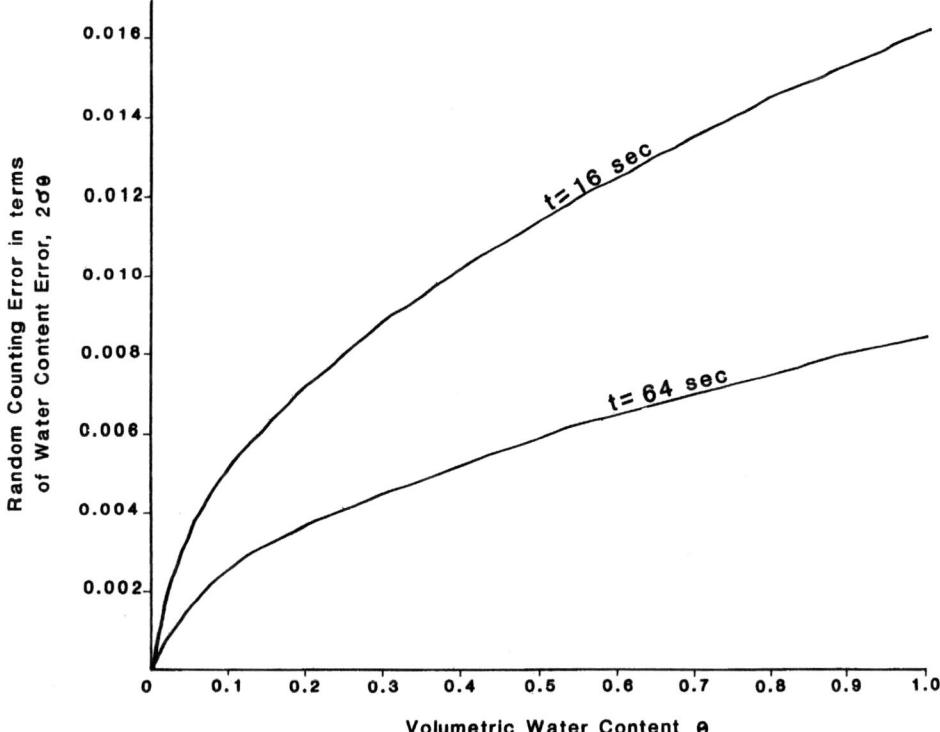

FIG. 2 Relationships between water content error $2\sigma_\theta$, resulting from the random counting error, and water content, for counting periods of 16 and 64 seconds. Bell [32].

This approach will provide a better estimate of water content variability in plot and catchment experiments than more precise data collected from fewer tubes.

The equivalent depth of water in a layer of soil in which θ has been measured is obtained by multiplying by the layer depth. Similarly, σ_θ is multiplied by the layer depth to give the corresponding error. The error associated with a profile water content value is the square root of the sum of squares of the errors corresponding to the individual layer values.

4. Standard Counts

It should be regular practice to record standard neutron probe count rates R_s in a standard medium, and calibrations should be made in terms of count rate ratio R/R_s, where R is the neutron count in the

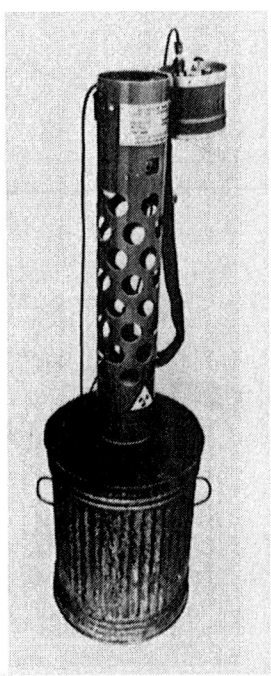

FIG. 3   Neutron probe mounted on a water drum while recording a
standard count.

soil. This ensures data compatibility if a probe requires repair,
after which it may have a different sensitivity, if slow aging of
components occurs, or if more than one probe of the same type is
in use. The sensitivity of probes of the same type is not identical
because of differences in source strengths and detector efficiencies.
However, use of count rate ratios enables their use with the same
calibrations. Data from probes of different designs cannot be normalized in this way, but a method for intercalibrating different probes
has been published by Nakayama and Reginato [50].

Weekly standard counts are recommended, but if a probe is used
less frequently, a standard count should be made before or after
each reading occasion. The count time should be an hour if possible. Standard counts should always be made in water. Water is
preferable to other hydrogen-rich media such as plastics because
the count rate is almost independent of the water temperature [32]
and there is no possibility of moisture absorption from the atmosphere. The container containing the water and access tube must

be large enough to ensure that there is a negligible difference between the count rate obtained in it and in an infinite volume of water. These conditions are met with a water-filled drum at least 0.6 m deep and 0.5 m in diameter having a watertight access tube fixed axially in the center, if the probe is lowered to about the midpoint of the drum. Figure 3 shows a water count being taken. The optimum depth for making the standard count is determined initially by taking a profile of measurements within the water drum.

Some manufacturers suggest taking standard counts in the probe transport shield. This is not advisable because of the influence of temperature and humidity on count rates and because transport shields are small and cannot be regarded as representing infinite media; the counts are easily influenced by surrounding neutron-moderating materials—the floor, for example. Hauser [51] describes precautions that may be used to overcome some of these shortcomings, but his work serves to emphasize the simplicity and reliability of use of a water standard.

D. Field Measurements

Before any measurement is made with a depth probe, access tubes must be installed and the depths at which measurements are to be made selected. A policy regarding surface measurements also must be decided and procedures for data handling established. The question of soil calibration (see Section III.E) should be examined at an early stage.

Depth intervals of 0.1 or 0.2 m, perhaps increasing to 0.3 m at depth, are generally appropriate. The detail of a measured water content profile improves if counts are made at many points within it. However, when total profile water content information is of interest, it is preferable to take counts at a few depths in each of several tubes and so allow for soil variability, rather than to make many measurements in one tube. Once a set of measurement depths has been established, it should be adhered to. If a change is made, the two sets of data will not be strictly comparable because different parts of the soil will have been measured. For the same reason, it is important that the probe be accurately relocated on every measuring occasion. Confusion sometimes arises because the depth of the center of sensitivity registered on the equipment is measured relative to the top of the access tube; to obtain the source depth below ground level, it is necessary to subtract the height by which the access tube protrudes from the ground. A further adjustment may be necessary to allow for the distance that the probe must be lowered until the center of sensitivity is at the top of the access tube.

1. Access Tubes

Aluminum, aluminum alloy, brass, stainless steel, and plastic are all suitable for access tubes. The factors to consider in selecting which material to use are transparency to neutrons, mechanical strength, resistance to corrosion in the soils to be investigated, cost, and availability. In addition to the tubing, stoppers to seal the top end and end pieces to close the bottom of each access tube are required.

Aluminum tubing is preferable because it has almost no effect on neutrons. Aluminum alloy tubing, which is stronger and cheaper, is often used. Either may corrode in very alkaline or saline soils, where plastic or stainless steel tubing might be more suitable. Higher neutron counts are obtained with most plastic tubes relative to aluminum because of the hydrogen content of the former. However, count rates in polyvinyl chloride (PVC) tubes are considerably reduced because of the chlorine content and their use is not advised. Stainless steel and brass have a similar, but less serious, disadvantage because of their iron and copper contents. However, for some applications, the strength of steel is required and the 10-15% reduction in count rate must be accepted.

The internal diameter of the tubing should be sufficient to allow free movement of the probe; a difference of 2-4 mm between the outside diameter of the probe and the inside diameter of the tubing ensures this. A wall thickness of 1.5-5 mm is appropriate. In fact most probes are designed for use with either 44.5 mm (1.75 in.) or 50.8 mm (2.0 in.) outer diameter tubing. If tubing of the correct diameter is not available, an adaptor can easily be made to allow the probe to be fitted on to the top of the access tube.

Suitable tube can normally be obtained from stock from tubing suppliers, as can rubber stoppers to close the exposed end, but bottom end-pieces usually must be specially made. They should be turned or cast of the same material as the tube and should be conical with a shank that can be fitted into the end of the access tube (Fig. 4). Waterproof glue is used to seal the join.

Whatever the material that is chosen, it is very important that all calibration work and all water counts be made using tubes of the same material and diameter used in the field.

2. Access Tube Installation

If precautions are taken to prevent damage to the surface of the surrounding soil in the course of making readings in access tubes, the only disturbance to the soil system will be that caused by the installation of the access tube. During the installation procedure, disturbance to the soil, the soil surface, and vegetation in the vicinity must be minimized, to ensure that the measurements are truly representative of the surrounding area. At the same time,

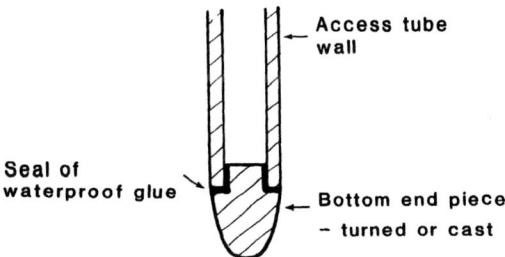

FIG. 4 Turned or cast end-piece fitted in to the bottom end of an access tube.

the access tube must fit tightly into the soil to prevent both the formation of voids adjacent to it and the channeling of water down beside it. When an access tube has been poorly installed, all the readings obtained from it will be biased. McGowan and Williams [36] have considered the effects of poor access tube installation.

If there is any doubt about how well a tube has been installed, or if difficulties are encountered in the process, it is best to withdraw it, backfill the hole, and resite it nearby. It is preferable to make a little more effort at this stage than to collect suspect readings over a long period. Thus, plenty of time should be allowed for installation work. Bell [32] states that two people working together in favorable conditions can be expected to install only three or four access tubes, 2 m deep, per day, using the method described briefly below. Longer tubes or problematic soils may permit completion of only one per day. The time of year can be important. In heavy clay soils, installation is liable to be difficult both when the soil is very wet (because of soil sticking to equipment) and when it is very dry (if the soil becomes very hard). Sand may present problems in dry conditions, as it tends to be unstable, making augering difficult; also, the sides of the reamed hole may collapse.

The access tube installation method originally described by Eeles [52] and Bell [32] is widely used; adaptations of it to meet specific circumstances are common, too. A hole is made by using a steel guide tube of the same outside diameter as the access tube. A screw auger that moves easily within the guide tube is used to drill out soil to about 0.1 m below the guide tube; the tube is then hammered down 0.1 m. The process is repeated until the required depth is reached, at which stage the guide tube is withdrawn and the access tube slid into place. If this procedure is followed, the guide tube will not be hammered down until a hole, of slightly smaller diameter, has been augered below it. The guide tube is not used for making the hole in the first instance because this could result in soil compaction around the hole.

If access tubes are being installed to more than 1 m depth, a series of guide tubes 1.15, 2.15, 3.15, and even 4.15 and 5.15 m in length is used successively with an auger having an extendable shaft. It is convenient to use lengths in one-meter increments to avoid the need for a raised platform to reach the top of the guide tube. However, tubes longer than 3 m are unwieldy, and an alternative is to use 1 m sections, which can be screwed onto the first tube of 1.15 m length. A removable collar is necessary to protect the top of the screw thread while hammering.

It is important that the cutting end of any guide tube be beveled on the inside; this ensures that the soil cuttings move into the tube as it is driven downward, preventing compaction of the adjacent soil. Well-sharpened, smooth tubes are both easier to use and ensure a better fit; a sharpener and a file, to remove any buckling of the cutting edge caused by stones, should be part of the installation kit.

The guide tube should not be driven in too far; at least 0.15 m should be left protruding so that a clasp may be attached if mechanical means are required to extract it. Clayey soils can be very difficult in this respect. Automobile jacks can be used, and powerful rod-pullers are available from drilling equipment suppliers. It is essential that the pull be exerted along the axis of the guide tube, both to reduce effort and to avoid creating air gaps during extraction. Use of a base plate with a central hole to guide the guide tube is recommended unless it is likely to damage the crop. This accessory provides a firm base when using tube extractors and prevents surface soil compaction and enlargement of the top of the hole. Several duckboards should be placed around the tube to protect the surroundings during installation work.

Once the access tube has been inserted into the prepared hole, gentle tamping may be necessary to drive it fully home. The tube should then be cut off so that the required length protrudes from the ground. In a field study with several tubes, it is advisable to cut all tubes off to the same height above ground and/or to mark each indelibly with the height. Otherwise, confusion easily arises when staff unfamiliar with the arrangement have to do the field work.

Adaptations to this method include use of the access tube itself to ream the hole, in which case there is no need to withdraw it; the strength of a stainless steel access tube may then be required. This may be preferable in heavy soils, where considerable effort is necessary to extract the guide tube and may lead to overenlarging of the hole near the surface. It is essential in soils with running sand and/or shallow water tables. Methods for sealing the bottom end of tubes installed in this fashion are available [53]. Bell et al. [54] describe a new technique developed for the capacitance probe, which can also be adapted to install neutron probe access tubes and to provide samples of known volume continuously down the profile from within the access tube. This could be a great help in calibration.

A power-driven hammer may be used to drive in guide tubes, but in most soils little time or effort will be saved, for the power device should be used only to drive the tube about 0.10 m down between augering. In very dense or stony soils where augering is not possible, or where tubes are being installed to great depth, such equipment may be useful, however. Installation work in soils with stones is often difficult; several attempts may be required to install a tube. Unfortunately, there is a greater tendency for success in less stony places, which may result in measurements that are not representative of the soil as a whole. Prebble et al. [53] have addressed this installation problem and have described a variety of methods that may be needed in other situations. Jarvis and Leeds-Harrison [55] have considered the effect of time of installation in swelling clays, on the susceptibility of soil adjacent to access tubes to cracking, thus leading to misleading measurements.

3. Site Maintenance

Field sites must be maintained so that vegetation and soil around access tubes are not damaged or modified after successive measurements. Duckboards should always be used on, or between rows of, short crops. With tall crops such as cereals, raised walkways such as that illustrated in Fig. 5 are necessary to enable the operator to get to the site over the vegetation. Grazing animals, particularly cattle, are very curious and may cause poaching immediately around a tube; they are also adept at removing stoppers from the top of a tube, and protective caps are necessary. Fencing off a small plot to enclose a set of tubes may be considered preferable, in which case the grass within should be cut frequently to emulate the grazing of the surrounding area. Similarly, the soil and crop around tubes in arable fields should be treated in the same way as the surrounding field. Access tubes with detachable top sections may be used so that the plot can be cultivated and harvested along with the surrounding field [56]. The tubes are disconnected about 0.25 m below ground and the lower part is sealed until the farm operation has been completed. Finding the buried tubes can be difficult unless their positions have been carefully recorded. Soil above the top of the tube is then carefully excavated, the top section is reconnected, and the soil is replaced.

4. Measurements Near the Soil Surface

Several methods have been tried in attempts to overcome the problem of making measurements with a subsurface probe, close to the soil surface. (The use of surface neutron meters is described in Section III.F.) These include (1) attempts to prevent the escape of neutrons

FIG. 5 Use of an elevated walkway to make neutron probe readings at an experimental plot.

into the atmosphere by placing neutron reflectors on the soil surface, or artificially extending the soil medium by placing a soil-filled
extension tray on the soil [32], (2) corrections to counts made in
the surface layer, and (3) use of calibrations specific to the surface
layer.

Neutron reflectors are not very useful because a separate calibration is normally required [32]; also the surrounding vegetation
may be damaged when they are used. The principle of surface extension trays is sound, but the trays are difficult to maintain in
practice and are rarely used. Parkes and Siam [57] evaluated their
use and that of correction methods and recommended the latter, in
particular Grant's method, which has a sound theoretical basis [58].
Correction methods require some field readings, which must be made
when the water content distribution is uniform in the top part of
the soil, a condition that can be difficult to satisfy.

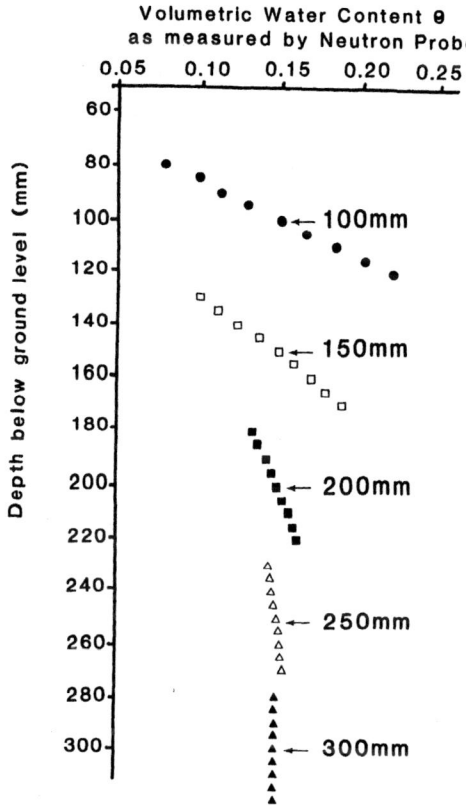

Volumetric Water Content θ
as measured by Neutron Probe

FIG. 6 Effect of depth location of probe on water content measurements at shallow depths. (After Karsten and Van der Vyver [59].)

Corrections require less effort to establish than conducting separate calibrations for surface layers. However, use of specific calibrations appears to be the most satisfactory means of dealing with them [8]. The field calibration procedures described below (Section III.E.2.c) are usually used. Hauser [51] suggests an excavation method to obtain surface calibrations, which obviates the need for core sampling. However, it is conditional upon a uniform moisture distribution to 0.9 m depth and is not appropriate for soils with marked horizons.

When repeated measurements are required at shallow depths, accurate placement of the probe at those depths is very important. The example shown in Fig. 6 indicates that at 0.10 m depth, a misplacement of only 0.01 m can introduce a 4% error in the volumetric water content [59]. Because this error is proportional to the variation of neutron count with depth, it is generally greatest nearer to the soil surface.

5. Data Handling

Large sets of data accumulate when neutron probes are used to monitor soil water content. A routine for dealing with the data as they are collected should be organized at the outset of any project. Manual recording calls for a printed field sheet that is simple to complete, having spaces for the access tube number, date, time, serial number of the rate-scaler and probe, observer, count time used, the reading depths, and the readings. In addition, there should be space for the observer to note any comments regarding the state of the field site, as well as the standard water count corresponding to the set of readings and the time used to obtain it.

The basis of neutron probe data processing is as follows.

1. Using the appropriate calibration and standard count rate, the volumetric moisture content is calculated.
2. Normally a reading is assumed to represent a layer extending halfway to the reading points above and below it; multiplication of the volumetric water content by the layer depth gives the equivalent depth of water in that layer.
3. Summation of these equivalent depths gives total profile water content.

The calculations involved are straightforward and easily undertaken by hand, but to facilitate subsequent analyses, and because of the volume of data produced, most data processing is done using computers. It is important, when designing a data-processing system for a computer, to consider carefully how to organize the data storage to economize on space while permitting easy retrieval. Neutron probe data should be stored in the form of neutron counts accompanied by the appropriate standard count so that, if doubts regarding a calibration or standard count arise, the data can be reprocessed. A variety of data-processing systems have been developed (see, e.g., Refs. 60-62). Alternatively some commercial database and spreadsheet packages are suitable.

There are two potential sources of error when using manual recording procedures and transferring data to a computer: mistakes by the operator in copying the ratescaler readings and mistakes when the data are keyed into the computer. Checking procedures can be used to eliminate the latter. Careful scrutiny of the data in relation to other sets of readings at that tube will throw up most anomalies caused by user error; plotting the data as either a series of profiles or as time series for each depth is very useful for this purpose but may not be practicable if there are numerous tubes. More automated methods of quality control have been described [60].

Automatic logging of neutron counts as they are made at an access tube is becoming increasingly common. The data that are logged should be the same as those that would have been recorded manually. The data-logging facilities being produced by different manufacturers vary in sophistication. Some simply store the data for later retrieval via microcomputer. At the other end of the range are units that allow interrogation of the memory in the field, have semipermanent memories to store information about individual access tubes (e.g., the measuring depths) that can be used to prompt the user, have options to calculate water contents and integrated profile water contents, and have sophisticated communications facilities to transfer data via telephone line to a computer, with error checking and retransmission facilities. These facilities can be very useful, although the calculation of water content requires that the user establish the calibrations required and know the standard count for the probe at the time. Often the data are to be compared with numerous other sets and transfer to a computer is necessary anyway. For most purposes in research it seems preferable to store only raw count rate data in the field, converting this to water content by the processing software "back at base." This avoids problems of operators using incorrect calibrations for different tubes/depths and enables reappraisal if calibrations are updated.

E. Calibration

1. Principles

Recent reviews of neutron probe calibration have been provided by Greacen et al. [63] and Hauser [51]. The most usual method of calibrating a soil is to compare neutron count ratios with water content measurements determined by the conventional oven-drying method. A linear relationship between count rate ratio and volumetric water content of the form

$$\theta = \frac{R}{R_s} a + b \tag{16}$$

where R is the neutron count rate in soil, R_s is the count rate in a standard medium, a is the slope of the line, and b its intercept, is obtained for most probes. The type of access tubing used influences the calibration, and it is essential to use the same tube diameter and material for calibration and field purposes. Calibrations should always be in terms of count-rate ratio (Section III.C).

The neutron count is a reflection of all the hydrogen in the soil: that is, hydrogen contained in free water molecules and in water

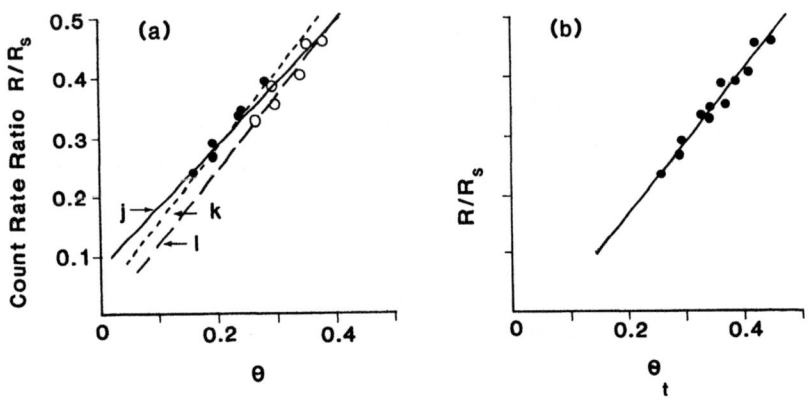

Volumetric Water Content

FIG. 7 Advantage of using calibrations based on total water content θ_t, rather than water released at 105°C, θ. (After Greacen et al. [63].) (a) Calibrations determined using all data from 0.4–0.8 m layer (j), for 0.4–0.6 m layer (k), and for 0.6–0.8 m layer (l). (b) Calibration using same data but expressed as θ_t.

molecules and hydroxyl groups in clay minerals and inorganic matter (i.e., adsorbed and structural water), as well as hydrogen in other compounds. The adsorbed and structural water hydrogen, and the "nonwater" hydrogen can be expressed in terms of an equivalent water content, since the hydrogen behaves similarly to that in free water [63]. It does not change with time and is not removed by conventional water content determinations by oven-drying, so is normally incorporated into the intercept of the calibration.

Greacen et al. [63] advocate that calibrations be made in terms of total water content θ_t (i.e., the sum of the contents of free water θ and equivalent water θ_e). The following example demonstrates how the need for individual calibrations for different soil layers can sometimes be obviated by use of a calibration made in terms of θ_t. Figure 7a illustrates the calibration line for the 0.4–0.8 m horizon of an Australian red-brown earth. The slope of the line is 1.008, compared with slopes of about 0.82 obtained if the same data are plotted separately for the two layers 0.4–0.6 and 0.6–0.8 m. The first equation appears quite plausible but gives changes in water content 25% greater than the field value. This can be overcome if the data are plotted against total water content (Fig. 7b). θ is obtained from $\theta_t - \theta_e$, where θ_e is the equivalent water content of the particular soil layer. Thus instead of having to establish two calibrations, one will suffice if θ_e is determined for each layer. Greacen et al.

[63] describe both a laboratory method for determining θ_e and a means of estimating it from soil clay content.

The calibration of a soil depends on its macroscopic cross-sections to both neutron scattering and capture. These are influenced by isotopic composition (Section III.A) and soil bulk density. It is important to be aware of the presence of strong neutron absorbers such as iron or chlorine and less usual elements that are stronger absorbers. The effect of gadolinium concentrations of only 1 to 36 mg kg^{-1} in Tasmanian soils on probe calibrations has been shown to be considerable [64].

An increase in bulk density causes an increase in the number of nuclei close to the source, resulting in more neutron scattering close to it and a consequent increase in the detection of thermal neutrons by the detector. This tendency for an increase in count rate with an increase in soil density is reinforced if the equivalent water content of the soil is large, because of the greater concentration of hydrogen close to the source. However, the concentration of neutron absorbers close to the source is also increased, and this counteracts the tendency toward a higher count rate. This has led to disagreement in the literature as to the importance of bulk density, but recent studies [65-67] have agreed that it is important in two respects. First, in establishing a calibration, by whatever method, it is essential that a correct measure of field soil bulk density be made and, where necessary, that calibrations be adjusted to this. Second, seasonal density changes may be significant in swelling clay soils. Difficulties in neutron probe work in swelling clay soils have been considered by Greacen and Hignett [66] and Jarvis and Leeds-Harrison [55].

Greacen and Schrale [65] have demonstrated that a count rate R_i measured in soil at one density (ρ_i) may be corrected to the count rate R which would be obtained at some mean field bulk density value ρ using

$$R = R_i \sqrt{\frac{\rho}{\rho_i}} \qquad (17)$$

Stones can present a problem when deriving calibrations. Lal [68], having investigated the effect of stone concentration and stone size on probe calibrations, concluded that both are important and that detailed calibrations are necessary for heterogeneous stony soils.

2. Methods

There are three main techniques for calibrating soil water content against neutron count-rate ratio: theoretical calibrations, laboratory calibrations using drums of soil, and field calibrations. Brief accounts of each are given.

Theoretical Calibration. Theoretical models based on diffusion theory have been designed to simulate neutron fluxes in soils of which the macroscopic neutron interaction cross-sections are known. The interaction cross-sections are determined either from a detailed chemical analysis of the concentration of at least 20 elements in representative samples of the soil concerned [69] or by direct measurement in a reactor [70,71]. With the former method it is possible unknowingly to omit analysis of a crucial neutron absorber; accuracy is also very important, but analyses of the high accuracy required are very expensive. Small traces of gadolinium or boron, which are particularly difficult to measure, can have a very large effect on the resulting calibration. The direct measurement method therefore offers great advantages.

Moutonnet et al. [72] recorded marked heterogeneity in neutron macroscopic cross-sections in samples from a homogeneous field soil and concluded that to represent their soil adequately, 22 samples were necessary. Assumptions have to be made in the theoretical calibration regarding soil density, the calibration is then corrected to allow for field soil bulk density, which must be carefully determined.

Greacen and Schrale [65] concluded that Olgaard's theoretical calibration method [69] was satisfactory for light-textured soils with low neutron capture cross-sections over a wide range of water contents. However, because of overestimation of the neutron absorption effect, it was not appropriate for soils with macroscopic absorption cross-sections of greater than 0.004 barn. Greacen et al. [63] noted that a semiempirical correction was used in Couchat's model [70] to counteract a similar problem. Vachaud et al. [73] compared theoretical calibrations obtained by the method of Couchat et al. [70] with those measured using the conventional field method for sand, chalk, silt, and chalky clay soils, and found good agreement. They particularly recommended this method for use in heavy soils, where obtaining a full range of water contents is difficult, and for soils with marked layering, as it enables "isolation" of the layers from one another for calibration purposes.

In the main, theoretical calibrations are not widely used. This is in part due to cost and problems of access to equipment as well as to some doubts regarding the models and their assumptions. However, improvements in services such as those of the Laboratoire d' Hydrologie Agricole du Centre d'Etudes Nucleaire de Cadarache (13108 Saint Paul Lez Durance, France), where calibrations can be provided at a modest cost, within 2 weeks of receiving suitable samples, make this method one that might be considered on the basis of speed and economics.

Drum Calibration. The drum calibration method requires the uniform packing of soil of known water content into a large drum. An access tube is then installed centrally using the field method, and neutron probe readings are made at the midpoint of the drum. Douglass [74] suggested that since the water content-neutron count relationship is linear, only two points are required, preferably those of air-dry and saturated soil packed to the original field bulk density. Use of the bulk density correction [65] obviates the need to pack the soil to field density, which is extremely difficult. Nevertheless, reliance on a two-point calibration is unwise, because a single error will cause permanent bias in the subsequent data. Also, the effect of any nonlinearity in the calibration curve may not be eliminated. It is preferable to determine several calibration points over a range of both water content and bulk density, although achieving intermediate, uniform, and constant values of water content is difficult. The drum must be large, to ensure that, with a dry soil of low neutron absorption, the difference between counts made within it and what would be obtained in an "infinite" medium of the same soil is negligible. A drum of at least 1.50 m diameter and 1.20 m depth is necessary [32].

The drum calibration technique is laborious. It requires the collection of very large representative quantities of soil from the field. However, with care, good calibrations with high correlation coefficients can be obtained for a wide variety of soils [63]. A combination of the laboratory and field calibration procedures has been suggested by Williamson and Turner [75].

Field Calibration. The field calibration method is simpler but results in greater scatter in the calibration points due to soil heterogeneity and sampling errors. This method has the advantage that the calibrations derived allow for the presence of site-specific features such as marked textural boundaries at certain depths in a profile. In contrast, the drum and theoretical calibration procedures assume a homogeneous soil, which rarely occurs in the field. Temporary access tubes are installed as near as practicable to the permanent ones for which calibrations are required. Neutron counts are first recorded for the required depth(s) of calibration; then five or six undisturbed samples of known volume are taken close to the access tube at these depths. The samples should be of adequate size; cores about 0.15 m long and 0.06 m in diameter are suitable. It is necessary to excavate a pit around the access tube if calibration samples are to be taken from much below 0.60 m. The volumetric water content of the samples is then determined thermogravimetrically (Section II.A). Data from different depths should be segregated, even if the soil appears to be homogeneous. If differences are present but the data are pooled, an incorrect calibration can

easily result, as demonstrated in the example described above in
connection with Fig. 7.

Several points are obtained on the calibration curve by repeating the method at other temporary access tubes when the soil is at
different water contents. Irrigation of a plot may be necessary to
obtain data for the soil when wet. Soil drying can be encouraged
by placing a polyethylene sheet over an area of ground or erecting
a temporary cover with open sides to allow good ventilation. Stocker
[76] has reported a field calibration method for use in stony soils
where obtaining volumetric samples with a corer is very difficult.
One large sample of soil is excavated from around the access tube
at the required depth. The volume of the sample is determined by
filling the void so created, with dry sand from a measuring cylinder,
so that the volume of sand can be determined. The soil sample is
returned to the laboratory for water content determination by the
thermogravimetric method as before.

Variations in the bulk density of the soil samples probably contributes to the scatter of data points in the calibration. Greacen
and Schrale [65] have shown that use of the bulk density correction
equation will reduce the scatter of the points and so improve the
correlation coefficient. Gamma soil density probes have been used
to measure bulk density in situ [6,73,77], rather than on core samples. This obviates the need to make bulk density determinations
on soil samples of known volume, probably the most difficult part of
field calibration. However, this technique is dependent on the calibration of the gamma probe.

An alternative in situ procedure has been presented recently by
Carneiro and De Jong [78]. The total equivalent depth of water in
a soil profile of depth z is given by

$$\int_0^z \theta \, dz = az + b \int_0^z \frac{R}{R_s} \, dz \qquad (18)$$

where a and b are the slope and intercept of Eq. 16. Addition of
water to the soil profile results in an increase in the integrated neutron count-rate ratio according to

$$I_{i+1} - I_i = a \int_0^z \left(\frac{R_{i+1}}{R_{s_{i+1}}} - \frac{R_i}{R_{s_i}} \right) dz \qquad (19)$$

where I is the total equivalent depth of water after the ith addition
of water. The value of a can therefore be found by making successive additions of known amounts of water to a soil, each followed by
recording a set of neutron count readings at intervals down the soil
profile. The added depth of water is plotted against the integrated

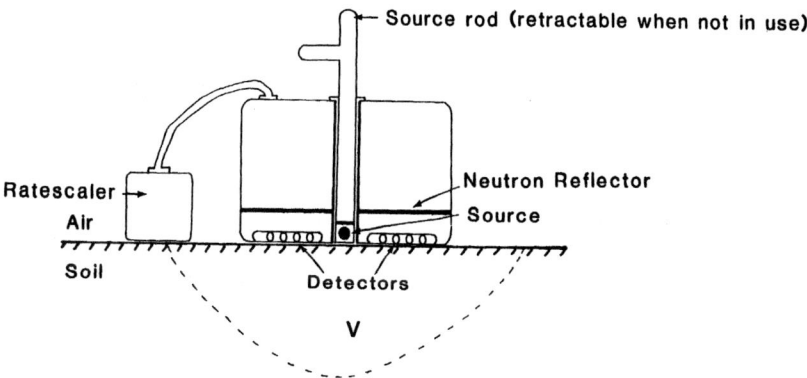

FIG. 8 Neutron probe for measurement of water content near soil surface. V is the volume of soil that principally contributes to soil moisture measurement.

count-rate ratio for the soil profile. The slope of that relationship is the slope of the neutron probe calibration line. It is assumed that there is no loss of water from the profile by evaporation or drainage. The method is therefore most applicable in dry, light-textured soils with a high infiltration rate. If the infiltration rate is slow, a plastic sheet may be used to prevent evaporation.

The intercept b of the calibration equation is then obtained by taking samples in the conventional manner and substituting the count-rate ratio and volumetric water content into the calibration equation. Calibrations obtained by this method are particularly good for accurate estimation of changes of water content.

F. Surface Neutron Meters

Surface neutron meters have been available for as long as their sub-surface counterparts, but the absence of descriptions of their application in the agricultural and hydrological literature implies that they have been little used in these disciplines until very recently. They, and combined neutron moisture and gamma density probes, are widely used in the fields of civil engineering and soil mechanics, for monitoring various types of earth and road works to determine whether their construction gives the required water content and density conditions.

Figure 8 illustrates the basic design of surface neutron meters. A surface meter is not as efficient as a depth probe having the same size of source and detector, because of the loss of neutrons to the atmosphere. To overcome this, a greater counting facility is incorporated, either by using a larger detector or several detectors. Plastic or iron neutron reflectors placed above the source and detector are also usual to contain the loss of neutrons. Generally, the source unit, which is placed on the soil surface when readings are made, can be withdrawn into the shield when not in use, so limiting the radiation hazard.

With a surface meter, measurements are made in a semi-infinite medium because of the presence of the soil-air interface. The soil water measurement represents the mean of a hemisphere of soil whose size changes with water content. Measurements represent a depth as great as 0.35 m in dry soil but only about 0.15 m when the soil is wet [4]. Evidently, if there is a shallow wetting front at, for example, 0.10 m, the measurement will be difficult to interpret.

It is important to prepare the soil surface before a series of measurements is commenced and to place the meter in the same position, in close contact with the soil, on each occasion, to avoid the introduction of errors. Use of a standard count, as with subsurface probes, is preferable in case of shifts in sensitivity of the detector. Evaluation of the relative merits of using the plastic references provided by manufacturers, and alternative standards such as water, has not been reported.

Farah et al. [79] have conducted very comprehensive calibrations with two makes of probe. Their analysis showed that for their soils, using either probe, only two calibrations were necessary to represent satisfactorily all or part of the layers 0-0.10 and 0-0.30 m deep. Alternatively, multiple linear regression could be used to obtain a single calibration that included a depth factor, with only a slight loss of accuracy.

Ahuja and Williams [80] and Ahuja et al. [81] have used surface gamma-neutron meters in two studies aimed at the characterization of soil properties, rather than monitoring of changes of water content. They concluded that for the purpose of describing field capacity water content variations over a field, the meters were very useful.

G. Radiological Safety

Neutron probes contain radioactive sources, and therefore their acquisition, use, and eventual disposal present a potential health hazard to the operator and to the public alike. Consequently most countries and organizations apply rigorous regulations governing the transport, storage, and use of such devices. Several publications consider this in detail and they should be referred to [31,32,82,83].

Before a probe is purchased, the appropriate national regulations governing its use and storage should be consulted, so that preparations for the training of staff and arrangement of storage facilities can be made. Notification of the appropriate authorities of possession of a neutron probe is usually mandatory. Radiological safety regulations vary somewhat between countries, and determining which regulations specifically apply to neutron probe usage can be difficult. The guidelines of the International Atomic Energy Agency [31] can be helpful in this respect, as most regulations are based on them. The relevant regulations have tended to become more stringent, for example, the British Ionising Radiations Regulations Act 1985 has reduced some permitted annual dose limits for operators.

Improvements in the efficiency of neutron detectors have enabled the use of sources that are less radioactive than those used in the past—for example, 1.85 GBq or smaller americium-beryllium sources. However, larger ones are still used in some types of neutron probe. Americium-beryllium sources are most common; their principal hazard is the fast neutrons, but slow neutrons and gamma radiation also contribute.

With sensible usage, the radiation hazard to regular neutron probe users should be only a little greater than that permitted for members of the public. The aim should always be to minimize the amount of exposure to radiation. Precautions such as maximizing one's distance from the source when carrying it or transporting it in a vehicle, and permitting only appropriately qualified personnel to remove the probe from its transport shield, are readily followed if the individuals concerned have been trained and so understand the reasons. Loss of a source would be very serious, and therefore probes should never be left unattended except when locked in their designated storage place.

Regular tests to check for leakage from sealed radioactive sources are advisable and are mandatory in some countries. For example, in Britain such tests must be conducted once every 2 years. Americium-beryllium sources have a half-life of 458 years; this is much longer than both the useful life of most other parts of the probe and the time over which the integrity of the source container can be expected to be maintained. When a source is no longer required, it must be properly disposed of at a designated repository for radiological waste.

IV. METHODS BASED ON MEASUREMENT OF THE DIELECTRIC CONSTANT OF SOIL

Both the capacitance and the time domain reflectometry (TDR) method for soil water content determination depend on measurement of K, the dielectric constant of the soil. The dielectric constant of water is about 80 and that of most soil material is between 3 and 5, so the dielectric constant of moist soil is a sensitive measure of the volumetric water content. Both methods are influenced to some extent by the dielectric constant of the soil matrix components, hence by the

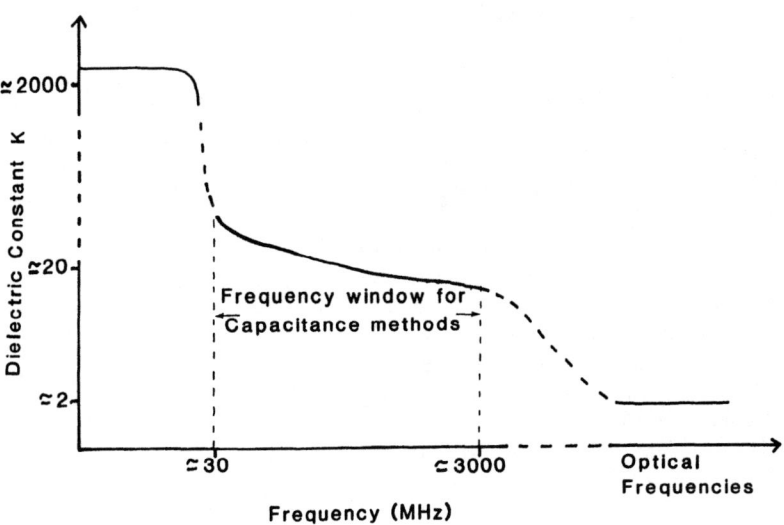

FIG. 9 Typical variation in dielectric constant with frequency for moist soil.

type of soil. For example, at a water content of 20%, a change in the dielectric constant of the soil matrix from 3 to 5 is equivalent to a change in water content from 20% to 21.5%.

A practical definition of the dielectric constant K of a material is the ratio of the value of a capacitor with the material between the plates, compared with the value with air between the plates. A dielectric material is an insulator, as distinct from a metal, which is a conductor. Under the influence of an electric field, the positive and negative charges in a dielectric material are displaced with respect to each other and tiny electric dipoles are produced. Some materials, such as water, also have permanent dipoles. The electric dipoles are aligned by the electric field, and the dielectric medium as a whole becomes polarized. The dielectric constant as defined above turns out to be a measure of the polarization; as a consequence, a material whose molecules have a permanent dipole moment and are free to align with the electric field has a very large dielectric constant.

The dipoles can arise from a number of physical mechanisms [84], many of which are present in moist soil, which is a composite material with varying proportions of air, water, and mineral material. The behavior of K for wet soil is correspondingly complex (Fig. 9).

At low frequencies all electric dipoles respond to the frequency of the applied electric field, and values for K of 100 or 1000 or even

higher have been reported for soils [85]. These extreme values
are due to interfacial polarization, with dipoles created from induced
charges on the surfaces of air voids in the soil. Such values for K
reflect the proportion and configuration of air voids present rather
than the proportion of soil water. Above about 30 MHz these relatively macroscopic dipoles can no longer follow the field reversals
and they cease to contribute to the dielectric constant, which drops
to the value determined by the proportions of water, air, and soil.
This value is dominated by the permanent dipole moment of the water
molecule and underlies the high sensitivity to the proportion of water
present.

At sufficiently high frequencies, of the order of 3000 MHz, the
water dipole fails to follow the field reversals and the dielectric constant falls still further, ultimately approaching near unity at frequencies in the optical region.

A. Capacitance Method

1. Introduction

The capacitance method includes the moist soil as part of the dielectric of a capacitor. Measurement of the capacitance gives the dielectric constant, hence the water content of the soil. The method has
several attractive features: it is cheap and entirely safe, and its
use is free from legal constraints. The instrumentation is stable,
rapid in manual operation, and amenable to automatic logging. Only
recently, however, have breakthroughs in design and electronics led
to two commercially available instruments [86,87]. Consequently the
practical aspects of the method are not yet well established. Both
these systems operate within an access tube and to this extent are
similar in use to the neutron probe.

2. Instrument Design

A variety of capacitor electrode geometries have been described in
the literature [3]. Wedge-shaped probes [88], parallel rods, and
parallel knife blades (Fig. 10a) have been used, but the disturbance
caused by direct insertion into the soil leads to difficulties. If the
electrodes comprise part of a probe inserted into an access tube in
the soil, then the soil surrounding the probe occupies the fringe
field of the capacitor (Fig. 10b). In this case the system must be
designed to minimize unpredictable and variable air gaps between
the access tube and the soil, yet avoid undue disturbance and compression of the soil.

With the use of simple "push-in" electrodes, which do not utilize
an access tube, air gaps created in installation and air gaps created
by shrink/swell soils, are avoided. However, this type of system
permits shallow measurements (adequate for many purposes) only,

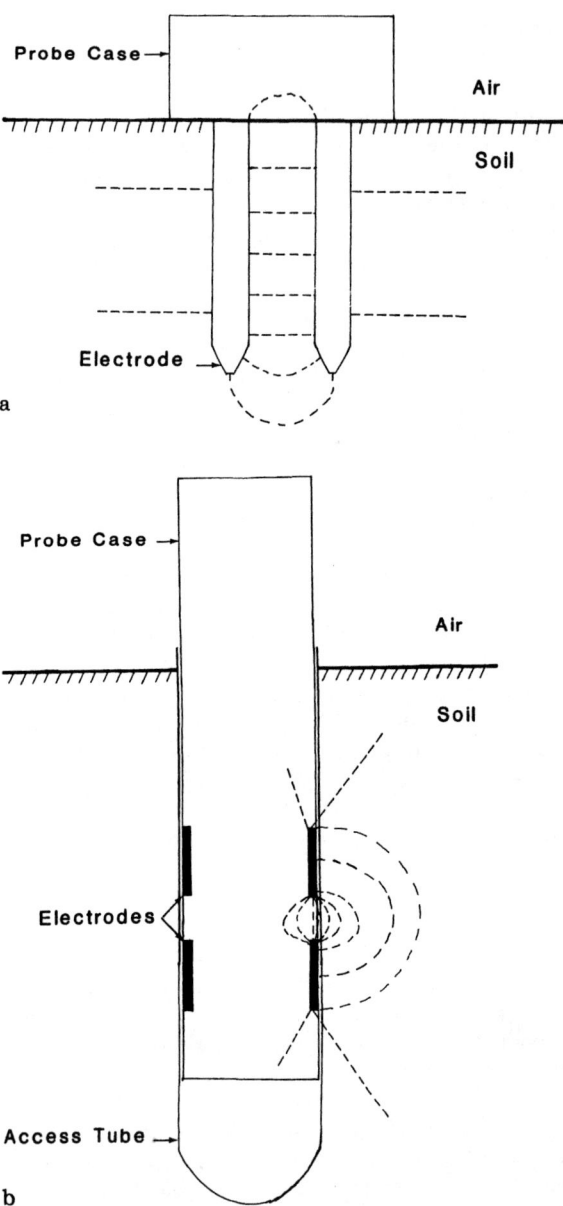

FIG. 10 Basic capacitance probe designs showing the approximate form of the associated electrical field lines. (a) Parallel electrodes inserted directly into the soil. (b) Electrodes installed within a probe that can be lowered down an access tube.

and normally each determination would be unique and unrepeatable, hence introducing the need to deal with soil heterogeneity, which tends to be large in the surface zone. The best method for accurate measurements is probably the use of an access tube installed by a controlled and reproducible technique [54] but at the expense of a lengthy installation time. Many of the practical details described in the following sections relate to use of a probe within an access tube rather than the use of direct-insertion electrodes.

The capacitance may be measured by a bridge method [88] at a frequency in the 30-3000 MHz range. An alternative and more straightforward method is to arrange for the capacitor to be part of an oscillator circuit so that the frequency of oscillation is a direct measure of the capacitance [87].

Interference from acidity and salinity of the soil water is reduced by operating at higher frequencies than the minimum of 30 MHz, as soil conductivity effects are inversely proportional to frequency. At 100-150 MHz, the problems are sufficiently reduced without encroaching into the very high frequencies that are operationally difficult for circuit technology.

3. Access Tube Installation and Calibration

A metallic access tube, as used for the neutron probe, is not suitable because it acts as a barrier to the electric field and would shield the probe from the soil outside the access tube. A plastic access tube must be used, and this necessitates some modification to the method of installation used for the neutron probe.

The soil volume comprising the capacitor is physically small, with dimensions of a few centimeters, and since the dielectric constant of air is unity, any variation in the volume of air "seen" by the capacitor is potentially a source of error. With fringe field geometries, the field decays over a few centimeters, and care during access tube installation is very important. For accurate work, the gap between access tube and soil should be less than 0.5 mm [87].

Because the lateral penetration of the electric field is small, the best field calibration procedure appears to be the taking of soil samples of known volume from within the access tube hole as each access tube is inserted [54], the water content being determined thermogravimetrically. Repetition of the procedure at each site under driest and wettest possible soil conditions, using additional nearby temporary access tubes, provides calibration data for different depths. This assumes that lateral homogeneity of the soil is likely to be better than homogeneity as a function of soil depth.

For a composite material such as moist soil, the dielectric constant is not a simple function of the values for the individual components. The relationship between K and the volumetric proportion of water present is, therefore, not linear. A theoretical treatment

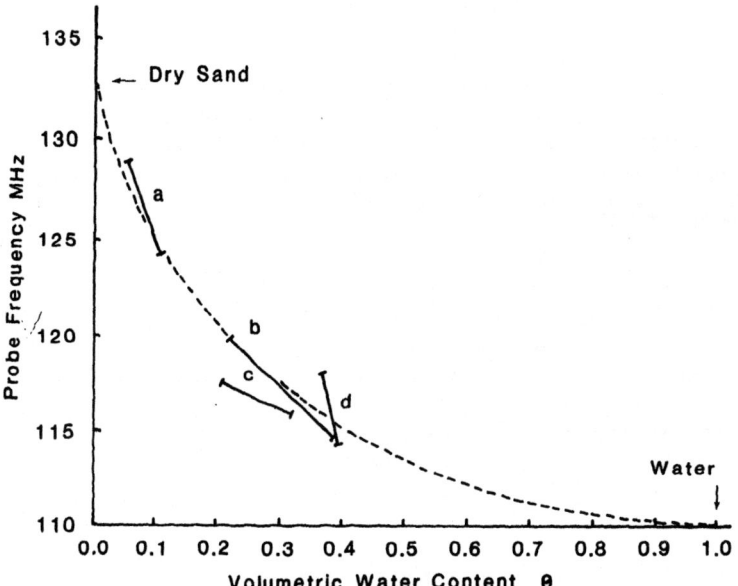

FIG. 11 Generalized calibration for capacitance probe. Linear re-
gressions show variability of calibrations for examples of different
soils: a, medium-fine sand overlying Jurassic sandstone; b, chalky
silt overlying poorly sorted fine gravel (repacked within a lysimeter);
c, sandy clay drift overlying Lower Greensand; and d, chalk overlain by topsoil. Bell et al. [54].

has recently been published based on a statistical analysis, but a
somewhat arbitrary assignment of parameters is required to match
the theoretical curve with the available data [89]. The method of
measuring the capacitance may also be nonlinear. This is the case
if a resonant frequency technique is used which has the general
form

$$F = \frac{1}{2\pi \sqrt{LC}} \qquad (20)$$

where F is the resonant frequency, L is a circuit inductance, and
C is a total capacitance that includes the soil component together
with some constants. Figure 11 illustrates a typical response, which
exhibits enhanced sensitivity at low water contents. However, over
the normal range of moisture variation for each soil, field calibrations
can be regarded for most purposes as linear. While falling in general

terms on an overall curve, these lines sometimes have gradients that appear to differ significantly from the slope of the overall curve relevant to the soil water content range concerned.

On theoretical grounds it seems likely that, as variation in composition and density of the soil have only second-order effects on the dielectric constant of the soil, there may be a family of curves which are quite closely spaced and converge on the frequency for water. Anomalous calibration gradients may therefore be derived inadvertently by combining all the data from a given soil profile that represents more than one soil layer. A regression on such a data set would lead to derivation of a curve of inappropriate gradient.

A further factor that could contribute to this effect is that the water that responds to the reversals of the applied field of the instrument is only that which is held loosely in the soil (i.e., water held in the pores by surface tension). Water held more strongly by surface adsorption forces or in chemical association with humus, iron hydroxide, etc. would be expected to be less free to respond to the probe, hence would not be seen as "water." Both types of soil water are, however, expelled without distinction by oven drying at 105°C (Section I.A.1). Different proportions of firmly held water in different soil layers could therefore distort calibration curves derived by the thermogravimetric reference method.

4. Resolution and Speed of Use

The radius of the soil zone around the electrodes, which dominates the response, depends on electrode geometry but in general is about 30-40 mm. Thus one of the main disadvantages of the capacitance method is that the effects of small-scale lateral heterogeneity may be considerable. A more serious disadvantage is that minor cavities, which can easily be created when installing an access tube, can give rise to significant anomalies. Installation of the access tube must be done with extreme care; the access tube should be regarded as part of the instrument. With the Institute of Hydrology probe, annular air gaps between access tube and soil produced a change in frequency corresponding to approximately 7% volumetric water content per millimeter [87]. Future research may lead to a means of designing electrode geometries to produce more lateral penetration of the field and so reduce the influence of air gaps. Air gaps can also arise after access tube installation, due to soil shrinkage, which is a particular problem in clay soils.

The depth resolution of a capacitance system (i.e., the ability of the sensor to resolve between soil layers of differing water content) depends on the size and geometry of the electrodes. It is evident, from work carried out at the Institute of Hydrology [54,87], that the capacitance method is capable of much better resolution than the neutron probe. The sensor used in this work is comprised of a

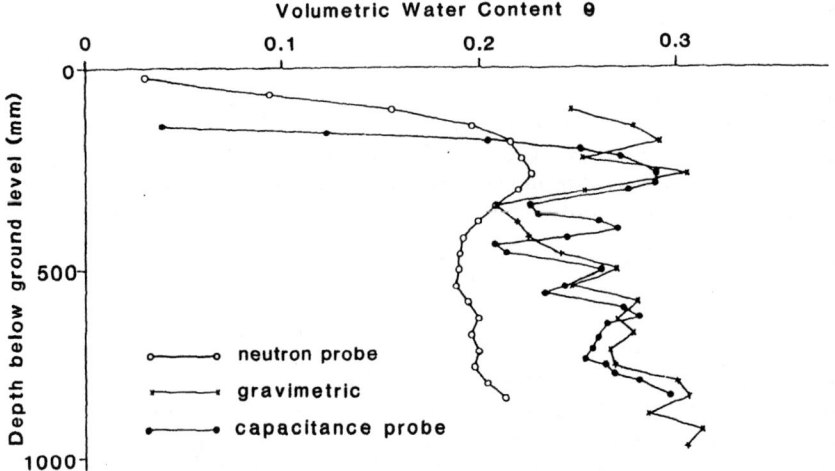

FIG. 12 Soil water content profiles measured by capacitance probe, neutron probe, and gravimetry at the same access tube. The neutron probe profile is based on a standard calibration curve appropriate to the soil, resulting in a minor displacement of the plotted profile. Bell et al. [54].

pair of annular electrodes 44 mm in diameter and 30 mm apart, operating at a frequency of 140 MHz. Detailed variations in water content determined by this capacitance probe closely match variations in the gravimetric water content profile (Fig. 12), which was determined on 40 mm depth increments. The neutron probe profile matches the other two curves in general form, but the detail is much less well resolved.

Determination of water content by the capacitance method is fast; each reading takes no longer than the time required to write down or otherwise record the frequency. A complete 1.5 m profile can be read at 20 mm intervals in 3 minutes or less. This contrasts with 20 minutes required for the neutron probe to read a 1.5 m profile using, for example, a 64-second integration time at 100 mm depth increments.

In addition, there is no random counting error. Hence resolution of water content changes is more precise, by a factor of about 4, than with the neutron probe.

5. Conclusions

The capacitance method offers a practical alternative to the neutron probe, being safe, cheaper, and quicker to use. Thus it can be installed and left in situ not only to monitor soil water content changes but also to be linked, for example, with flood warning or irrigation control systems. It has better depth resolution than the neutron probe but in its present form lacks lateral penetration and is probably more affected by small-scale soil heterogeneity and air gaps around the access tube. Now that commercial instruments are becoming available, further development will doubtless lead to improvements.

B. Time Domain Reflectometry

1. Introduction

The time domain reflectometry (TDR) technique for measurement of soil water content is relatively new and is only beginning to gain acceptance as a routine technique, although it is a little better established than the capacitance probe method. It is an alternative method for measuring the dielectric constant of the soil. The principle of the method is that a high-frequency electromagnetic pulse is fed into the soil between two metal rods. Part of the pulse is reflected back up through the soil from the bottom of the rods, and the time interval between the incident and reflected pulses is measured. TDR differs from the capacitance probe method in three main ways.

1. It measures dielectric constant over a broad band of frequencies, typically between 100 and 1000 MHz.
2. It uses a pair of parallel rods inserted in the ground for measurement.
3. It measures an averaged dielectric constant over the length of the emplaced rods.

A recent advance in the method has been to measure electrical conductivity and dielectric constant simultaneously, so that solute concentration may be estimated at the same time as water content. (See also, Note Added in Proof—p. 63.)

2. Theory

The propagation velocity v of an electromagnetic wave down a transmission line in a nonmagnetic medium, such as soil, is equal to

$$v = \frac{c}{\sqrt{K}} \tag{21}$$

where c is the velocity of light in free space (3×10^8 m s^{-1}) and K is the dielectric constant of the medium.

Thus, in pure water, with a dielectric constant of about 80, the speed of propagation will be about 3.3×10^7 m s^{-1}, and in moist

soil, with a typical dielectric constant of 16, it will be about 7.5 ×
10^7 m s^{-1}. The corresponding wavelengths in these three cases
(free space, water, and soil) will be 3, 0.33, and 0.75 m at 100
MHz and 0.3 m, 33 mm, and 75 mm at 1000 MHz.

In practice, the medium is not purely capacitive and inductive,
but there is also some conduction within the medium. This can be
taken into account by writing the dielectric constant as a complex
quantity

$$K = K' - jK'' \qquad\qquad\qquad\qquad (22)$$

where K' is the real part of the dielectric constant (i.e., the capacitive part), K'' is the imaginary part, giving rise to energy losses,
and j is $\sqrt{-1}$. The imaginary part of the dielectric constant arises
from dc conduction as well as radio-frequency effects associated
with, for instance, energy dissipation, as the dipoles of the water
molecules reverse their orientation with the electric field.

In principle, the imaginary part of the dielectric constant affects
the propagation velocity of the wave, but in practice the effect has
been found to be negligible, even when solute concentrations are
very high [90,91]. The effect does, however, affect the amplitude
of the wave as it propagates in the medium, causing it to be attenuated down the transmission line. This effect can be used to estimate the electrical conductivity of the medium [91-93].

3. Practical Arrangement for Measurement of Water Content

Electronic Arrangement. Figure 13 is a block diagram of the
parts of a TDR setup. A timer supplies synchronizing information
to a pulse generator and receiver.

The pulse generator supplies a voltage step with a very fast
rise time, effectively feeding a train of high-frequency electromagnetic waves with a wide frequency distribution into the sample.
The detector circuit measures the sum of the input voltage and the
reflected pulse. Because the times involved are very short, a few
nanoseconds, the time dependence of the output voltage is determined by sampling the voltage at a series of times after the initial
pulse. Pulses are sent repeatedly every millisecond or so, and one
voltage sample is measured after each pulse cycle. Thus, a voltage-
time curve can be reconstructed from these separate measurements.

An example of an output trace is shown in Fig. 14. To interpret this, it is important to remember that the curve is the sum of
a step input and the reflected voltage.

Transmission Line Arrangement. Topp et al. [90] demonstrated
the feasibility of measuring the water content of a sample of soil

TDR INSTRUMENT

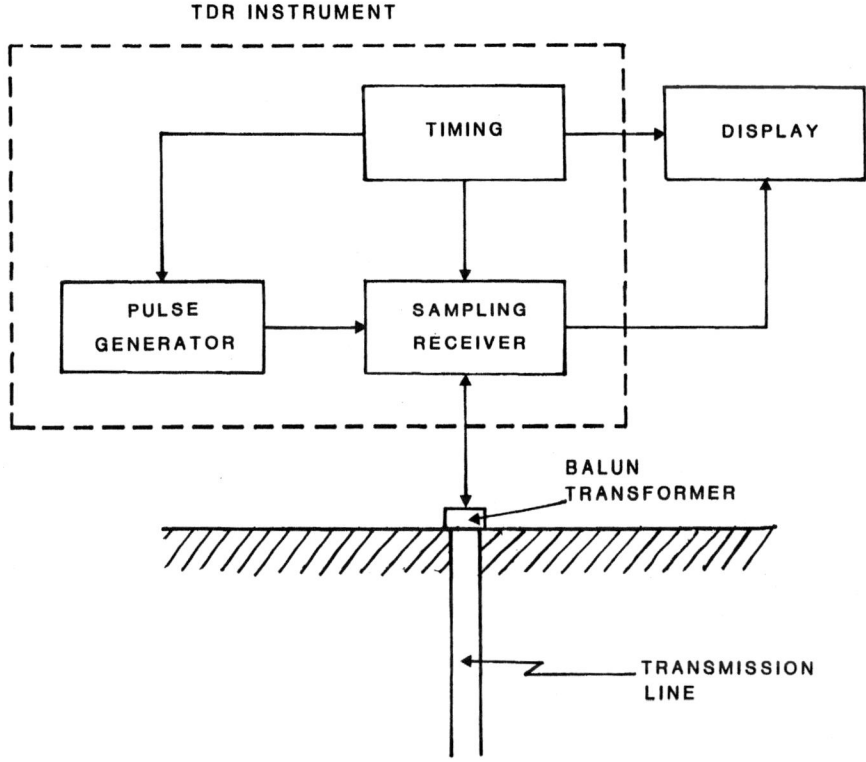

FIG. 13 Block diagram of a TDR instrument and its display unit.

packed in a metal cylinder with a central conductor. This arrangement is clearly not practical for field use, and the coaxial arrangement has been superseded by one using parallel-wire conductors
pushed into the soil. Usually, a coaxial cable is used to connect
the transmission line to the TDR instrument, and a "balun" transformer is needed to provide impedance matching (which prevents
too much of the wave energy from being reflected at this junction)
and to isolate electrically the grounded outer conductor of the cable
from the twin probes.

    In some soils, it has been found possible to insert 6 mm diame-
ter probes to a depth of 1 m or more. The parallelism of the probes
has been found not to be critical for water content measurement [94],
but it would be important if electrical conductivity were to be measured as well. Nevertheless, careful installation of the probes is
necessary to avoid air gaps between the rods and the soil, which

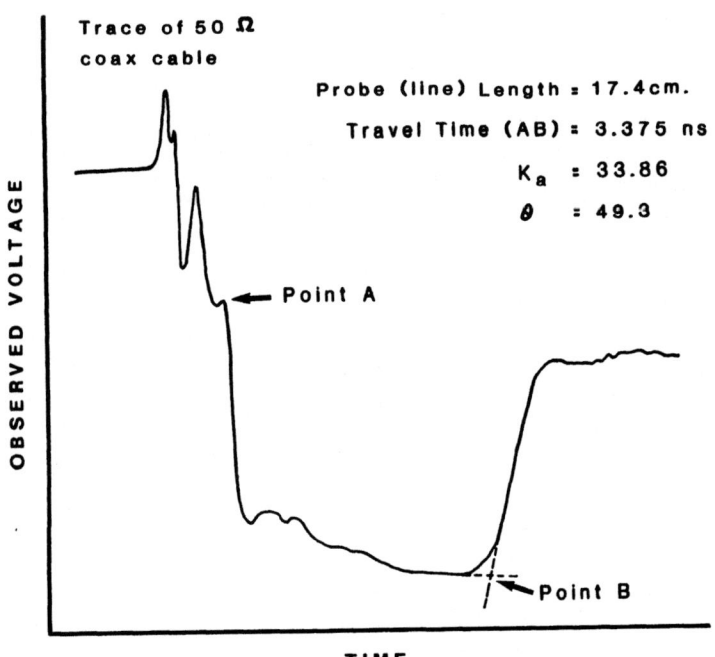

FIG. 14 Typical TDR trace in wet soil from parallel transmission lines, with an open circuit at the end. Point A: start of parallel transmission lines; point B: end of parallel transmission lines. (Reproduced by permission of the National Research Council of Canada from the *Canadian Geotechnical Journal* [104].)

may allow water to run down them and also may lead to the recording of artificially low water contents [95].

The metal probes inserted into the soil form part of a transmission line, and the method gives an average value of water content over the length of this transmission line (Section IV.B.4). Three methods are used to obtain details of variations in water content with depth. One is to insert a line horizontally in the soil at each of a number of depths, thus sampling water content at a single depth with each line. Second, "steps" can be made in the probes by, for instance, reducing their diameter in places and installing them vertically in the soil. These provide points at which the pulse reflects back and act as depth markers on the output trace. Third, lines of different length can be installed vertically from the surface in nests. Water content in the depth interval between two lines can then be calculated by difference.

4. Interpretation of a TDR Trace

Interpretations of the various features of the typical TDR trace presented in Fig. 14 are shown in the figure. Difficulties arise because there are often several reflections along the line, caused, for instance, by minor inhomogeneities in the soil. An appreciable water content gradient may lead to what is effectively a distributed reflection, giving a trace with a marked gradient. In some cases, the signal may be so attenuated by the end of the line that no reflection is visible. Multiple reflections may also lead to interpretation difficulties. In the simplest case, the reflected wave from the end of the line will itself be reflected as it returns to the discontinuity at the beginning of the line. If there are intermediate reflections along the line, these will complicate the issue, possibly to the point that interpretation of the trace becomes ambiguous.

It is necessary to identify the precise position of the reflection arising from the end of the line, and this can present a problem. In the ideal situation, the reflection is sharp, but in practice, as shown in Fig. 14, it is distributed over a range of times. The reasons for this are not completely understood, but probably they involve some dispersion of the pulse and penetration of the pulse field beyond the end of the transmission line, making its electrical end "fuzzy."

It has been found that the method shown diagrammatically in Fig. 14, where the end of the line is estimated from the interaction of two tangents to the line near the end, is in practice a reliable measure, enabling estimation of the time of propagation to within about 0.08 ns, equivalent to a precision of volumetric water content of about 1% [90]. The rationale for this method is that it gives a measure of the first incidence of the wave train in the pulse with the end of the line [Topp, personal communication].

5. Practical Use of TDR

Several examples of the practical use of time domain reflectometry in both field and laboratory have been published.

Calibration. Topp et al. [90] calibrated a coaxial TDR apparatus with several soils in the laboratory (Fig. 15). There were some differences between soils, but these were surprisingly small. Overall, the data fitted a curve

$$\theta = 5.3 \times 10^{-2} + 2.92 \times 10^{-2}K - 5.5 \times 10^{-4}K^2 + 4.3 \times 10^{-6}K^3$$

$$(23)$$

There was some evidence that sandy soils gave a more curved line than clay ones, which is consistent with the presence of water bound to the clay lattice in an icelike structure, and consequent

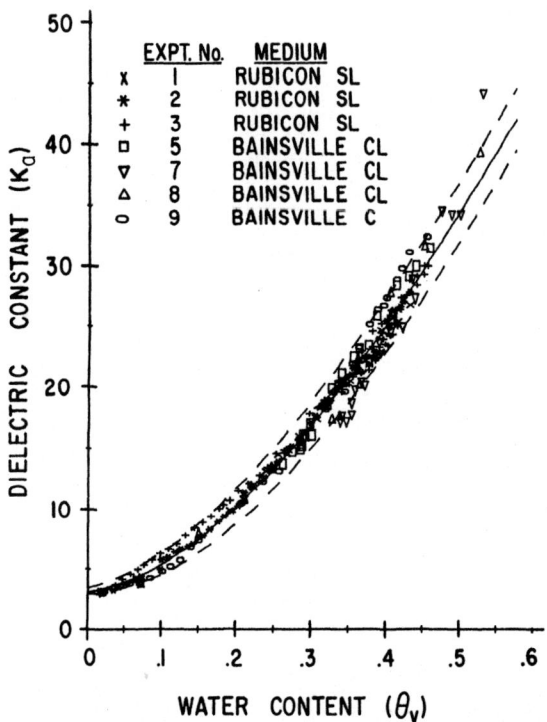

FIG. 15 Relationship between apparent dielectric constant K_a and volumetric water content θ_v, for several mineral soils measured in the laboratory [90]. (Copyright by American Geophysical Union.)

restraint of the dipoles. Topp et al. found very little evidence for the effect of bulk density or temperature, but Ledieu et al. [96] were able to define corrections for both variables, though the effects were very small and of little practical consequence. Overall, their calibration line was very close to that of Topp et al. [90].

Use in Soils. Topp et al. [97] investigated the effect of wetting fronts, steep gradients, and impedance steps in the lines. They found that TDR gave a well-averaged value for water content along the length of the line, even when there was considerable variation of water content. The reason for this is that inversion of Eq. 23 gives

$$K_a = 3.03 + 9.3\theta + 146.0\theta^2 + 76.7\theta^3 \qquad (24)$$

Dialectric constant is, therefore, very nearly proportional to the square of water content, and the speed of travel is consequently inversely proportional to water content. Hence, the time taken for the pulse to propagate through the medium is proportional to the water content. The total travel time is thus a good measure of the total water content of the medium integrated over the length of the line.

Sharp wetting fronts can give rise to reflections [97], and their position can be detected by examination of the TDR trace. Although initial results were encouraging [95,98,99], the use of "markers" on the line, to allow some indication of depth in the profile by forcing partial reflection of the waves at particular places, has not met with general acceptance. This is probably partly because of difficulties in interpreting the traces, especially if there are other reflections along the line, caused, perhaps, by stones or wetting fronts. A better method for gaining depth-specific information is to bury lines horizontally at several different depths, or to use a set of lines in incremental depth intervals.

Dalton and coworkers [91-93] have used TDR to estimate water content and electrical conductivity simultaneously in rhizotrons. They found that correspondence between TDR and the thermogravimetric method for water content and between TDR and electrical conductivity of saturation extracts was as good as between the neutron probe and thermogravimetric methods and between four-point conductivity probes and saturation extracts. This implies that the imaginary part of the dielectric constant is dominated by the dc conductivity of the soil solution. Other data [100,101] suggest that this applies to the conductivity of solutions, but the relationship is much less good when measuring the bulk conductivity of soil.

Aubic and Bottreau [102] were able to reconstruct depth profiles of dry bulk density and water content in an almost uniform sand column in the laboratory, using a single transmission line with no artificial depth markers. They succeeded in this by carefully analyzing the shape of the reflected pulse. Whether it would be possible to perform a similar analysis in the much less ideal conditions of a field soil appears not to have been tested.

An example of the use of TDR in the field is the study of Topp and Davis [98], in which TDR probes were inserted into a cracked soil, both in the middle of peds and near to the cracks. Depth resolution was obtained by reducing the diameter of the rods at intervals. They were able to show that water penetrated the soil near the cracks in the 0.1-0.3 m depth interval after 2 hours of water application. Away from the cracks, water did not penetrate deeper than 0.1 m over a 9-hour experiment. The estimate of 2 hours before water started to flow into the cracks and infiltrate from them was consistent with both visual observation and estimates of the infiltrability of the soil. Other field tests have demonstrated good agreement between TDR and thermogravimetric measurements [99,103].

Use in Frozen Soil and Snow. The difference in the dielectric constant of water between the frozen and unfrozen (about 3.15 and 80, respectively) states makes dielectric methods attractive to distinguish between those states. TDR has been used in soil [94,104, 105] to determine the depth of freezing and in snow [94] to determine the unfrozen water content. There is clearly scope for use in such studies of capacitance methods in conjunction with others (e.g., the neutron probe), which are sensitive to total water content.

6. Automatic Water Content Measurement by TDR

Automatic instruments have been developed for the measurement of water content by TDR which eliminate the time-consuming manual examination of many traces. They work by performing an analysis, similar to that described in Section IV.B, using a microprocessor. One instrument is known as IRAMS (Instrument for Reflectometry Analysis of Moisture in Soils) and is available from CPN Corporation (2830 Howe Road, Martinez, CA 94553). Each measurement takes about one minute. Clearly, there are situations in which it will fail to give reliable answers, as in some cases the interpretation requires considerable skill and experience. Nevertheless, the saving in labor is very great, which probably outweighs these problems in many cases. Another device is marketed by Soil Moisture Equipment Corporation (P.O. Box 30025, Santa Barbara, CA 93105). A fuller review of the TDR method can be found in the publication of Topp and Davis [106].

V. ALTERNATE APPROACHES TO FIELD MEASUREMENT

A. Measurement of Soil Water Potential

An alternative approach to water content determination is to measure soil water matric potential (see Chap. 2) and to translate the data into water contents using the relationship between water content and matric potential (the water release characteristic). This relationship is subject to hysteresis, and consequently precise determination of water content is not possible. The method can be satisfactory, however, in the absence of water content measurements, and if used cautiously. Determination of soil water release characteristics is described in Chap. 3.

B. Gamma-Ray Attenuation

When gamma rays are passed through soil, scattering and absorption of the radiation occur; soil density is the principal determinant of these interactions. Where there is no shrink-swell soil behavior, temporal variations in total bulk density are entirely due to variation in water content and gamma-ray attenuation can be used to monitor water content. The technique is used principally for soil bulk density measurements, and its theory and practice are detailed in Chap. 7 (Section II.B.5 covers water content determination).

The technique has been used in laboratory studies of water movement in soil columns [107-109] but only occasionally in the field [110,111], the better-established neutron method being preferred. Ferraz [112] and Morrison [4] have reviewed its use, and the latter publication usefully summarizes its merits and limitations for soil water content determinations.

C. Thermal Conductivity

When a heat source is placed in a soil, the rate of heat dissipation from it is a function of the soil's thermal diffusivity. This increases with increasing water content, and thus the rate of heat dissipation can be used to measure soil water content. The technique was originally proposed by Shaw and Baver [113]. A heating element and a temperature sensor are placed in the soil; the element is warmed by passing an electric current through it, and the sensor indicates the time taken for a certain temperature to be attained.

There are two principal designs: direct-contact probes [114, 115] and those in which the element and sensor are encased in a porous material [116]. However, measurements made with the latter type are a reflection of the water content of the block when its matric potential is at equilibrium with the surrounding soil, and hence respond to matric potential (See Chap. 2, Section II.C). The limitations in attempting to convert this to water content have already been discussed (Section V.A) in this chapter.

The commonest problems with direct-contact probes arise from poor contact between the instrument and the surrounding soil; hence the method is not successful in swelling soils. In fact, to date it has not been exploited to any great extent in any type of soil.

D. Nuclear Magnetic Resonance Spectrometry

The use of nuclear magnetic resonance (NMR) techniques for measurement of soil water content was investigated in the past [117,118], but recent improvements in NMR technology have made its routine use a much more practical proposition [119]. NMR techniques depend on the influence that a magnetic field has on nuclear dipoles. When

subjected to an oscillating electromagnetic field, atomic nuclei change
their energy levels. Different frequencies affect different nuclei,
causing absorption of energy by nuclear magnetic dipole systems.
The NMR signal is produced by electronic detection of either the
energy absorption or the concomitant nuclear dipole excitation. The
signal depends on the nuclei concerned, the local molecular environment, and the physical state of the sample. Hydrogen nuclei give
the strongest response, so the method is well suited to measurement
of the amount of water present in a medium.

Prototype field instruments for measurements on soil samples and
for remote measurements of surface soil water contents using tractor-
mounted equipment have been described [120]. Calibration against
volumetric water content for different soils is required, and a temperature correction may also be necessary. The method is still in
its infancy but could prove to be a useful complement to other
techniques.

VI. REMOTE SENSING OF SOIL WATER CONTENT

A. General Considerations

The development of remote sensing, which was given considerable
impetus through the advent of the Soviet and U.S. space programs
in the early 1960s, is now a flourishing subdiscipline with a wide
range of proven applications in the appraisal and monitoring of many
aspects of the environment. In remote sensing several methods are
used to convey data from the object of interest, called the "target,"
to the sensor. Electromagnetic energy has proved to be the most
useful for studying soil water content. Sensors may be mounted
just above ground level (e.g., on a moving vehicle), on an aircraft,
or on a satellite. In the latter case data are purchased from the
relevant space agency for processing by the user.

Figure 16 shows the electromagnetic spectrum, with an indication
of the sensing technologies that have been most usefully applied in
each portion of the spectrum. Remote sensing studies of soil water
have exploited a wide range of wavelengths from gamma rays (< 0.003-
10 nm), to long-wavelength microwave radiometry and radar (1-800
mm). Both "passive" and "active" remote sensing techniques have
been successfully employed. In the passive techniques the sensor
monitors radiation that either is emitted by the target or is reflected,
refracted, or polarized by it, having originated from a natural source
such as the sun. Active remote sensing uses an artificial source of
radiation. This radiation is detected after being reflected from the
target; sonar, radar, and monochromatic lidar are examples of active
systems. Further technical introductions to remote sensing data acquisition and data-processing techniques are available in a variety of

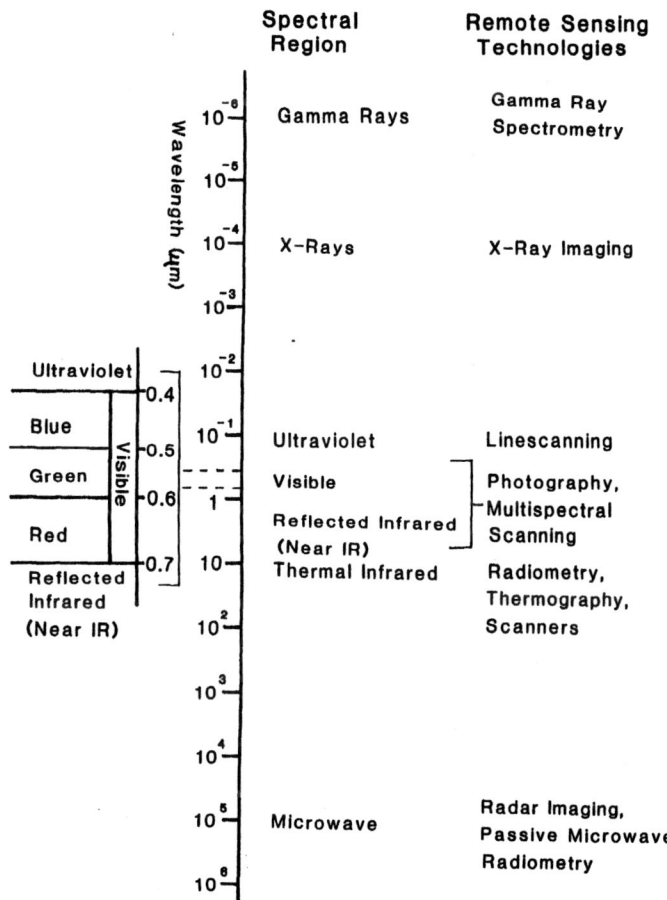

FIG. 16 The electromagnetic spectrum, indicating the principal spectral regions exploited in remote sensing and applicable sensing technologies.

introductory texts [121,122]. Table 2 indicates the principal remote sensing approaches that have been used in the estimation of soil water content.

Remote sensing techniques for measurement of soil water content have several important features in common. First, all the available techniques are capable of measuring water content in surface soil only. Measurements of the average water content to a maximum depth of 0.3 m are possible using gamma-ray spectrometry, but other techniques measure a much shallower soil layer. Second,

TABLE 2 The Principal Remote Sensing Approaches Used in the Estimation of Soil Water Content

Spectral region	Sensing technology	Comments
1. Gamma rays	Spectrometry	Limited potential for ground-based studies but unlikely to compete effectively with other technologies
2. Visible and reflected infrared	Airborne and satellite multispectral photography or scanning of reflectance properties	Few indications of viability in early studies; little promise for future
3. Visible	Polarization photography from aircraft	Promising early results; possible potential for future
4. Thermal infrared	Thermal radiometry or thermal scanners	Considerable potential for further development
5. Microwave	Passive microwave radiometry (airborne or ground-based)	Feasible; some future potential
6. Microwave	Satellite microwave radiometry	Feasible, but spatial resolution may be unacceptably coarse for practical purposes
7. Microwave	Active microwave (radar)	Early results equivocal; possibly useful

because the measurements are made at a distance from the soil, they are subject to interference from objects positioned between the soil and the sensor. Vegetation and cloud are the commonest causes of interference. All but the radar method are affected by cloud cover, while the gamma-radiation and the reflectance and polarization techniques are not suitable for use in vegetated areas. Third, spatial resolution varies with the technique as well as the location of the sensor. The limited spatial resolution means that these techniques are best applied to measurement of water content changes over large areas. There are many applications for such areal estimates, and it is in this context that remote sensing has most potential. However, practical considerations, such as the cost and availability of appropriate data and imagery (particularly if aircraft- or satellite-mounted sensors are necessary), and difficulties of calibration, will often necessitate recourse to point estimation procedures; for although the technology of remotely sensed soil water assessment is now well established, it is still far from a routinely available operational tool for everyday use by agriculturalists and hydrologists.

B. Measurement Techniques

Many of the techniques discussed in this section have been reviewed more fully by Myers [123].

1. Techniques Based on Naturally Occurring Gamma Radiation

The natural gamma radiation field has been widely used with terrestrial and airborne sensors in mineral prospecting [124]. All rock and soils are inherently radioactive and emit gamma radiation. Since soil water effectively attenuates such natural gamma radiation, it is possible to deduce changes in soil water content by repeated gamma-ray spectrometry of areas of interest. Early Soviet work [125] indicates that average near-surface soil water content, in the 0-0.3 m zone can be measured to an accuracy of 10%. A comparable level of accuracy was achieved in a study in the American Great Plains [126], where the gamma-radiation flux was assessed from repeated overpasses by low-flying aircraft. The risk of noise from atmospheric gamma-ray emissions necessitates a very low aircraft altitude, often as low as 100-200 m [127], and consequently the technique can be used only in areas of low relief. Even at such low altitudes, the "ground footprint" of gamma-radiation attenuation techniques may be unacceptably large; no studies have achieved a spatial resolution better than about 200 m. It may well be that the most promising future application of gamma-ray spectrometry for soil water assessment lies in ground-based studies [128].

2. Reflectance and Polarization Techniques in the Visible and Near-Infrared Regions

The interactions that occur between visible or near-infrared radiation and the ground surface are, in part, a function of soil water content. The spectral reflectance of soil generally decreases at higher water contents [129] (i.e., wet soil is darker in color). However, soil spectral properties are influenced by a variety of other factors such as soil texture, structure, illumination geometry, and atmospheric conditions, and care must be taken before ascribing any change in reflectance to water content variation. Both airborne [130] and satellite [131] imagery have been evaluated as a means of monitoring spectral reflectance changes for soil water studies, but the results have been mixed. It is not likely that direct-reflectance studies offer an immediately viable method of soil water measurement.

The polarization characteristics of visible light are significantly affected by soil water content. Early work established that the degree of polarization of reflected visible sunlight is a function of soil water content and soil type [132]. This relationship has since been exploited to assess surface soil water contents from light aircraft, using cameras with polarizing filters [133].

3. Techniques Using Thermal Infrared Radiation

Surface soil temperatures are influenced by a number of factors, one of which is the water content of the soil below. Empirical work in the United States has established how diurnal variations in observed soil temperature can be related to soil water contents at various depths [134,135]. This property has been exploited in ground-based, airborne, and satellite remote sensing studies of soil water, usually employing passive radiometers or line scanning techniques in the 8-14 μm portion of the electromagnetic spectrum, where atmospheric attenuation is at a minimum. It has been demonstrated that even under a straw mulch cover, near-surface water content was the most important variable influencing diurnal soil temperature variation measured by an airborne thermal scanner [136]. Other airborne studies have been similarly encouraging but have highlighted the noise factors introduced by the atmosphere [137].

There have been several attempts to monitor soil water status over a wide area from space, using thermal data. Moore et al. [131] successfully used Skylab S-192 imagery in the 10.2-12.5 μm band. More recently, the Heat Capacity Mapping Mission satellite, launched in 1978, provided data twice daily from two single-channel thermal radiometers, which was used to develop the TELLUS model to estimate thermal inertia, daily evaporation, and water content [138].

4. Passive and Active Microwave Techniques

The solar reflectance and infrared emission techniques already discussed are of use for assessing water contents only in the upper few centimeters of the soil. Considerable promise for sensing the water content of a greater thickness of soil is shown by studies in the microwave portion of the spectrum. The technique depends on the influence of water on the dielectric properties of the soil (see Section IV), causing a decrease in emitted microwave energy at higher water contents [139,140]. Microwaves have the advantage of being scarcely affected by atmospheric conditions.

Experiments over the past few years have demonstrated that airborne microwave radiometers, typically operating in the L-band (~ 0.2 m wavelength), show considerable sensitivity to soil water content variation [141]. For example, Schmugge [142] reports results from a variety of vegetated and unvegetated environments showing how radiometer response correlates well with water content in the 0-0.05 m soil layer. The technique has also been successful

with similar instruments mounted on vehicles [143]. Some researchers [143] have used the shorter wavelength (about 0.06 m) C-band,
although three factors have combined to favor studies in the L-band;
these are the broader range of dielectric constant, and the larger
sampling depth and the reduced attenuation by vegetation canopies
at longer wavelengths. The evidence of current research is that
L-band passive microwave radiometry offers an expensive but reliable method of soil water content determination over large areas.
The size of the radiometer footprint is necessarily large, which can
be very useful, but does present particular difficulties when calibrating such data. This is especially true with satellite-borne microwave radiometers; the Earth Resources Experiment Package (EREP)
on the Skylab mission, for example, afforded a footprint of 115 km
diameter, which was used to generate areal estimates of soil water
status over the southwestern United States [144].

Active microwave techniques offer a better spatial resolution
than passive microwave techniques. Early work focused on determining optimal combinations of frequency and "look angle" [145].
Truck-mounted active microwave sensors have been evaluated under
a variety of conditions [146], and these have recently been extended
to airborne studies [147]. The information content in active microwave data is still uncertain, although the approach offers promise
for the future. Microwave methods liberate the researcher from
operational restrictions such as cloud cover and, with better calibration of radar data, active techniques may yet afford high-resolution estimation of soil water content. It has also been suggested
that a combination of active and passive microwave approaches may
be particularly appropriate [148].

VII. CONCLUSIONS AND SUMMARY

As the preceding pages have shown, the methods available for soil
water content measurement are very diverse. Table 3 summarizes
the features of the methods that have been described, to facilitate
comparisons between them. When planning a program of soil water
content measurement, the following points should be considered before a particular technique is selected.

1. Is the conventional definition of soil water as the water expelled
   on heating to 100-110°C appropriate to the study?
2. Are absolute measures of water content, or water content
   changes, more important?
3. The majority of methods are indirect and so require calibration.
   Are the calibrations already available? If not, calibration will
   increase the cost of the project.

TABLE 3 Available Methods for Measurement of Soil Water Content[a]

| | Applicability | | | | | Maximum measurement depth (m) | Suitable for surface | Depth resolution[b] | Accuracy[c] (% volumetric water content) | Cost | | Advantages | Disadvantages |
	Labor-atory	Field	Saline soils	Organic soils	Stony soils					Initial[c]	Recurrent[d]		
Thermogravimetric	x	x	x	x*	x*	>6[f]	x	Good	±0.1	Low	High	Direct measurement of water content; simple procedure; equipment readily available	Destructive sampling; bulk density required for volumetric determinations; large numbers of samples needed if monitoring changes
Neutron probe		x	x*	x	x*	>6	x*	Poor	±1	High	Low	Established method for monitoring water content in situ; reliable; well researched; can be computerized in some versions	High equipment cost; radiological safety procedures required; special measures necessary to deal with readings in surface soil; care required in access tube installation
Capacitance probe		x	x	x	x*	<2	x*	Good	±2	Mod-erate	Low	Good depth resolution; readings are instantaneous; can be logged automatically	Particular care required for access tube installation; poor sensitivity at surface
TDR	x	x		x	x*	<1.5	x	Poor	±2	High	Low	Ease of installation (except in stony soils)	High cost of equipment; limited depth range
Soil water potential	x	x	x	x	x	>3	x	Good	±1	Low	Low	Soil water potential data provide additional information such as the direction of fluxes in soil profile	Predetermination of soil water characteristics essential; inaccuracies due to hysteresis of water content/potential relationship

Method							Accuracy[c]	Cost[d]	Cost[e]	Remarks	Remarks
Gamma (attenuation)	x	x	x	x*	<1	x*	Good	±2	High	Low	Very good depth resolution with attenuation method but poor with backscatter techniques; measurement of density possible	Double probes are unwieldy; radiological safety procedures necessary; responds to total bulk density, including water content
Gamma (backscatter)		x	x	x*	>3	x*	Poor	±2	High	Low		
Thermal conductivity	x	x	x	x	<3	x	Good	±5	Moderate	Low	Measurements possible in extremely saline soils	Good thermal contact between soil and instrument essential but difficult to ensure; difficulties in shrinking clay soils due to contact problems
NMR	x	x	x*	x	>3	x		±2	High	Low	Possibilities of method still being explored	
Remote sensing	x	x	x*	x	<0.3	x	N/A	±10	High	High	Provides areal estimates of soil water status	Measurement relates to soil surface only; interference from cloud and vegetation cover

[a] Asterisk indicates applicable with special care or modifications to usual methods.

[b] Depth resolution: good < 50 mm, moderate 50–200 mm, poor > 200 mm.

[c] Potential accuracy of absolute water content measurements. Precision of measurements of water content change by in situ methods is better than this.

[d] Initial cost: low < £50; moderate £50–£5000, high > £5000.

[e] Recurrent cost per measurement: low < £1, high > £1.

[f] Limited bulk density determinations required to obtain volumetric water content.

4. Are the measurements required to be areally representative?
    If they are, a spatial sampling study will be necessary, or a
    remote sensing approach might be appropriate.

5. Are there any particular factors (e.g., soil salinity, pH, the
    presence of particular elements, bulk density changes) to be
    taken into account?

6. Many methods require placement of equipment in the soil. The
    equipment and the installation procedure must be designed and
    executed in such a way that they do not influence the soil
    water regime. Equally, a good contact between the soil and
    the equipment is usually essential.

7. Equipment durability and longevity vary enormously, but an
    implicit assumption of any method used in situ is that all equip-
    ment will operate reliably despite extremes of temperature and
    humidity and complete wetting, and all should be free from
    corrosion problems when placed in soil.

8. The presence of stones can present problems because of instal-
    lation and calibration difficulties. Also the stones may have a
    specific effect on the procedure itself.

9. Most methods make a measurement over a volume of soil and so
    can achieve only a limited depth resolution, which must be de-
    fined. Depending on the method, measurement in the vicinity
    of pronounced water content gradients, soil boundaries and,
    in particular, close to the soil-air interface, can give mislead-
    ing results unless a particular method of dealing with such
    measurements is adopted.

10. In most situations, it is essential to avoid short term or long-
    term progressive damage to the soil surface and the vegetation
    growing there. Special precautions may be required, such as
    overhead access.

11. Animals can damage installations. Provision to protect the site
    (without modifying conditions there) may be necessary. Van-
    dalism can also be a problem.

It will be evident from Table 3 that no single method is appropriate for all applications but that some are less versatile than
others. The conventional thermogravimetric procedure will probably always have a place despite its disadvantages. It is possible
that the dominance of the neutron method for in situ field measurements may be reduced in the future when the capacitance probe and
TDR approaches have been fully evaluated. However, they are likely to complement one another, for each has its own particular merits.
For measurements of changes of soil water status over large areas,
remote sensing techniques offer great potential and in the future
are likely to become a more practical proposition.

NOTE ADDED IN PROOF

Since this chapter was written, several significant developments of the TDR technique have been published. Malicki and Shierucha [149] investigated the use of needle pulses rather than the "step" voltages normally employed and found that this simplifies and reduces the cost of the electronics, and allows the use of multiple lines of different length. Zegelin et al. [150] used lines constructed of 3 and 4 wires, which were found to approximate a coaxial line well, eliminating the need for a balancing transformer. Baker and Allmaras [151] described systems for automating and multiplexing a conventional TDR arrangement using a portable data logger and personal computer.

REFERENCES

1. Gardner, W. H., Water content, in *Methods of Soil Analysis*, Part 1 (C. A. Black, Ed.). Am. Soc. Agron., Madison, WI, 1965, pp. 82-187.
2. Cope, F., and E. S. Trickett, Measuring soil moisture, *Soils Fertil.*, 28: 201-208 (1965).
3. Schmugge, T. J., T. J. Jackson, and H. L. McKim, Survey of methods for soil moisture determination, *Water Resour. Res.*, 16: 961-979 (1980).
4. Morrison, R. D., *Groundwater Monitoring Technology: Procedures, Equipment and Applications*, Timco Mfg. Inc., Prairie du Sac, WI, 1983.
5. Gardner, W. H., Water content, in *Methods of Soil Analysis*, Part 1, 2nd ed. (A. Klute, Ed.), Am. Soc. Agron., Madison, WI, 1986, pp. 493-544.
6. Haverkamp, R., M. Vauclin, and G. Vachaud, Error analysis in estimating soil water content from neutron probe measurements: 1. Local standpoint, *Soil Sci.*, 137: 78-90 (1984).
7. Rijksdienst voor de Ijsselmeerpolders, *Explanatory Memorandum on Methods of Analysis for Soil, Organic Matter, Water and Soil Moisture*, Rijksdienst voor de Ijsselmeerpolders, Lelystad, The Netherlands, 1983.
8. Avery, B. W., and C. L. Bascomb (Eds.), *Soil Survey Laboratory Methods*, Soil Survey Tech. Monogr. No. 6, Harpenden, U.K., 1974.
9. Reynolds, S. G., The gravimetric method of soil moisture determination: 1. A study of equipment and methodological problems, *J. Hydrol.*, 11: 258-273 (1970).
10. Liu, C., and J. B. Evett, *Soil Properties: Testing, Measurement and Evaluation*, Prentice-Hall, Englewood Cliffs, NJ, 1984.

11. British Standards Institution, *Methods of Testing Soils For Civil Engineering Purposes*, BS 1377, British Standards Institution, London, 1975.

12. Australian Water Resources Council, *Soil Moisture Measurement and Assessment*, AWRC Hydrological Series No. 9, Australian Dept. Environ. Conserv., Canberra, 1974.

13. Reynolds, S. G., The gravimetric method of soil moisture determination: 2. Typical required sample sizes and methods of reducing variability, *J. Hydrol.*, 11: 274-287 (1970).

14. American Society for Testing and Materials, *Annual Book of ASTM Standards*, ASTM, Philadelphia, 1981.

15. Reinhart, K. G., The problem of stones in soil moisture measurement, *Soil Sci. Soc. Am. Proc.*, 25: 268-270 (1961).

16. Flint, A. L., and S. Childs, *Physical Properties of Rock Fragments and Their Effect on Available Water in Skeletal Soils*, Soil Sci. Soc. Am. Special Publ., No. 13, 1984, pp. 91-103.

17. Hanson, C. T., and R. L. Blevins, Soil water in coarse fragments, *Soil Sci. Soc. Am. J.*, 43: 819-820 (1979).

18. Hewlett, J. D., and J. E. Douglass, A method for calculating error of soil moisture volumes in gravimetric sampling, *For. Sci.*, 7: 265-272 (1961).

19. Bouyoucos, G. J., The alcohol method for determining moisture content of soil, *Soil Sci.*, 32: 173-179 (1931).

20. Bouyoucos, G. J., Evaporating the water with burning alcohol as a rapid means of determining moisture content of soils, *Soil Sci.*, 44: 377-383 (1937).

21. Standards Association of Australia, *Determination of the Moisture Content of a Soil: Microwave-Oven Drying Method*, AS 1289.B1.4-1986 (1986).

22. Roundy, B. A., J. A. Young, G. J. Cluff, and R. A. Evans, *Measurement of Soil Water on Rangelands*, U.S. Dept. Agric. Agricultural Research Results, Western Series No. 31, Oakland, CA, 1983.

23. Baver, L. D., *Soil Physics*, 3rd ed., Wiley, New York, 1964, p. 291.

24. Gupta, U. S., and R. K. Gupta, Evaluation of a sulphuric acid procedure for the determination of soil moisture, *J. Indian Soc. Soil Sci.*, 29: 156-159 (1981).

25. Brummer, E., and E. S. Mardock, A neutron method for measuring saturation in laboratory flow measurements, in *Proc. Am. Inst. Mining Metal Eng.*, Los Angeles, 1945.

26. Pieper, G. F., The measurement of soil moisture by the slowing down of neutrons, Ph.D. thesis, Cornell University, Ithaca, NY, 1949.

27. Belcher, D. J., *The Measurement of Soil Moisture by Neutron and Gamma Ray Scattering*, Civil Aeronautics Admin. Tech. Dev. Report No. 127, 1950, pp. 98-110.
28. Underwood, N., C. H. M. Van Bavel, and R. W. Swanson, A portable slow neutron flux meter for measuring soil moisture, *Soil Sci.*, 77: 339-340 (1954).
29. Stone, J. F., D. Kirkham, and A. A. Read, Soil moisture determination by a portable neutron scattering moisture meter, *Soil Sci. Soc. Am. Proc.*, 19: 419-423 (1955).
30. Holmes, J. W., Calibration and field use of the neutron scattering method of measuring soil water content, *Aust. J. Appl. Sci.*, 7: 45-58 (1956).
31. International Atomic Energy Agency, *Neutron Moisture Gauges*, IAEA Tech. Report Series, No. 112 (1970).
32. Bell, J. P., *Neutron Probe Practice*, Report No. 19, Inst. Hydrol., Wallingford, Oxfordshire, U.K., 1976.
33. Greacen, E. L. (Ed.), *Soil Water Assessment by the Neutron Method*, CSIRO, East Melbourne, Victoria, Australia, 1981.
34. Bautista, I., G. Cruz Romero, J. R. Castel, and C. Ramos, Spatial and time variability of soil moisture in citrus orchards as measured by neutron probe, *Acta Hortic.*, 171: 61-73 (1985).
35. Cooper, J. D., *Measurement of Moisture Fluxes in Unsaturated Soil in Thetford Forest*, Report No. 66, Inst. Hydrol., Wallingford, Oxfordshire, U.K., 1980.
36. McGowan, M., and J. B. Williams, The water balance of an agricultural catchment: 1. Estimation of evaporation from soil water records, *J. Soil Sci.*, 31: 217-230 (1980).
37. McGowan, M., J. B. Williams, and J. L. Monteith, The water balance of an agricultural catchment: III. The water balance, *J. Soil Sci.*, 31: 245-262 (1980).
38. Sophocleus, M., and C. A. Perry, Experimental studies in natural groundwater—Recharge dynamics: The analysis of observed recharge events, *J. Hydrol.*, 81: 297-332 (1985).
39. Gardner, C. M. K., and M. Field. An evaluation of the success of MORECS, a meteorological model, in estimating soil moisture deficits, *Agric. Meteorol.*, 29: 269-284 (1983).
40. Jensen, K. J., Application of soil water flow theory in field simulation, *Nord. Hydrol.*, 12: 167-184 (1981).
41. Wellings, S. R., and J. D. Cooper, The variability of recharge of the English Chalk, *Agric. Water Manage.*, 6: 243-253 (1983).
42. Hodnett, M. G., The neutron probe for soil moisture measurement, in *Advanced Agricultural Instrumentation* (W. G. Gensler, Ed.), Martinus Nijhoff, Dordrecht, The Netherlands, 1986, pp. 148-192.
43. Goldberg, I., N. A. MacGillivray, and R. R. Ziemer, Effects of neutron source type on soil moisture measurement, *Trans. Am. Nucl. Soc.*, 10: 20-21 (1967).

44. Long, I. F., and B. K. French, Measurement of soil moisture in the field by neutron moderation, *J. Soil Sci.*, 18: 149-166 (1967).

45. Sicamois, D., Etude comparative des sources [241]Am-Be, [244]Cm-Be et [252]Cf pour équiper les humidimétres à neutrons, *Bull. Fr. Humidimétres Neutronique*, 8: 35 (1980).

46. Lorch, E. A., The concept of the recommended working life applied to radiation sources, *Radiol. Prot. Bull.*, 34: 20-22 (1980).

47. Cameron, J. F., Nucleonic soil density and moisture gauges, in *The Use of Nuclear Techniques in the Prospecting and Development of Mineral Resources*, Proc. Symposium in Argentina, IAEA, Vienna, 1968.

48. Nicolls, K. D., J. L. Honeysett, and M. W. Hughes, Instrument design, in *Soil Water Assessment by the Neutron Method* (E. L. Greacen, Ed.), CSIRO, East Melbourne, Victoria, Australia, 1981, pp. 24-34.

49. Van Vuuren, W. E., Problems involved in soil moisture determinations by means of a neutron depth probe, in *Recent Investigations in the Zone of Aeration*, Proc. Int. Symposium in Munich, 1984, pp. 281-293.

50. Nakayama, F. S., and R. J. Reginato, Simplifying neutron moisture meter calibration, *Soil Sci.*, 133: 48-52 (1982).

51. Hauser, V. L., Neutron meter calibration and error control, *Trans. Am. Soc. Agric. Eng.*, 27: 722-728 (1984).

52. Eeles, C. W. O., *Installation of Access Tubes and Calibration of Neutron Moisture Meters*, Report No. 7, Inst. Hydrol., Wallingford, Oxfordshire, U.K., 1969.

53. Prebble, R. E., J. A. Forest, J. L. Honeysett, M. W. Hughes, D. S. McIntyre, and G. Schrale, Field installation and maintenance, in *Soil Water Assessment by the Neutron Method* (E. L. Greacen, Ed.), CSIRO, East Melbourne, Victoria, Australia, 1981, pp. 82-98.

54. Bell, J. P., T. J. Dean, and M. G. Hodnett, Soil moisture measurement by an improved capacitance technique: II. Field techniques, evaluation and calibration, *J. Hydrol.*, 93: 79-90 (1987).

55. Jarvis, N. J., and P. B. Leeds-Harrison, Some problems associated with the use of the neutron probe in swelling/shrinking clay soils, *J. Soil Sci.*, 38: 149-156 (1987).

56. Howse, K. R., A technique for using permanent neutron meter access tubes in cultivated soils, *Exp. Agric.*, 17: 265-269 (1981).

57. Parkes, M. E., and N. Siam, Error associated with the measurement of soil moisture change by the neutron probe, *J. Agric. Eng. Res.*, 24: 87-93 (1979).

58. Grant, D. R., Measurement of soil moisture near the surface using a neutron moisture meter, *J. Soil Sci.*, 26: 124-129 (1975).

59. Karsten, J. H. M., and C. J. Van der Vyver, The use of a neutron moisture meter near the surface, *Agrochemophysica*, 11: 45-49 (1979).

60. Roberts, G., *The Processing of Soil Moisture Data*, Report No. 18, Inst. Hydrol., Wallingford, Oxfordshire, U.K., 1972.

61. Hignett, C. T., *A Computer Program to Process Field Data Collected with the Neutron Moisture Meter*, Soil Technical Paper No. 42, CSIRO, East Melbourne, Victoria, Australia, 1980.

62. Hulsman, R. B., The neutron probe and the microcomputer, *Soil Sci.*, 140: 153-157 (1985).

63. Greacen, E. L., R. L. Correl, R. B. Cunningham, G. G. Johns, and K. D. Nicolls, Calibration, in *Soil Water Assessment by the Neutron Probe Method* (E. L. Greacen, Ed.), CSIRO, East Melbourne, Victoria, Australia. 1981, pp. 50-81.

64. Nicolls, K. D., J. T. Hutton, and J. L. Honeysett, Gadolinium in soils and its effects on the count rate of the neutron moisture meter, *Aust. J. Soil Res.*, 15: 287-291 (1977).

65. Greacen, E. L., and G. Schrale, The effect of bulk density on the neutron meter calibration, *Aust. J. Soil Res.*, 14: 159-169 (1976).

66. Greacen, E. L., and C. T. Hignett, Sources of bias in the field calibration of a neutron meter, *Aust. J. Soil Res.*, 17: 405-415 (1979).

67. Rahi, G. S., and S. F. Shih, Effect of bulk density on calibration of neutron moisture probes for organic soils, *Trans. Am. Soc. Agric. Eng.*, 24: 1230-1233 (1981).

68. Lal, R., Concentration and size of gravel in relation to neutron moisture and density probe calibration, *Soil Sci.*, 127: 41-50 (1979).

69. Olgaard, P. L., *On the Theory of the Neutronic Method for Measuring the Water Content in Soil*, Riso Report No. 97, Danish Atomic Energy Commission, 1965.

70. Couchat, P., C. Carre, J. Marcesse, and J. le Ho, The measurement of thermal neutron constants of the soil: Application to the calibration of neutron moisture gauges and to the pedological study of the soil, in *Nuclear Cross-sections and Technology, October 1975* (R. A. Schrack and C. D. Bowman, Eds.), U.S. Dept. Commerce, Washington, DC, 1975, pp. 516-579.

71. McCulloch, D. B., and T. Wall, A method of measuring neutron absorption cross-sections of soil samples for calibration of the neutron moisture meter, *Nucl. Instrum. Meth.*, 137: 577-581 (1976).

72. Moutonnet, P., P. Perrochet, and P. Couchet, Variabilité spatiale des characteristiques neutroniques d'un sol, in *Isotope and Radiation Techniques in Soil Physics and Irrigation Studies*, IAEA, Vienna, 1983, pp. 41-54.

73. Vachaud, G., J. M. Royer, and J. D. Cooper, Comparison of methods of calibration of a neutron probe by gravimetry or neutron capture model, *J. Hydrol.*, 34: 343-356

74. Douglass, J. E., Volumetric calibration of neutron moisture probes, *Soil Sci. Am. Proc.*, 30: 541-544 (1966).

75. Williamson, R. J., and A. K. Turner, Calibration of a neutron moisture meter for catchment hydrology, *Aust. J. Soil Res.*, 18: 1-11 (1980).

76. Stocker, R. V., Calibration of neutron moisture meters on stony soils, *J. Hydrol. N.Z.*, 23: 34-46 (1984).

77. Karsten, J. H. M., and F. J. Haasbroek, A new method of calibrating a neutron moisture meter and a correction for the effect of soil bulk density, *Agrochemophysica*, 5: 35-40 (1973).

78. Carneiro, C., and E. de Jong, In situ determination of the slope of the calibration curve of a neutron probe using a volumetric technique, *Soil Sci.*, 139: 250-254 (1985).

79. Farah, S. M., R. J. Reginato, and F. S. Nakayama, Calibration of a soil surface neutron moisture meter, *Soil Sci.*, 138: 235-239 (1984).

80. Ahuja, L. R., and R. D. Williams, Use of a surface gamma-neutron gauge to measure bulk density, field capacity and macro-porosity in the topsoil, in *Isotope and Radiation Techniques in Soil Physics and Irrigation Studies*, IAEA, Vienna, 1983, pp. 469-478.

81. Ahuja, L. R., R. D. Williams, G. C. Heathman, and J. W. Naney, Use of a surface gamma-neutron gauge to measure effects of tillage, cropping and erosion of soil properties, *Soil Sci.*, 140: 278-286 (1985).

82. Gee, G. W., J. F. Stiver, and H. R. Borchert, Radiation hazard from americium-beryllium neutron probes, *Soil Sci. Soc. Am. J.*, 40: 492-494 (1976).

83. Watson, C. L., and J. L. Honeysett, Safety aspects, in *Soil Water Assessment by the Neutron Method* (E. L. Greacen, Ed.), CSIRO, East Melbourne, Victoria, Australia, 1981, pp. 126-137.

84. Hasted, J. B., *Aqueous Dielectrics*, Chapman and Hall, London, 1973.

85. Hoekstra, P., and A. Delaney, Dielectric properties of soils at UHF and microwave frequencies, *J. Geophys. Res.*, 79: 1699-1708 (1974).

86. Galfy, J., *High-Frequency Moisture Probe*, Commercial Brochure, Vituki Research Centre, Budapest, Hungary, 1984.

87. Dean, T. J., J. P. Bell, and A. J. B. Baty, Soil moisture measurement by an improved capacitance technique: I. Sensor design and performance, *J. Hydrol.*, 93: 67–78 (1987).

88. Thomas, A. M., In situ measurement of moisture in soil and similar substances by fringe capacitance, *J. Sci. Instrum.*, 43: 21–27 (1966).

89. Ansoult, M., L. W. De Backer, and M. Declerq, Statistical relationship between apparent dielectric constant and water content in porous media, *Soil Sci. Soc. Am. J.*, 49: 47–50 (1985).

90. Topp, G. C., J. L. Davis, and A. P. Annan, Electromagnetic determination of soil water content: Measurements in coaxial transmission lines, *Water Resour. Res.*, 16: 574–582 (1980).

91. Dalton, F. N., and M. T. van Genuchten, The time-domain reflectometry method for measuring soil water content and salinity, *Geoderma*, 38: 237–250 (1986).

92. Dalton, F. N., W. N. Herkelrath, D. S. Rawlins, and J. D. Rhoades, Time-domain reflectometry: Simultaneous measurement of soil water content and electrical conductivity with a single probe, *Science*, 224: 989–990 (1984).

93. Dasberg, S., and F. N. Dalton, Time-domain reflectometry field measurements of soil water content and electrical conductivity, *Soil Sci. Soc. Am. J.*, 49: 293–297 (1985).

94. Stein, J., and D. L. Kane, Monitoring the unfrozen water content of soil and snow using time-domain reflectometry, *Water Resour. Res.*, 19: 1573–1584 (1983).

95. Topp, G. C., J. L. Davis, and A. P. Annan, Electromagnetic determination of soil water content using TDR: II. Evaluation of installation and configuration of parallel transmission lines, *Soil Sci. Soc. Am. J.*, 46: 678–684 (1982).

96. Ledieu, J., J. De Ridder, P. De Clerk, and S. Dautrebande, A method of measuring soil moisture by time-domain reflectometry, *J. Hydrol.*, 88: 319–328 (1986).

97. Topp, G. C., J. L. Davis, and A. P. Annan, Electromagnetic determination of soil water content using TDR: I. Applications to wetting fronts and steep gradients, *Soil Sci. Soc. Am. J.*, 46: 672–678 (1982).

98. Topp, G. C., and J. L. Davis, Detecting infiltration of water through soil cracks by time-domain reflectometry, *Geoderma*, 26: 13–23 (1981).

99. Topp, G. C., J. L. Davis, W. G. Bailey, and W. D. Zebchuk, Measurement of soil water content using a portable TDR hand probe, *Can. J. Soil Sci.*, 64: 313–321 (1984).

100. Yanuka, M., G. C. Topp, S. Zegelin, and W. D. Zebchuk, Multiple reflection and attenuation of TDR probes: Theoretical considerations and applications to soil and water, *Water Resour. Res.*, 24: 939–944 (1988).

101. Topp, G. C., M. Yanuka, W. D. Zebchuk, and S. Zegelin, The determination of electrical conductivity using TDR: Soil and water experiments in coaxial lines, *Water Resour. Res.*, 24: 345-352 (1988).

102. Aubic, J., and A. M. Bottreau, Utilisation de la réflecto-métrie en domaine temporal pour l'étude des transports d'humidité en milieu poreux perméable, *J. Hydrol.*, 57: 337-357 (1982).

103. Topp, G. C., and J. L. Davis, Measurement of soil water content using time-domain reflectometry (TDR); a field evaluation, *Soil Sci. Soc. Am. J.*, 49: 19-24 (1985).

104. Patterson, D. E., and M. W. Smith, The measurement of unfrozen water content by time-domain reflectometry: Results from laboratory tests, *Can. Geotech. J.*, 18: 131-144 (1981).

105. Hayhoe, H. N., G. C. Topp, and W. G. Bailey, Measurement of soil water contents and frozen soil depth during a thaw using time-domain reflectometry, *Atmos. Ocean*, 21: 299-311 (1983).

106. Topp, G. C., and J. L. Davis, Time-domain reflectometry (TDR) and its application to irrigation scheduling, *Adv. Irrig.*, 3: 107-127 (1985).

107. Gurr, C., Use of gamma rays in measuring water content and permeability in unsaturated columns of soil, *Soil Sci.*, 94: 224-229 (1962).

108. Ferguson, H., and W. H. Gardner, Water content measurement in soil columns by gamma ray absorption, *Soil Sci. Soc. Am. Proc.*, 26: 11-18 (1962).

109. Wood, B., and N. Collis-George, Moisture content and bulk density measurements using dual-energy beam gamma radiation, *Soil Sci. Soc. Am. J.*, 44: 662-663 (1980).

110. Giesel, W., S. Lorch, and M. Renger, Water flow calculations by means of gamma absorption and tensiometer field measurements in the unsaturated soil profile, in *Isotope Hydrology*, IAEA, Vienna, 1970, pp. 663-672.

111. Reginato, R. J., and R. D. Jackson, Field measurement of soil water content by gamma ray transmission compensated for temperature fluctuations, *Soil Sci. Soc. Am. Proc.*, 35: 529-533 (1971).

112. Ferraz, E. S. B., Gamma-ray attenuation to measure water contents and/or bulk densities of porous materials, in *Isotope and Radiation Techniques in Soil Physics and Irrigation Studies*, IAEA, Vienna, 1983, pp. 449-460.

113. Shaw, B., and L. Baver, An electrothermal method for following moisture changes of the soil in situ, *Soil Sci. Soc. Am. Proc.*, 4: 78-83 (1939).

114. Fritton, D., W. Busscher, and J. Alpert, An inexpensive but durable thermal conductivity probe for field use, *Soil Sci. Soc. Am. Proc.*, 38: 854-855 (1974).

115. Sophocleus, M., A thermal conductivity probe designed for easy installation and recovery from shallow depths, *Soil Sci. Soc. Am. J.*, 43: 1056-1058 (1979).

116. Bloodworth, M., and J. Page, Use of thermistors for the measurement of soil moisture and temperature, *Soil Sci. Soc. Am. Proc.*, 21: 11-15 (1957).

117. Andreyev, S. V., and B. K. Martens, Soil moisture determination by the method of nuclear magnetic resonance, *Sov. Soil Sci.*, 10: 1129-1132 (1960).

118. Prebble, R. E., and J. A. Currie, Soil water measurement by a low-resolution nuclear magnetic resonance technique, *J. Soil Sci.*, 21: 273-288 (1970).

119. Paetzold, R. F., and G. A. Matzkanin, *NMR Measurement of Water in Clay*, ILRI Publication No. 37, U.S. Dept. Agric., Soil Conserv. Serv., Beltsville, MD, 1984, pp. 316-319.

120. Paetzold, R. F., G. A. Matzkanin, and A. De Los Santos, Surface soil water content measurement using pulsed nuclear magnetic resonance techniques, *Soil Sci. Soc. Am. J.*, 49: 537-540 (1985).

121. Colwell, R. N., D. S. Simonett, and J. E. Estes (Eds.), *Manual of Remote Sensing*, 2nd ed. (2 vols.), Am. Soc. Photogrammetry, Falls Church, VA, 1983.

122. Barret, E. C., and L. F. Curtis, *Introduction to Environmental Remote Sensing*, 2nd ed., Chapman and Hall, London, 1982.

123. Myers, V. I., Remote sensing applications in agriculture, in *Manual of Remote Sensing*, Vol. 2 (R. N. Colwell, D. S. Simonett, and J. E. Estes, Eds.), Am. Soc. Photogrammetry, Falls Church, VA, 1983, pp. 2111-2228.

124. Grasty, R. L., Applications of gamma radiation in remote sensing, in *Remote Sensing for Environmental Sciences* (E. Schanda, Ed.), Springer-Verlag, Berlin, 1976, pp. 257-276.

125. Zotimor, N. V., Use of the gamma field of the earth to determine the water content of the soil, *Sov. Hydrol.*, 4: 313-320 (1971).

126. Carroll, T. R., Airborne soil moisture measurement using natural terrestrial gamma radiation, *Soil Sci.*, 132: 358-366 (1981).

127. Salomonsen, V. V., Water resources assessment, in *Manual of Remote Sensing*, Vol. 2 (R. N. Colwell, D. S. Simonett, and J. E. Estes, Eds.), Am. Soc. Photogrammetry, Falls Church, VA, 1983, pp. 1497-1570.

128. Loijens, H. S., Determination of soil water content from terrestrial gamma radiation measurements, *Water Resour. Res.*, 16: 565-573 (1980).

129. Condit, H. R., The spectral reflectance of American soils, *Photogramm. Eng.*, 36: 955-966 (1970).

130. Evans, R., Air photos for soil survey in lowland England: Factors affecting the photographic images of bare soils and their relevance to assessing soil moisture content and discrimination of soils by remote sensing, *Remote Sensing Environ.*, 8: 39-63 (1979).

131. Moore, F. G., M. L. Horton, J. J. Russell, and V. I. Myers, Evaluation of thermal X/5 detector Skylab S-192 data for estimating evapotranspiration and thermal properties of soils for irrigation management, in *Proc. NASA Earth Resources Survey Symp.*, NASA Report TM-X-58168, Houston, TX, 1975, pp. 2561-2583.

132. Stockhoff, E. H., and R. T. Frost, Polarisation of light reflected by moist soils, in *Proc. 7th Symp. Remote Sensing of Environment*, Environ. Res. Inst. Michigan, Ann Arbor, MI, 1972.

133. Curran, P. J., Remote sensing: The use of polarized visible light to estimate soil moisture, *Appl. Geog.*, 1: 41-53, 1981.

134. Idso, S. B., T. J. Schmugge, R. D. Jackson, and R. J. Reginato, The utility of surface temperature measurement for the remote sensing of soil water status, *J. Geophys. Res.*, 80: 3044-3049 (1975).

135. Reginato, R. J., S. B. Idso, J. F. Vedder, R. D. Jackson, M. B. Blanchard, and R. Goettelman, Soil water content and evaporation determined by thermal parameters obtained from ground based and remote measurements, *J. Geophys. Res.*, 81: 1616-1620 (1976).

136. Cihlar, R., T. Sommerfeldt, and B. Patterson, Soil water content estimation in fallow fields from airborne thermal scanner measurements, *Can. J. Remote Sensing*, 5: 18-32 (1979).

137. Elkington, M. D., and J. Hogg, The characterisation of soil moisture content and actual evapotranspiration from crop canopies using thermal infrared remote sensing, in *Geological and Terrain Analysis Studies by Remote Sensing* (J. A. Allen and M. Bradshaw, Eds.), Remote Sensing Soc., Reading, U.K., 1981, pp. 69-90.

138. Gurney, R. J., The estimation of soil moisture content and actual evapotranspiration using thermal infrared remote sensing, in *Remote Sensing and National Mapping* (J. A. Allen and R. Harris, Eds.), Remote Sensing Soc., Reading, U.K., 1979, pp. 101-109.

139. Schmugge, T. J. P. Gloersen, T. Wilheit, and F. Geiger, Remote sensing of soil moisture with microwave radiometers, *J. Geophys. Res.*, 79: 317-323 (1974).

140. Njoku, E. G., and J. A. Kong, Theory for passive microwave sensing of near-surface soil moisture, *J. Geophys. Res.*, 82: 3108-3114 (1977).

141. Estes, J. E., M. R. Mel, and J. O. Hooper, Measuring soil moisture with an airborne imaging passive microwave radiometer, *Photogramm. Eng. Remote Sensing*, 43: 1273-1281 (1977).

142. Schmugge, T., Remote sensing of soil moisture with microwave radiometers, *Trans. Am. Soc. Agric. Eng.*, 26: 748-753 (1983).

143. Newton, R. W., and J. W. Rouse, Microwave radiometer measurements of moisture content, *IEEE Trans. Antennae Propagation*, AP-28: 680-684 (1980).

144. Eagleman, J., and W. Lin, Remote sensing of soil moisture by a 21 cm passive radiometer, *J. Geophys. Res.*, 81: 3660-3666 (1976).

145. Ulaby, F. T., Radar measurement of soil moisture content, *IEEE Trans. Antennae and Propagation*, AP-22: 257-265 (1974).

146. LeToan, T., Scatterometer measurements on crop and soil surfaces, Report No. ESA-SP-166, in *Proc. ESA Workshop on Coherent and Incoherent Radar Scattering from Rough Surfaces and Vegetated Areas*, European Space Agency, Paris, 1981.

147. Jackson, T. J., and P. E. O'Neil, Aircraft scatterometer observations of soil water on rangeland watersheds, *Int. J. Remote Sensing*, 6: 1135-1162 (1985).

148. Ulaby, F. T., M. C. Dobson, and D. R. Brunfeldt, Improvement of moisture estimation accuracy of vegetation covered soil by combined active/passive microwave remote sensing, *IEEE Trans. Geoscience Remote Sensing*, GRS-21: 300-307 (1983).

149. Malicki, M. A., and W. M. Skierucha, A manually controlled TDR soil moisture meter operating with 300 ps rise-time needle pulse, *Irrig. Sci.*, 10: 153-163 (1989).

150. Zegelin, S. J., I. White, and D. R. Jenkins, Improved field probes for soil water content and electrical conductivity measurement using time domain reflectometry, *Water Resour. Res.*, 25: 2367-2376 (1989).

151. Baker, J. M., and R. R. Allmaras, System for automating and multiplexing soil moisture measurement by time domain reflectometry, *Soil Sci. Soc. Am. J.*, 54: 1-6 (1990).

2
Matric Potential

CHRIS E. MULLINS *Department of Plant and Soil Science,
University of Aberdeen, Aberdeen, Scotland*

I. INTRODUCTION

The total potential ψ_t of soil water refers to the potential energy
of water in the soil with respect to a defined reference state. Various components of this total potential control water flow in the soil
(Chaps. 4 and 5), from the soil into the plant root, and through
the plant. Matric potential refers to the tenacity with which water
is held by the soil matrix [1] and, in the absence of high concentrations of solutes, is the major factor that determines the availability of water to plants. Differences in the value of matric potential
between different parts of the soil also provide the driving force
for the unsaturated flow of soil water after any differences in elevation have been allowed for (Chap. 5).

A. Definition

Precise definitions of total potential and its various component potentials were provided by a committee of the International Soil Science
Society (ISSS) in 1963 [2] and slightly modified in 1976 [3]. A
brief summary is provided here, but for a more detailed discussion
of the meaning and significance of these definitions the reader is
referred to books on soil physics such as those by Hillel [4,5],
Marshall and Holmes [6], and Rose [7].

Total potential of soil water can be divided into three components:

$$\psi_t = \psi_p + \psi_g + \psi_o \tag{1}$$

75

The pressure potential ψ_p is defined as "the amount of useful work that must be done per unit quantity of pure water to transfer reversibly and isothermally to the soil water an infinitesimal quantity of water from a pool at standard atmospheric pressure that contains a solution identical in composition to the soil water and is at the elevation of the point under consideration" [6]. Similar definitions have been given for gravitational, ψ_g, and osmotic potential, ψ_o, which refer to the effects of elevation (i.e., the gravitational field) and of solutes on the energy status of soil water. The sum of gravitational and pressure potential is called the hydraulic potential ψ_h. Differences between the hydraulic potential at different places in the soil provide the driving force for the movement of soil water. Matric potential ψ_m is a subcomponent of pressure potential and is defined as the value of ψ_p, where there is no difference between the pressure of air or gas in the soil and the gas pressure on the water in the reference state.

The ISSS definition of pressure potential already given includes (1) the positive hydrostatic pressure that exists below a water table, (2) the potential difference experienced by soil that is under a gas pressure different from that of water in the reference state, and (3) the negative pressure or suction experienced by soil water as a result of its affinity for the soil matrix. Some authors [8,9] depart from this definition and use the term "pressure potential" to refer only to subcomponents 1 and 2. Fortunately, all authors agree on equivalent definitions for matric potential, which is subcomponent 3. Matric potential can have only a zero or negative value, the magnitude of which increases as water is held more and more tightly in the soil. Matric or soil water suction or tension refers to the same property but takes the opposite sign (to matric potential). In a swelling soil, overburden pressure can cause a slight error in applications where it is intended to relate matric potential to soil water content [10].

The sum of matric and osmotic potential is referred to as the water potential ψ_w and is directly related to the relative humidity of vapor in equilibrium with the liquid phase in soils and in plants. ψ_w is an important measure of plant water status and is also important in saline soils, where the osmotic potential of the soil solution is of sufficient magnitude to influence plant water uptake.

B. Units

Since potentials are defined as energy per unit mass, they have units of joules per kilogram in the SI system. However, it is also possible to define potentials as energy per unit volume or per unit weight. Thus, since the dimensions of energy per unit volume are identical to those of pressure, the appropriate SI unit of pressure

TABLE 1 Summary of Methods for Measuring Matric ψ_m or Water ψ_w Potential[a]

Method	Range	Accuracy ($\pm$)[a]	Commercial equipment	Approx. cost (U.S. $)
Tensiometers (ψ_m)	0 to -85 kPa	0.25 kPa	Bourdon gauge Portable Bourdon gauge Mercury manometer[b] Puncture tensiometer Portable electronic readout	100 1000 300 50 3000
Filter paper (ψ_m/ψ_w)	-1 kPa to -100 MPa	~ $\pm 30\%$ at -10 kPa	Paper	1
Electrical resistance sensors (ψ_m) Fiberglass Gypsum	-0.1 to -100 kPa -50 to -1.5 MPa	Low, depends on calibration, salinity, and temperature correction	Fiberglass sensor Gypsum sensor Meter	60 20 1000
Heat dissipation sensors (ψ_m)	0 to -1 MPa	0 to -100 $\pm$ 10 kPa 0 to -300 $\pm$ 20 kPa 0 to -600 $\pm$ 100 kPa		
Psychrometers (ψ_w) Richards Spanner	0 to -300 MPa 0 to -7 MPa	Laboratory, 4 kPa Field, 50 kPa	Laboratory sample changer psychrometer Soil psychrometer Nanovoltmeter thermometer Dew-point microvoltmeter	2000 50 2000 3500

[a]Accuracy figure represents the best reliable reported values, but see text for more detail since accuracy may depend on a number of factors.
[b]Multiple units are also available.

TABLE 2 Addresses of Some Manufacturers (M), and Other
Suppliers (S) of Equipment

Equipment[a]	Address of manufacturer
	United States
(M) T, E	Soilmoisture Equipment Corp., P.O. Box 30025, Santa Barbara, CA 93105
Pun	Soil Measurement Systems, 1906 South Espina Street, Suite Six, Las Cruces, NM 88001
(M) Ps, Pf, D	Wescor, Inc., 459 South Main, Logan UT 84321
(M) Pr	Decagon Devices Inc., P.O. Box 835, Pullman, WA 99163
(M) Pf	J. R. D. Merrill Equipment, RFD Box 140A, Logan, UT 84321
(M) D	Campbell Scientific, Inc. Box 551, Logan, UT 84321
	United Kingdom
(S) T, E, D	ELE International, Agronomics Division, Eastman Way, Hemel Hempstead, HERTS HP2 7HB
(S) T, Pun	Van Walt Agricultural Services, 108 Blackdown Rural Industries, Haslemere, Surrey GU 27 3BR
(S) T, E	Vinten Instruments Ltd., Jessamy Road, Weybridge, Surrey KT13 8LE
	France
(M) T, Pun	Nardeux Humisol, 4 Avenue du Quebec—Batiment Saint Laurent, Z.A. de Courtaboef, 91940 Les Ulis
	The Netherlands
T	Eijkelkamp, Nijverheidsstraat 14, 6987 EM Giesbeck
	West Germany
(M) Pun	Adolf Thies GmbH, Postfach 3536 + 3541, Hauptstrasse 76, D-3400 Göttingen

[a]Key: T, tensiometers; Pun, puncture tensiometers; E, electrical
resistance sensors. Laboratory psychrometers: Pr, Richards type;
Ps, Spanner type; Pf, field psychrometers; D, data loggers.

is the pascal (at present the bar, equal to 10^5 Pa, is more commonly used). Similarly, the dimensions of energy per unit weight are
identical to those of length, so the appropriate SI unit is the meter.
Because it is common to refer to the pressure due to a height h of
a column of water as a pressure head (or simply head) h, this terminology is often used to describe potential energy per unit weight.
The relation

$$\psi(\text{J kg}^{-1}) = \gamma\psi(\text{Pa}) = \frac{\psi}{g} \text{ (m)} \qquad (2)$$

where γ is the density of water and g is the acceleration due to
gravity ($\sim$ 1000 kg m^{-3} and 9.81 m s^{-2}, respectively), is used to
convert potentials from one system of dimensions to another. A
logarithmic pF scale for expressing matric potential proposed by
Schofield [11] is also in use, where

$$\text{pF} = \log_{10} \quad \text{(negative pressure head, in cm)} \qquad (3)$$

II. MEASURING INSTRUMENTS

A. An Overview

The main features of the available methods for measuring matric potential are summarized in Table 1. Methods have been classified
according to the measurement principle involved and are discussed
in detail in the following sections. Thus tensiometers (Section II.B)
consist of a porous cup attached via a liquid-filled column to a manometer. Porous material sensors (Section II.C) consist of a porous
material whose water content varies with matric potential in a reproducible manner; a physical property of the material that varies with
water content is measured and related to matric potential, using a
calibration curve. Psychrometers (Section II.D) measure the relative humidity of water vapor in equilibrium with the soil solution.
Because psychrometers measure the sum of matric and osmotic potentials, they are also readily applicable for measurement of the potential in various parts of plants.

Many of these instruments are available in versions that may be
attached to a data logger or some kind of continuous recording device in the field. The addresses of some manufacturers and suppliers are listed in Table 2. There are many earlier reviews of the
design and use of measuring instruments [1,12-15].

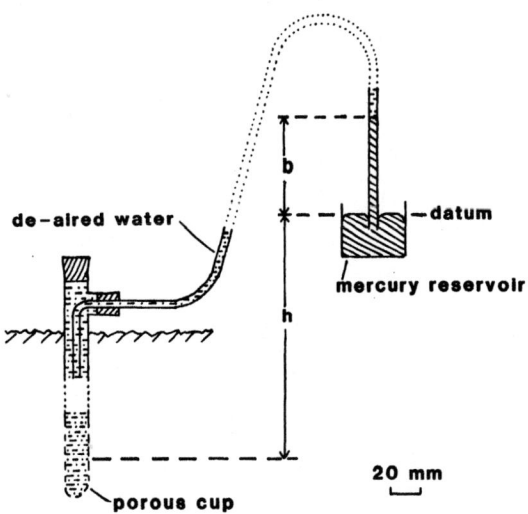

FIG. 1   Mercury manometer tensiometer.

B.   Tensiometers

A tensiometer consists of a porous cup connected to a manometer
with all parts of the system water-filled (Fig. 1).   When the cup is
in contact with the soil, water films make a hydraulic connection between soil water and the water within the porous cup via the pores
in the cup wall.   Water then moves into or out of the cup until the
(negative) pressure inside the cup equals the matric potential of the
soil water outside.

The following equations are used to obtain matric and hydraulic
potential from the readings obtained from the mercury manometer
shown in Fig. 1.

$$\psi_m = \frac{h - 12.6b - c}{g}$$

$$\psi_h = -\frac{12.6b + c}{g}$$ (4)

The factor of 12.6 appears as the difference between the relative
density of mercury and that of water.   c is a correction factor to
allow for the capillary depression that occurs at the mercury-water
interface.   If g is omitted from these two equations, they will give
the potentials in head units.

Tensiometers are also available with Bourdon vacuum gauges or transducer pressure sensors. The latter are used when automatic recording is required. Finally, portable tensiometers are also available. Discussion of the limitations common to most of these designs precedes a consideration of each type of tensiometer.

1. Design Limitations

Trapped Air. All water-filled tensiometers have a lower measuring limit of about -85 kPa (-0.85 bar) because, at more negative potentials, there is a tendency for air bubbles to nucleate at microscopic irregularities within the instrument. At such a low pressure relative to atmospheric pressure these bubbles grow in size, augmented by dissolved air coming out of solution, and can eventually form within the instrument blockages that make further readings unreliable. To counteract this tendency it is usual to fill tensiometers with (deaired) water, which has had some of its dissolved air removed by boiling or by leaving the water for a period of hours in an evacuated container. Despite this procedure, even at potentials greater than -85 kPa there is a tendency for dissolved air to move in through the porous cup and come out of solution within the instrument under the reduced pressure that exists there. Consequently tensiometers are designed with an air trap positioned to allow air to collect within without interfering with operations (Fig. 1). Because accumulation of air within the instrument will eventually lead to deterioration in its response time, especially at low potentials, it is normal to "purge" tensiometers at regular intervals (~ weekly; less often under cool wet conditions) [13]. During this operation, which involves replacing the trapped air with deaired water, there is a temporary release of any suction that exists within the tensiometer, causing some water to pass out of the porous cup into the surrounding soil. Therefore readings taken shortly after purging are unreliable.

Response Time. Any tensiometer will take a finite time to register a change in matric potential, since this will cause a change in the volume of liquid within the tensiometer, and consequently water needs time to move from the soil into the tensiometer or vice versa. The rate at which water can transfer is influenced by the conductance of the porous cup and by the unsaturated hydraulic conductivity of the soil, while the amount of water that must move for a given change in potential (known as the "gauge" sensitivity) depends on the type of pressure sensor. Mercury manometer and Bourdon vacuum gauges are much less sensitive than pressure transducers; in practice, however, because most tensiometers operate under conditions in which there is some air within the instrument, and because the instrument tubing is not entirely rigid, the difference in

sensitivity between pressure transducer and other types of tensiometer is much less than would be predicted from the sensitivity of
the transducer alone.

If the tensiometer response time is not influenced by soil properties but is determined by the cup conductance and gauge sensitivity alone, then the tensiometer behavior is said to be "tensiometer-
limited;" otherwise it is "soil-limited." The tensiometer-limited response time is inversely proportional to cup conductance and to
gauge sensitivity [16], and high conductance cups with 100 times
greater conductivity than normal cups are available for specialized
applications. Towner [17] has demonstrated that it is not difficult
to attain tensiometer-limited conditions in practice, although in some
soils in the drier state many tensiometers will be soil-limited. According to Towner's theory, it should always be possible to obtain
a tensiometer-limited response by increasing the sensitivity sufficiently.

It is advantageous to operate under tensiometer-limited conditions,
since the instrument behavior is then not dependent on variable soil
conditions but is predictable [18]. This is particularly important
when the potential is changing fast, and a small response time is in
any case desirable. However, where tensiometers are used to monitor conditions in the field over long periods and are read only at
infrequent intervals, obtaining a tensiometer-limited response is not
an overriding design consideration. Furthermore, tensiometers of
high sensitivity will be more responsive to factors other than matric
potential, which can also cause small changes in the liquid-filled volume, such as changes in temperature [19] and bending of the tensiometer tubing; thus such systems must be treated with care and
are found to operate better in a laboratory environment. In field
use, even with mercury manometer tensiometers, it is important to
keep all water-filled parts of the system out of direct sunlight insofar as is possible. Otherwise, a sudden change in temperature can
cause volume changes in the tubing, the water it contains, and in
any air in the air trap, thus producing a temporary perturbation
of the reading.

Porous cups, usually made of a ceramic material, must have
pores small enough to prevent air from entering the cup when its
walls are saturated. The cups must also have a conductance high
enough to allow the tensiometer to respond quickly to changes in
potential. Ceramic tensiometer cups for field use have a conductance of about 3×10^{-9} m^2 s^{-1}, and a mercury manometer tensiometer constructed with such a cup will have a (tensiometer-limited)
response time of about one minute in the absence of any trapped
air [13], more than adequate for most field uses.

2. Mercury Manometer and Bourdon Gauge Tensiometers

Commercial mercury manometer and Bourdon gauge tensiometers are readily available (Table 2). A mercury manometer scale can easily be read to the nearest millimeter so that such tensiometers have a scale resolution of about ±0.1 kPa. However, with the smallest (1.7 mm diameter) nylon tubing commonly used for the manometer, there is a significant capillary correction of about 0.8 kPa to be made (see Eq. 4) and hysteresis, caused by the mercury meniscus sticking to the walls of the nylon tube, limits the accuracy to ±0.25 kPa even when the nylon tube is agitated to produce a small fluctuation in mercury level before measurement [20]. Without this precaution, much larger hysteresis errors can occur. An accuracy of ±0.25 kPa is still an order of magnitude better than Bourdon vacuum gauges which, typically, have a scale division every 2 kPa. Furthermore, friction within the gauge mechanism and the difficulty of obtaining an accurate zero setting further limit the accuracy of such gauges. Consequently, although Bourdon gauge tensiometers do not present the environmental hazard of mercury and avoid the inconvenience of a 1 m manometer post, they are of little use in applications calling for high accuracy (e.g., measurement of vertical gradients in hydraulic potential).

It is possible to construct mercury manometer tensiometers without the use of workshop facilities. These probably represent the cheapest accurate soil physics equipment in use. Construction, installation, and maintenance details have been given by Webster [21] and Cassell and Klute [13]. Where several tensiometers are to be used in the same vicinity, it is common to share a single mercury reservoir among 6-30 tensiometers. Because the mercury withdrawn from the reservoir will cause a slight drop in the reservoir level, for high accuracy, it is advisable to measure the reservoir level each time a reading is taken or to have a reservoir with a cross-section very much greater than the sum of the cross-sections of all the manometer tubes that dip into it. It is also advisable to check each tensiometer for air leaks before installation. This is done by soaking the porous cup in water, then applying an air pressure of 100 kPa to the inside of the tensiometer while it is immersed in water [13]. To minimize thermal effects, the manometer tubing should be shielded from direct sunlight (e.g., by facing the manometer post away from the midday sun). With prolonged outside use, some plasticizer may gradually come out of the nylon tubing and collect as a white deposit, which can cause problems and may eventually block the tubing. This has not been found to be a problem over a single growing season, but when a manometer is used for long periods, replacement of 1.7 mm (i.d.) tubing is occasionally necessary.

3. Pressure Transducers and Automatic Systems

Because of the high gauge sensitivity of pressure transducers, the use of these devices is appropriate wherever a short response time is important. They are also required where automatic recording is used. Until recently, temperature-compensated strain gauge pressure transducers have been the main form of pressure transducer available, and the high cost of the transducer and associated circuitry has severely limited the extent of their use. Low-cost solid-state transducers, which have become available more recently, should transform this situation. Long [22] describes the use of a temperature-compensated transducer with integral circuitry that gives a change in dc output of 48 mV per kilopascal from a 15 V dc supply. Mullins et al. [20] have used a piezoresistive silicon transducer and have described associated circuitry, including a portable digital readout, which can be assembled at a cost of a few hundred dollars. These transducers are not sensitive to temperature effects and have a precision of ±100 Pa, although if not periodically corrected for, long-term drift may limit their accuracy. Whatever the design of transducer used, it should be a type that is vented to the atmosphere to avoid the necessity for making adjustments for changes in atmospheric pressure.

In the unusual case that matric potentials are required at a considerable depth (say 10 m) in the profile, a pressure transducer located close to the measuring depth is essential because a hanging water column will break once the tension in it approaches 100 kPa (i.e., 10 m of water or atmospheric pressure).

Automatic Recording Systems. Automatic recording systems are required both at remote sites, where measurements are required more often than the systems can be visited, and to study laboratory or field situations in which many measurements are required over a period of hours or days (e.g., drainage studies). In the former case, a provision for periodic automatic purging of the tensiometers may also be necessary. Many systems have been described (see, e.g., Refs. 23-25) that use a motor-driven fluid-scanning switch to allow a number of tensiometers to be connected each in turn to a single pressure transducer. In this way, up to 24 tensiometers may be connected to a single chart recorder or data logger, although in practice some switch positions are used for pressure calibration checks [25].

These systems are not appropriate where a very short measurement interval is required because a small fluctuation in pressure cannot be avoided when the transducer is switched between tensiometers at different potentials. Reequilibration can take 2 minutes [25] or more [68]. Temperature fluctuations can also influence the output, especially where nylon tubing is exposed above ground [19,25,

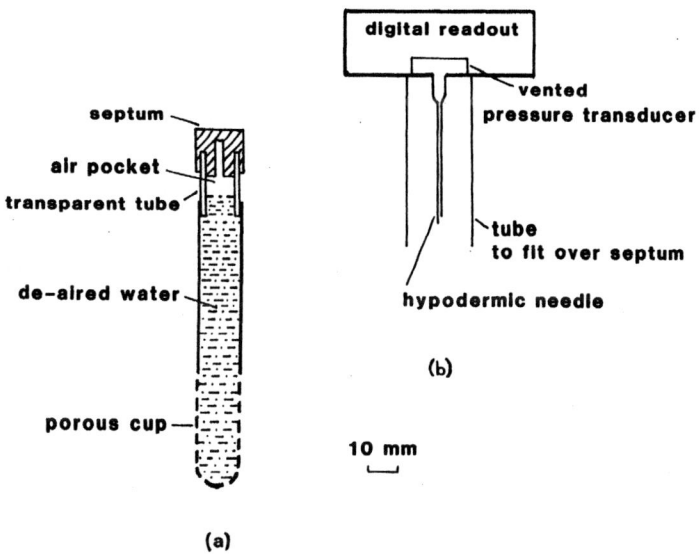

septum

air pocket

transparent tube

de-aired water

porous cup

digital readout

vented
pressure transducer

tube
to fit over septum

hypodermic needle

(b)

10 mm

(a)

FIG. 2 Puncture tensiometer: (a) tensiometer and (b) portable pressure sensor. (Adapted from Ref. 26.)

68]. Both these problems are avoided if a transducer is attached to each tensiometer and, because the cost of a suitable pressure transducer is now less than $100, it may be more simple and economical with small numbers of tensiometers to use a transducer with each tensiometer, connected in turn to a millivoltmeter or automatic recording system [22].

Portable Transducer Systems. Portable pressure transducer systems, which can be temporarily connected to permanently installed tensiometers by puncturing a septum on the tensiometer with a hypodermic needle attached to the pressure transducer are commercially available (Table 2). This arrangement allows one pressure transducer to be used to monitor a large number of tensiometers. A small air pocket is deliberately left at the top of each tensiometer (Fig. 2) to reduce the small pressure surge caused by introducing a hypodermic needle through the septum. In the design of Marthaler et al. [26] the needle-pressure transducer assembly contains an air-filled dead volume, which should result in a fractional error roughly equal to the dead volume divided by the volume of the air pocket (1/20 in their example). In practice they observed considerably larger systematic errors of about 0.7 kPa in values of potentials close to zero (2-3.6 kPa) although the overall relation between

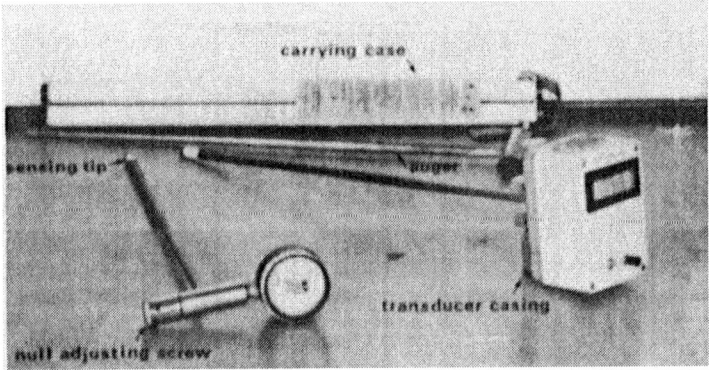

FIG. 3 Portable tensiometers with a Bourdon vacuum gauge or
transducer and digital display [20].

mercury manometer readings and those from the puncture tensiometer was good (r = 0.99). In the design of Frede et al. [27], the
dead volume is kept filled with water, but the accuracy of such a
system in practice has not been reported.

4. Portable Tensiometers

A portable tensiometer with a Bourdon vacuum gauge is commercially
available, and Mullins et al. [20] provide a design whereby this can
be modified to take a pressure transducer and digital readout (Fig.
3), thus increasing the accuracy to ±100 Pa. Another transducer
tensiometer with a 2.3 mm diameter probe was specifically designed
for use in peat blocks used for raising seedlings [28] but would be
applicable wherever minimal soil disturbance is required, provided
sufficient hydraulic contact between porous sensor and the soil
could be obtained. Because portable tensiometers are stored with
their sensing tips immersed in water when not in use, there is little
accumulation of air within the tensiometer and the devices rarely
need to be refilled. Such tensiometers should be of use where single or occasional measurements are required, such as when checking on the performance of irrigation or drainage systems, or in
studying the spatial variation of potential before choosing a representative site for a permanent tensiometer installation. Unfortunately, even when the hole into which the tensiometer is inserted has
been made with an auger of the same diameter, insertion of the tensiometer causes deformation of the soil around the sensing tip, and

it takes some time for this soil to reequilibrate with the undisturbed soil (irrespective of use of the null-point device supplied on the commercial model). In the sandy loam used by Mullins et al. [20], the investigator's transducer tensiometer equilibrated within a few minutes at matric potentials greater than -5 kPa but required more than 2 hours at potentials less than -30 kPa.

5. Osmotic Tensiometers

The design and performance of an osmotic tensiometer has been described by Peck and Rabbidge [29]. This device consists of a cell containing a high-molecular-weight (20,000) polyethylene glycol solution confined between a pressure transducer and a semipermeable membrane supported on a porous ceramic. The cell is pressurized so that, when immersed in pure water, it registers a pressure of about 1.5 MPa; this allows the transiometer to measure matric potentials in the range of 0 to -1.5 MPa. Problems of leakage of solution or solute from within the tensiometer, solution depolymerization, and strain gauge creep were all suggested as possible explanations for the observed rate of pressure loss of between 1 and 2% per month [29]. This loss limits the practical application of such tensiometers which have been superseded by sensors of cheaper types (see Sections II.C, II.D).

C. Porous Material Sensors

Sensors of this type consist of a porous material whose water content varies with matric potential in a reproducible manner. A physical property (of the material) that varies with water content is measured and related to matric potential using a calibration curve. Sensors based on measurement of the weight of filter paper, electrical conductivity, and heat dissipation are discussed.

Irrespective of the method used to measure water content of the porous material, the physical properties of the porous material itself determine the range of matric potentials over which the sensor will be sensitive and accurate. Sensitivity depends on the rate of change of water content with matric potential, hence on the pore size distribution of the porous material. A major restriction to accuracy can be the amount of hysteresis the material displays. The porous material is calibrated by equilibrating it at a set of known matric potentials. The reliability of published calibration curves or those supplied by manufacturers depends on how closely the water release characteristic of the sensor resembles that of the sensor used in the original calibration. For greater accuracy, users should calibrate all, or at least a representative sample, of their sensors at potentials at either end of the range of interest and at least one point in between.

1. The Filter Paper Method

The filter paper method, originally used in 1937 as a simple procedure for obtaining the soil water release characteristic [30], is a
cheap and simple method for measuring matric potential that has not
received the use it deserves. The method consists of placing filter
paper in contact with a soil sample (> 100 g) in a sealed tin or plastic container at constant temperature until an equilibrium is reached.
The gravimetric water content of the filter paper is then determined,
and this is converted to matric potential using a calibration curve.
Apart from calibrated filter papers, this technique requires only a
home-made lagged sample-equilibration box, an oven set at 105°C,
and a balance weighing to ±1 mg.

The water retention characteristic of a filter paper (which is its
calibration curve) can usefully cover a very wide range of potentials
from -1 kPa to -100 MPa [31]. At the moister end of this range, as
equilibration will occur by water flow between soil and the filter paper; it is important that the soil sample be trimmed to make good
contact with the filter paper and that it rest on top of the paper
and fully cover it. Alternatively, the paper may be sandwiched between the two halves of a core. For dryer soil, vapor equilibration
is more likely so that the paper responds to the sum of both matric
and osmotic potential. Vapor equilibration is a much slower process,
and equilibration times from 3 to 7 days have been used [31-33].
Small temperature fluctuations during vapor equilibration can disturb
the process and may even cause distillation (i.e., the condensation
of vapor on the walls of the container) [34]. To avoid these problems, the sealed containers should be kept thermally insulated in
Styrofoam (expanded polystyrene) containers, out of direct sunlight,
and in a room or cupboard that does not have a large diurnal temperature variation [15].

Since the potential of a sample can be changed by deformation,
it is important to use an undisturbed sample or soil core and to
transport it with a minimum of vibration or to equilibrate it in situ
[33]. Hamblin [33] has also used the technique in situ by introducing papers into slits cut with a spatula in field soils, or by placing
papers between sections of long cores to obtain a matric potential
profile. However, as Mullins et al. [20] have demonstrated, even
the small disturbance likely to be incurred during sampling or placing filter paper in contact with a cut surface can change the potential of the cut surface and, at potentials less than -30 kPa, this
can take over an hour to reequilibrate with the bulk of the soil.
Hence the shortest acceptable equilibration time may be set by the
behavior of the soil; more information is needed on this topic.

Most authors have found it necessary to impregnate their filter
papers to avoid fungal degradation during equilibration. Both

0.005% HgCl$_2$ [31] and 3% pentachlorophenol in ethanol [32] have been successfully used by moistening the filters, which are allowed to dry again before use. This procedure has not been found to affect the calibration curve [31,32]. Hamblin [33] found that when using shorter equilibration times of 3 days or less, the fungicide treatment was unnecessary. Because the equilibrated filter paper is likely to have some soil attached to it, various methods have been found to account for this. Gardner [30] corrected for the mass of soil adhering to the filter paper by obtaining its oven-dry mass (when it was brushed off the paper) and then back-calculating what its moist mass would have been from a knowledge of the water content of the soil sample. Another successful procedure is to sandwich three layers of filter paper between two halves of the sample and use only the unsoiled central paper [31], although this may increase the equilibration time [33]. Hamblin [33] suggests that, with short equilibration times, the errors resulting from soil that cannot be quickly tapped off a single filter paper are small and no correction is required.

Calibration and Accuracy. Filter papers have a measurable hysteresis [31,32]; therefore it is preferable to bring the filter paper to equilibrium in the same way when it is calibrated and when it is used. Either dry filter papers should be used together with a calibration obtained on their wetting curve [31,33] or moist filter papers should be used together with a drying calibration curve. Comparisons between six different batches of the same Whatman No. 42 filter paper have shown that between- and within-batch differences in the water content of equilibrated papers were relatively small (CV < 3%) at potentials less than -50 kPa but were larger and could introduce some error (CV 7%) at -1 kPa. Published calibration curves are available for Schleicher and Schuell No. 589 White Ribbon filters [32,34] and for Whatman No. 42 filters [31,33]. The latter was obtained on a wetting curve and can be approximated by

$$\ln \psi = -4.700 - 3.683 \ln m \qquad \text{for } \psi > -3 \text{ MPa} \qquad (5)$$

where ψ is the potential in megapascals and m is the water content of the filter paper in kilograms per kilogram on an oven-dry basis [33: an error in the published equation has been corrected]. For more accurate work, a batch of papers can be calibrated using a tension table, pressure plate, psychrometer, and/or vapor equilibration to cover different parts of the calibration [15].

Error Analysis. Fawcett and Collis-George [31] have presented an error analysis of the filter paper method. From the variation within sets of equilibrated papers, they found that in the range of -2 kPa to -160 MPa, the coefficient of variation for the potential was between 7 and 20%. The actual error of any single reading will be somewhat greater than this, depending on the accuracy of the calibration curve and any inaccuracy due to poor equilibration.

2. Electrical Resistance

Electrical resistance sensors, which consist of two electrodes with a porous material in between them, have been used since the 1940s to provide a crude measure of soil water content. This method, which requires a separate calibration for each soil, is hardly ever used today, but such sensors are widely used to measure matric potential. At equilibrium, the matric potential of the solution within the sensor is equal to that of the surrounding soil. Commercial sensors can be purchased cheaply (Table 1), and it is also not difficult to construct large numbers of sensors for very little cost. Suitable portable meters for measuring sensor resistance can also be purchased or constructed for a modest price. However, the method is subject to a series of limitations, which restrict the accuracy that can be obtained.

The potential of the sensor is obtained by measuring the electrical resistance between the two electrodes, which is a function of the water content of the porous material, hence of its matric potential. Unfortunately, the resistance is also a function of temperature and of the concentration of solutes in the soil solution. Temperature effects can be corrected for, but the sensors cannot be used in saline soils unless the electrical conductivity of the soil solution is also known. Scholl [35] has reported on the construction and use of a combined salinity-matric potential sensor designed to overcome this limitation. More commonly, the sensor is cast from gypsum, which slowly dissolves and should maintain a saturated solution of calcium sulfate within itself. At 20°C, the solubility of the calcium sulfate is about 1 g/l, which should be more than 10 times greater than the concentration of the soil solution in most nonsaline soils, rendering gypsum sensors insensitive to the electrical conductivity of the soil solution in such cases. Fiberglass and nylon, the other common sensor materials, are sensitive to the electrical conductivity of the soil solution and therefore are potentially less accurate.

The measured resistance R_s at temperature T_s can be converted to the resistance R_c at the calibration temperature T_c, using the following approximation [15]

$$R_c = R_s[1 + 0.03(T_s - T_c)]   \qquad (6)$$

or [36]

$$R_c = R_s{}^{[1 + 0.002(T_s - T_c)]}   \qquad (7)$$

Sensor Materials and Construction. Many authors have given construction details for gypsum [37-39], nylon [40], and fiberglass [41] sensors. Both gypsum and fiberglass sensors are commercially available, and it is possible to include a thermistor within the sensor [41] to allow for temperature correction. The geometry of the electrodes depends on the material used but must aim to minimize any contribution to the current between the electrodes arising from electrical conduction through the soil. Sensors constructed from nylon or fiberglass fabric usually consist of two flat-mesh electrodes, between and around which are one or more layers of fabric, the whole being held in a rigid arrangement such as a perforated metal casing (typically 25 × 40 × 3 mm thick). Because gypsum ($CaSO_4 \cdot 2H_2O$) sensors are cast from a mixture of plaster of Paris ($CaSO_4 \cdot \frac{1}{2}H_2O$) and water, a greater variety of electrode geometries is possible. A popular design consists of two concentric cylindrical mesh electrodes set in a cylindrical gypsum block (typically 30 × 25 mm diameter) [37,38].

Gypsum sensors have a limited lifetime because they slowly dissolve in the soil and their calibration can change with time [42,43]. Bouyoucos [42] was able to extend the lifetime of gypsum sensors by about 10 times, without altering their electrical characteristics, by dipping them in a solution of nylon resin in alcohol. He suggested that ordinary gypsum sensors may last more than 10 years in dry soil, whereas their useful life in very wet soil where they are close to the water table may not exceed 1 year. Aitchison et al. [36] report that gypsum sensors degenerate at a much faster rate in saline soils. Both the durability and the calibration curve of gypsum sensors depend on the source of the plaster of Paris used in their construction and the ratio of plaster to water used in casting [36,44]. Considerable calibration drift has also been observed in fiberglass sensors by England [45]. He found that calibration changes were fast in the first few years after installation and that calibration stabilized after 5 or 6 years. These changes were related to deposition of colloidal and amorphous soil solids within the sensor and to decomposition of the fiberglass, and England suggests a maximum sensor life expectancy of under 15 years.

Irrespective of the sensor material, it seems likely that the calibration curve may change significantly well before the sensor shows obvious signs of wear. Thus the only guarantee of consistent results is to recheck at regular intervals (say 1 year) the calibration of a sample set of sensors taken from the whole range of soil conditions in which the sensors are installed.

Range and Response Time. The selection of published calibration characteristics shown in Fig. 4 illustrates the influence of the water release characteristic of the porous material. Sensors

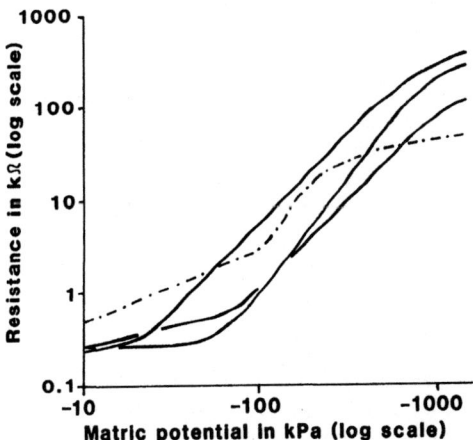

FIG. 4 Calibration characteristics of a range of electrical resistance sensors. Solid lines represent three different gypsum sensors; the dash-dot line represents a fiberglass sensor. (From Ref. 36.)

constructed of nylon or fiberglass cloth undergo rapid desaturation at potentials below about -100 kPa and, on a log-log scale, they have a linear range from near saturation down to this value. In contrast, gypsum sensors, which do not usually contain large pores, do not lose much water until the potential drops below -30 to -80 kPa (depending on the sensor) [36,44] but can then display an approximately linear calibration (on a log-log scale) to potentials down to -1 MPa or less. By firing a mixture of ground charcoal and clay, Scholl [35] was able to produce a porous material that can be used over a wide range of potentials. The range of sensors has also been extended by encasing nylon or fiberglass sensors in gypsum [46].

It is not possible to generalize about sensor response time because this can depend on the unsaturated hydraulic conductivity of the soil and the goodness of the soil-sensor contact as well as the potential toward which the sensor is equilibrating and the physical properties of the sensor itself. Gypsum sensors require about a week to equilibrate fully on a pressure plate at potentials between -0.1 and -1.5 kPa, but most of the equilibration has occurred within the first 48 hours [43,47]. Thus such sensors cannot be expected to respond any faster in the soil. In practice, fast changes in potential in the field are associated with rewetting events to which sensors are found to respond quickly [e.g., 48], whereas it is unlikely that sensors will lag much behind the rate at which soils dry out, except near to the soil surface.

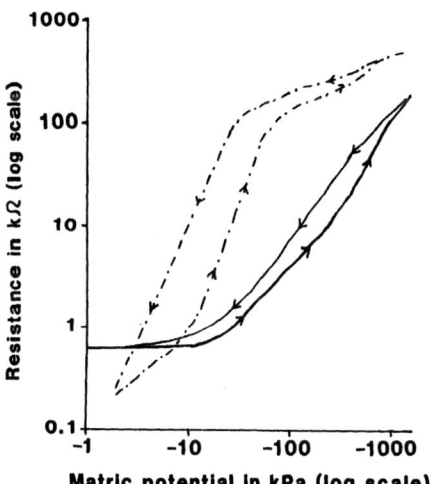

FIG. 5 Hysteresis in the calibration curves of gypsum (solid lines) and nylon (dash-dot lines) electrical resistance sensors. (From Ref. 46.)

Hysteresis and Uniformity. A study by Tanner et al. [49] of the effect of method of wetting on the resistance of saturated gypsum sensors showed that vacuum saturation gave lower resistances than saturation by immersion, whereas capillary wetting resulted in an intermediate value. These authors suggest that vacuum wetting is the most appropriate wetting method for testing a set of sensors for uniformity, since other wetting methods resulted in greater variation between the resistances of saturated sensors. The influence of saturation procedure on sensor resistance is related to air trapped within the sensor during saturation. Capillary saturation in which a sensor is allowed to wet up slowly from one end is suggested as the most appropriate procedure before field installation, since this is closest to the way in which the sensor is likely to be rewet in the field.

The effect of method of wetting is only one aspect of the hysteresis in resistance exhibited by sensors, whereby resistance values on a drying curve are less than those on a wetting curve (Fig. 5). Since sensors are calibrated by desaturation and since they are often installed at the start of a growing season into a wet soil that subsequently dries out, it has often been argued that hysteresis problems may not be serious. However, in nearly all applications there are likely to be transient rewetting events (rain or irrigation) that result in partial rewetting of the soil profile so that

some inaccuracy due to hysteresis is unavoidable. Laboratory measurements of hysteresis [46,50] indicate that, with gypsum sensors
operating in the range -30 kPa to -1 MPa, calibration based on a
drying curve can typically result in an overestimate of matric potential measured during rewetting by a factor of 2. Other sensors
tested by Bourget et al. [46] (nylon, nylon encased in gypsum, and
fiberglass encased in gypsum) exhibited similar or greater hysteresis
and a greater within-batch variability in the behavior of different
sensors. Uniformity in behavior is important, since it may be inconvenient to calibrate a large number of sensors, and calibration must
then depend on a representative subset.

Calibration. Detailed methods for the calibration of gypsum sensors using a pressure membrane [47] or pressure plate [43] have
been given. Care must be taken to ensure that good contact is maintained between the sensors, which are initially saturated, and the
membrane or plate. This may be obtained by attaching sensors to
the membrane with plaster of Paris [47] or by embedding sensors
into a paste of ground chalk on top of a pressure plate [43]. Electrical connection to the sensors through the wall or lid of the pressure chamber is made via metal-through-glass or metal-through-cer-
amic insulated connectors (commercially available with some pressure
chambers). Separate pairs of lead-through connections must be made
for each sensor, to avoid a reduction in the measured resistance due
to current flow from adjacent sensors. A limitation to the use of simple parallel-electrode sensors, where a significant contribution to the
current flow path can occur outside the sensor, was noted by Tanner
and Hanks [50], who found that the sensor resistance in air was between 1.3 and 1.6 times greater than the resistance measured on a
pressure membrane. With the latter, current can flow down through
the membrane, along the metal supporting mesh, and back up. Electrical leads within the pressure chamber need to be covered to prevent condensation from providing an additional conducting pathway
between the electrodes. This is most important if all the wires enter
the chamber through a multipin connector. Sealing the wires at the
connector with silicone rubber to avoid this problem is suggested
[43].

When calibrating nylon or fiberglass sensors, it is desirable to
obtain hydraulic contact with the membrane or plate by placing sensors in saturated soil taken from the location at which they are to
be used, so that the calibrating soil solution has a similar conductivity to that in the field. The calibration curve is often nearly linear
over the useful range when plotted using log-log axes (Fig. 4) so
that calibration can be approximated by

$$\log \psi_m = x \log R_c + k \tag{8}$$

where x and k are constants for any given sensor.

Meters. To avoid polarization effects, sensor resistance must be measured with an alternating current. In the past, low-frequency (~ 1 kHz) ac bridge circuits were used to measure this resistance and, because the sensor also has a capacitance that varies with its water content, it was necessary to balance out this capacitance to obtain a satisfactory null. Modern measuring circuits operate on a different principle, in which a voltage output is produced that is proportional to the sensor's resistance [43] and can be read off directly from a meter or fed into a data logger or automatic recording system. Strangeways [51] describes a very simple circuit for use in conjunction with a data-logging system. Goltz et al. [48] describe a more sophisticated circuit, which avoids errors due to the capacitance of sensors and their leads and in which the output is proportional to the logarithm of the sensor resistance. This should represent the most accurate system, especially where the resistance is likely to vary over 2 or more orders of magnitude.

Summary. Electrical resistance sensors represent a cheap method of limited accuracy for measuring matric potential. Nylon or fiberglass sensors, which cover a range similar to that of tensiometers, are comparable in price to puncture tensiometer units but are far inferior in terms of accuracy and should not be used except for applications in which a low and uncertain level of accuracy is acceptable. Gypsum sensors, which can cover a range of potentials down to about -1.5 MPa, the approximate limit for water extraction by roots, are a useful complement to the use of tensiometers. They offer an attractive proposition in nonsaline soils for applications requiring large numbers of sensors and/or data logging. A major limitation to these sensors is the uncertainty in the limited level of accuracy that can be expected and the time required for laboratory calibration and recalibration, to ensure that this uncertainty remains within acceptable limits. Temperature correction of readings is necessary to obtain the best accuracy. They are useful for applications such as irrigation scheduling, where a limited degree of accuracy may be acceptable [48].

3. Heat Dissipation

This technique involves sensing the heat dissipation in a porous material sensor, to the center of which a short (150 s) heat pulse has been applied. The thermal diffusivity of the sensor determines its rate of heat dissipation and is itself related to the water content, hence to the matric potential of the sensor. Heat dissipation is measured as the difference between the temperature at the center of the sensor before and after the heat pulse has been applied. Sensor performance is unaffected by the thermal diffusivity of the surrounding soil because the device is large enough to contain the heat pulse. The sensor is constructed of a germanium junction diode (used to measure temperature) around which is wrapped a heating coil, the whole being encased in a cylinder of plaster of Paris or of a ceramic material about 50 mm long by 20 mm diameter. Unlike electrical resistance sensors, it is not responsive to the salinity of the soil solution.

The sensor is calibrated by equilibrating it at various matric potentials in the same way as described for electrical resistance sensors (Section II.C.2). Theory, design, and constructional details are given by Phene et al. [52], who have also compared the performance of these sensors (with and without temperature compensation) against that of psychrometers [53]. Sensors with their associated electronics and data loggers are also commercially available.

Sensor performance depends on the porous material that is used. Phene et al. [53] report a calibration accuracy of ±20 kPa for matric potentials from 0 to -300 kPa and ±100 kPa from -300 to -600 kPa for home-made ground ceramic/Castone sensors. Campbell and Gee [15] estimate a precision of ±10 kPa in the range 0 to -100 kPa for commercially available sensors. As with electrical resistance sensors, accuracy will be further restricted by hysteresis of the porous material, and contamination with fine particles and microbial debris may alter the calibration after long periods in the soil. Although sensors are well suited for data-logging applications, they cannot be interrogated too frequently because each heat pulse must have time to dissipate before the next reading can be taken [15]. Their limited and uncertain accuracy does not commend them to experimental applications, but they appear well suited to irrigation scheduling [54].

D. Psychrometers

Thermocouple psychrometers (hereafter referred to as psychrometers) sense the relative humidity of vapor in equilibrium with the liquid phase in the soil or plant. They measure water potential in a range that overlaps the lower limit of tensiometer response (~ -80 kPa) at one end, and extends well beyond the limits of available water (< -1.5 MPa) at the other. The technique is widely used not only to measure soil but also to determine plant water status, and equipment has been commercially available for more than 10 years (Table 2).

Because psychrometers cover an important range of potentials in which there is a lack of accurate measurement techniques and are in theory capable of high accuracy, they have sometimes been used as a standard against which to compare other methods. Factors that can restrict their accuracy or cause systematic errors have therefore received considerable attention, and these have been summarized in an excellent review by Rawlins and Campbell [14].

1. Mode of Operation and Range

In principle, the simplest psychrometer to be developed is based on
the design of Richards and Ogata [55] (often referred to as the
Richards psychrometer) in which a water droplet is placed on a silver loop attached to the measuring junction of a thermocouple thermometer. Water evaporates from the droplet at a rate controlled by
its temperature and the relative humidity of the surrounding air.
This evaporation causes the temperature of the loop to drop slightly
below that of its surroundings. Eventually a steady rate of evaporation is reached when the loop has a constant temperature difference
ΔT from its surroundings, such that the heat lost by evaporation is
balanced by the heat gained in various ways (radiation, conduction
along the thermocouple wires, etc.) from the surroundings. ΔT is
measured simply by connecting two thermocouple junctions in opposi-
tion, a reference junction attached to some thermal ballast (e.g., a
piece of metal whose mass is very large compared to that of the loop)
and in good thermal contact with the soil and surroundings, and the
sensing junction attached to the loop. In a commercially available
version of the Richards psychrometer, the sensing thermocouple
junction is coated with porous ceramic to form a bead that is wetted
before measurement by immersing it in water.

Most currently used psychrometers are based on a modification
of this technique introduced by Spanner [56]. In this system, a
reverse current passed through a miniature sensing junction causes
it to cool—an effect known as the Peltier effect. Water condenses
onto the tip of the cooled junction, and the psychrometer leads are
then connected to a nanovoltmeter to measure ΔT as already described.
In a further modification, the temperature of the sensing junction is
controlled by an electronic feedback mechanism that switches the cooling current on for just the correct proportion of the time so that the
junction is held at the dew point [57,58]. The dew point is the temperature at which water vapor first starts to condense on the junction and, when operated in this way, the psychrometer is often referred to as a dew-point hygrometer. The dew-point temperature
depression is measured during the intervals when the cooling current
is switched off.

Spanner psychrometers are limited to a range of potentials exceeding roughly -7 MPa, whether operated in the dew-point mode or
not, because a larger cooling current is necessary at lower potentials to cool the sensing junction sufficiently, and this results in
Joule heating of the thermocouple wires. The appropriate size and
duration of the cooling current depends on the dimensions of the
psychrometer [59]. Unlike Spanner psychrometers, Richards psychrometers are not easily adapted for installation in the soil for monitoring purposes because of the necessity to wet the junction before

measurement; they can, however, be used with a specially designed sample changer on soil samples in the laboratory or, with suitable precautions in the field, to extend the measuring range down to less than -300 MPa (relative humidity < 0.1). All psychrometers are limited at the wet end of the range by the smallest temperature difference that can be meaningfully detected. Modern portable nano-voltmeters have a readability of ±10 nV, corresponding to a potential of ±2 kPa; however, the problems associated with measuring such small temperature differences (~ 0.0002°C) probably limit the useful range of current field psychrometers to potentials below -100 kPa (see Section 3).

Spanner psychrometers are commercially available both as small devices that are inserted into the soil, and in the form of a labora-tory sample changer [60] that can be used on small (~ 2 cm^3) soil samples. The commercial version of this design (Model SC-10A, Decagon Devices Inc.) can also be used with a Richards-type psy-chrometer with a ceramic-covered sensing tip. It will hold up to 10 samples, permitting a number of samples to be thermally equili-brated with the psychrometer at one go. Nanovoltmeters made for use with psychrometers, dew-point control systems, and systems that can automatically scan and record from a number of psychrom-eters [61,62], are all commercially available (Table 2). Wiebe et al. [63] give instructions for construction of home-made psychrometers.

2. Theory

The relative humidity p/p_0 of vapor in equilibrium with the liquid phase in the soil (or plant) is related to its water potential γ_w (in Pa) by

$$\psi_w = \frac{\gamma RT}{M} \ln\left(\frac{p}{p_0}\right) \tag{9}$$

where γ is the density of water (kg m^{-3}), R is the ideal gas con-stant (8.31 K^{-1} mol^{-1}), T is the temperature (K) of the liquid phase, M is the molecular weight of water (0.018 kg mol^{-1}), p is the vapor pressure of water in equilibrium with the liquid phase (Pa), p_0 is the saturated vapor pressure of the liquid phase (Pa), and ψ_w is the sum of the matric and osmotic potentials. Except in saline soils matric potential is likely to be the dominant component of ψ_w in the working range of the psychrometer. In the range of potentials that water is available to plants ($\psi_w > \sim -1.5$ MPa), Eq. 9 can be approx-imated by

$$\psi_w \simeq \frac{\gamma RT}{M}\left(\frac{p}{p_0} - 1\right) \tag{10}$$

to an accuracy of better than 1%.

The relative humidity sensed by a psychrometer is related to the temperature depression of the sensing junction ΔT by [14,64]

$$\frac{p}{p_0} = 1 - \left(\frac{s + \gamma^*}{p_0}\right) \Delta T \tag{11}$$

where s is the slope of the saturation vapor pressure curve and γ^* is the apparent psychrometer constant ($\sim$ 60-90 Pa K^{-1}). Combining Eqs. 10 and 11 gives

$$\psi_w = -\frac{\gamma RT}{M}\left(\frac{s + \gamma^*}{p_0}\right) \Delta T \tag{12}$$

Thus, under ideal conditions, where all other terms in this equation are held constant, ψ_w is proportional to ΔT for $\psi_w > -1.5$ MPa.

3. Calibration and Accuracy

A psychrometer is usually calibrated empirically by placing it over a range of salt solutions of known potentials, in a sealed chamber in a constant temperature bath. Published values are available for the potentials of KCl [65] and NaCl [66] solutions at a range of temperatures. The detailed procedure for calibration using a laboratory sampleholder is given in the manufacturer's instructions. Merrill and Rawlins [59] describe calibration of field psychrometers and, for both laboratory and field psychrometers, recommended calibration procedures are given by Rawlins and Campbell [14]. If the sample temperature is not the same as the temperature at which calibration was performed, and the psychrometer is not used in the dew-point mode, it is necessary to make a temperature correction. Temperature correction can be made either by calibrating at a series of temperatures and interpolation of the corrected calibration curve or by a theoretical correction procedure [14,59].

The major factors that influence the accuracy of psychrometer results and can cause large systematic errors are mainly associated with temperature, as shown in Table 3. This is understandable when it is realized that to sense a potential of -10 kPa requires accurate measurement of a ΔT of about 0.001°C. A detailed review of the factors in Table 3 can be found in Rawlins and Campbell [14].

Laboratory Sample Changer. The best observed precision of psychrometer readings under near-ideal laboratory conditions is ±4 kPa [14], but this value is of relevance only when the psychrometer is used to measure the potential of samples in laboratory experiments such as the determination of the water release characteristic. Precautions to minimize temperature gradients include use in a room where temperature changes are not rapid and there is little air

TABLE 3 Factors That Can Introduce Systematic Errors in Soil Psychrometer Readings[a]

Factor and source	Effect	Remedy
1. Temperature gradients (variation in temperature of surroundings, electrical heating of thermocouple wires, absorption of external radiation)	Temperature difference between reference and sensing junction [30]	i. (L) Use thermal insulation and/or a water bath to avoid gradients, allow 1/2 hour for samples to equilibrate in sample-holder. ii. (Ps & Pd) If reference junction is isolated from sample, measure temperature difference before Peltier cooling and subtract it from the reading. iii. (F) Align psychrometer, with reference and sensing junctions parallel to isotherms (i.e., insert parallel to soil surface). iv. (F) Use a thermally shielded psychrometer with shield attached to the reference junction.
2. Temperature fluctuations with time (Variation in temperature of surroundings)	a. See 1, above b. Variation of relative humidity within chamber [10]	v. Arrange sample to surround the sensing junction as nearly as possible. vi. As for remedy v [20].
3. Vapor pressure gradient (L) only (extraneous sources or sinks of water vapor, especially where samples are warmer than the chamber and water condenses on chamber walls)	Relative humidity in chamber is not controlled by the sample and reading is erroneous	vii. Ensure that sample and holder have reached the same temperature before moving under the sensing junction; do not insert samples that are warmer than the holder into it.

4. Contamination of sensing junction or chamber walls		viii. Clean junction and chamber and recalibrate.
5. Zero offset	Nonzero output when calibrated over water	ix. Subtract offset reading before converting it to a potential.
6. Temperature correction (calibration temperature was not the same as measurement temperature)	Not important for Pd; incorrect readings for Pr and Ps	x. Calibrate at more than one temperature and interpolate to measurement temperature or use a theoretical correction procedure [40].
7. Insufficient equilibration time (L) only	Incorrect reading	xi. Plot psychrometer reading versus time to gain familiarity with its performance and use an adequate time. Equilibration time will be reduced by remedy v.

aKey: L, laboratory sample changer arrangement; F, field psychrometer; Pr, Richards psychrometer; Ps, Spanner psychrometer; Pd, dew-point mode.

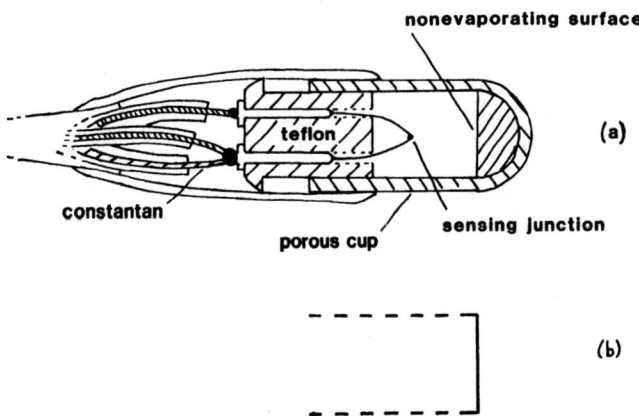

FIG. 6 (a) Three-wire Spanner psychrometer. (Adapted from
Ref. 14.) A stainless steel screen (b) can be used in place of
the porous cup.

movement, minimizing hand contact with the sample changer, and
encasing the sample changer in polyurethane foam or other thermal
insulation. For samples with a high relative humidity (e.g.,
ψ_w < -6 MPa), samples should be transferred to and loaded into the
sample changer in a humid atmosphere (e.g., a box lined with wetted paper towels and with limited access, ideally a glove box).
Manufacturer's instructions (Decagon Devices, Table 2) indicate
that soil samples require 30 minutes to attain thermal equilibrium
with the sample changer. A further period of 10 minutes or more
is necessary after each sample has been moved under the sensing
junction for vapor equilibrium to be established for potentials of
less than -0.4 MPa, although shorter periods are required at higher
potentials.

   Use of the laboratory apparatus on samples that are removed in
the field, transported in thermally insulated sealed containers, and
then subsampled to fill the sample holder will depend on factors
such as water loss by distillation onto the container walls, variation
of sample potential with temperature, and the effects of mechanical
disturbance of the samples. Because both the sample changer and
measuring electronics are portable, there is no reason not to perform measurements in the field if a suitable, stable thermal environment for the sample changer can be arranged and samples are cooled
to below the chamber temperature before loading (to avoid condensation on the chamber).

Psychrometers for Insertion Into the Soil. Unless used in the dew-point mode, these psychrometers require temperature correction for accurate results [59]. Three-wire Spanner psychrometers (Fig. 6), in which a second wire is attached to the reference junction, can be used for this purpose to measure ambient temperature. Diurnal soil temperature variations depend on climate and are greatest when the surface is bare of vegetation. They impose an important limitation on the accuracy of psychrometer readings made near the soil surface (< 0.25 m depth). Merrill and Rawlins [59] have discussed the installation and calibration stability of soil psychrometers. They observed errors in excess of 50% for Wescor ceramic-enclosed psychrometers installed vertically at 0.25 m depth, in soil with a bare surface. Diurnal temperature variation at this depth was ±1.3°C and, when psychrometers were installed horizontally to minimize the influence of soil temperature gradients, the variation in psychrometer readings was reduced to about 10%. Improved design can further reduce the sensitivity to soil temperature gradients [67]. In addition to horizontal placement, Merrill and Rawlins [59] recommend that 50-100 mm of the lead wire adjacent to the psychrometer be horizontally oriented also. They also observed a 5.3% median change in calibration sensitivity of 33 Wescor ceramic-enclosed psychrometers after an 8-month period of use in the field; only one psychrometer changed by more than 15%. They considered that field psychrometers were able to distinguish day-to-day water potential variations to within ±50 kPa.

Merrill and Rawlins [59] have discussed the relative merits of the commercially available ceramic cup and wire screen-shielded psychrometers (Fig. 6). The ceramic cup excludes contamination by fungal hyphae and prevents flooding of the chamber if it is below the water table for short periods, whereas the wire screen may be more suitable in soils that are likely to shrink away from the psychrometer upon drying.

III. APPLICATIONS

Measurements of soil matric, hydraulic, and water potential are so fundamental for studying water movement and plant growth that the literature is full of references to their use. Examples of some of the major applications are given here.

Irrigation scheduling can be based on data from tensiometers [13,69], electrical resistance [48], or heat dissipation sensors [54], all of which can be adapted to continuous automatic monitoring and to automatic irrigation control. Tensiometers, with their greater accuracy but restricted lower limit, are most suited for applications, such as the irrigation of vegetable and greenhouse crops where it is intended to keep the soil permanently at a high potential and fairly accurate control is required to avoid overwatering. Small portable tensiometers can be used for testing the suitability of conditions for germination and establishment in seedbeds, peat blocks, and other materials used to raise plants [28].

For monitoring the potential in the root zone under nonirrigated conditions, the best accuracy will be obtained with a combination of tensiometers and psychrometers. If there is little recharge of the soil profile during the growing season, it is possible to identify a zero flux plane, where there is zero hydraulic potential gradient. This plane represents an imaginary watershed, above which water moves upward to plant roots and below which drainage may occur [70-72]. By following the movement of the zero flux plane down the profile during the growing season, it is possible to follow changes in the maximum depth of root water extraction and to obtain improved estimates of the soil water balance. Psychrometers designed for attachment to leaves or the stems [73] of plants may be used in combination with soil sensors to obtain detailed information on the diurnal pattern of the plant water regime [67].

For measuring matric and hydraulic potential under wet conditions, there is no substitute for the accuracy of tensiometers, especially as they will function equally well below the water table. Tensiometers can be used to study the water regime in relation to restrictions on soil aeration and root growth [74,75] and to follow the pattern of water flow that determines the water regime on hillsides and hollows [23]. Under wet (ψ_m > -10 kPa) conditions, portable tensiometers can be used to study spatial variation of matric potential [20], and the recent introduction of cheap puncture tensiometers should see a further increase in the use of tensiometers. Cassell and Klute [13] provide a good discussion of methods for installing and maintaining tensiometers.

In addition to spatial variations resulting from plant water uptake, the soil water regime may be heterogeneous in structured soils. Sensors that connect with cracks or biopores, which form preferred pathways for infiltration, may then give readings different from sensors installed within structural units. In such cases there is no single representative value, and the positioning of sensors must be related to the object of the particular investigation. Superimposed on such structure-related variability there is also likely to be longer range variability in the soil water regime. Thus, even in a relatively uniform soil, Greminger et al. [76] observed significant spatial variability between tensiometer readings at a separation of greater than 10 m.

Use of matric potential sensors for in situ determination of the water release characteristic [76] and for determination of unsaturated hydraulic conductivity is discussed in Chaps. 3 and 5, respectively.

REFERENCES

1. Marshall, T. J., Relations between water and soil, *Tech. Commun.*, *No. 50*, Commw. Bur. Soils, Harpenden, U.K., 1959.
2. Aslyng, H. C. (Chairman), Soil physics terminology, *Int. Soc. Soil Sci. Bull.*, 23: 7-10 (1963).
3. Bolt, G. H. (Chairman), Soil physics terminology, *Int. Soc. Soil Sci. Bull.*, 49: 16-22 (1976).
4. Hillel, D., *Fundamentals of Soil Physics*, Academic Press, New York, 1980.
5. Hillel, D., *Introduction to Soil Physics*, Academic Press, New York, 1982.
6. Marshall, T. J., and J. W. Holmes, *Soil Physics*, 2nd ed., Cambridge University Press, Cambridge, 1988.
7. Rose, C. W., *Agricultural Physics*, Pergamon Press, Oxford, 1969.
8. Taylor, S. A., and G. L. Ashcroft, *Physical Edaphology*, Freeman, San Francisco, 1972.
9. Hanks, R. J., and G. L. Ashcroft, *Applied Soil Physics*, Springer-Verlag, Berlin, 1980.
10. Towner, G. D., The correction of in situ tensiometer readings for overburden pressures in swelling soils, *J. Soil Sci.*, 32: 499-504 (1981).
11. Schofield, R. K., The pF of the water in soil, *Trans. 3rd Int. Congr. Soil Sci.*, 2: 37-48 (1935).
12. Rawlins, S. L., Measurement of water content and the state of water in soils, in *Water Deficits and Plant Growth*, Vol. 4 (T. T. Kozlowski, Ed.), Academic Press, New York, 1976, pp. 1-55.
13. Cassell, D. K., and A. Klute, Water potential: Tensiometry, in *Methods of Soil Analysis*, Part 1, 2nd ed. (A. Klute, Ed.), Am. Soc. Agron., Madison, WI, 1986, pp. 563-596.
14. Rawlins, S. L., and G. S. Campbell, Water potential: Thermocouple psychrometry, in *Methods of Soil Analysis*, Part 1, 2nd ed. (A. Klute, Ed.), Am. Soc. Agron., Madison, WI, 1986, pp. 597-618.
15. Campbell, G. S., and G. W. Gee, Water potential: Miscellaneous methods, in *Methods of Soil Analysis*, Part 1, 2nd ed. (A. Klute, Ed.), Am. Soc. Agron., Madison, WI, 1986, pp. 619-633.
16. Richards, L. A., Methods of measuring soil moisture tension, *Soil Sci.*, 68: 95-112 (1949).
17. Towner, G. D., Theory of time response of tensiometers, *J. Soil Sci.*, 31: 607-621 (1980).
18. Klute, A., and W. R. Gardner, Tensiometer response time, *Soil Sci.*, 93: 204-207 (1962).
19. Watson, K. K., and R. D. Jackson, Temperature effects in a tensiometer-pressure transducer system, *Soil Sci. Soc. Am. Proc.*, 31: 156-160 (1967).

20. Mullins, C. E., O. T. Mandiringana, T. R. Nisbet, and M. N. Aitken, The design, limitations, and use of a portable tensiometer, *J. Soil Sci.*, 37: 691-700 (1986).

21. Webster, R., The measurement of soil water tension in the field, *New Phytol.*, 65: 249-258 (1966).

22. Long, F. L., A new solid-state device for recording tensiometers, *Soil Sci.*, 133: 131-132 (1982).

23. Anderson, M. G., and T. P. Burt, Automatic monitoring of soil moisture conditions in a hillslope spur and hollow, *J. Hydrol.*, 33: 27-36 (1977).

24. Lee-Williams, T. H., An automatic scanning and recording tensiometer system, *J. Hydrol.*, 39: 175-183 (1978).

25. Blackwell, P. S., and M. J. Elsworth, A system for automatically measuring and recording soil water potential and rainfall, *Agric. Water Manage.*, 3: 135-141 (1980).

26. Marthaler, H. P., W. Vogelsanger, F. Richard, and P. J. Wierenga, A pressure transducer for field tensiometers, *Soil Sci. Soc. Am. J.*, 47: 624-627 (1983).

27. Frede, H. G., W. Weinzerl, and B. Meyer, A portable electronic puncture tensiometer, *Z. Pflanzenernaehr. Bodenk.*, 147: 131-134 (1984).

28. Goodman, D., A portable tensiometer for the measurement of water tension in peat blocks, *J. Agric. Eng. Res.*, 28: 179-182 (1983).

29. Peck, A. J., and R. M. Rabbidge, Design and performance of an osmotic tensiometer for measuring capillary potential, *Soil Sci. Soc. Am. Proc.*, 33: 196-202 (1969).

30. Gardner, R., A method of measuring the capillary tension of soil moisture over a wide moisture range, *Soil Sci.*, 43: 277-293 (1937).

31. Fawcett, R. G., and N. Collis-George, A filter-paper method for determining the moisture characteristics of soil, *Aust. J. Exp. Agric. Animal Husb.*, 7: 162-167 (1967).

32. McQueen, I. S., and R. F. Miller, Calibration and evaluation of a wide-range gravimetric method for measuring moisture stress, *Soil Sci.*, 106: 225-231 (1968).

33. Hamblin, A. P., Filter-paper method for routine measurement of field water potential, *J. Hydrol.*, 53: 355-360 (1981).

34. Al-Khafaf, S., and R. J. Hanks, Evaluation of the filter paper method for estimating soil water potential, *Soil Sci.*, 117: 194-199 (1974).

35. Scholl, D. G., A two-element ceramic sensor for matric potential and salinity measurements, *Soil Sci. Soc. Am. J.*, 42: 429-432 (1978).

36. Aitchison, G. D., P. F. Butler, and C. G. Gurr, Techniques associated with the use of gypsum block soil moisture meters, *Aust. J. Appl. Sci.*, 2: 56-75 (1951).

37. Pereira, H. C., A cylindrical gypsum block for moisture studies in deep soils, *J. Soil Sci.*, 2: 212-223 (1951).

38. Cannell, G. H., and C. E. Asbell, Prefabrication of mould and construction of cylindrical electrode-type resistance units, *Soil Sci.*, 97: 108-112 (1964).

39. Fourt, D. F., and W. H. Hinton, Water relations of tree crops. A comparison between Corsican pine and Douglas fir in southeast England, *J. Appl. Ecol.*, 7: 295-309 (1970).

40. Bouyoucos, G. J., Nylon electrical resistance unit for continuous measurement of soil moisture in the field, *Soil Sci.*, 67: 319-330 (1949).

41. Colman, E. A., and E. A. Hendrix, The Fiberglas electrical soil-moisture instrument, *Soil Sci.*, 67: 425-438 (1949).

42. Bouyoucos, G. J., More durable plaster of Paris moisture blocks, *Soil Sci.*, 76: 447-451 (1953).

43. Wellings, S. R., J. P. Bell, and R. J. Raynor, *The Use of Gypsum Resistance Blocks for Measuring Soil Water Potential in the Field*, Report No. 92, Inst. Hydrol., Wallingford, Oxfordshire, U.K., 1985.

44. Perrier, E. R., and A. W. Marsh, Performance characteristics of various electrical resistance units and gypsum materials, *Soil Sci.*, 86: 140-147 (1958).

45. England, C. B., Changes in fiberglass soil moisture-electrical resistance elements in long-term installations, *Soil Sci. Soc. Am. Proc.*, 29: 229-231 (1965).

46. Bourget, S. J., D. E. Elrick, and C. B. Tanner, Electrical resistance units for moisture measurements: Their moisture hysteresis, uniformity, and sensitivity, *Soil Sci.*, 86: 298-304 (1958).

47. Haise, H. R., and O. J. Kelly, Relation of moisture tension and electrical resistance in plaster of Paris blocks, *Soil Sci.*, 61: 411-422 (1946).

48. Goltz, S. M., G. Benoit, and H. Schimmelpfennig, New circuitry for measuring soil water matric potential with moisture blocks, *Agric. Meteorol.*, 24: 75-82 (1981).

49. Tanner, C. B., E. Abrams, and J. C. Zubriski, Gypsum moisture-block calibration based on electrical conductivity in distilled water, *Soil Sci. Soc. Am. Proc.*, 13: 62-65 (1948).

50. Tanner, C. B., and R. J. Hanks, Moisture hysteresis in gypsum moisture blocks, *Soil Sci. Soc. Am. Proc.*, 16: 48-51 (1952).

51. Strangeways, I. C., Interfacing soil moisture gypsum blocks with modern data-logging system using a simple, low-cost, dc method, *Soil Sci.*, 136: 322-324 (1983).

52. Phene, C. J., G. J. Hoffman, and S. L. Rawlins, Measuring soil matric potential in situ by sensing heat dissipation within a porous body. I. Theory and sensor construction, *Soil Sci. Soc. Am. Proc.*, 35: 27-33 (1971).

53. Phene, C. J., S. L. Rawlins, and G. J. Hoffman, Measuring soil matric potential in situ by sensing heat dissipation within a porous body. II. Experimental results, *Soil Sci. Soc. Am. Proc.*, 35: 225-229 (1971).

54. Phene, C. J., and D. W. Beale, High-frequency irrigation for water nutrient management in humid regions, *Soil Sci. Soc. Am. J.*, 40: 430-436 (1976).

55. Richards, L. A., and G. Ogata, Thermocouple for vapor-pressure measurement in biological and soil systems at high humidity, *Science*, 128: 1089-1090 (1958).

56. Spanner, D. C., The Peltier effect and its use in the measurement of suction pressure, *J. Exp. Bot.*, 11: 134-168 (1951).

57. Neumann, H. H., and G. W. Thurtell, A Peltier cooled thermocouple dewpoint hygrometer for in situ measurement of water potentials, in *Psychrometry in Water Relations Research* (R. W. Brown and B. P. van Haveren, Eds.), Utah State University, Logan, 1972, pp. 103-112.

58. Campbell, E. C., G. S. Campbell, and W. K. Barlow, A dewpoint hygrometer for water potential measurement, *Agric. Meteorol.*, 12: 113-121 (1973).

59. Merrill, S. D., and S. L. Rawlins, Field measurement of soil water potential with thermocouple psychrometers, *Soil Sci.*, 113: 102-109 (1972).

60. Campbell, G. S., W. D. Zollinger, and S. A. Taylor, Sample changer for the thermocouple psychrometers: Construction and some applications, *Agron. J.*, 58: 315-318 (1966).

61. Lang, A. R. G., and E. S. Trickett, Automatic scanning of Spanner and droplet psychrometers having outputs up to 30 μV, *J. Sci. Instrum.*, 42: 777-782 (1965).

62. Hoffman, G. J., W. N. Herkelrath, and R. S. Austin, Simultaneous cycling of Peltier thermocouple psychrometers for rapid water potential measurements, *Agron. J.*, 61: 597-601 (1969).

63. Wiebe, H. H., G. S. Campbell, W. H. Gardner, S. L. Rawlins, J. W. Cary, and R. W. Brown, *Measurement of Plant and Soil Water Status*, Bull. No. 484, Utah State University, Logan, 1971.

64. Campbell, G. S., Improved thermocouple psychrometers for measurements of soil water potential in a temperature gradient, *J. Phys. E: Sci. Instrum.*, 12: 739-743 (1979).

65. Campbell, G. S., and W. H. Gardner, Psychrometric measurement of soil water potential: Temperature and bulk density effects, *Soil Sci. Soc. Am. Proc.*, 35: 8-12 (1971).

66. Lang, A. R. G., Psychrometric measurement of soil water potential in situ under cotton plants, *Soil Sci.*, 106: 460-464 (1967).

67. Bruini, O., and G. W. Thurtell, An improved thermocouple hygrometer for in situ measurements of soil water potential, *Soil Sci. Soc. Am. J.*, 46: 900-904 (1982).

68. Rice, R., A fast response, field tensiometer system, *Trans. Am. Soc. Agric. Eng.*, 12: 48-50 (1969).

69. Hagan, R. M., H. R. Haise, and T. W. Edminster (Eds.), *Irrigation of Agricultural Lands*, Am. Soc. Agron., Madison, WI, 1967.

70. Arya, L. M., D. A. Farrell, and G. R. Blake, A field study of soil water depletion patterns in presence of growing soybean roots: I. Determination of hydraulic properties of the soil, *Soil Sci. Soc. Am. Proc.*, 39: 424-430 (1975).

71. McGowan, M., Depths of water extraction by roots: Application to soil-water balance studies, in *Isotopes and Radiation Techniques in Soil Physics and Irrigation Studies*, IAEA, Vienna, 1974, pp. 435-445.

72. Cooper, J. D., *Measurement of Moisture Fluxes in Unsaturated Soil in Thetford Forest*, Report No. 66, Inst. Hydrol, Wallingford, Oxfordshire, U.K., 1980.

73. McBurney, T., and P. A. Costigan, Plant water potential measured continuously in the field, *Plant Soil*, 97: 145-149 (1987).

74. King, J. A., K. A. Smith, and D. G. Pyatt, Water and oxygen regimes under conifer plantations and native vegetation on upland peaty gley soil and deep peat soils, *J. Soil Sci.*, 37: 485-497 (1986).

75. Nisbet, T. R., C. E. Mullins, and D. A. MacLeod, The variation of soil water regime, oxygen status and rooting pattern with soil type under Sitka spruce, *J. Soil Sci.*, 40: 1-15 (1989).

76. Greminger, P. J., Y. K. Sud, and D. R. Nielsen, Spatial variability of field-measured soil-water characteristics, *Soil Sci. Soc. Am. J.*, 49: 1075-1082 (1985).

3

Water Release Characteristic

MALCOLM J. REEVE and ANDRÉE D. CARTER *Soil Survey and Land Research Centre, Shardlow, Derby, England*

I. INTRODUCTION

The water release characteristic is the relationship between water content and matric potential (soil water or matric suction) in a drying soil. The water release characteristic is one of the most important measurements for characterizing soil physical properties, since it can (1) indicate ability of the soil to store water that will be available to growing plants, (2) indicate the aeration status of a drained soil, and (3) be interpreted in nonswelling soils as a measure of pore size distribution.

There are a range of methods used for measurement of the water release characteristics of soil. This chapter describes the physical properties that determine the release characteristic, outlines the most common methods used to measure it and their suitability for a range of analytical environments, and briefly illustrates the ways in which the results can be presented and applied so that we can make more beneficial use of dwindling soil resources.

Soil water that is in equilibrium with free water is by definition at zero matric potential. As the soil dries, matric potential decreases and large pores empty of water. Progressive decreases in matric potential will continue to empty narrower pores until eventually water is held in only the finest pores. Not only is water removed from soil pores, but the films of water held around soil particles are reduced in thickness. Therefore a decreasing matric potential is associated with a decreasing soil water content. Laboratory or field measurements of these two parameters can be made and the relationship plotted as a curve, called the soil moisture characteristic by Childs [1]. Soil water retention characteristic, soil moisture characteristic curve, pF curve, and soil water release characteristic have all been used as synonymous terms. However, the use of *water release characteristic* is preferred because this term refers specifically to the unique relationship obtained for a soil that is progressively dried from zero matric potential. Because of hysteresis (Section III.A), this may differ from a relationship derived otherwise.

II. ENERGY LEVELS OF SOIL WATER

Because matric potential can have only zero or negative values, the terms *soil water potential* or *matric suction* (which represent the same quantity but take the opposite sign) are often used as convenient alternatives. The units used to express the energy levels of soil water are diverse, and Table 1 provides a conversion for some of those more commonly used. The kilopascal is the most commonly applied SI unit, but the bar is still in common use, and suctions of 0.1 bar or less are often expressed as centimeters of water. Schofield [2] proposed the pF scale, which is the logarithm of the soil water suction expressed in centimeters of water. The scale is analogous to the pH scale and is designed to avoid the use of very large numbers, but has not been universally adopted.

TABLE 1 Conversion Table for Energy Levels of Soil Water

Soil water potential			
kPa[a]	bar	cm H_2O at 20°C	pF
−1	−0.01	−10.2	1.01
−5	−0.05	−51.1	1.71
−10	−0.1	−102.2	2.01
−20	−0.2	−204.3	2.31
−30	−0.3	−306.5	2.49
−40	−0.4	−408.6	2.61
−50	−0.5	−510.8	2.71
−60	−0.6	−613.0	2.79
−70	−0.7	−715.1	2.85
−80	−0.8	−817.3	2.91

TABLE 1 (continued)

-90	-0.9	-919.4	2.96
-100	-1.0	-1,021.6	3.01
-200	-2	-2,043.2	3.31
-1,500	-15	-15,324	4.18

[a]1 kPa = 10 hPa = 0.001 MPa
= 1 cbar = 10 mbar = 0.01 bar
= 4.022 in H_2O = 10.216 cm H_2O = 7.5 cm Hg
= 0.145 lb$_f$ in^{-2}

III. THE SOIL WATER RELEASE CHARACTERISTIC

The soil water release characteristic is not generally unique for either a group of similar soils or even an individual soil type. The shape and position of the curve depend not only on the soil properties but also on the wetting and drying history of the soil because of hysteresis.

A. Hysteresis

As discussed in Section I, the term "water release characteristic" implies a measurement made by desorption from saturation or a low suction. However, this curve is different from the sorption curve, obtained by gradually rewetting a dry sample. Both curves are continuous, but not identical, forming a hysteresis loop (Fig. 1). Partial drying followed by rewetting, or partial wetting followed by drying, can result in intermediate curves known as scanning curves, which lie within the hysteresis loop. Because the same soil can hold different quantities of water at the same matric potential, a knowledge of the wetting and drying history of a soil is essential for interpretation of results. The phenomenon of hysteresis [3] has been frequently documented, more lately by Poulovassilis [4] and Shcherbakov [5].

The main reasons for the hysteresis phenomena, described in detail by Hillel [6], are:

1. *Pore irregularity.* Pores are generally irregularly shaped voids interconnected by smaller passages. This results in the "ink-bottle" effect, illustrated in Fig. 2.
2. *Contact angle.* The angle of contact between water and the solid walls of pores tends to be greater for an advancing meniscus than for a receding one. A given water content will tend therefore to exhibit greater suction in desorption than in sorption.

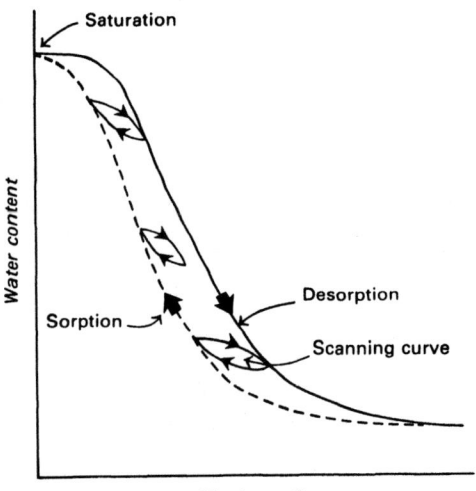

FIG. 1 The hysteresis loop, showing intermediate scanning curves. (After Ref. 6.)

3. *Entrapped air.* This can decrease the water content of newly wetted soil.
4. *Swelling and shrinking.* Volume changes cause changes of soil fabric, structure, and pore size distribution, with the result that interparticle contacts differ on wetting and drying.

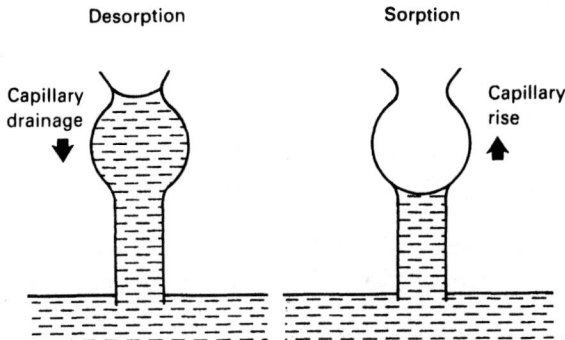

FIG. 2 The "inkbottle" effect, determining the equilibrium height of water in irregularly shaped pores 'n response to the same suction. (After Ref. 6.)

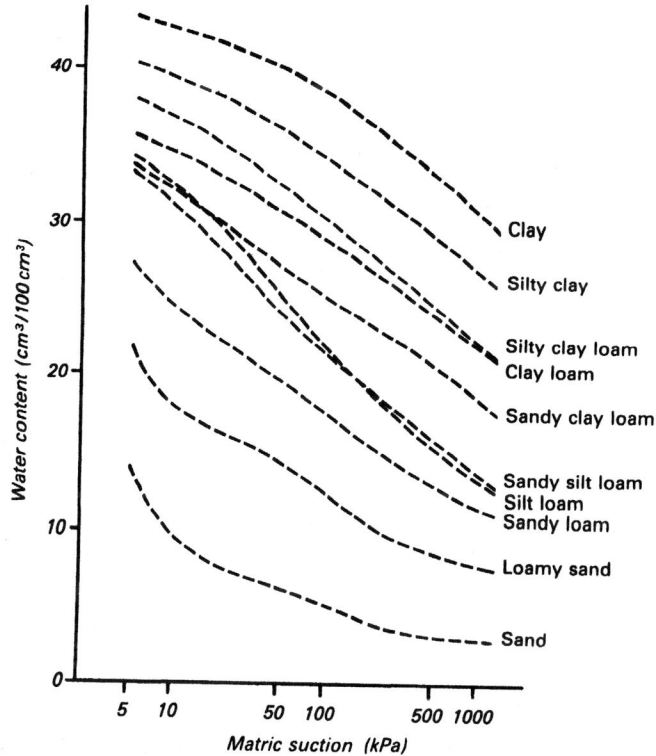

FIG. 3 The water release characteristic for subsoils of different texture. (After Ref. 7.)

Poulovassilis [4] adds that the rate of wetting or drying may also cause hysteresis. Although an understanding of hysteresis is central to any explanation of soil water release characteristics, the overriding influence on the shape of the water release curve is soil composition.

B. Effect of Soil Properties

The amount of water retained at low suctions (0-100 kPa) is strongly dependent on the capillary effect and, in nonshrinking soils, on pore size distribution. Sandy soils contain large pores, and the majority of water is released at low suctions (Fig. 3). Clay soils release small amounts of water at low suctions and retain a large proportion of their water even at high suctions, where retention is attributable to adsorption. Clay mineralogy is also important,

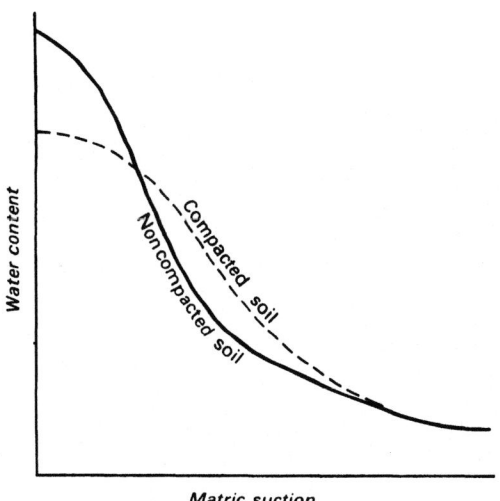

FIG. 4   The effect of compaction on the water release characteristic
of an aggregated soil.

smectitic clays with high cation-exchange capacity and specific surface having larger adsorption forces than kaolinitic clays [8].   Organic matter increases the amount of water retained, especially at
low suctions, but at higher suctions soils rich in organic materials
release water rapidly.   The presence of free iron oxides and calcium carbonate has also been shown to affect the release characteristic [9,10], though the effect of free iron is difficult to separate
from the effect of the high clay contents and good structural conditions with which it is often associated [11].

Soil structure and density have significant effects.   For example,
compaction decreases the total pore space of a soil [12], mainly by
reducing the amount of large pores, which retain water at low suctions (Fig. 4).   Whereas the amount of fine pores remains largely
unchanged, the amount of intermediate size pores is sometimes increased, and this can increase the amount of water retained between
specific matric suctions of agronomic importance [12].

C.   Relevance to Land Use

An important aspect of the water release characteristic for land use
is the shape or form of the curve at or between suction limits found
to be critical to plant functions or to water movement through the
soil.   Early research on plant response to low soil moisture contents
[13] indicated that sunflowers wilt permanently at a suction of about
1500 kPa (15 bar).   This suction was widely adopted as the general

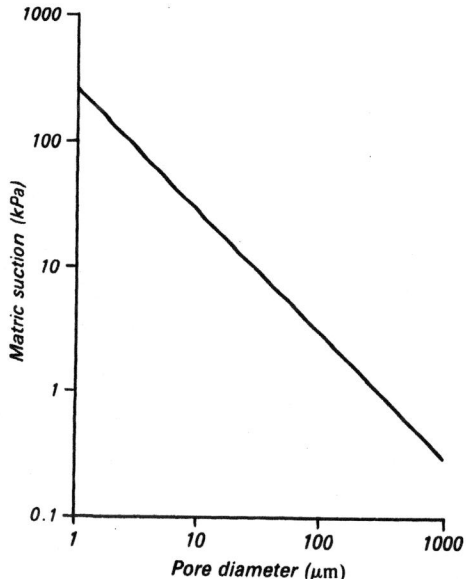

FIG. 5   The relationship between pore diameter and matric suction.

limit of water available to plants and is still in use today.   However, as research has furthered our knowledge of water flow to plant roots and the importance of unsaturated hydraulic conductivity, the shape of the curve at lower suctions has assumed more importance.   For example, the slope of the curve, termed the differential or specific water capacity, is an important function in calculating soil water diffusivity (Chap. 5) used in modeling water use by crops.   At low suctions the release of water by the soil is significant in designing drainage [14] and irrigation systems [15].   Further applications are in relating the soil water release curve to other physical parameters such as bearing capacity [16] and soil shrinkage [17], both of which are important in construction and in agriculture.

The water release characteristics of soil can also be used to define the structure of the soil in terms of the shape and size of the pores that make up its pore space.

D.   Suction and Pore Size

In a simple situation of a rigid soil containing uniform cylindrical pores, the applied suction is related to the pore size by the equation

$$d = \frac{4\sigma}{\rho g h} \tag{1}$$

where d is the diameter of pores, σ is the surface tension, ρ is the density of water, h is the soil water suction, and g is the acceleration due to gravity. At 20°C Eq. 1 gives d = 3.06h, where h is in kilopascals and d is in micrometers (Fig. 5).

The volume of water released by an increase in matric solution from h_1 to h_2 therefore equals the volume of pores having an effective diameter between d_1 and d_2, where d and h are related by Eq. 1. This simple relationship will operate only in nonshrinking soils and where the pore space consists of broadly circular pores with few "blind ends" or random restrictions (necks). Real soils can contain planar voids, pores with blind ends, and/or restrictions. If a void of 200 μm diameter has a neck exit of only 30 μm, water in the void will be released only when the suction exceeds 10 kPa. Thus the water release characteristic is at best only a general indicator of the effective pore size distribution.

The size distribution of pores in a soil can be used as a means of quantifying soil structure [7] or to give a general indication of saturated hydraulic conductivity, the value of which is largely determined by the volume of larger pores. Aeration is also largely a function of larger pores. Whereas larger pores may be defined as macropores and related to the water released at an arbitrary low suction, other pore sizes may be termed meso- or micropores [18], the latter being related to the water release characteristic at higher suctions.

In clay soils, shrinkage results in pores reducing in size as water is withdrawn, and thus a relation between water release and pore size does not hold.

IV. MEASUREMENT METHODS

There are two distinct ways to obtain a release characteristic. The usual procedure is to equilibrate samples at a chosen range of potentials and then determine their moisture contents. Suction tables, pressure plates, and vacuum desiccators are examples of this approach. In the second procedure, samples are allowed to dry out progressively and their potential and moisture content are both directly measured. In principle, any of the methods for measuring potential (see Chap. 2) can be used before determining moisture content, gravimetrically or by other means (see Chap. 1).

For laboratory measurements of the release characteristic, it is necessary initially to wet the sample to a suction lower than that at which it is to be equilibrated (see Section V.B for wetting fluid and wetting procedure).

TABLE 2   Methods of Determining Soil Water Release Characteristics in the Laboratory

Method	Approximate range of application (kPa, suction)	Type of potential measured	Early reference to method
Büchner funnel	0-20	Matric	3
Porous suction plate	0-70	Matric	20
Sand suction table	0-10	Matric	21
Sand-kaolin suction table	10-50	Matric	21
Porous pressure plate (including Tempe cell)	0-1500	Matric	22, 23
Pressure membrane	10-10,000	Matric	24, 25
Centrifuge	10-3000	Matric	26
Osmosis	30-2500	Matric and osmotic	27, 28
Consolidation	1-1000	Matric	29
Vapor pressure (vacuum desiccator)	3000-1,000,000	Matric and osmotic	29
Sorption balance	3000-1,000,000	Matric and osmotic	30
Filter paper	1000-10,000,000	Matric	31

A. Main Laboratory Methods for Potentials of 0 to -1500 kPa

Diverse methodologies for the determination of water release characteristics have evolved since Buckingham [19] introduced the concept of using energy relations to characterize soil water phenomena. The most important techniques of measuring water release characteristics in the laboratory and the ranges of suction for which each method can be used are shown in Table 2.

1. Vacuum or Suction Methods for Measurement at High Potentials (< 100 kPa suction)

The basis of these methods is that soil is placed in hydraulic contact with a medium whose pores are so small that they remain in a saturated state up to the highest suction to be measured. The suction can be applied by using either a hanging water column or a pump and suction regulator. The soil in contact with the medium loses or gains water depending on whether the applied suction is greater or less than the initial value of soil water suction. Because it is more

common to carry out such measurements on the desorption segment
of the hysteresis curve, we are usually concerned with the loss of
water. Attainment of equilibrium with the applied suction can be
determined by regularly weighing the soil sample or by measuring
the outflow of water until either the weight loss or outflow ceases
or becomes minimal. The main restriction to such methods is the
bubbling pressure of the medium used. The bubbling pressure
(which is negative) is the suction applied to the medium that empties
the largest pores, thus allowing air to pass through the pores and
causing a breakdown in the applied suction. Various experimental
arrangements to apply the suction are discussed in the following
sections.

Büchner Funnel. In the simplest application of the suction principle, a Büchner funnel and a filter paper support the soil. The
apparatus, introduced by Bouyoucos [32] and later adapted by Haines
[3] to demonstrate hysteresis effects, is still occasionally referred to
as the Haines apparatus, even in installations where the funnel is
fitted out with a porous ceramic plate [33-35].

One type of installation is illustrated in Fig. 6. One end of a
flexible PVC tube is connected to the base of a funnel and the other
end to an open buret. The tubing should be flexible but resistant
to collapse, which can result in measurement errors. The tubing
and funnel are filled with deaerated water and the buret adjusted
until the water is level with the ceramic plate or filter paper. Air
bubbles trapped within the funnel can be expelled upward by tap-
ping the funnel while applying a gentle air pressure through the
end of the buret. If a porous ceramic plate is used, as in Fig. 6,
deaerated water will need to be drawn through the plate by applying a vacuum to the open end of the buret while the funnel is inverted in the water. Once the system is air-free, a prewetted soil
sample (normally a soil core) is placed in contact with the filter paper or ceramic plate. The water level is maintained level with the
base of the sample until it is saturated, whereupon the volume in
the buret is recorded. A suction h cm of water can then be applied
by adjusting the buret so that the water level in it is h cm below
the midpoint of the sample. Water that flows out of the sample in
response to the applied suction can be measured by the increase in
volume of the water in the buret after the water level has stopped
rising.

No detectable change in buret water level within 6 hours is suggested as a satisfactory definition of equilibrium [36], but a shorter
period without change might be acceptable. Small evaporative losses
through the open end of the buret can be suppressed by adding a
few drops of liquid paraffin to the water in it. Evaporative losses
from the sample can be minimized by covering the open top of the

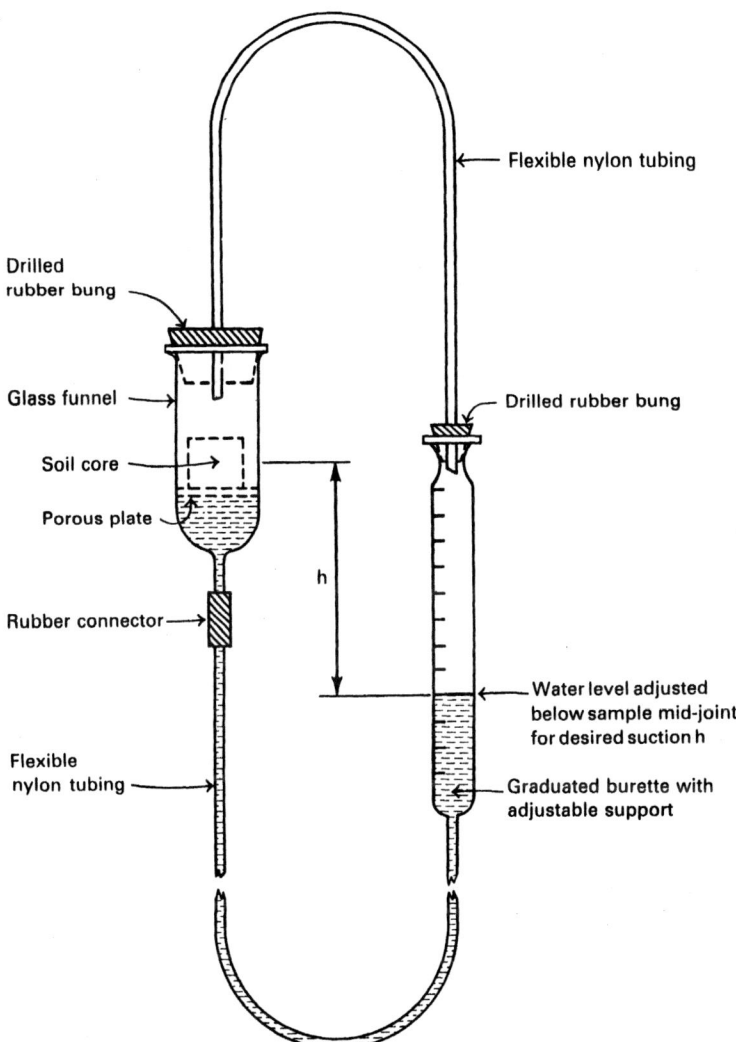

FIG. 6 Büchner funnel or Haines apparatus tension method.

funnel or creating a closed system as in Fig. 6. If the final level in the buret is h', then the final suction applied is h', rather than h. However, by altering the level of the free water surface to h at each inspection, the desired suction can be maintained. By repeating the exercise at successively increasing suctions, a soil moisture characteristic curve can be plotted by calculating back from

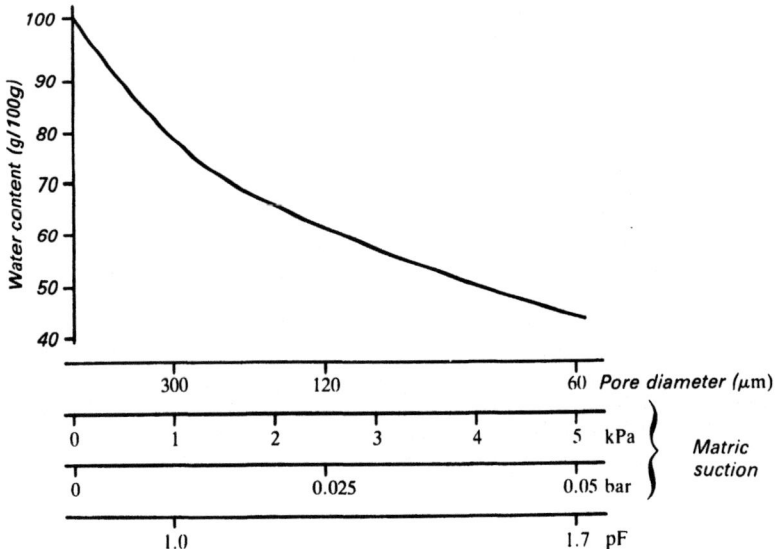

FIG. 7 A soil water release curve obtained using the Büchner funnel method.

the final moisture content of the soil sample (determined gravimetric-ally) using the volumes of water extracted between successive applied suctions.

Using a filter paper, the maximum suction that can be applied is only 50-70 cm of water before air entry occurs around the sides of the paper; but using a porous ceramic plate, the maximum suction attainable is much higher, depending on the air-entry (bubbling) pressure of the plate. In practice, the maximum suction applied using a ceramic insert is restricted by the distance to which the leveling buret can be lowered below the funnel (i.e., typically < 200 cm of water).

The Büchner funnel technique is not only very suitable as a teaching method, it is also trouble-free. Even with the limitations of using filter paper, a curve can be obtained that can be used to interpret the soil pore size distribution in a range important for soil drainage (Fig. 7). The volume of water extracted from some soils between successive suctions might be small and difficult to measure accurately in the buret. An alternative, possible only if a ceramic plate is used in the Büchner funnel, is to determine the water content of the soil sample gravimetrically after each successive equilibrium is reached [34]. Because the Büchner funnel method requires a separate piece of apparatus for each soil sample, it lends itself to small research and/or teaching laboratories, where large numbers of samples are not normally analyzed. However, the method should not be disregarded for other situations, as accuracy is claimed to be good and material costs are low [34].

Porous Suction Plate. The Büchner funnel method has been adapted in a variety of ways [29, 37], but most assemblies retain the common property of accommodating only one sample at a time. Czeratzki [38] described the construction and use of a ceramic suction plate 500 mm by 350 mm, capable of taking several samples, and several European institutions were reported as using the method [39]. Loveday [20] describes three designs of ceramic suction plate extractor, although noting that only one was commercially available in Australia. One design consists of a large ceramic plate sealed onto a clear, water-filled acrylic container with outlet. The space between the plate and container is kept water-filled, and air bubbles trapped below the plate can be readily seen and removed. A cover to the whole assembly reduces evaporative losses and, depending on the size of the plate, several soil cores can be brought to equilibrium at one time. The suction can be applied either by using a hanging water column (as for the Büchner funnel) attached to a leveling bottle or buret, or by a vacuum pump and regulator. If several contrasting soils are being analyzed at the same time, some might reach equilibrium much more quickly than others. Then, if water outflow were used as a criterion of equilibrium, the samples could not be removed until the last sample had reached equilibrium. Because the water extracted from each sample cannot be measured by the outflow and must be determined from the equilibrium weight, it is easier to determine equilibrium of each individual sample by regular weighing, as for sand suction tables (see next section). Problems can be encountered in regaining hydraulic contact between samples and plate after weighing. Solutions include using plaster of Paris in the bottom of the sample and using a fine layer of silt on the plate, but care must then be taken to remove silt adhering to the sample before it is weighed.

The requirement for regular weighing means that porous suction plates must be maintained at working height, thus limiting the height available below the plate for a suspended water column (unless in multifloor buildings it can be extended into an underlying story). For suctions in excess of 10 kPa, a complex sequence of bubbling towers [20] or an accurately controlled mechanical vacuum system [29] is then required, and this has probably limited the widespread adoption of the porous suction plate.

Sand Suction Tables. The use of sand suction tables is fully described by Stakman et al. [21], who refer to them as the sandbox apparatus. Instead of applying a suction to a ceramic plate or

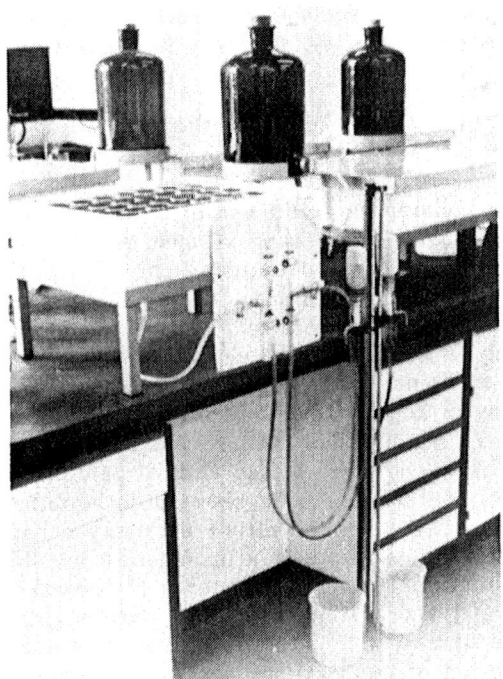

FIG. 8 Sand suction tables. (Reproduced by permission of the
Institute for Land Reclamation and Improvement, The Netherlands.)

filter paper, suction is applied to saturated coarse silt or very fine
sand held in a rigid container, and core samples are then put into
contact with it. The maximum suction that can be applied before
air entry occurs is related to the pore size distribution of the packed
fine sand or coarse silt and is thus related to its particle size distribution. The original design (Fig. 8) has been adapted, sometimes
with minor modifications, elsewhere (Fig. 9). A version is available
commercially, but one of the attractions of sand suction tables is
that they can be constructed easily and cheaply from readily available materials, although care must be taken during assembly. They
are thus well suited to laboratories in locations where supplies of
more sophisticated equipment are available only at great cost as im-
ports, or not at all. The container need not be a ceramic sink,
though such receptacles are very suitable. Any rigid, watertight,
nonrusting container, with a cover to prevent evaporative losses,
will suffice, and slightly flexible plastic stacking storage bins can
be used successfully, provided the sides cannot flex away from the

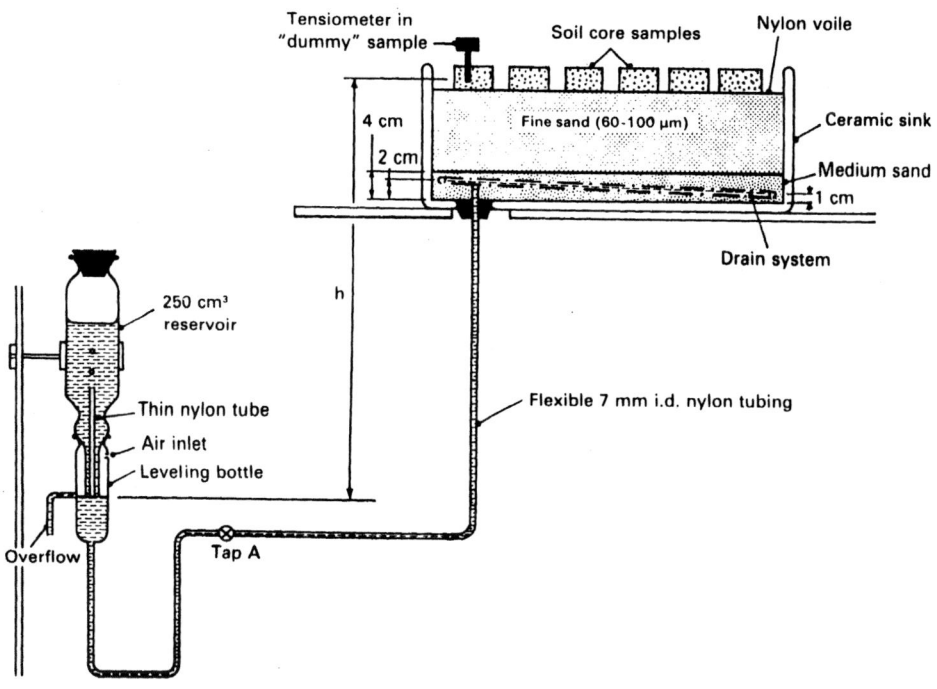

FIG. 9 Components of a sand suction table. The suction applied is equivalent to the difference in height h. (After Ref. 7.)

sand to allow air entry. Industrial sands with a narrow particle size distribution are most suitable because they contain few fines; the particle size distribution of some suitable grades available commercially in Britain is given in Table 3. In practice, local sources of sediments, such as from rivers, estuaries, coastal flats [21], or the washing lagoons of aggregate plants, can often provide a suitable particle size distribution. Fine glass beads and aluminum oxide powder have been shown to have adequately high air-entry values and hydraulic conductivities for use as tension media [40], but these materials cost considerably more than sand. Ball and Hunter [41] report a shallower design of suction table, which utilizes a strengthened Perspex tray with integral drainage channels overlain by glass microfiber paper and a thin layer of commercially available silica flour with particles mainly of 10-50 μm.

It follows that sand suction tables can be of a variety of designs and sizes. Typically though, each should hold 30-50 undisturbed

TABLE 3 Industrial Sands and Silica Flour for Suction Tables[a]

Type	Chelford 60	Redhill 110	Redhill HH	HPF2
Use:	Base of suction tables	Surface of suction tables (< 50 cm suction)	Surface of suction tables (< 110 cm suction)	Surface of suction tables (< 210 cm suction)
Typical particle size distribution				
> 600 µm	1	1	1	0
200-600 µm	63	8	1	0
100-100 µm	37	69	11	1
60-100 µm	1	20	30	2
20-60 µm	1	3	53	39
< 20 µm	0	0	5	59

[a]All samples available in U.K. from Hepworth Minerals and Chemicals Ltd, Moneystone Quarry, Oakamoor, Stoke-on-Trent STIO 3DZ.

presaturated soil cores. The upper face of the core is kept covered by a lid, while the lower face is covered by a piece of nylon voile secured with an elastic band. Vomocil [36] considers that the voile interferes with hydraulic contact only if a suction of more than 15 kPa is applied. By placing tensiometers beneath the surface of the sand and in the samples, we have confirmed that hydraulic contact is maintained to suction of at least 10 kPa. Sand baths up to 10 kPa suction are fairly reliable and maintenance-free. The applied suction can be monitored by a tensiometer embedded in a "dummy" sample and connected to a mercury manometer [7] or by a standard nondegradable porous sample weighed at regular intervals. The occasional air locks that do occur can be cured by temporarily flooding the bath with deaerated water and drawing it through under vacuum.

For full characterization of the water release at high potentials, samples on sand baths need to be brought to equilibrium at a series of increasing suctions [21]. Regular alteration of the tension applied to a single suction table can result in more frequent air locks, and furthermore, all samples must reach equilibrium before the tension can be changed. A more practical solution is to wait until samples have reached equilibrium, then transfer them to tables set at progressively higher suctions [7].

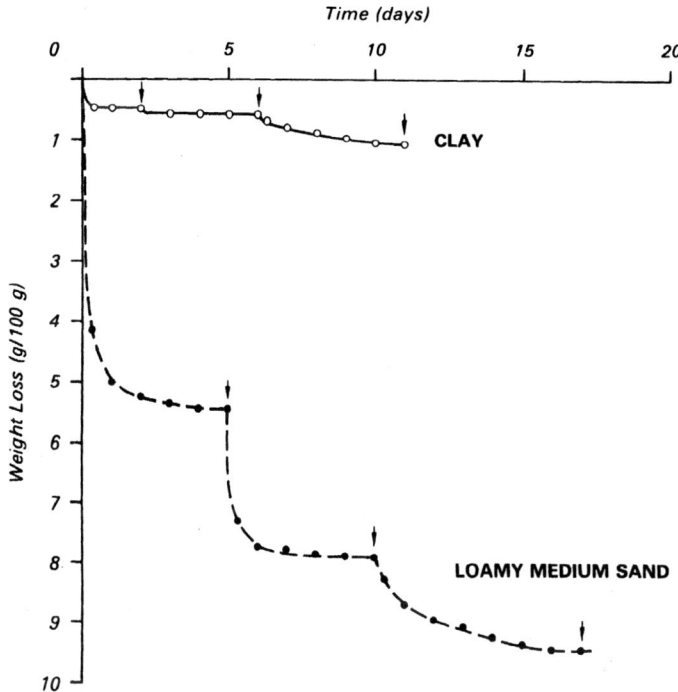

FIG. 10 Outflow curves for two soils equilibrated from natural saturation at three successive suctions (2.5, 5, and 10 kPa) on sand suction tables.

The attainment of equilibrium at a given suction is determined by weighing the samples at 2-3 day intervals. If the decline in weight does not follow the general shape of the curves in Fig. 10 but continues at the same magnitude, hydraulic contact is likely to have been lost. Weight loss criteria for equilibrium depend on sample size and accuracy required, and thus quoted equilibration times [38,41] are not of general use. By recording the equilibrium weight, the moisture content at any given suction can be calculated after the sample has been oven-dried. The time taken to reach equilibrium depends on the particle size distribution of the sample, its organic matter content, and the suction being applied. For example, equilibration times for sandy soils are often longer than those for clayey soils (Fig. 10). This is because a loamy sand subsoil that has the same unsaturated hydraulic conductivity as a clay loam subsoil at 1 kPa suction has an unsaturated hydraulic conductivity of only one-tenth that of the clay loam at 10 kPa [42].

The air-entry value of fine sand precludes the use of sand suction tables at suctions above about 10 kPa. Stakman [21] extended the range of the sand suction table by first applying layers of a sand-kaolin mixture and then pure kaolin to the top of a sand suction table. The required suction was maintained by a vacuum pump. The kaolin-sand suction table has been reported to be in use elsewhere [7], but it is more difficult to construct than a sand suction table. It also suffers from problems of entrapped air [40] and capillary breakdown and thus requires more maintenance than a sand suction table. The kaolin used has a low hydraulic conductivity; hence samples require a long time to reach equilibrium. Ball and Hunter [41] reported achieving suctions of 20 kPa with their silica flour assembly but did not report an air-entry value for it. Such a medium might be usable up to 33 kPa and might result in fewer problems than the sand-kaolin combination.

Because sand or silt suction tables provide an excellent low-cost method of measuring the soil water characteristic for a large number of samples at high potentials, they have been adopted by many researchers (see, e.g., Refs. 7 and 10). Their main limitation is capillary breakdown as larger suctions are applied, and for this reason, pressure methods are more commonly adopted for suctions in excess of 10 kPa.

2. Gas Pressure Methods (0 to -1500 kPa potential)

As with the vacuum or suction methods, soils are placed on a porous medium, but they are brought into equilibrium at a given matric potential by applying a positive gas pressure. To maintain this pressure, the porous medium and samples are contained within a pressure chamber while the underside of the porous medium is maintained at atmospheric pressure. Various designs of pressure chamber have been reported [7,20] since Richards [22,24] developed the original designs. All use either a porous plate or a cellulose acetate membrane as the porous medium. The pressure is supplied via regulators and gauges, by bottled nitrogen, or by a mechanical air compressor. Most designs of pressure chamber can take soils in a variety of physical states, but as equilibration times in pressure cells depend on the height of the soil sample, core samples in excess of 5 cm high are undesirable. At -1500 kPa, a sample height of 1 cm is convenient. Because the water in samples equilibrated at low potentials is held in small pores, it is acceptable to use disturbed samples, provided the soil is not compressed or remolded.

Pressure Plate Extractor. With the development of porous ceramics, pressure plate extractors have become available to cover a range of potentials down to -1500 kPa (Fig. 11) and have been widely used [8,43-47] for measurement of the water release characteristic,

FIG. 11   A 5-bar pressure plate extractor.   (Reproduced by permission of Soil Moisture Equipment Corp.)

although recent research [48] casts some doubt over their accuracy. Most are designed to accommodate several samples contained within soil sample retaining rings in contact with the porous plate.   Once the extractor has been sealed, a gas pressure is applied to the air space above the samples, and water moves downward from the samples through the plate, for collection in a buret or measuring cylinder.   Equilibrium is judged to have been attained when outflow of water ceases.   The samples can then be removed and their moisture content determined gravimetrically.   Burke et al. [34] report that 2-14 days is necessary to establish equilibrium.   Precision of the method is good, a coefficient of variation of 1-2% being attainable [49].   However, clogging of the ceramic plates by soil particles or algal growth can occur after repeated use, reducing the efficiency of the extractor.   Furthermore, Chahal and Yong [50] discovered that because of air bubbles trapped or nucleated in the water-filled pores, the soil water characteristic curve obtained with the pressure plate apparatus at high potentials (low suction) is higher than that obtained using the suction method of Haines.   Thus pressure plate extractors are best suited to suctions of 33 kPa or greater.

Pressure Membrane Apparatus.   In contrast with pressure plate extractors, in the pressure membrane apparatus the soil sample sits in contact with a semipermeable cellulose acetate (Visking) membrane. This allows passage of water from the sample but retains the air pressure applied to the upper surface of the membrane.   Since the first pressure membrane cell was developed [24], designs have

FIG. 12 Pressure membrane extractors: (a) individual cells capable of taking 7.5 cm diameter samples and (b) large cell with several small samples and an internal diaphragm in the lid. (b reproduced by permission of Soil Moisture Equipment Corporation.)

varied (Fig. 12), and the technique has been used in many parts of the world (7,10,45,51,52]. Larger cells take several small disturbed samples contained in retaining rings, and some designs incorporate in the lid a diaphragm that expands during use to hold the soil samples in firm contact with the cellulose membrane (Fig. 12). As with pressure plate extractors, outflow from large cells is measured in a single container, and thus all samples must have reached equilibrium before any can be removed for gravimetric determination of moisture content. Because gas diffuses slowly through the membrane and is replaced by drier gas from the pressure source, samples that reach equilibrium several days before others may start to dry by evaporation [53] and give erroneous results. This is likely to be a more serious problem with systems powered by bottled dry nitrogen gas than with those using laboratory or outdoor air compressed mechanically to a high humidity. Evaporation is also less likely to be a problem with smaller cells, designed to take only one sample [7] from which the outflow is monitored by a single collection device. With these, the sample can be removed as soon as equilibrium is reached. Texture-related equilibrium times for pressure membrane analysis are given by Stakman and van der Harst [54]. The pressure membrane apparatus gives moisture contents comparable to those from pressure plate extractors at the same applied pressure [55] but is found by some [49,55] to be prone to membrane leaks due to microbial action, iron rust from the chamber, or sand

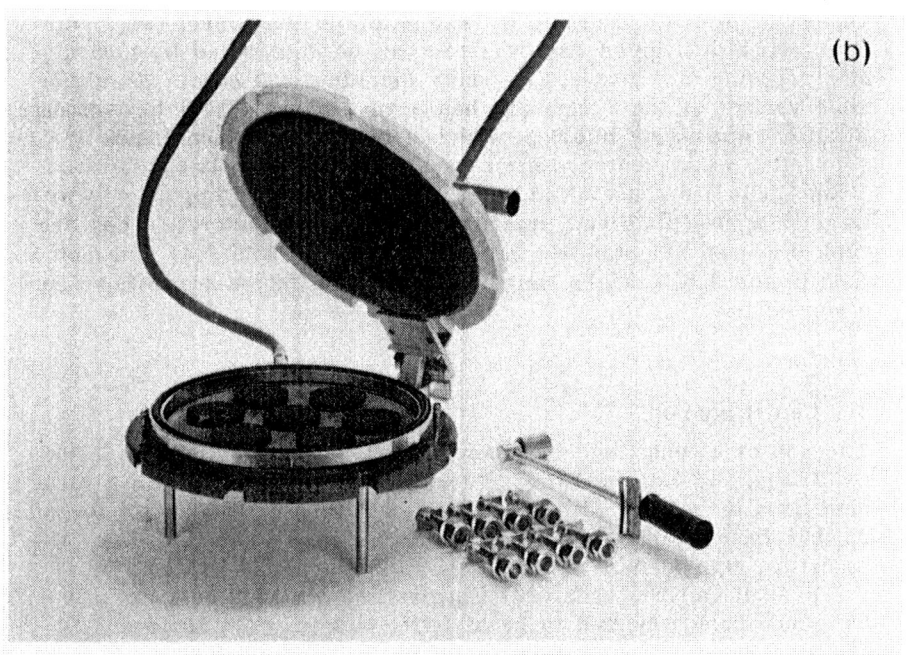

(b)

FIG. 12 (continued)

grains trapped near the gasket seals. These problems are a greater nuisance with a large cell containing many samples, and we find that such problems are rare when we use brass or stainless steel pressure cells and two membranes for high pressures (> 1000 kPa), and exercise care in operation.

Tempe Cells. Most pressure membrane and pressure plate extractors have been designed to extract moisture from small disturbed soil samples and are thus not suitable for characterizing the low suction range, where soil structure is all-important. Because of this, an individual cell, similar to the individual pressure membrane cells described by Hall et al. [7], but of lightweight construction, has been developed for measurement on undisturbed soil cores using pressures of 0-100 kPa. The commercially available design is a de-

velopment of that described by Reginato and Van Bavel [23], and
equilibrium at a given gas pressure can be determined by periodically weighing the complete assembly including soil core. A submersible variant of the Tempe cell has been developed [56] to overcome
problems due to air bubbles, which can result in inaccuracies in
volumetric water content measurements and porous plate failure.
Tempe cells are a useful addition to installations equipped only with
large pressure plate and pressure membrane extractors. They are
typically used at potentials between 0 and -100 kPa [44]; for potentials in the 0 to -20 kPa range, sand suction tables are cheaper and
easier to use.

3. Centrifugation

The use of a centrifuge to extract water from soils was introduced
by Briggs and McLane [57]. These investigators centrifuged saturated soils in perforated containers at a speed that exerted a force
of 1000 times gravity and termed the resulting moisture content the
"moisture equivalent."

   Russell and Richards [26] improved on the technique, and it
has since been reported to be in fairly wide use [14,29,58,59] for
measuring moisture retained at a variety of applied suctions. The
soil sample is commonly supported on a porous medium in a cup containing a water table at the opposite end from the soil. The force
exerted by the centrifuge during spinning is related to the angular
velocity and the distances of the water table and sample from the
center of rotation, given by

$$\log_{10} h = \log_{10}\left(\frac{r_2^2 - r_1^2}{2} \cdot \frac{w^2}{g}\right) \tag{2}$$

where h is the suction in centimeters of water, r_1 and r_2 are the
distances (cm) between the center of rotation and the midpoint of
the sample and of the water table, respectively, w is the angular
velocity, and g is the acceleration due to gravity.

   Thus, by varying the angular velocity, different suctions can
be applied to the soil sample. Odén [59] recommends the following
centrifugation times for equilibrating saturated soils 3 cm high and
with a volume of 50 cm^3, though the precise time will depend also
on the sample composition.

   The advantages of centrifugation as a method are that it can
quickly produce a soil water release curve. However, as Childs

Matric potential (kPa)	Centrifugation time (min)
< 1	5
1–20	15
20–200	30
200–2500	60

[60] has pointed out, the suction actually varies over the thickness of the sample, and other methods give better accuracy. While the centrifuge stops spinning and before the sample can be removed for weighing, the sample might reabsorb some moisture from the porous medium on which it sits.

Furthermore, in saturated compressible samples thicker than 0.5 cm, consolidation during centrifugation can introduce further errors [29].

B. Main Laboratory Methods for Potentials of Less than –1500 kPa

Although it is uncommon to measure the water release characteristic to less than –1500 kPa potential, several methods are available to extend the curve to higher soil water suctions. Some methods, such as the pressure membrane apparatus, can be considered direct, while others are indirect (vapor pressure and sorption balance), involving the thermodynamic relationships between the suction of retained water and freezing point and vapor pressure depressions.

1. Pressure Membrane

By using strengthened assemblies, the usefulness of the pressure membrane apparatus can be extended to extract water held at potentials less than –1500 kPa. Richards [25] measured moisture retention in soils to –10,000 kPa potential, while the apparatus of Coleman and Marsh [61] can accept pressures of almost 150,000 kPa. Even though pressure membranes measure matric potential, while a sorption balance measures matric and osmotic potential, Coleman and Marsh [61] found good agreement between results from the two methods applied to a clay soil at around –10,000 kPa.

2. Vapor Pressure

The relationship between relative humidity at 20°C and soil water suction h (cm H_2O) is expressed by

$$\log h = 6.502 + \log(2 - \log H) \tag{3}$$

where H is the relative humidity in percent [2].

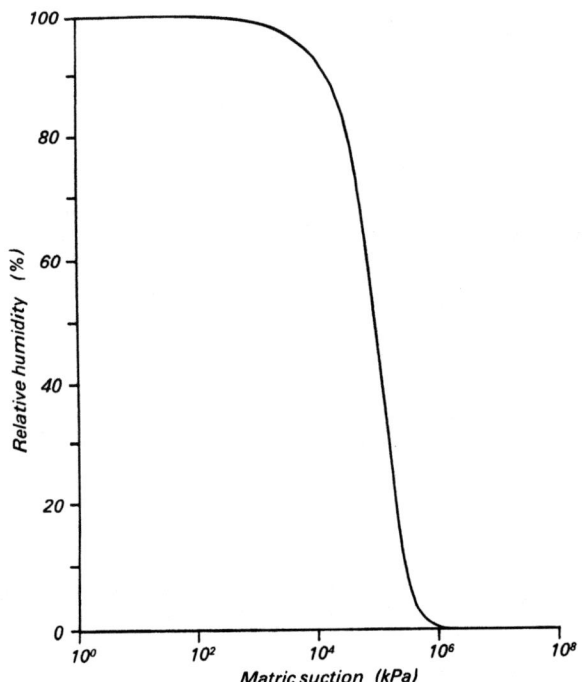

FIG. 13 The relationship between matric suction and relative humidity at 20°C. (After Ref. 29.)

This relationship, shown in Fig. 13, can be used in two ways to determine the water release characteristic at high suctions.

Vacuum Desiccator. By placing soil that has been broken into small aggregates (passed through a 2 mm sieve) on a petri dish, in constant-humidity atmospheres in a vacuum desiccator or other sealed container, soil can be equilibrated at a chosen potential before its moisture content is determined gravimetrically. Aqueous sulfuric acid solutions have been used, but Loveday [20] recommends the use of several easily available neutral or acid salts to achieve a range of vapor pressures (Table 4). Although equilibrium times are long (5-15 days), the accuracy of the method is claimed to be good [34]. To minimize errors due to temperature fluctuations, however, it is essential that the vapor pressure method be used only in an environment (room or insulated container) with temperature control to better than 1°C, especially for potentials higher than -10,000 kPa [61].

TABLE 4 Saturated Salt Solutions and Vapor Pressures at 20°C [20]

Salt	Relative humidity (%)	Potential (kPa)
$CaSO_4 \cdot 5H_2O$	98	−2730
$Na_2SO_3 \cdot 7H_2O$	95	−6935
$ZnSO_4 \cdot 7H_2O$	90	−14245
NaCl	75	−38893
$Ca(NO_3)_2 \cdot 4H_2O$	56	−78389
$CaCl_2 \cdot 6H_2O$	32	−15404

Sorption Balance. The sorption balance also uses the relationship between the soil water potential and the vapor pressure of the atmosphere with which the soil is in equilibrium. In the sorption balance, water from the sample is allowed to evaporate into a previously evacuated chamber and the potential deduced from measurements of the vapor pressure [29]. The sample is weighed continuously by a sensitive spring balance as the vapor pressure is reduced. It is important to maintain a constant temperature, but Coleman and Marsh [61] found the sorption balance less prone than the vacuum desiccator to temperature-induced errors.

C. Other Laboratory Methods

1. Osmosis

Zur [27] was the first to present a method of analysis based on the osmotic pressure of different solutions. A polyethylene glycol solution is separated from a soil-water system by a membrane that is permeable to water and ions but impermeable to certain solute molecules and soil particles. The water in the solution has a lower partial free energy than that of the water in the soil, and this tends to move water from the soil to the glycol solution until an equilibrium is established. Since the membranes are permeable to ions, the osmotic system controls the matric soil water potential only. By using solutions of different concentrations, calibrated to apply given matric potentials, a water release characteristic can be determined. Pritchard [28] developed the apparatus and extended the method to cover a range of potentials from −30 to −1500 kPa but encountered problems with microbial breakdown of membranes. Although there is fairly good agreement between water release characteristics obtained by the osmotic method and those by pressure membrane [27],

the osmotic method has not been applied widely because of long
sample equilibration times [62].

2. Consolidation

Measurement of the water release characteristic by applying a direct
load to the soil was described by Croney et al. [29]. A saturated
soil sample, laterally confined and sandwiched top and bottom between
two porous disks, is loaded with successive weights on a consolidation
frame (oedometer) [63]. The excess pore water pressure induced by
each load is dissipated through the porous disks at a rate dependent
on the hydraulic conductivity of the soil, and the soil compresses to
a new state of equilibrium in which the load is equaled by the matric
potential of the new soil-water system. When compression ceases for
any given load, the equilibrium moisture content can be calculated
from reduction in sample thickness (measured by micrometer) and
plotted against applied pressure. The method is applicable only to
compressible soils such as shrinking clays and only over the primary
consolidation phase [63]. Croney et al. [29] point out that the friction between the sample and the containing ring can affect accuracy
at low suctions. However, our research on disturbed clays indicates
that the method gives a water release characteristic for clays comparable to that obtained by a combination of sand suction tables and
pressure membrane apparatus (Fig. 14). The consolidation method
is also faster than most others (the curves in Fig. 14 were obtained
in 6 days), but it is mainly likely to find application in laboratories
with an interest in the engineering application of soil physical data
and already possessing the necessary equipment.

3. Filter Paper

The filter paper method is based on the assumption that the matric
potential of moist soil and the potential of filter paper in contact
with it will be the same at equilibrium. The method assumes also
that if the soil sample is large compared with the filter paper (Al-
Khafaf and Hanks [64] used about 100 g of soil), then contact with
the filter paper will result in minimal change in soil water potential.
By separately calibrating the water content of the filter paper
against matric potential, the matric potential in the soil can be derived. The technique was originated by Gardner [65] and was further developed [66,67] for use with Whatman No. 42 filter paper,
which is comparatively thick and has a small pore size. Above all,
the technique requires a good contact between the sample and the
soil. After an equilibration period of 5-7 days in a sealed container
at constant temperature, the filter paper is removed carefully with
tweezers; adhering soil must be brushed free, and the paper is
quickly weighed before oven-drying.

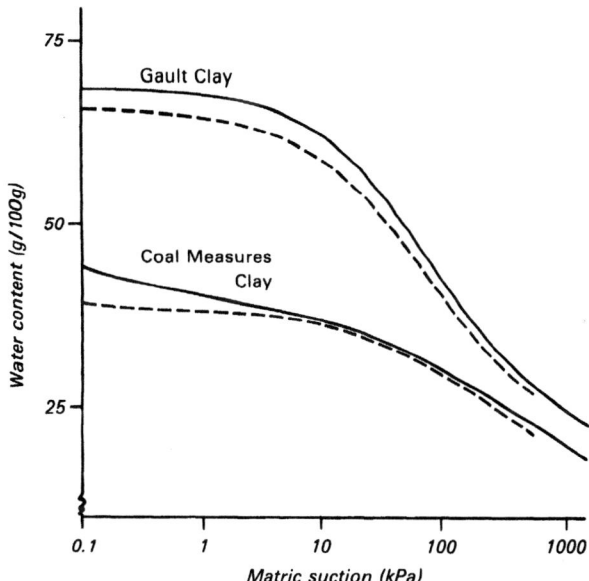

FIG. 14 Comparison of water release characteristics obtained by consolidation (---) and by sand suction table-pressure membrane apparatus (——) for two sieved and rewetted subsoil clays.

To plot the water release characteristic, however, soil samples uniformly dried to a range of moisture contents are required. These are best obtained by successive sampling of field soils as they dry out, though the climate and the season will then determine the scope of the water release characteristic obtained. One of the main interests in the filter paper method is for quick measurements of soil water potential, which, in fine-grained soils, controls soil strength [68].

4. Rapid (Null) Method

In the methods described in Section IV.A, the moisture content of the sample is measured after equilibrium has been reached in response to a constantly applied suction or pressure. It may take several days to reach this equilibrium. Croney et al. [29] described apparatus in which the suction applied to the sample is adjusted to be equal to the suction of the soil without change in moisture content. The method is best described by analogy to the Büchner funnel (Section IV.A.1). Instead of being connected via a PVC tube to a buret, the porous plate is connected via a fine horizontal tube to a vacuum system. The plate, reservoir, and part of the flow

tube contain air-free water, which can be colored to increase the
visibility of the meniscus (which, at the beginning of the test, is
half-way along the flow tube). When a soil sample is placed in contact with the porous plate, water will be drawn from the reservoir,
causing the meniscus to recede. By applying a suction to the open
end of the tube so that the movement of the meniscus is stopped,
the suction can be related to the moisture content of the sample determined gravimetrically. Dumbleton and West [69] extended the
range of the method but, as with filter paper, soil samples at a
range of moisture contents are necessary to plot the water release
characteristic. The method is currently being reinvestigated for
the measurement of suction-moisture content relationships in samples
that are continually disturbed as they are dried. This can provide
data of relevance to studies of critical state soil mechanics and for
comparison with simple engineering index tests.

5. Psychrometry

The application of, and equipment for, thermocouple psychrometry
was described in Chap. 2. However, provided samples uniformly
dried to a suitable range of moisture contents are available, laboratory psychrometers such as those described by Rawlins and Campbell [70] can also be used to determine the water release characteristic.

D. Field Methods

It is relevant briefly to discuss field methods of determining the
soil water release characteristic, as these are done in situ and consequently are more representative than laboratory measurements.
Laboratory measurements often deviate significantly from the field-
measured water release curve, especially in fine-grained compressible soils where there is the influence of overburden load in the
field [71]. Thus Ratliff et al. [72] recommend that if absolute
accuracy is required—for example, in soil water balance calculations—field-measured curves should be taken. By installing tensiometers at different depths in the field, suction readings of potential can be related to water content measurements determined
either gravimetrically (hence destructively) or by a neutron probe
[34,73]. The method is limited by the range of tensiometers (0 to
-80 kPa) and, although use of electric resistance sensors [74] or
thermocouple psychrometers can extend this range, there can be
calibration problems, and a long time is needed before a soil water
characteristic curve can be obtained. If the soil rewets between
readings, hysteresis can be a problem, and fluctuations in soil temperature cause further complications through their effect on the
viscosity of soil water. For these reasons, field methods are less

commonly used than laboratory methods. Bruce and Luxmoore [75] provide a useful summary of references describing measurement of the release characteristic in the field.

E. Choice of Method

Having reviewed the various methods available to measure the soil water release characteristic, it is pertinent to consider external factors that might influence the choice of method in any particular situation. Many of the factors are unrelated to the scientific merits or accuracy of the method but, in practical terms, are of great, if not overriding, importance.

An analysis might be carried out either to demonstrate the principle of the water release characteristic or to obtain data for further interpretation. Even in the latter case, some situations require a large volume of fairly accurate data, while others may require highly accurate data from a small number of samples. The Büchner funnel or Tempe cell might have considerable advantage in demonstrating principles in an educational environment inasmuch as each allows individual analysis of individual samples. However, the same methods may not be cost-effective or the most efficient when many analyses are required. The equipment requirements for measurement of the water release characteristic will thus vary considerably from a small research laboratory to a laboratory serving, for example, an active national soil survey organization.

Some small analytical installations in universities might have to serve a dual function in research and teaching. Use of the measurement equipment for hands-on learning might be more effective if the sample belonging to each student or student group is contained in an individual pressure cell or in an individual core on a sand bath rather than enclosed in a larger pressure extractor along with samples belonging to other students or student groups. In a laboratory responsible for measurements on a large number of undisturbed soil samples, a requirement for measurements mainly at high potentials is more likely to be satisfied by sand baths or by large ceramic suction plates than by the limited capacity of a pressure plate or pressure membrane extractor.

1. Analysis Time

Most methods of measuring the water release characteristic involve leaving samples until their potential reaches equilibrium with an applied suction or pressure. Because of this, the time taken for full characterization can be considerable when compared, for example, with many methods of soil chemical analysis. Samples can take 4-12 days to reach each successive equilibrium on sand suction tables and in pressure cells [41]. Thus, determination of five or six equilibrium points using one sample can result in a total analysis time of 3-4 months, once peripheral laboratory tasks such as oven-drying and data calculation have been taken into account. This time scale might not be a problem for a laboratory servicing a large strategic soil resource survey, but it is totally unacceptable for short-term, customer-oriented projects. Analysis time in such situations can be shortened by careful division of samples so that different equilibrium points can be determined simultaneously on sub-samples or by taking a large number of replicate undisturbed samples. However, a reliable soil water characteristic curve is not likely to be obtained from subsamples unless the original sample is homogeneous. Any requirement for more rapid analysis is likely to be met only by methods such as those using a centrifuge, and meeting it will entail any inaccuracies inherent in such methods.

2. Equipment Availability and Price

Perhaps the major influence on methods adopted in soil physics laboratories around the world is the availability of an extensive range of soil moisture extractors manufactured by the Soil Moisture Equipment Corporation (Santa Barbara, CA). Smaller ranges of similar equipment are available in the United Kingdom, Australia, and the Netherlands, but they are not in use widely outside their country of origin. A list of suppliers is given in Table 5. In many developing countries, however, acquisition of imported equipment is strongly discouraged by fiscal policies. Thus, although a range of suitable equipment may be available, it is not easily obtainable, and alternative supplies or methodologies may need to be adopted. Under such circumstances, it might be pertinent to consider adopting methods that are less capital-intensive or manufacturing equipment locally. It must be remembered though that whereas a commercially available system such as a pressure plate extractor and peripherals comes well documented with a complete set of instructions, a proven methodology for measurement, and a single source of replacement parts, self-designed installations require staff with the necessary aptitude for construction and maintenance and often necessitate considerable effort in locating and obtaining component parts. Whatever the degree of sophistication of the equipment used, the usefulness of the data will be affected by many other factors including the quality of available staff. Accurate temperature control of the analytical environment is also important because of the effect of temperature changes on the viscosity of water [76,77].

3. Safety and Statutory Requirements

The most common techniques used to characterize the low-potential part of the water release characteristic employ high air pressures.

TABLE 5 Some Equipment Suppliers and Typical Prices

Equipment	Typical unit cost ex works 1989 (U.S. $)	Suppliers[a]	Remarks
Sintaglass funnel (Buchner)	40	E	Available from general labora- tory suppliers
Suction plate	75	H	(e.g., 0.35 × 0.5 × 0.08 m, 1 μm pore diameter)
Sand suction tables	1800–2100	B, C, I	Can be hand-made
Sand-kaolin tables	3500–4200	B, C	Can be hand-made
Pressure plate, 500 kPa	1175–2000	A, B, I	
Ceramic plate, 1500 kPa	1400–4750	A, B	
Pressure membrane cells	150[b]–1100[c]	A[c], D[b]	
Pressure membrane (with gauges and regulators)	2550[c]–2750[d]	C[c], D[d]	
Tempe cell	125	A	
Centrifuge	3500–4500	B, E	
Laboratory psychrom- eters	2000–2500	F, G	
Sample corers	550–925	A, C, D, I	
Corer sets	950–1550	B, C	
Laboratory com- pressor	2000–3500	A, B	

[a]Most suppliers have agents around the world.

[b]Individual sample cell.

[c]Multisample cell.

[d]Module with eight individual cells.

Key: A. Soil Moisture Equipment Corporation, P.O. Box 30025, Santa Barbara, CA 93105. B. ELE International Ltd., Eastman Way, Hemel Hempstead, Herts, HP2 7HG, U.K. C. Eijkelkamp, Nijverheidsstraat 14, 6987EM Giesbeek, The Netherlands. D. Westbrook Engineering, 34/36 Brook Street, Derby, DE1 3PF, U.K. E. Gallenkamp, Belton Road West, Loughborough, Leics, LE11 OTR, U.K. F. Decagon Devices Inc., P.O. Box 835, Pullman, WA 99163. G. Wescor, 459 South Main Street, Logan, UT 84321. H. KPM, Wegelystrasse 1, 1000 Berlin 12, West Germany. I. Irricrop Technological Pry. Ltd., P.O. Box 487, Narrabri, N.S.W. 2390, Australia.

Thus it is essential that the equipment used and the peripheral supply lines be designed not only to withstand safely the pressure ranges
applied but to do so within an acceptable safety margin. This is an
important consideration not only for equipment made locally according
to laboratory specifications but can also apply to internationally available standard pieces of equipment. Different countries interpret
safety criteria differently and apply different safety margins. In
the United Kingdom, for example, the design, operation, and maintenance of air receivers comes under the control of the Factories Act
of 1961. The act is normally interpreted as including pressure plate
and pressure membrane extractors. These devices are subject to initial inspections and pressure tests to ensure that their design incorporates a sufficient safety margin against failure, and then to regular (26-month) inspections to ensure that they are maintained in a
safe condition. The same rules apply to the air receivers of com-
pressors, which may be used to pressurize the extractors.

The application of these stringent safety regulations in the early
1980s prevented many U.K. laboratories from using the pressure
plate extractors with which they were already equipped, thus dis-
rupting research programs and incurring considerable costs for re-
equipping. We do not detail here the safety regulations in other
countries, but clearly it is advisable to be aware of the statutory
or local constraints on the use of pressure apparatus before equipping a laboratory for measurement of the soil water characteristic.

4. Standardization

In certain allied disciplines, such as in soil analysis for engineering
purposes, there are well-documented standard methods [78] using
equipment of standard design. There have been attempts at some
degree of standardization for methods of determining the water release characteristic, most recently by Burke et al. [34], and an
International Standard is being prepared. However, a variety of
analytical methods are still in use worldwide and will continue to be
used as long as individual requirements differ. In practice, errors
arising from the execution of laboratory procedures are likely to exceed the differences resulting from the use of alternative methods.
Given the wide variety of physical states in which samples are tested,
any attempt at standardization should start with sampling procedure
and sample preparation. These are major factors in analytical differences, and a correct choice of sample state and sample size will
largely decide the analytical technique used.

V. SAMPLING METHODOLOGY AND PRETREATMENT FOR ANALYSIS

A. Field Sampling

Soil samples taken for water release analysis should be isolated with minimal disturbance so that they are closely representative of the in situ soil property. McKeague [79] states that the quality of samples depends on the judgment and ingenuity of the sampler, and the re-liability of the physical data depends on the original soil sample more than any other factor. Burke et al. [34] list the following as impor-tant factors that should be carefully considered to obtain a represen-tative sample: the method to be used, the sample dimension, the sampling location within the field and within the soil profile, the number of replicates, and the time of sampling.

Loveday [20] provides a comprehensive discussion on sampling technique and sampler design.

1. Location

If soil samples are to be taken to represent an area of land such as a field or soil mapping unit, they should be taken from several soil pits to characterize the natural variability. The location of each pit should be carefully chosen with regard to soil and landscape condi-tions. Burke et al. [34] state that random sampling is suitable only when soil differences are not evident; these authors recommend the use of soil survey information in site selection. Greminger et al. [73] present field-measured water release data for 100 locations, demonstrating variability attributable to soil changes along a 100 meter line.

2. Sampling From a Soil Profile

Samples should be taken from representative locations within a fresh-ly dug soil profile (e.g., the midpoints of discrete soil layers or horizons), taking special note of such management-induced boundar-ies as plow pans, deep loosening, and drainage treatments. Where obvious differences occur within a soil horizon or layer, each dis-crete area should be sampled. Detailed profile descriptions (whether in soil science [80,81] or geotechnical [82] terminology and particle size analysis are essential aids to the interpretation of analytical results.

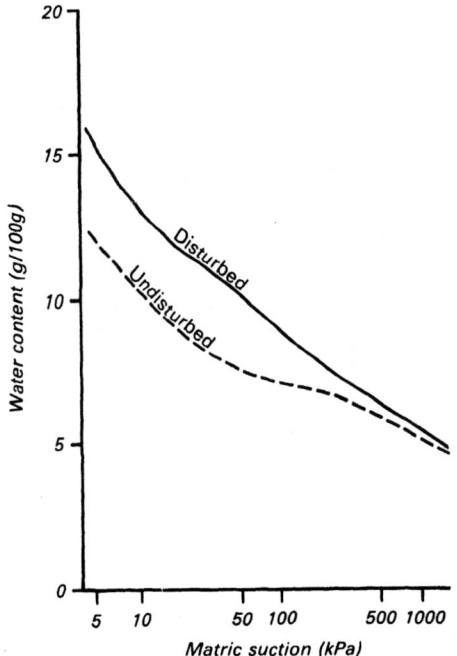

FIG. 15 The effect of sample disturbance on the water release
characteristic of a loamy sand subsoil.

3. Time of Sampling

To standardize procedures and to minimize the effect of hysteresis,
water release samples should be taken when the soil is fully wetted.
In humid areas of northwestern Europe and the United States, the
period from late October to early April, when the soil moisture deficit is usually zero, is suitable. Where clay soils with shrink-swell
properties are being investigated, it is preferable to sample a few
months after the soil has returned to and remained close to zero
soil moisture deficit, to ensure that maximum soil expansion has
occurred.

4. Sample Type and Dimensions

Disturbed Versus Undisturbed. It was shown in Section III.B
that the shape of the water release curve at high potentials is largely dependent on soil structure and the associated pore size distribution. Thus if a sample is disturbed or sieved it cannot reflect the
true properties of a relatively undisturbed field soil because its pore
size distribution will have been greatly altered. Figure 15 shows the

effect of sample disturbance on the water release curves of a loamy medium sand. Unger [83], who made comparative water retention analyses using core and sieved samples, found that disturbance generally decreased and increased water storage in coarse- and fine-textured soils, respectively, but that organic matter content and structural development in the undisturbed soil affected this general trend. Similar results have been recorded by others [84,85].

Disturbed samples, provided they have not been crushed, compressed, or in any other way remolded, may however be acceptable for measurements at potentials less than -100 kPa, and remolded samples might be used for certain geotechnical applications.

Sample Size. The minimum sample volume required to represent a given soil layer without producing unacceptable variation is termed the representative elementary volume [34]. For each soil type this is largely dependent on soil structure, being smaller for sandy soil with a single grain structure than for a clay soil with larger natural aggregates or peds. Although samples of different size should be taken for different representative elementary volumes, many workers use a standard sample size because of the fixed dimensions of the sampler and increase the volume sampled by replication. In practice the number of replicates is often limited by the time and expense of field work and laboratory analysis. Generally cores with diameters of at least 7.5 cm but preferably 10 cm are the most practical for measurements at potentials in the 0 to -100 kPa range. The core length should ideally be equal to the diameter, but since a long sample takes a long time to reach equilibrium during analysis, a length of 2-3 cm is most common.

Coring Devices. The core method normally uses a cylindrical metal sampler that is pressed or driven into the soil to the desired depth and is carefully removed to preserve a known volume of soil as it existed in situ. Dagg and Hosegood [86] devised a sampler incorporating several existing designs which, with further slight modifications, is used on a routine basis in England and Wales [7]. A tin-plated sleeve 7.5 cm diameter and 5.0 cm high is placed in a machined steel barrel and a cutting ring attached. The coring device is driven carefully, using an integral 3.5 kg sliding hammer, into a flat, horizontal surface prepared in the relevant soil layer. Compaction of the sample is avoided by not coring beyond a level marked in the barrel of the corer. The corer is dug out with a trowel and the core ejected by means of a spring-loaded plunger. Various other designs are available internationally, and the suppliers of some of these are listed in Table 5.

Stony Soils. Many soils are difficult to sample because of stones, and although specially designed corers have been recommended [87,88], sample disturbance will be unavoidable in many soils. Rimmer [89], working on reclaimed colliery spoil heaps with large stone contents, filled cans with disturbed material. Alternatively, water release data can be derived from sieved soil repacked to field density and the results corrected using a stone content measured in the field [81]. Where it is not possible to obtain core samples, or expansion or excessive shrinking of a sample is expected, a clod sample can be taken. Loveday [20] describes a method in which natural clods are immersed in Saran resin; after initial measurements of the sample volume, the Saran coating is removed from one flat face and the clods can be equilibrated at various potentials.

B. Sample Preparation

In the field, the soil core should be trimmed roughly with a knife before being fitted with lids at each end and labeled clearly. Samples should be double-wrapped in plastic bags to prevent drying and if necessary packed in foam or polystyrene to avoid damage in transit. Cores taken to the laboratory should be stored in a refrigerator at 1-2°C to reduce evaporation and suppress biological activity. Biotic activity in soil cores can make the determination of equilibrium conditions difficult, and where activity is evident, samples should be treated with an inhibitor, such as a 0.05% solution of copper sulfate or copper chloride. Freezing of samples is to be avoided at all costs because it is likely to alter the pore size distribution, hence the release curve. Preparation for water retention measurements varies between laboratories and the following is the procedure followed by Hall et al. [7].

The core is trimmed flush with the ends of the sleeve and one lid is replaced by a circle of nylon mesh or voile (fine net curtain is suitable) secured with an elastic band. The other lid is sprayed with a dry film lubricant to ease subsequent removal, as the tins can become corroded after a few weeks in the moist atmosphere during equilibration. When trimming the cores, small projecting stones may be carefully removed and the cavity filled with surplus soil or a smaller stone. Samples with large projecting stones may need to be discarded.

The samples are wetted by standing on a sheet of saturated foam rubber to ensure that they are brought to a suction of less than 0.05 bar (the first equilibration point). The time required for wetting varies with particle-size class, being a day or two for sands and as much as two weeks for clayey soils. Care is taken not to leave sandy soils wetting for too long since they may slake. Low-density subsoil sands without the stabilizing influence of organic matter or roots are the most susceptible.

Klute [62] suggests that a wetting solution of deaerated 0.005 M $CaSO_4$ is preferable to either deionized or tap water. The former promotes dispersion of clays in the sample, and dissolved air in the latter can come out of solution, affecting the water content at a given potential.

Fast wetting such as by submergence is not recommended for swelling soils or those with a fragile structure. Klute [62] points out that wetting in the fashion described by Hall et al. [7] brings the sample to natural saturation rather than total saturation because of the presence of trapped air. The water release characteristic will then follow a different curve initially from that from total saturation. It will be representative of field situations but, for detailed studies of pore size distribution, vacuum saturation may be necessary. Too great a vacuum should be avoided, as the water can boil under the reduced pressure and disrupt the sample.

A final point concerns the representativeness of measurements on unconfined swelling clays. In situ they are subjected to an overburden load. To mimic this situation, a similar external load should be applied in the laboratory before wetting and subsequent measurement, but routine techniques for this have not been developed.

VI. APPLICATIONS OF WATER RELEASE MEASUREMENTS

Knowledge of the amount of water held at various matric potentials is required for many purposes and is frequently used in agronomic, engineering, and environmental applications. In agronomic applications three soil moisture constants are regularly used, as discussed below.

A. Soil Moisture Constants

1. Field Capacity

Field capacity as defined by Veihmeyer and Hendrickson [90] is the amount of water held in the soil after excess gravitational water has drained away and after the rate of downward movement has materially decreased. The value of this concept is that it specifies approximately the maximum amount of water that can be held in the soil without further loss by drainage. The matric potential of the soil when in this state is also commonly referred to as field capacity (as an abbreviation for "matric potential at field capacity").

Many problems arise with the assumption of a single static value for a dynamic process such as field capacity. The redistribution of draining water in a soil profile is a continuous process, which may be influenced by many factors [91-93], including antecedent moisture conditions, depth of wetting, soil texture, type of clay present, organic matter, presence of slowly permeable horizons, and the rate of evapotranspiration. Consequently matric potential can be different in deep horizons of less permeable soils than in an overlying topsoil The field capacity concept is most acceptable for the coarser and loamy textured soils, where a static or equilibrium state is more easily defined because of the sharp decrease in unsaturated hydraulic conductivity with a comparatively small drop in matric potential.

Values ranging from -3 to -8 kPa have been reported in a range of freely draining soils for the matric potential at which water is held at field capacity [92,94-96].

Ideally, field capacity should be determined in the field by monitoring soil water content. Alternatively the matric potential measured at field capacity should be used, together with a water release characteristic, to deduce a field capacity value. However, both these procedures are time-consuming, so in most applications published values of matric potentials that are thought to approximate to field capacity are used, together with a water release characteristic. Such values vary from -5 to -50 kPa [92].

2. Permanent Wilting Point

The permanent wilting point is defined as the soil moisture content at which the leaves of a growing plant first reach a stage of wilting from which they do not recover. Different plants have different values of matric potential at wilting point, with values between -800 and -3000 kPa being reported [20]. Since the change in moisture content is so small across this range for most soils, a potential of -1500 kPa, based on wilting studies with dwarf sunflowers [97], is generally taken to be an approximation of permanent wilting point.

3. Available Water Capacity

The amount of water held in the soil between field capacity and permanent wilting point is generally defined as being available for plant growth. The concept is widely used, although it is subject to many limitations [91]. In particular it must be pointed out that available water capacity represents the maximum amount of water that is available to the plant under conditions of low evaporative demand [92]. The soil water release curve provides a means of obtaining the volumetric available water capacity (θ_A) for any soil horizon:

$$\theta_A = \theta(5) - \theta(1500)$$

where $\theta(5)$ is the volumetric water content at a potential of -5 kPa and $\theta(1500)$ is the volumetric water content at a potential of -1500 kPa.

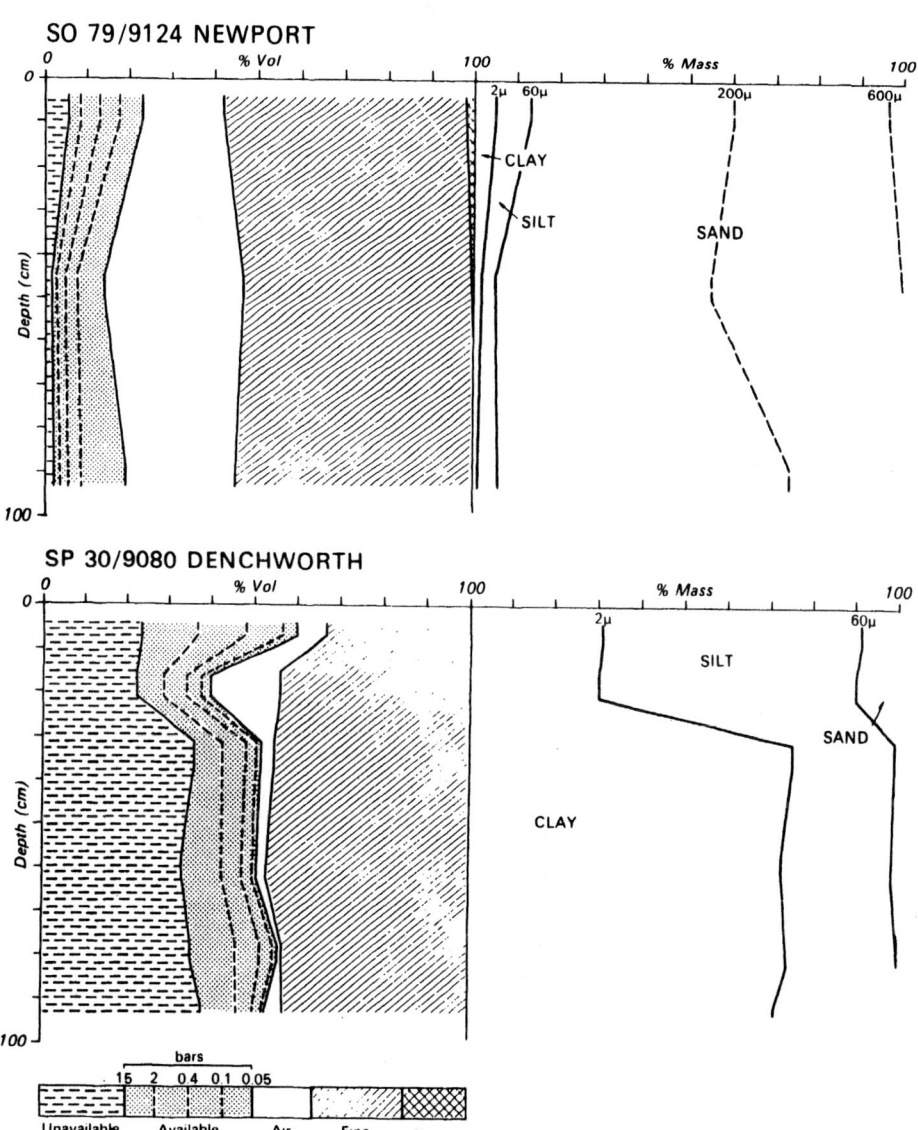

FIG. 16 Water release characteristics of two contrasting soil profiles. (From Ref. 7.)

Available water for the soil horizon is then the product of the horizon thickness and θ_A, while that for the whole profile is the sum of such values down to a specified depth or a barrier to rooting.

4. Air Capacity

The terms "air capacity" and "coarse porosity" usually refer, interchangeably, to the air-filled porosity of soil equilibrated at a suction approximating to field capacity. Air capacity is obtained as the difference between the total porosity and the volumetric water content at field capacity. Such pores are normally air-filled except during short periods following heavy rainfall. Because air capacity is a measure of the fractional volume of large pores in the soil, it also provides a reasonable indication of saturated hydraulic conductivity where the large pores are continuous [98].

B. Diagrammatic Presentation and Interpretation of Data

The relationship between soil air, soil water, and the soil solids can be obtained from the water release characteristic and can be presented diagrammatically for a complete soil profile (Fig. 16). The horizontal axis is divided into unavailable water, available water (at stated suctions), air capacity, fine earth (< 2 mm), and stones, all on a percentage volume basis. The vertical axis represents depth below the soil surface, and mean results for each sampling depth are plotted. The lines for each sampling depth are then connected by an oblique line added solely for diagrammatic clarity and having no analytical basis. Soil horizons or a change to bedrock can be shown where appropriate.

Advantages of this style of representation are that data for a soil profile can be presented concisely and that changes in air-water-solid relationships down the profile can be seen at a glance. Particle size distribution can be presented in a similar format for easy comparison.

The Newport series profile in Fig. 16 is a haplumbrept with a large amount of fine sand (60-200 µm) in all horizons. It has a large air capacity and negligible unavailable water, both typical features of sandy soils; neither bulk density nor the available water content changes much with depth. This soil is relatively dense (1.5 g cm^{-3}), but the air capacity is also large. The fine sand content influences the volume of water released between -5 kPa (-0.05 bar) and -10 kPa (-0.10 bar) and gives a volume in this range of up to 11%, well above average for sandy soils. A precise field measurement of the field capacity may be very difficult in this soil, since water content would change markedly between 1 and 5 days after saturation.

The chief features of the Denchworth series profile (Fig. 16), a haplaquept formed on Mesozoic clay shales, are the marked reduction in air capacity and a sharp increase in unavailable water at the boundary to clay at the base of the A horizon. The very large volume of available water (35%) in the surface is attributable to a moderate organic matter content (4.7% organic carbon) and good structural condition under old pasture. The associated low bulk density does not necessarily give a large air capacity, however, as the surface layer is susceptible to compaction by traffic or stock, the increase in density occurring initially at the expense of coarser pores. The deeper horizons are very clayey, with typical properties: a large volume of unavailable water, small air capacity, and little water released at potentials more than -40 kPa (-0.40 bar). The slight increase in air capacity below 90 cm depth is associated with the graduation to clay shale.

It is clear that downward percolation of water will be severely impeded below 30 cm in this soil. The ability of the upper horizons to accept rainfall is accordingly very limited. Accumulated air capacity above 30 cm is equivalent to about 35 mm of rainfall, and the rainfall of two wet winter days could saturate the soil to the surface. Vertical disposal of water is rather slow; thus this soil is prominently mottled within 40 cm depth and has gray colors on ped faces.

C. Agronomic Applications

1. Crop Water Supply

For annual crops, the amount of available water that is genuinely accessible varies with crop and soil. Different crops have different rooting habits, some rooting to well below 1 m depth, others rooting only to 50 cm. Similarly they have different tolerances to water stress, and many crops have water-sensitive stages of growth when, for optimum yield, they need water held at low suctions. Thus different approaches have been taken to assess long-term moisture limitations to optimum crop production. On the broad scale, one can classify profile available water according to climatic moisture regime [95]. At a more detailed level, the available water range can be split into easily and less easily available portions, and empirical models can be set up (Table 6) to obtain crop-adjusted profile available water values. These can then be used with data on potential soil moisture deficit to assess soil droughtiness in a given area [99].

At a field scale, water retention data are important when considering a soil for irrigation requirements. A full water release curve is required for each soil type to assess available water capacity, critical deficits, and optimum frequency and volume of water applications [94]. Reeve [15] has explained the relevance of water

TABLE 6   Available Water Models for Common U.K.
Field Crops [99]

Crop	Depth for summation within soil profile if no rooting barrier (cm)	Water held between potentials of (kPa)
Wheat/barley ⎱	0-50	-5 to -1500
Temporary grass ⎰	50-120	-5 to -200
Potatoes	0-70	-5 to -1500
Sugar beet	0-80	-5 to -1500
	80-130	-5 to -200
Permanent grass	0-70	-5 to -1500
	70-100	-5 to -200

retention measurements to irrigation planning in New Zealand in
terms of the ability of a soil to sustain crop transpiration during
drought or between irrigation events, the ability of soil to absorb
irrigation water when dry, potential losses of irrigation water by
drainage, the possibility of waterlogging caused by slowly permeable subsoils, and the existence of dense or compact layers, which
may restrict rooting.

2.   Porosity and Structure

Values of air capacity have been used as a guide to the recognition
of impermeable horizons [100], and values integrated down to the
top of an impermeable horizon have been used to represent the storage capacity of soils for irrigation water [15] and for rainwater in
flood response studies.

In addition, the water release characteristic can be used as a
measure of soil structure in an undisturbed situation [7], or to
record the recovery of land after damage [101].

D.   Other Applications

A knowledge of the water release characteristic is useful in various
engineering applications such as off-road trafficability and stability
of earthworks formed from clay.   In the latter situation the shape
of the curve can be important.   From Fig. 14, a small water loss
over the middle section of the characteristic represents a much larger
strength increase in the Mesozoic (Gault) Clay than in the Paleozoic
(Coal Measures) Clay.

Many physically based models depend on the use of water release data. These models include assessments of soil suitability for restoring damaged land and accepting municipal sewage sludge [95], predictions of nitrate leaching [102], aquifer vulnerability measurements [103], and descriptions of the residual behavior of pesticides [104] in the profile. Substances such as nitrate and certain pesticides are readily soluble in water, and their movement in the profile is largely controlled by the water release characteristic of that soil.

Regional simulations of moisture availability and soil water fluxes often incorporate soil water release data. Predictions of the effect of groundwater lowering on crop production may require water release data and hydraulic conductivities for all soil horizons [105,106].

Many of these applications require a large amount of data, which may present a formidable barrier to progress. In these cases, rapid measurement methods (see, e.g., Ref. 106) or estimations may be necessary. Estimations can be based on tables relating soil moisture constants to texture classes and horizon types [107] or on multiple regression equations [7,108-110]. Bouma et al. [105] describe how transfer functions (mathematical expressions relating different characteristics or properties with one another or to land qualities) also can be used in the absence of measured data.

REFERENCES

1. Childs, E. C., The use of soil moisture characteristics in soil studies, *Soil Sci.*, 50: 239-252 (1940).
2. Schofield, R. K., The pF of the water in soil, *Trans. 3rd Int. Congr. Soil Sci.*, 2: 37-48 (1935).
3. Haines, W. B., Studies in the physical properties of soils, *J. Agric. Sci.*, 20: 97-116 (1930).
4. Poulovassilis, A., The uniqueness of the moisture characteristics, *J. Soil Sci.*, 25: 27-33 (1974).
5. Shcherbakov, R. A., Model of the hysteresis of water retention by soils, *Pochvovedeniye*, 8: 54-60 (1985).
6. Hillel, D., *Soil and Water*, Academic Press, New York, 1971.
7. Hall, D. G. M., M. J. Reeve, A. J. Thomasson, and V. F. Wright, *Water Retention, Porosity and Density of Field Soils*, Soil Survey Tech. Monog. No. 9, Harpenden, U.K., 1977.
8. Lambooy, A. M., Relationship between cation exchange capacity, clay content and water retention of Highveld soils, *S. Afr. J. Plant Soil*, 1: 33-38 (1984).
9. Williams, J., R. E. Prebble, W. T. Williams, and C. T. Hignett, The influence of texture, structure and clay mineralogy on the soil moisture characteristic, *Aust. J. Soil Res.*, 21: 25-32 (1983).

10. Stakman, W. P., and B. G. Bishay, Moisture retention and plasticity of highly calcareous soils in Egypt, *Neth. J. Agric. Sci.*, 24: 43-57 (1976).

11. Prebble, R. E., and G. B. Stirk, Effect of free iron oxide on range of available water in soils, *Soil Sci.*, 88: 213-217 (1959).

12. Archer, J. R., and P. D. Smith, The relation between bulk density, available water capacity, and air capacity of soils, *J. Soil Sci.*, 23: 475-480 (1972).

13. Veihmeyer, F. J., and A. H. Hendrickson, Methods of measuring field capacity and permanent wilting percentage of soils, *Soil Sci.*, 68: 75-94 (1949).

14. Scullion, J., A. R. A. Mohammed, and G. A. Ramshaw, Statistical evaluation of drainage treatments in simple field trials with special reference to former opencast coal mining land, *J. Agric. Sci. Camb.*, 107: 515-520 (1986).

15. Reeve, M. J., Water retention, porosity and composition interrelationships of alluvial soils in mid-Hawke's Bay and their relevance in irrigation planning, *N.Z. J. Agric. Res.*, 29: 457-468 (1986).

16. Mullins, C. E., and A. Fraser, Use of the drop-cone penetrometer on undisturbed and remoulded soils at a range of soil-water tensions, *J. Soil Sci.*, 31: 25-32 (1980).

17. Reeve, M. J., D. G. M. Hall, and P. Bullock, The effect of soil composition and environmental factors on the shrinkage of some clayey British soils, *J. Soil Sci.*, 31: 429-442 (1980).

18. Beven, K., Micro-, meso-, macroporosity and channeling flow phenomena in soils, *Soil Sci. Soc. Am. J.*, 45: 1245 (1981).

19. Buckingham, E., Studies on the movement of soil moisture, Bur. Soils, Bull No. 38, U.S. Dept. Agric., 1907.

20. Loveday, J. (Ed.), *Methods for Analysis of Irrigated Soils*, Tech. Commun. No. 54, Commonwealth Bureau of Soils, Farnham Royal, U.K., 1974.

21. Stakman, W. P., G. A. Valk, and G. G. van der Harst, *Determination of Soil Moisture Retention Curves: I. Sand-Box Apparatus—Range pF 0 to 2.7*, 3rd rev. ed., Institute for Land and Water Management Research, Wageningen, The Netherlands, 1969.

22. Richards, L. A., Porous plate apparatus for measuring moisture retention and transmission by soil, *Soil Sci.*, 66: 105-110 (1948).

23. Reginato, R. J., and C. H. M. van Bavel, Pressure cell for soil cores, *Soil Sci. Soc. Am. Proc.*, 26: 1-3 (1962).

24. Richards, L. A., A pressure-membrane extraction apparatus for soil solution, *Soil Sci.*, 51: 377-386 (1941).

25. Richards, L. A., Methods of measuring soil moisture tension, *Soil Sci.*, 68: 95-112 (1949).

26. Russell, M. B., and L. A. Richards, The determination of soil moisture energy relations by centrifugation, *Soil Sci. Soc. Am. Proc.*, 3: 65-69 (1938).

27. Zur, B., Osmotic control of the matric soil-water potential: I. Soil-water system, *Soil Sci.*, 102: 394-398 (1966).

28. Pritchard, D. T., An osmotic method for studying the suction/moisture content relationships of porous materials, *J. Soil Sci.*, 20: 374-383 (1969).

29. Croney, D., J. D. Coleman, and P. M. Bridge, *The Suction of Moisture Held in Soil and Other Porous Materials*, Road Research Tech. Paper No. 24, Road Research Laboratory, HMSO, London, 1952.

30. Wadsworth, H. A., An interpretation of the moisture content-surface force curve for soils, *Soil Sci.*, 58: 225-242 (1944).

31. McQueen, I. S., and R. F. Miller, Calibration and evaluation of a wide-range gravimetric method for measuring moisture stress, *Soil Sci.*, 106: 225-231 (1968).

32. Bouyoucos, G. J., A new, simple and rapid method for determining the moisture equivalent of soils, and the role of soil colloids on this moisture equivalent, *Soil Sci.*, 27: 233-241 (1929).

33. Russell, M. B., Pore size distribution as a measure of soil structure, *Soil Sci. Soc. Am. Proc.*, 6: 108-112 (1941).

34. Burke, W., D. Gabriels, and J. Bouma (Eds.), *Soil Structure Assessment*, A. A. Balkema, Rotterdam, 1986.

35. Danielson, R. E., and P. L. Sutherland, Porosity, in *Methods of Soil Analysis*, Part 1 (A. Klute, Ed.), Am. Soc. Agron., Madison, WI, 1986, pp. 443-461.

36. Vomocil, J. A., Porosity, in *Methods of Soil Analysis*, Part 1 (C. A. Black, Ed.), Am. Soc. Agron., Madison, WI, 1965, pp. 299-314.

37. Jamison, V. C., Structure of a Dunkirk silty clay loss in relation to pF moisture measurements, *J. Am. Soc. Agron.*, 34: 307-321 (1942).

38. Czeratzki, W., Eine keramische Platte zur Serienmaszigen Untersuchung von Porengroszen in Boden in Spannungsbereich bis ca. -1 Atm., *Z. Pflanzenernaehr. Dueng. Bodenkd.*, 81: 50-56 (1958).

39. de Boodt, M. (Ed.), *West European Methods for Soil Structure Determination*, State Faculty of Agricultural Sciences, University of Ghent, Belgium, 1967.

40. Topp, G. C., and W. Zebchuk, The determination of soil-water desorption curves for soil cores, *Can. J. Soil Sci.*, 59: 19-26 (1979).

41. Ball, B. C., and R. Hunter, *Improvements in the Routine Determination of Soil Pore Size Distribution from Water Release Measurements on Tension Tables and Pressure Plates*, Departmental Note No. SIN/314, Scottish Inst. Agric. Eng., Penicuik, 1980.

42. Carter, A. D., and A. J. Thomasson, Data to feed and calibrate land evaluation models, in *Application of Computerized EC Soil Map and Climate Data* (H. A. J. van Lanen and A. K. Bregt, Eds.), Soil Survey Institute, Wageningen, The Netherlands, 1989, pp. 35–43.

43. Lal, R., Physical properties and moisture retention characteristics of some Nigerian soils, *Geoderma*, 21: 209–223 (1979).

44. Puckett, W. E., J. H. Dane, and B. F. Hajek, Physical and mineralogical data to determine soil hydraulic properties, *Soil Sci. Soc. Am. J.*, 49: 831–836 (1985).

45. Gradwell, M. W., The available water capacities of North Auckland soils, *N.Z. J. Agric. Res.*, 14: 253–287 (1971).

46. Datta, B., and O. P. Singh, Comparative study of moisture release behaviour of soils, soil clays and pure clays, *Aust. J. Soil Res.*, 19: 79–82 (1981).

47. Kumar, S., R. S. Malik, and I. S. Dahiya, Water retention, transmission and contact characteristics of Ludas Sand as influenced by farmyard manure, *Aust. J. Soil Res.*, 22: 253–259 (1984).

48. Madsen, H. B., C. R. Jensen, and T. Boysen, A comparison of the thermocouple psychrometer and the pressure plate methods for determination of soil water characteristic curves, *J. Soil Sci.*, 37: 357–362 (1986).

49. Richards, L. A., Physical condition of water in soil, in *Methods of Soil Analysis*, Part 1 (C. A. Black, Ed.), Am. Soc. Agron., Madison, WI, 1965, pp. 128–152.

50. Chahal, R. S., and R. N. Yong, Validity of the soil water characteristics determined with the pressurized apparatus, *Soil Sci.*, 99: 98–103 (1965).

51. Heinonen, R., On the pre-treatment of samples of heavy clay soil for determinations by the pressure membrane apparatus, *J. Sci. Agric. Soc. Finland*, 33: 153–158 (1961).

52. Kuznetsova, I. V., and Vinogradova, G. B., Wilting moisture of plants in compacted soil horizons, *Pochvovedeniye*, 5: 58–64 (1982).

53. Collis-George, N., A note on the pressure plate-membrane apparatus, *Soil Sci.*, 74: 315–322 (1952).

54. Stakman, W. P., and G. G. van der Harst, *Pressure Membrane Apparatus Range pF 3.0 to 4.2*, Institute for Land and Water Management Research, P.O. Box 35, Wageningen, The Netherlands, 1969.

55. Waters, P., Comparison of the ceramic plate and the pressure membrane to determine the 15 bar water content of soils, *J. Soil Sci.*, 31: 443-446 (1980).

56. Constantz, J., and W. N. Herkelrath, Submersible pressure outflow cell for measurement of soil water retention and diffusivity from 5 to 95°C, *Soil Sci. Soc. Am. J.*, 48: 7-10 (1984).

57. Briggs, L. J., and J. W. McLane, *The Moisture Equivalent of Soils*, Bur. Soils Bull. No. 45, U.S. Dept. Agric., 1907.

58. Kyuma, K., Y.-S. Suh, and K. Kawaguchi, A method of capability evaluation for upland soils: I. Assessment of available water retention capacity, *Soil Sci. Plant Nutr.*, 23: 135-149 (1977).

59. Odén, S., An integral method for the determination of moisture retention curves by centrifugation, *Grundförbättring*, 27: 137-143 (1975/76).

60. Childs, E. C., *An Introduction to the Physical Basis of Soil Water Phenomena*, Wiley, London, 1969.

61. Coleman, J. D., and A. D. Marsh, An investigation of the pressure-membrane method for measuring the suction properties of soil, *J. Soil Sci.*, 12: 343-362 (1961).

62. Klute, A., Water retention: Laboratory methods, in *Methods of Soil Analysis*, Part 1 (A. Klute, Ed.), Am. Soc. Agron., Madison, WI, 1986, pp. 635-662.

63. Head, K. H., *Manual of Soil Laboratory Testing*, Vol. 2, Pentech Press, London, 1982.

64. Al-Khafaf, S., and R. J. Hanks, Evaluation of the filter paper method for estimating soil water potential, *Soil Sci.*, 117: 194-199 (1974).

65. Gardner, R., A method of measuring the capillary tension of soil moisture over a wide moisture range, *Soil Sci.*, 43: 277-283 (1937).

66. Fawcett, R. G., and N. Collis-George, A filter paper method for determining the moisture characteristics of soil, *Aust. J. Exp. Agric. Animal Husb.*, 7: 162-167 (1967).

67. Hamblin, A. P., Filter-paper method for routine measurement of field water potential, *J. Hydrol.*, 53: 355-360 (1981).

68. Chandler, R. J., and C. I. Gutierrez, The filter-paper method of suction measurement, *Geotechnique*, 36: 265-268 (1986).

69. Dumbleton, M. J., and G. West, Soil suction by the rapid method: An apparatus with extended range, *J. Soil Sci.*, 19: 40-46 (1968).

70. Rawlins, S. L., and G. S. Campbell, Water potential: Thermocouple psychrometry, in *Methods of Soil Analysis*, Part 1 (A. Klute, Ed.), Am. Soc. Agron., Madison, WI, 1986, pp. 597-618.

71. Yong, R. N., and B. P. Warkentin, *Soil Properties and Behaviour*, Elsevier, Amsterdam, 1975.
72. Ratliff, L. F., J. T. Ritchie, and D. K. Cassel, Field-measured limits of soil water availability as related to laboratory-measured properties, *Soil Sci. Soc. Am. J.*, 47: 770-775 (1983).
73. Greminger, P. J., Y. K. Sud, and D. R. Nielsen, Spatial variability of field-measured soil-water characteristics, *Soil Sci. Soc. Am. J.*, 49: 1075-1082 (1985).
74. Campbell, G. S., and G. W. Gee, Water potential: Miscellaneous methods, in *Methods of Soil Analysis*, Part 1 (A. Klute, Ed.), Am. Soc. Agron., Madison, WI, 1986, pp. 619-633.
75. Bruce, R. R., and R. J. Luxmoore, Water retention: Field methods, in *Methods of Soil Analysis*, Part 1 (A. Klute, Ed.), Am. Soc. Agron., Madison, WI, 1986, pp. 663-686.
76. Hopmans, J. W., and J. H. Dane, Effect of temperature-dependent hydraulic properties on soil water movement, *Soil Sci. Soc. Am. J.*, 49: 51-58 (1985).
77. Hopmans, J. W., and J. H. Dane, Combined effect of hysteresis and temperature on soil-water movement, *J. Hydrol.*, 83: 161-171 (1986).
78. British Standards Institution, *Soils for Civil Engineering Purposes*, BS1377, British Standards Institution, London, 1975.
79. McKeague, J. A. (Ed.), *Manual on Soil Sampling and Methods of Analysis*, Can. Soc. Soil Sci., Ottawa, Ontario, 1978.
80. Soil Survey Staff, *Soil Survey Manual*, U.S. Dept. Agric. Handbook No. 18, Government Printing Office, Washington, DC, 1951.
81. Hodgson, J. M. (Ed.), *Soil Survey Field Handbook*, Tech. Monogr. No. 5, Harpenden, U.K., 1976.
82. Carter, M., *Geotechnical Engineering Handbook*, Pentech Press, London, 1983.
83. Unger, P. W., Water retention by core and sieved soil samples, *Soil Sci. Soc. Am. Proc.*, 39: 1197-1200 (1975).
84. Elrick, D. E., and C. B. Tanner, Influence of sample pretreatment on soil moisture retention, *Soil Sci. Soc. Am. Proc.*, 19: 279-282 (1955).
85. Young, K. K., and J. D. Dixon, Overestimation of water content at field capacity from sieved sample data, *Soil Sci.*, 101: 104-107 (1966).
86. Dagg, M., and P. H. Hosegood, Details of hand sampling tool for taking undisturbed soil cores, *E. Afr. Agric. For. J.*, *Special Issue*: 129-131 (1962).
87. McLintock, T. F., A method for obtaining soil-sample volumes in stony soils, *J. For.*, 57: 832-834 (1959).
88. Jurgensen, M. F., M. J. Larsen, and A. E. Harvey, *A Soil Sampler for Steep, Rocky Sites*, U.S. Dept. Agric. Forest Service Research Note No. INT-217, Government Printing Office, Washington, DC, 1977.

89. Rimmer, D. L., Soil physical conditions on reclaimed colliery spoil heaps, *J. Soil Sci.*, 33: 567-579 (1982).

90. Veihmeyer, F. J., and A. H. Hendrickson, The moisture equivalent as a measure of the field capacity of soils, *Soil Sci.*, 32: 181-193 (1931).

91. Hillel, D., *Introduction to Soil Physics*, Academic Press, New York, 1982.

92. Cassel, D. K., and D. R. Nielsen, Field capacity and available water capacity, in *Methods of Soil Analysis*, Part 1 (A. Klute, Ed.), Am. Soc. Agron., Madison, WI, 1986, pp. 901-926.

93. Beukes, D. J., The effect of certain soil properties on matric potential at, and time duration to, field capacity, *S. Afr. J. Plant Soil*, 1: 126-131 (1984).

94. Dent, D. L., and R. P. Scammell, Assessment of long-term irrigation need by integration of data for soil and crop characteristics and climate, *Soil Sur. Land Eval.*, 1: 51-57 (1981).

95. *National Soils Handbook*, U.S. Dept. Agric., Washington, DC, 1983.

96. Webster, R., and P. H. T. Beckett, Matric suctions to which soils in South Central England drain, *J. Agric. Sci. Camb.*, 78: 379-387 (1972).

97. Richards, L. A., and L. R. Weaver, Fifteen-atmosphere percentages as related to the permanent wilting percentage, *Soil Sci.*, 56: 331-339 (1943).

98. Ahuja, L. R., J. W. Naney, R. E. Green, and D. R. Nielsen, Macroporosity to characterize spatial variability of hydraulic conductivity and effects of land management, *Soil Sci. Soc. Am. J.*, 48: 699-702 (1984).

99. Thomasson, A. J., Assessment of soil droughtiness, in *Soil Survey Applications* (M. G. Jarvis and D. Mackney, Eds.), Tech. Monogr. No. 13, Harpenden, U.K., 1979, pp. 43-50.

100. Avery, B. W., *Soil Classification for England and Wales*, Soil Survey Tech. Monogr. No. 14, Harpenden, U.K., 1980.

101. Bullock, P., A. C. D. Newman, and A. J. Thomasson, Porosity aspects of the regeneration of soil structure after compaction, *Soil Tillage Res.*, 5: 325-341 (1985).

102. Addiscott, T. M., A simple computer model for leaching in structured soils, *J. Soil Sci.*, 28: 554-563 (1977).

103. Carter, A. D., R. C. Palmer, and R. A. Monkhouse, Mapping the vulnerability of groundwater to pollution from agricultural practice, particularly with respect to nitrate, in *Vulnerability of Soil and Groundwater to Pollutants* (W. van Duijvenbooden and H. G. van Waegeningh, Eds.), CHO-TNO, The Hague, The Netherlands, 1987, pp. 333-342.

104. Nicholls, P. H., Predicting the availability of soil-applied pesticides, *Aspects Appl. Biol.*, 21: 173-184 (1989).

105. Bouma, J., H. A. J. Van Lanen, A. Breeusma, H. J. M. Wosten, and M. J. Kooistra, Soil survey data needs when studying modern land use problems, *Soil Use Manage.*, 2: 125-130 (1986).

106. Wosten, J. H. M., J. Bouma, and G. H. Stoffelsen, The use of soil survey data for regional water simulation, *Soil Sci. Soc. Am. J.*, 49: 1238-1245 (1985).

107. McKeague, J. A., R. G. Eilers, A. J. Thomasson, M. J. Reeve, J. Bouma, R. B. Grossman, J. C. Favrot, M. Renger, and O. Strebel, Tentative assessment of soil survey approaches to the characterization and interpretation of air-water properties of soils, *Geoderma, 34*: 69-100 (1984).

108. Petersen, G. W., R. L. Cunningham, and R. P. Matelski, Moisture characteristics of Pennsylvania soils: I. Moisture retention as related to texture, *Soil Sci. Soc. Am. Proce.*, 32: 271-275 (1968).

109. Gupta, S. C., and W. E. Larson, Estimating soil water retention characteristics from particle size distribution, organic matter percent and bulk density, *Water Resour. Res.*, 15: 1633-1635 (1979).

110. Rawls, W. J., D. L. Brakensiek, and K. E. Saxton, Estimation of soil water properties, *Trans. Am. Soc. Agric. Eng.*, 25: 1316-1328 (1982).

4

Hydraulic Conductivity of Saturated Soils

EDWARD G. YOUNGS* *Soils Division, Rothamsted Experimental Station, Harpenden, Hertfordshire, England*

I. INTRODUCTION

The physical law describing water movement through saturated porous materials in general and soils in particular was proposed by Darcy [1] in 1856 in a work concerned with the water supplies for the town of Dijon. Darcy established the law from the results of experiments with water flowing down columns of sands in an experimental arrangement shown schematically in Fig. 1. He found that the volume of water Q flowing per unit time was directly proportional to the cross-sectional area A of the column and to the difference Δh in hydraulic head, measured by the level of water in manometers, causing the flow, and inversely proportional to the length L of the column; thus

$$Q = \frac{KA \; \Delta h}{L} \tag{1}$$

where the proportionality constant K with units of velocity (dimensions LT^{-1}) is now known as the *hydraulic conductivity* of the porous material. Typical values of K for soils of different textures are given in Table 1. Conversion factors relating various units are given in Table 2. Since the hydraulic conductivity of a soil is inversely proportional to the viscous drag of the water flowing between the soil particles, its value increases by about 3% per degree Celsius rise as the viscosity of water decreases.

Current affiliation: Silsoe College, Cranfield Institute of Technology, Silsoe, Bedfordshire, England.

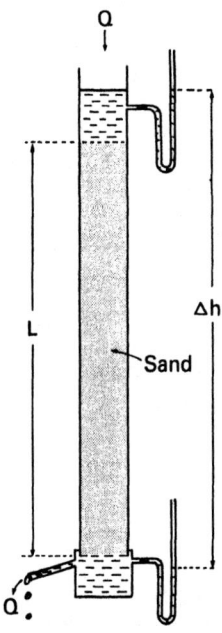

FIG. 1 Darcy's experimental arrangement.

The hydraulic head measured directly by the level of water in the manometers in Darcy's experiment is the water potential expressed as the work done per unit *weight* of water in transferring it from a reference source. The potential may also be defined as the work done per unit *volume* of water, in which case the potential difference in Darcy's law would be $\rho g \, \Delta h$, where ρ is the density of water and g is the acceleration due to gravity. K then has dimensions $M^{-1}L^3T$. In this chapter we will adopt the usual convention of defining the potential as the work done per unit weight, that is, as a head of water, so that K is simply expressed in units of a velocity.

TABLE 1 Hydraulic Conductivity Values of Saturated Soils

Soil	Hydraulic conductivity (mm d^{-1})
Fine-textured soils	< 10
Soils with well-defined structure	10–1000
Coarse-textured soils	> 1000

TABLE 2   Conversion Factors for Units of Hydraulic Conductivity[a]

m d^{-1}	cm h^{-1}	cm min^{-1}	mm s^{-1}
1	4.17	0.0694	0.0116
0.24	1	0.0167	0.00278
14.4	60	1	0.167
86.4	360	6	1

[a]*Example.* To convert x cm min^{-1} to meters per day, find 1 in the cm min^{-1} column. Numbers on the same horizontal row are values in other units equivalent to 1 cm min^{-1}, so that 1 cm min^{-1} ≡ 14.4 m d^{-1} and x cm min^{-1} ≡ 14.4x m d^{-1}.

This is very convenient when computing water flows in soils, but it has the disadvantage that the value of the hydraulic conductivity of a porous material depends on g. This means that the hydraulic conductivity of a given porous material depends on altitude and is smaller at the top of a mountain than at sea level, but this is of little importance in most practical situations concerned with groundwater movement.

Equation 1 describes the flow of water in porous materials at low velocities when viscous forces opposing the flow are much greater than the inertial forces. The ratio of the viscous forces in the inertial forces is represented by the Reynolds number Re [2,3], which may be defined as

$$Re = \frac{vd\rho}{\eta} \qquad (2)$$

where v is the mean flow velocity, d a characteristic length that may be conveniently taken as the mean pore diameter, ρ the density of water as before, and η the viscosity of water. When Re exceeds a value of about 1.0, Darcy's law no longer describes the flow of water. Under field conditions this is unlikely to occur, except in some situations of flow in gravels and in structural fissures and worm holes.

Darcy's work was concerned with one-dimensional flow. Generally, however, flows in soil are two- or three-dimensional, and Eq. 1 has to be extended to take into account multi-dimensional flow. Slichter [4] in 1899 argued that the flow of water in soil described by Darcy's law is analogous to the flow of electricity and heat in conductors, and so generally Darcy's law may be written in vectorial notation as

$$v = - K \text{ grad } h \tag{3}$$

where v is the flow velocity and h is the hydraulic potential of the soil water expressed as the hydraulic head. If the water can be considered to be incompressible and the soil does not shrink or swell, the equation of continuity is

$$\text{div } v = 0 \tag{4}$$

so that h is described by Laplace's equation

$$\nabla^2 h = 0 \tag{5}$$

Thus it is only a matter of solving Eq. 5 for the hydraulic head h with the given boundary conditions to obtain a complete solution to a given problem of groundwater flow in one, two, or three dimensions [4]. With h known throughout the flow region from Eq. 5, flows can be found from Eq. 3 if K is known. Conversely, if flows and hydraulic heads are measured in the flow region, the hydraulic conductivity can be found. Measurement techniques for the determination of hydraulic conductivities of porous materials in general, including soils, make use of solutions of Laplace's equation with the prescribed boundary conditions imposed by the particular method.

The concept of hydraulic conductivity is derived from experiments on "uniform" porous materials. Methods of measuring hydraulic conductivity assume implicitly that Darcy's law applies to flow in the soil region concerned; that is, they presuppose that the soil is uniform. As discussed in Section II, soils can be far from uniform because of heterogeneities at various scales, and measurements need to be made on some representative volume of the whole flow region. Thus, although values of "hydraulic conductivity" for a soil in a given region can be obtained using any method, such values will be of little relevance in the context of predicting flows if the volume of soil sampled by the method is unrepresentative of the soil region as a whole.

In the discussion above it has been tacitly assumed that the hydraulic conductivity of the soil is the same in all directions. However, anisotropy in soil properties can occur because of structural development and laminations, giving different hydraulic conductivity values in different directions. Darcy's law then has to be expressed in tensor form [3], and the streamlines of flow are orthogonal to the equipotential surfaces only when the flow is in the direction of one of the three principal directions. The theory of flow in anisotropic soils [2,3,5] shows that Laplace's equation can still be used to obtain solutions to flow problems if a transformation incorporating the components of hydraulic conductivity in the principal directions is applied to the spatial coordinates. If the soil is anisotropic, the two- and three-dimensional flows usually used in hydraulic conductivity measurement techniques in the field require analysis using this theory to obtain values of the hydraulic conductivity in the principal directions.

II. FUNDAMENTAL CONSIDERATIONS OF WATER FLOW THROUGH SOILS

A. Soil Considered as a Continuum

The movement of water through soils takes place in the tortuous channels between the soil particles with velocities varying from point to point and described by the Stokes-Navier equations [3]. Darcy's law does not consider this microscopic flow pattern between the particles but instead assumes the water movement to take place in a continuum with a uniform flow averaged over space. It therefore describes the flow of water macroscopically in volumes of soil much larger than the size of the pores. In fact it can be used to describe only the macroscopic flow of water through soil regions of volume greater than some *representative elementary volume* that encompasses many soil particles.

The concept of representative elementary volume of a porous material is most easily illustrated by considering the measurement of the water content of a sample of unstructured "uniform" saturated soil, starting with a very small volume and then increasing the sample size. For very small volumes comparable to the size of the soil particles, the sample volume would include only solid matter if located wholly within a soil particle, giving zero soil-water content. If located wholly in a pore, it would contain only water, giving a soil-water content of one. All values between zero and one are possible when the sample is located partly within a soil particle and partly within the pore. As the volume size is increased, the lower limit of measured water content increases while the upper limit decreases, as shown in Fig. 2. When the sample is large enough, repeated measurements on random samples of the soil give the same value of soil-water content. The smallest sample volume that produces a consistent value is the representative elementary volume. Measurements of hydraulic conductivity and other soil properties must be made on volumes larger than this volume. While additive soil properties, such as the water content, can be obtained by averaging a large number of measurements made on smaller volumes within the representative elementary volume, the hydraulic conductivity cannot be obtained in this way because of the interdependent complex pattern of flows that this property embraces.

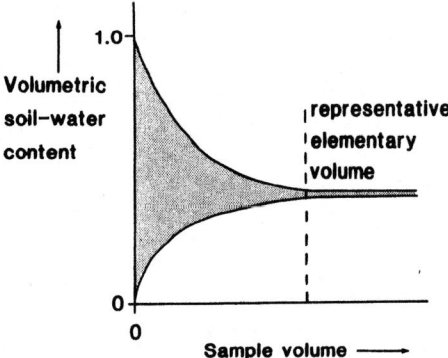

FIG. 2   Measurement of soil-water content of a saturated "uniform" soil.

Figure 2 illustrates the variability of a soil physical property that exists in all porous materials at a small enough scale because of their particulate nature.   Variability can also be present in soils at larger scales.   For example, in aggregated and structured soils where a distribution of macropores between the aggregates or peds is superimposed on the interparticle micropore space, the soil-water content would vary with sample size as shown in Fig. 3; only when the sample size encompasses a representative sample of macropore space do we have a representative volume.   This volume will be characteristic of the soil's structure that determines the hydraulic conductivity.

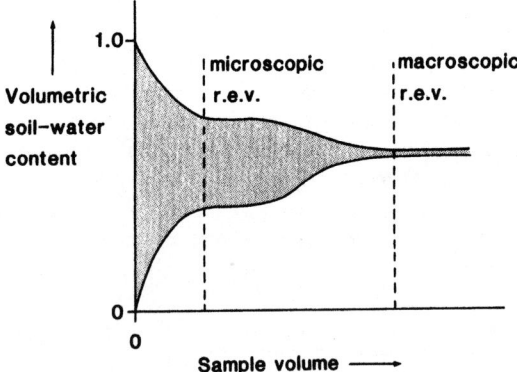

FIG. 3   Measurement of soil-water content of a saturated soil with superimposed macrostructure (r.e.v. = representative elementary volume).

It is only in materials that show behavior similar to that depicted in Fig. 2 that continuum physics, such as that implied by Darcy's law, can be applied macroscopically without difficulty to soil-water flow phenomena. In materials such as that illustrated in Fig. 3, boundary conditions at the surfaces of the aggregates and fissures affect the flow patterns throughout the soil region. However, for saturated conditions, as long as sufficiently large volumes are considered, continuum physics can still be applied to water flows at this larger scale using an appropriate value of hydraulic conductivity.

B. Heterogeneity

Because of the complex geometry of the pore system of soils, there is an inherent heterogeneity at pore size dimensions that is not observed when measurements are made on volumes containing a large number of pores. Soil heterogeneity usually implies variations of soil properties between soil volumes containing such a large number of pores. Such heterogeneity occurs at many scales in the following progression:

particle → aggregate → pedal/fissure → field → regional

The objective in making measurements of hydraulic conductivity is to enable quantitative predictions of soil water flows under given conditions. In a soil showing heterogeneity at various scales, different values of hydraulic conductivity apply at different spatial scales and need to be obtained by appropriate measurements. For example, the calculation of water movement to roots requires measurements at the scale of the soil aggregates, whereas the calculation of the flow to land drains in the same soil requires measurements on a much larger scale that take into account the flow through fissures. For hydrological purposes, measurements need to be made on even larger volumes to ensure consideration of flows at the field or regional scale.

The discussion so far has considered soil heterogeneity as stochastic so that measurements of physical properties can be made on a sample larger than some representative elementary volume. However, changes in soil occur often abruptly or as a trend, that is, in a deterministic manner. One particularly important aspect of soil variability occurs with the variation of the soil with depth. This has a profound effect on field soil-water regimes. There is often a gradual change of soil properties with depth that makes it impossible to define a representative elementary volume as previously described. In such cases it is assumed that Eq. 1 defines the hydraulic conductivity; hence with vertical flow in soils with a hydraulic conductivity $K(z)$ varying with the height z, we have

$$K(z) = - \frac{v}{dh/dz} \tag{6}$$

where v is the vertical flow velocity; that is, we assume the soil to be a continuum with properties varying with depth.

C. Equivalent Hydraulic Conductivity

As noted in Section I, the measurement of the flow that occurs with imposed boundary conditions in a uniform soil allows the determination of the hydraulic conductivity. For a nonuniform soil, the measurement gives an *equivalent hydraulic conductivity* value for the flow region with the imposed boundary conditions: that is, a value of hydraulic conductivity that would give the measured flow under the same conditions if the soil were uniform. The measurement of hydraulic conductivity by any method gives an equivalent value for the particular flow pattern produced by the boundary conditions used in the measurement. The value will be different for different boundary conditions if the soil varies spatially. Some examples will illustrate this.

First consider the flow between opposite faces of a composite rectangular slab of porous material $ABCD$ illustrated in Fig. 4, half of which has a hydraulic conductivity K_1 and the other half has a hydraulic conductivity K_2. For flow between the faces AD and BC the equivalent hydraulic conductivity K_e is given by

$$\frac{1}{K_e} = 2\left(\frac{1}{K_1} + \frac{1}{K_2} \right) \tag{7}$$

whereas for flow between the faces AB and DC, the equivalent hydraulic conductivity is

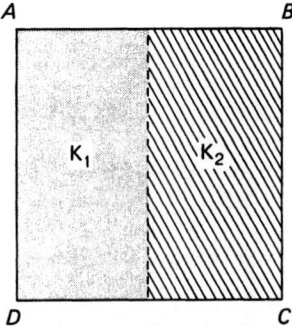

FIG. 4   Flow in a composite slab of soil.

$$K_e = \frac{K_1 + K_2}{2} \tag{8}$$

That is, for this arrangement of hydraulic conductivities the effective hydraulic conductivity can be either the harmonic mean or the arithmetic mean of the hydraulic conductivity values, depending on the direction of flow. It can be shown [6] that for any distribution of hydraulic conductivity values within a cube of soil, the equivalent hydraulic conductivity value K_e for flow between opposite faces lies between the arithmetic mean value K_a and the harmonic mean value K_h of the conductivities of the elements making up the region; that is,

$$K_a > K_e > K_h \tag{9}$$

and since

$$K_a > K_g > K_h \tag{10}$$

where K_g is the geometric mean value, this result is in keeping with the fact that the geometric mean is often taken as the equivalent hydraulic conductivity value for groundwater flow computations. For an isotropic soil, it can be argued [6] that

$$K_e = (K_a^2 K_h)^{1/3} \tag{11}$$

The dependence of the equivalent hydraulic conductivity value on the boundary conditions of the flow region has been further demonstrated in calculations of flow through an earth bank with a complex spatial variation of hydraulic conductivity [7].

Thus, while hydraulic conductivities obtained by methods employing any boundary conditions will give correct predictions when used in computations of groundwater flows in uniform soils, the accuracy of predictions in a nonuniform soil will depend on the relevance of the measured equivalent hydraulic conductivity. If the measurement imposes boundary conditions that produce flow patterns very different from those of the flows to be calculated, the predictions will lack accuracy. For accurate predictions, the pattern of flow in the measurement must approximate as nearly as possible those of the problem, since local variations of hydraulic conductivity can distort flows profoundly.

From the foregoing remarks it is obvious that the measurement of hydraulic conductivity is not a simple matter when the soil is nonuniform. Methods of measurement in nonuniform soils must be conditioned by the purpose for which they are made. Otherwise values obtained are of little relevance. Unless otherwise stated, the methods described in this chapter, as in other reviews of methods [3, 8-10], assume that the soil is uniform and isotropic; that is, it is assumed that the measurements are on flow regions made up of several representative elementary volumes with no preferred direction, so that the geometry of the boundaries and the orientation of the flows is of no import.

III. LABORATORY MEASUREMENTS

A. General Principles

Most laboratory measurements of hydraulic conductivity on saturated samples of soils essentially repeat Darcy's original experiments [1] that led to his formulation of the law for water movement through porous materials. The principles that apply for soil samples taken from the field are the same as those for the sands used by Darcy. The soil is removed from the field, hopefully undisturbed, to form a column on which measurements can be made with the sides enclosed by impermeable walls. With the column of soil standing on a permeable base, the soil is saturated and the surface ponded so that water percolates through the column. The soil-water pressure head is measured with piezometers at positions down the column and the rate of flow through the soil measured. The hydraulic conductivity is the rate of flow per unit cross-sectional area per unit hydraulic head gradient. An arrangement used for measuring hydraulic conductivity is known as a *permeameter*.

In addition to methods that involve measurements on a completely saturated material, there are other methods that involve wetting up an unsaturated sample from a surface maintained saturated at zero soil-water pressure. These methods utilize infiltration theory to obtain the hydraulic conductivity of the saturated soil from measurements on the rate of uptake of water.

B. Collection and Preparation of Soil Samples

For loosely bound soil materials such as the sands and sieved soils that are often used in various tests, care has to be taken to obtain uniform packing of columns on which measurements are to be made. If the material is not packed uniformly as the column is filled, separation of different-sized particles can occur, resulting in a column with spatially variable hydraulic conductivity; even columns of coarse sand can pack to give a twofold variation of hydraulic conductivity down the column [11]. In filling columns it is useful to attach a short extension length to the top of the column and fill above the top, pouring continuously but slowly while tamping to obtain a uniform density. The material in the top extension is then removed, leaving the bottom part for the measurement. For granulated materials with particles passing through a 2 mm sieve, the representative elementary volume is small enough to allow columns of small diameter, 100 mm or less, to be used.

The taking of samples in the field requires great care to ensure that samples as nearly representative of the field soil as possible are obtained. The size of sample required cannot easily be inferred from visual inspection because fine cracks in soils, which contribute largely to the hydraulic conductivity of a soil, may not be noticed. In poorly structured soils small samples of cross-sectional area 0.01 m^2 or less can be representative even for groundwater flow calculations. In highly structured soils the size of a sample that is representative for a measurement will depend on the purpose for which the measurement is required. Small samples of the size of those suitable for poorly structured soils might suffice for some purposes (e.g., for studies on water movement in the soil matrix between cracks in a fissured soil), but groundwater movement predictions generally call for a much larger sample that includes the highly conducting cracks and fissures. Cylindrical samples 0.4 m in diameter and 0.6 m high have been used [12,13]. For special purposes larger "undisturbed" samples can be obtained as for lysimeter studies [6, 14], typically 0.8 m in diameter.

Soil samples can be collected in large diameter PVC or fiberglass cylinders. A steel cutting edge is first attached to one end and the sample taken by jacking the cylinder into the soil hydraulically. While samples are usually taken vertically, horizontal samples can also be taken. As the sampling cylinder is forced into the soil, the surrounding soil is removed to lessen resistance to passage. When the required sample is contained in the cylinder, the surrounding soil is dug away to a greater depth to allow a cutting plate to be jacked underneath, separating the sample from the soil beneath. The sample is then removed to the laboratory, covered by plastic sheeting to retain moisture. In the laboratory the upper and lower faces are carefully prepared by removing any smeared or damaged surfaces before saturating the samples for the hydraulic conductivity measurements.

While taking and removing the sample, soil disturbance or shrinkage may occur, notably with the soil coming detached from the side of the sampling cylinder. A seal can be made by pouring liquid bentonite down the edge. The wetting of the sample will swell the soil and make the seal watertight.

An alternative method of preparing a sample for hydraulic conductivity measurements has been devised by Bouma [15]. A cylindrical column of soil is sculptured in situ so that the column is left in the middle of a trench. Plaster of Paris is then poured over it to seal the sides. The column can either be cut from the base and removed to the laboratory for measurements of hydraulic conductivity, both in saturated and unsaturated conditions, or alternatively left in place for measurements to be made in the field. A cube of soil is sometimes cut [16] so that flow measurements can be made in different directions after the removal of the plaster from the appropriate faces, allowing the components of hydraulic conductivity in the different directions to be obtained in anisotropic soils. In a modification of the method [17], a cube of soil is carved around a tile drain to permit measurements of hydraulic conductivity to be made in this sensitive region in drained lands.

C. Constant Head Permeameter

The constant head permeameter uses exactly the same arrangement as Darcy [1] used in 1856 (Fig. 1). The soil column is supported on a permeable base such as a wire gauze or filter, or sometimes a sand table. Water flows through the column from a constant head of water on the soil surface and is collected for measurement from an outlet chamber attached to the base. Slichter [4] recommended that soil water pressures be measured by piezometers within the soil column, since he noted that "there appears sudden reduction in pressure as the liquid enters the soil." The error arising from not accounting for this reduction is considered to be of no great importance today because of the recognition of the true degree of accuracy that can be expected in hydraulic conductivity values due to inhomogeneities in most soils. The hydraulic conductivity is given from the measurements by

$$K = \frac{QL}{A \; \Delta h} \tag{12}$$

where Q is the flow rate, L the length of the column, A its cross-sectional area, and Δh the head difference causing the flow. In Eq. 12, as with all formulas for K in this chapter, the units of K are the same as the units used for length and time for the quantities on the right-hand side of the equation. Care must be taken to use the same length unit for all variables (e.g., for the units of K to be cm s^{-1}, Q must be in cm of s^{-1}, L in cm, A in cm^2, and Δh in cm).

Errors often occur because of preferential boundary wall flow between the soil and the sides of the permeameter. This can be reduced by separately collecting and measuring the throughput from the central area of the sample [18], as illustrated in Fig. 5.

The measurements made using a constant head permeameter actually give the hydraulic conductance between the levels at which the measurements of head are made. These are interpreted as hydraulic conductivity values, assuming the soil to be uniform; that is, equivalent hydraulic conductivity values are obtained. Thus the measured

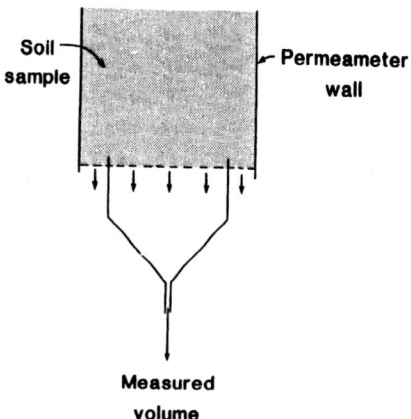

Soil sample

Permeameter wall

Measured volume

FIG. 5 Arrangement at base of soil sample to reduce permeameter boundary wall flow errors.

conductance C_{12} between two piezometers at heights L_1 and L_2 showing heads h_1 and h_2, respectively, with a steady flow Q is given by

$$C_{12} = \frac{Q}{h_2 - h_1} \qquad (13)$$

and the equivalent hydraulic conductivity K_e is given by

$$K_e = \frac{C_{12}}{A(L_2 - L_1)} \qquad (14)$$

Youngs [19] has described an alternative technique in which the piezometers act as interceptor drains, as illustrated in Fig. 6. With only one of the piezometers at a height Z above the base acting as a drain and removing water at a rate Q_Z, and with no flow through the base, the conductance C_{LZ} between the top of the column at height L and the height Z is given by

$$C_{LZ} = \frac{Q_Z}{h_L - h_0} \qquad (15)$$

where h_L is the measured head of the ponded water on the surface and h_0 is that measured at the base of the column. When the conductance profile is obtained by making measurements of flows from piezometers down the column, the hydraulic conductivity profile is given by

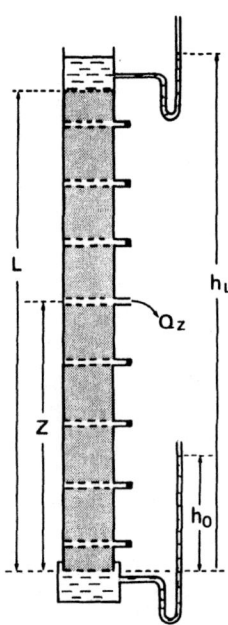

FIG. 6   Measurement of hydraulic conductivity profiles down soil monoliths using interceptor drains.

$$K(Z) = \left[A \frac{d}{dz} \left(\frac{1}{C_{LZ}} \right) \right]^{-1} \tag{16}$$

This technique can be used [6] to obtain the variation of hydraulic conductivity with depth on a soil monolith contained in a lysimeter.

D. Falling Head Permeameter

The falling head permeameter is similar to the constant head permeameter except that instead of maintaining a constant head of water on the surface of the soil sample, no more water is added and the changing level of the head is observed as the water percolates through the sample.   Such an arrangement is shown in Fig. 7. Magnification of the rate of fall of the standing head is achieved by containing it in a tube of smaller cross-sectional area A' than the cross-sectional area A of the soil sample.   With the initial height of the water level h_0 at time t_0 falling to h_1 at t_1, the hydraulic conductivity is given by

$$K = \frac{A'L \ \ln(h_0/h_1)}{A(t_1 - t_0)} \tag{17}$$

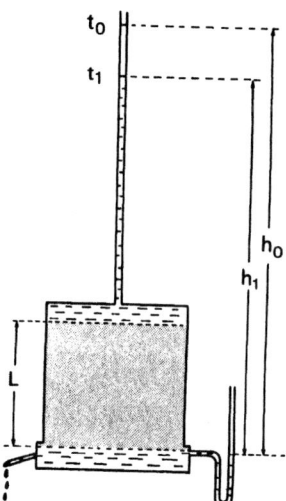

FIG. 7   Falling head permeameter.

E.   Oscillating Permeameter

A drawback of the constant head and falling head permeameters is that a fairly large volume of water percolates through the soil sample during the course of a measurement of hydraulic conductivity. If the material is surface-active, structural changes may occur during the test because of changes in chemical constitution, thus producing changes in the hydraulic conductivity of the soil sample.

A variation of the falling head permeameter is the oscillating permeameter [20]. This device utilizes the passage to and fro through the soil sample of a limited volume of water, very little in excess of that required to saturate the pore space. Such a small quantity of water quickly comes to equilibrium with the soil without affecting greatly its chemical composition, therefore remaining in equilibrium throughout the test, however long its duration.

In the permeameter the sample forms one arm of a U-tube system, as shown in Fig. 8a. The other arm contains a forcing head of water that oscillates sinusoidally about a mean position. This head and the head of water standing on the surface of the soil sample on the other side of the U-tube are recorded with time (e.g., with pressure transducers). After a few cycles, the two heads oscillate as shown in Fig. 8b, with the head above the sample being out of phase with the forcing head and of a smaller amplitude. If the amplitude of the forcing head is H_0 and that over the sample is h_0, the phase angle β is given by

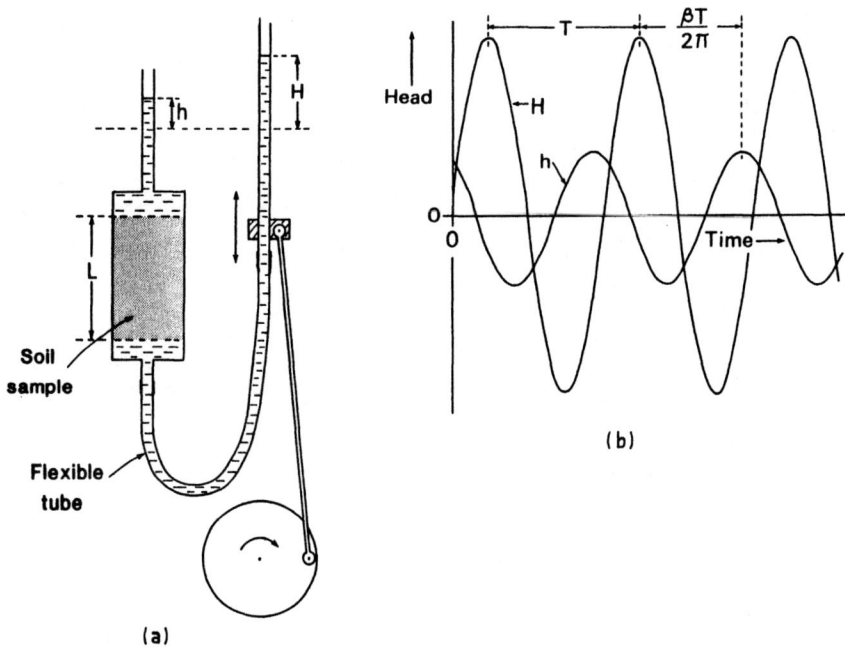

FIG. 8 Oscillating permeameter: (a) experimental arrangement and
(b) measured oscillating heads.

$$\tan \beta = \left(\frac{H_0^2}{h_0^2} - 1 \right)^{1/2} \tag{18}$$

and the hydraulic conductivity of the sample is given by

$$K = \frac{2\pi A'L}{AT \tan \beta} \tag{19}$$

where A is the cross-sectional area of the sample of length L, A' is
that of the tube containing the head above the sample, and T is the
period of one cycle.   The hydraulic conductivity can thus be found
from the phase angle obtained either by direct measurement or from
measurements of the amplitudes of the heads and the use of Eq. 18.

F. Infiltration Method

Infiltration theory shows that the infiltration rate from a ponded surface into a long vertical column of uniform porous material eventually

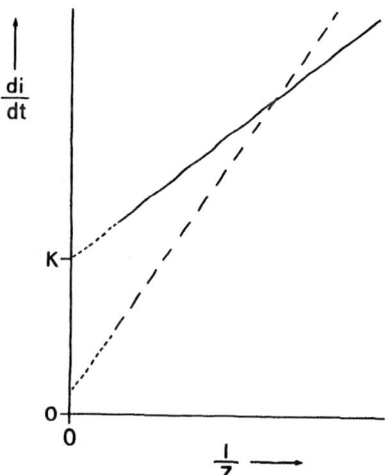

FIG. 9 Plot of the rate of infiltration di/dt against the reciprocal of the depth of wetting front 1/Z: ————, for a uniform soil; ------, that which might be obtained for a soil with hydraulic conductivity decreasing with depth.

approaches a constant rate, equal to the hydraulic conductivity of the saturated material. The approximate Green and Ampt theory of infiltration [21] gives the infiltration rate di/dt when the wetting front has advanced to a depth Z as

$$\frac{di}{dt} = K\left(\frac{h_f}{Z} + 1\right) \tag{20}$$

where $-h_f$ is the soil-water pressure head at the wetting front. Thus a plot of di/dt against 1/Z gives an intercept K on the di/dt axis, as sketched in Fig. 9. The hydraulic conductivity of saturated uniform porous materials can thus be obtained by observing the position of the wetting front while measuring the infiltration rate from a ponded surface. However, the fact that a linear plot is found when plotting di/dt against 1/Z should not be taken as proof that the column is uniform, since it has been found [22-24] that such a linear plot is obtained in certain situations characterized by a decrease in hydraulic conductivity with depth. The intercept in this case is less than if the soil were uniform, and can even become negative. The method is therefore reliable only if the soil profile is known to be uniform within the wetted depth, and this may be difficult to ascertain.

G. Varying Moment Permeameter

The varying moment permeameter [25], although originally used to measure the hydraulic conductivity of unsaturated soils, provides a quick method of measuring the hydraulic conductivity of soil samples that are initially unsaturated. Water is infiltrated horizontally at a positive pressure head into columns of the unsaturated soils, and the rate of change of moment of the advancing water profile about the plane through which infiltration takes place is measured. It can be shown that this rate of change of the moment is equal to the integral of the hydraulic conductivity with respect to the soil-water pressure along the column multiplied by the cross-sectional area A of the column. Thus:

$$\frac{dM}{dt} = A\left[\int_{p_i}^{p_0} \rho g K' \, dp\right] = A\left[\int_{p_i}^{0} \rho g K' \, dp + \rho g K p_0\right] \tag{21}$$

where M is the moment of the advancing soil-water profile at time t, p is the soil-water pressure head (with the subscripts 0 and i referring to that at the infiltrating surface and that in the soil not yet reached by the advancing water front, respectively), and $K'(p)$ is the hydraulic conductivity of the soil, which is a function of the soil-water pressure head p in unsaturated soils but equal to K for saturated soils. By measuring dM/dt for different pressure heads p_0 of infiltrating water, the hydraulic conductivity of the saturated soil can be obtained.

An arrangement of the varying moment permeameter for measuring the hydraulic conductivity of saturated soils is shown in Fig. 10a. The sample in the form of a column contained within impermeable walls is arranged on a balance with a fulcrum immediately below the surface through which water is infiltrated at a pressure p_0 from the reservoir. A counterbalance weight is used at the start of the experiment to tare the weight of the column so that the changing moment of the advancing soil-water profile can be easily measured by adding weights to the balance pan. With the pressure p_0 maintained constant, this moment varies linearly with time, so that dM/dt can simply be determined for a given p_0 by observing the time taken for the apparatus to balance for a known added weight. When measurements have been made at a given head, the head is increased and the procedure repeated. By plotting dM/dt against the pressure head p_0 as shown in Fig. 10b, the slope of the straight line obtained gives the hydraulic conductivity K.

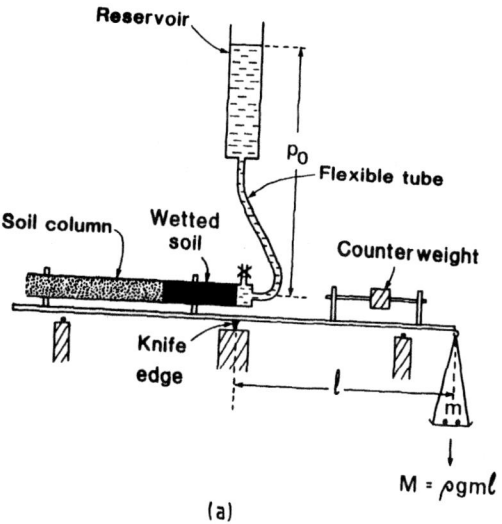

(a)

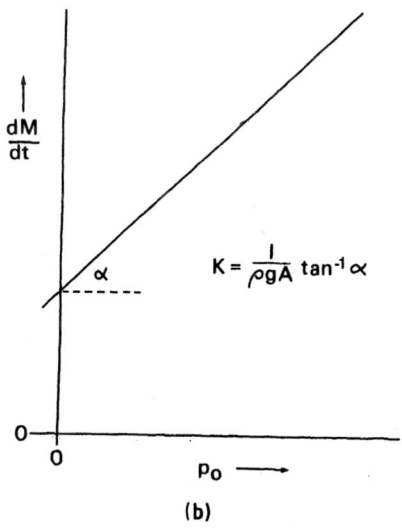

(b)

FIG. 10 Varying moment permeameter: (a) experimental arrangement and (b) plot of the rate of change of the moment dM/dt against the soil-water pressure head p_0 of the infiltrating water.

IV. FIELD MEASUREMENTS BELOW A WATER TABLE

A. General Principles

In situ measurements of hydraulic conductivity below the water table provide the most reliable values for use in calculations of ground-water flows, especially when they sample large volumes of soil. Techniques usually employ unlined or lined wells sunk below the water table and involve measurements of flow into or out of the wells when the water levels in them are perturbed from the equilibrium. The hydraulic conductivity values are calculated from the solution of the potential problem for the flow region with the imposed boundary conditions. If no analytical solution is available, recourse can be made to electrical analogs or numerical methods to obtain solutions.

For satisfactory measurements, wells must be large enough to allow a representative volume of soil to be sampled in the measurement. However, it is not easy to deduce the volume of soil sampled in a given measurement. Some indication of this volume might be obtained from the volume traced out by 90% (say) of the stream tubes for a 90% (say) reduction in head. It obviously increases with the size of well used. It will also depend on other geometrical factors of the flow system: for example, the area of the well walls through which water can flow, and the spacing of wells in a multi-well system.

Well radii of 100 mm or larger are typically used. The wells are made with post augers, and special tools are used to form the holes into an exact cylindrical shape. Some difficulties may be encountered doing this [26]. First, there is the common problem of making holes when the soil is stony; stones may have to be cut with chisels during the operation. Second, there is the possibility of unstable soils slumping below the water table; permeable liners can be used to alleviate this problem. And third, in clay soils there is the problem of smearing of the sides of the walls of the wells, thus creating surfaces of low conductivity that restrict flow entry; to lessen this effect, the wells are emptied to allow inflowing water to unblock the pores before measurements are made.

The various well techniques for measuring the hydraulic conductivity are described mostly in books on land drainage [8,9]. Since all gave satisfactory results in a comparison of well methods in a hydraulic sand tank [27], it would appear that the choice of method depends largely on site conditions, resources available, and individual preference. However, in some methods the flow is predominantly horizontal while in others it is vertical, so that if the soil is suspected of being anisotropic, the method to be employed must take into consideration the direction of flow in the region under investigation.

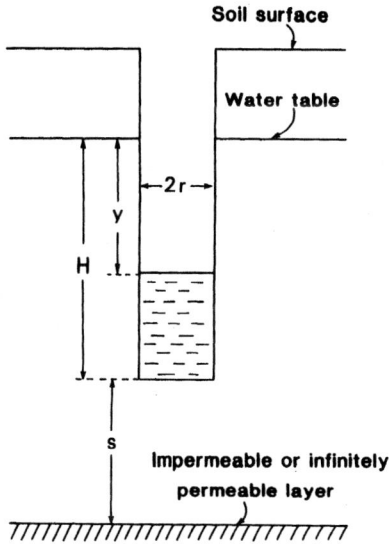

Soil surface

Water table

FIG. 11   Geometry of the auger-hole method.

While the use of wells gives a practical and convenient method of providing an arrangement of groundwater flows that can be analyzed to give hydraulic conductivity values, any arrangement of sinks and/or sources that produce flows that can be analyzed may be used for the purpose. For example, land drains themselves can be used as permeameters [28,29] that sample much larger volumes of soil than can be sampled using wells.

B. Auger-hole Method

In the auger-hole method of determining the hydraulic conductivity of a soil, an unlined cylindrical hole is made below the water table (Fig. 11). The position of the water table is found by allowing the water in the hole to return to its equilibrium water level. The water level in the hole is then lowered by removing water by pumping or bailing and its rate of rise observed as it returns to equilibrium. The hydraulic conductivity is calculated using the formula

$$K = C \frac{dy}{dt} \tag{22}$$

where y is the depth of the water level in the hole below the water table at time t and C is a factor that depends on the radius r of the hole, the depth s of an impermeable or infinitely permeable layer

TABLE 3   Values of the Shape Factor C × 10³ for Auger Holes

H/r	y/H	Impermeable layer at s/H								s/H = ∞	Infinitely permeable layer at s/H			
		0	0.05	0.1	0.2	0.5	1	2	5		5	2	1	0.5
1	1	518	490	468	435	375	331	306	296	295	292	280	247	193
	0.75	544	522	503	473	418	376	351	339	338	335	322	287	230
	0.5	643	623	605	576	521	477	448	441	440	437	416	376	306
2	1	215	204	193	178	155	143	137	135	133	133	131	123	106
	0.75	227	216	208	195	172	160	154	152	152	151	148	140	123
	0.5	271	261	252	240	218	203	196	194	194	193	190	181	161
5	1	60.2	56.3	53.6	49.6	44.9	42.8	41.9		41.5		41.2	40.1	37.6
	0.75	63.6	60.3	57.9	54.3		47.6	46.6		46.4		45.9	44.8	42.1
	0.5	76.7	73.5	71.1	67.4	62.5	60.2	59.1		58.8		58.3	57.1	54.1
10	1	21.0	19.6	18.7	17.5	16.4	15.8	15.5		15.5		15.4	15.2	14.6
	0.75	22.2	21.0	20.2	19.1	18.0	17.4	17.2		17.2		17.1	16.8	16.2
	0.5	27.0	25.9	24.9	23.9	22.6	22.0	21.8		21.7		21.6	21.3	20.6
20	1	6.86	6.41	6.15	5.87	5.58	5.45	5.4		5.38		5.36	5.31	5.17
	0.75	7.27	6.89	6.65	6.38	6.09	5.97	5.9		5.89		5.88	5.82	5.67
	0.5	8.90	8.51	8.26	7.98	7.66	7.52	9.5		7.44		7.41	7.35	7.16
50	1	1.45	1.37	1.32	1.29	1.24	1.22			1.21			1.19	1.18
	0.75	1.54	1.47	1.42	1.39	1.35	1.32			1.31			1.30	1.28
	0.5	1.90	1.82	1.79	1.74	1.69	1.67			1.66			1.65	1.61
100	1	0.43	0.41	0.39	0.39	0.38	0.37			0.37			0.37	0.36
	0.75	0.46	0.44	0.42	0.42	0.41	0.41			0.41			0.39	0.39
	0.5	0.57	0.54	0.53	0.52	0.51	0.51			0.51			0.50	0.50

Source: After Boast and Kirkham [34].

below the bottom of the hole, and the depth y, all measured relative to the depth H of the water in the hole when in equilibrium with the water table; thus we can write C = C(r/H, s/H, y/H).

Formulas for obtaining the factor C in Eq. 22 have been given by Diserens [30], Hooghoudt [31], Ernst [32], and Kirkham and van Bavel [33]. An exact mathematical solution was obtained for C in terms of a complex series by Boast and Kirkham [34]. Their results are presented in Table 3 in a form that gives the hydraulic conductivity K in the same units as those for the rate of rise of the water level dy/dt. These values of C can be used in Eq. 22 to give the hydraulic conductivity from experimental results. (Boast and Kirkham's original values for C require dy/dt values to have units of centimeters per second to give K in meters per day. This practice, also adopted in other formulas, can give rise to confusion.) Alternatively, Ernst's formulas

$$K = \frac{4.63}{(20 + H/r)(2 - y/H)} \frac{r}{y} \frac{dy}{dt} \quad \text{for} \quad s > 0.5H \quad (23)$$

and

$$K = \frac{4.17}{(10 + H/r)(2 - y/H)} \frac{r}{y} \frac{dy}{dt} \quad \text{for} \quad s = 0 \quad (24)$$

can be used when the hole is in effectively infinitely deep soil and when the hole extends down to an impermeable layer, respectively. These formulas give results that agree satisfactorily with Boast and Kirkham's values, especially when H ≈ 10r. Measurements are sometimes made using seepage into large holes below the water table, a method sometimes referred to as the "pit-bailing" method. Then shape factors are required for r > H, a situation not encountered with the normal use of auger holes. These shape factors for r > H have been given by Boast and Langebartel [35].

The auger-hole method has been developed, as have most other methods, for use in uniform soils, but both Hooghoudt [31] and Ernst [32] considered how estimates of hydraulic conductivity could be made with auger holes in layered soils. The flow during tests is primarily horizontal; thus in anisotropic soils the results obtained approximate to the horizontal component of the hydraulic conductivity.

C. Piezometer Method

The piezometer method uses lined wells, sunk below the water table, with or without a cavity at the bottom, as illustrated in Fig. 12. This cavity is usually cylindrical, although other shapes, for example hemispherical, can be used. As in the auger-hole method, after the water level in the well has come into equilibrium with the water

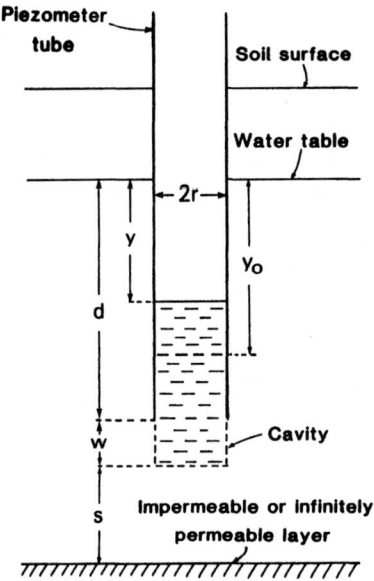

FIG. 12   Geometry of the piezometer method.

table, it is depressed by pumping or bailing and its rate of rise observed as it returns to equilibrium. The hydraulic conductivity is
then given by

$$K = \frac{\pi r^2 \ln(y_0/y)}{A(t - t_0)} \tag{25}$$

where y_0 and y are the depths of the water level in the well below
the equilibrium level at time t_0 and at time t, respectively, and A
is a shape factor that depends on the depth d of water in the well
at equilibrium, the length w of the cavity at the bottom of the well,
and the depth s of soil to an impermeable or infinitely permeable
layer, all measured relative to the radius r of the well; that is, $A = A(d/r, w/r, s/r)$.

Shape factors obtained with an electrical analog were given by
Frevert and Kirkham [36]. More accurate values were presented by
Smiles and Youngs [27] and a comprehensive table of accurate values,
reproduced in Table 4, was given by Youngs [37]. As shown by
these values, as long as the cavity is not less than about a radius
from an impermeable or permeable stratum, the results are very
nearly the same as for an infinitely deep soil and so are unaffected

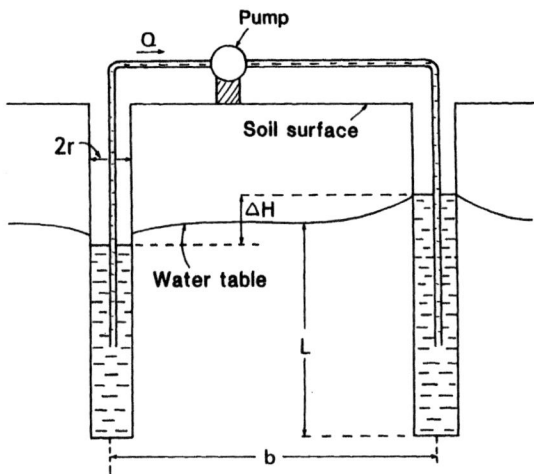

FIG. 13   The geometry of the two-well method.

by changes of hydraulic conductivity at this distance away. Thus accurate determinations of hydraulic conductivity can be made with this method in layered soils by making measurements at different depths with the cavity within the different layers. With cavities of small length, the flow is mainly vertical, so that values reflect the vertical component of hydraulic conductivity in anisotropic soils.

Piezometers installed for soil-water pressure measurements may also be used to measure hydraulic conductivity. For example, Goss and Youngs [38] used an existing installation of piezometers inserted horizontally from the walls of an inspection pit. Such piezometers may not have cavities that conform to those for which shape factors are available, with the result that shape factors for the particular piezometers must be determined with an electrical analog. An arrangement of piezometers located at intervals down the soil profile allows the hydraulic conductivity variation with depth to be determined; and when the installation is from an inspection pit, measurements can be made from one year to another in a soil that remains undisturbed at depth, with normal cultivation practices being carried out above.

D.   Two-Well Method

The two-well method of Childs [26,39,40] uses a pair of unlined wells sunk to the same depth below the water table, as illustrated in Fig. 13. Water is pumped at a constant rate from one well into the other, thus depressing the level in one and raising it in the

TABLE 4 Values of the Shape Factor A (Expressed as A/r) for Piezometers with Cylindrical Cavities

w/r	d/r	Impermeable layer at s/r							Infinitely permeable layer at s/r						
		∞	8.0	4.0	2.0	1.0	0.5	0	∞	8.0	4.0	2.0	1.0	0.5	0
0	20	5.6	5.5	5.3	5.0	4.4	3.6	0	5.6	5.6	5.8	6.3	7.4	10.2	∞
	16	5.6	5.5	5.3	5.0	4.4	3.6	0	5.6	5.6	5.8	6.4	7.5	10.3	∞
	12	5.6	5.5	5.4	5.1	4.5	3.7	0	5.6	5.7	5.9	6.5	7.6	10.4	∞
	8	5.7	5.6	5.5	5.2	4.6	3.8	0	5.7	5.7	5.9	6.6	7.7	10.5	∞
	4	5.8	5.7	5.6	5.4	4.8	3.9	0	5.8	5.8	6.0	6.7	7.9	10.7	∞
0.5	20	8.7	8.6	8.3	7.7	7.0	6.2	4.8	8.7	8.9	9.4	10.3	12.2	15.2	∞
	16	8.8	8.7	8.4	7.8	7.0	6.2	4.8	8.8	9.0	9.4	10.3	12.2	15.2	∞
	12	8.9	8.8	8.5	8.0	7.1	6.3	4.8	8.9	9.1	9.5	10.4	12.2	15.3	∞
	8	9.0	9.0	8.7	8.2	7.2	6.4	4.9	9.0	9.3	9.6	10.5	12.3	15.3	∞
	4	9.5	9.4	9.0	8.6	7.5	6.5	5.0	9.5	9.6	9.8	10.6	12.4	15.4	∞
1.0	20	10.6	10.4	10.0	9.3	8.4	7.6	6.3	10.6	11.0	11.6	12.8	14.9	19.0	∞
	16	10.7	10.5	10.1	9.4	8.5	7.7	6.4	10.7	11.0	11.6	12.8	14.9	19.0	∞
	12	10.8	10.6	10.2	9.5	8.6	7.8	6.5	10.8	11.1	11.7	12.8	14.9	19.0	∞
	8	11.0	10.9	10.5	9.8	8.9	8.0	6.7	11.0	11.2	11.8	12.9	14.9	19.0	∞
	4	11.5	11.4	11.2	10.5	9.7	8.8	7.3	11.5	11.6	12.1	13.1	15.0	19.0	∞

2.0	20	13.8	13.5	12.8	11.9	10.9	10.1	9.1	13.8	14.1	15.0	16.5	19.0	23.0	8
	16	13.9	13.6	13.0	12.1	11.0	10.2	9.2	13.9	14.3	15.1	16.6	19.1	23.1	8
	12	14.0	13.7	13.2	12.3	11.2	10.4	9.4	14.0	14.4	15.2	16.7	19.2	23.2	8
	8	14.3	14.1	13.6	12.7	11.5	10.7	9.6	14.2	14.8	15.5	17.0	19.4	23.3	8
	4	15.0	14.9	14.5	13.7	12.6	11.7	10.5	15.0	15.4	16.0	17.6	20.1	23.8	8
4.0	20	18.6	18.0	17.3	16.3	15.3	14.6	13.6	18.6	19.8	20.8	22.7	25.5	29.9	8
	16	19.0	18.4	17.6	16.6	15.6	14.8	13.8	19.0	20.0	20.9	22.8	25.6	29.9	8
	12	19.4	18.8	18.0	17.1	16.0	15.1	14.1	19.4	20.3	21.2	23.0	25.8	30.0	8
	8	19.8	19.4	18.7	17.6	16.4	15.5	14.5	19.8	20.6	21.4	23.3	26.0	30.2	8
	4	21.0	20.5	20.0	19.1	17.8	17.0	15.8	21.0	21.5	22.2	24.1	26.8	31.5	8
8.0	20	26.9	26.3	25.5	24.0	23.0	22.2	21.4	26.9	29.6	30.6	32.9	36.1	40.6	8
	16	27.4	26.6	25.8	24.4	23.4	22.7	21.9	27.4	29.8	30.8	33.1	36.2	40.7	8
	12	28.3	27.2	26.4	25.1	24.1	23.4	22.6	28.3	30.0	31.0	33.3	36.4	40.8	8
	8	29.1	28.2	27.4	26.1	25.1	24.4	23.4	29.1	30.3	31.2	33.8	36.9	41.0	8
	4	30.8	30.2	29.6	28.0	26.9	25.7	24.5	30.8	31.5	32.8	35.0	38.4	43.0	8

Source: From Ref. 37, E. G. Youngs, Shape factors for Kirkham's piezometer method for determining the hydraulic conductivity of soil in situ for soils overlying an impermeable floor or infinitely permeable stratum, *Soil Science, 106:* 235–237 (1968). © by Williams and Wilkins, Baltimore, MD.

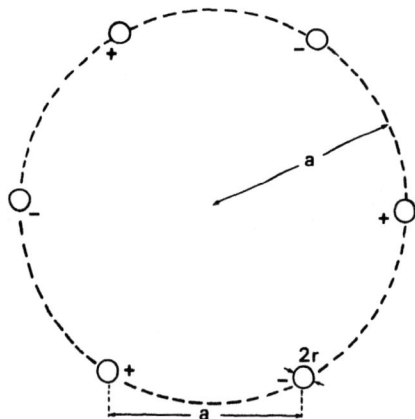

FIG. 14 Plan view of the multiple-well method using six wells:
—, pumped wells; +, receiving wells.

other. When a steady state ensues, the hydraulic conductivity of
the soil is given by

$$K = \frac{Q}{\pi \, \Delta H(L + L_f)} \, \cosh^{-1}\left(\frac{b}{2r}\right) \tag{26}$$

where Q is the steady flow rate, L the depth of the wells below the
water table, L_f an end correction to be added to take into account
the flow beneath the wells (if they do not reach to an impermeable
floor) and the flow in the capillary fringe, b the distance between
centers of the wells, r the radius of the wells, and ΔH the difference in water level in the two wells. The hydraulic conductivity
profile may be obtained when there is a soil variation with depth by
making measurements on wells sunk successively deeper. Alternatively, the seepage analysis of Youngs [41] can be used to measure
this variation with depth by making measurements using a range of
drawdowns in the pumped well.

    Childs' two-well method may be extended to a radial symmetri-
cal array of wells [42] such as that illustrated in Fig. 14, alternate
ones discharging and receiving the same rate of flow. The formula
for obtaining K for this case is

$$K = \frac{Q}{n\pi \, \Delta H(L + L_f)} \, \ln\left(\frac{4a}{hr}\right) \tag{27}$$

where n is the even number of wells of radius r, arranged symmetrically on the circumference of a circle of radius a and sunk to a

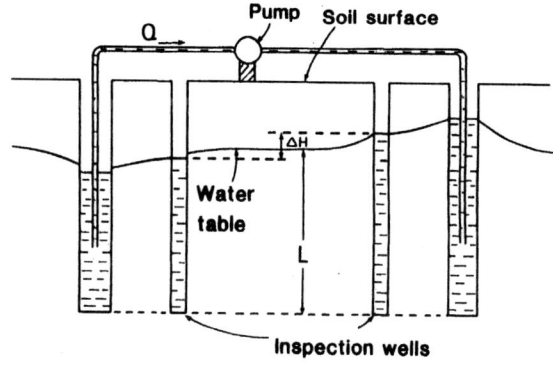

FIG. 15   Geometry of the four-well method.

depth L below the water table, L_f is an end correction as in the two-well method, and Q is now the total rate of water being pumped from the wells in the system when there is a head difference of ΔH between the levels of water in the pumped and receiving wells.

In uniform soils the depression of the water level in the pumped well is equal to the elevation in the receiving well. However, in field soils this is rarely found to be the case because of soil variation. Some indication of the variability of the soil is given by the differences between the elevations and depressions in the wells [40, 42].

A modification of the two-well method [43] employs two inspection wells symmetrically installed between the two wells to measure the heads in the flow system at these locations, as shown in Fig. 15. This arrangement overcomes difficulties associated with clogging of pores in the return well. The formula for calculating K is

$$K = \frac{BQ}{\Delta HL} \tag{28}$$

where B is a factor depending on the geometry of the system that is given by a set of graphs and ΔH is now the difference in level in the two inspection wells [44].

The flow produced in the unlined two-well and multiple-well methods is mainly horizontal, so that values obtained with these methods in anisotropic soils approximate the horizontal component of the hydraulic conductivity. The methods can be used in conjuntion with Kirkham's piezometer method at the same site to obtain both the vertical and horizontal components [40].

E. Pumped Wells

Pumped wells discharging at a constant rate are used extensively to
measure aquifer characteristics for groundwater supplies. They may
be employed to determine the hydraulic conductivity of the soil by
measuring the drawdown of the water table at some distance from
the pumped wells as a function of time. The transmissivity T, which
is the product of the hydraulic conductivity and the depth of the
aquifer, is given by Theis' formula

$$z = -\frac{Q}{4\pi T} \, Ei\left(-\frac{r^2 S}{4tT}\right) \tag{29}$$

where z is the drawdown at time t at a radial distance r from the
well pumped at a constant rate Q and S is the storage coefficient of
the aquifer [45]. In Eq. 29 Ei is the exponential integral of the expression within parentheses [8,46]. T and S are found by overlay-
ing the experimental results of z and r^2/t plotted on log-log graph
paper on top of a plot of the function Ei on identical graph paper,
keeping the axes on each parallel during matching. Values of $Q/4\pi T$
and $4T/S$ are the values of the coordinates z and r^2/t, respectively,
which superimpose unit values on the type curve. Some difficulties
in matching may arise because of delayed yield, with the value of S
varying with the time of pumping.

F. Land Drains Used as Permeameters

Drainage equations that give the relationship between water table
height and drain discharge for a particular drainage installation provide a means whereby land drains can be used as large permeameters
to give equivalent hydraulic conductivity values of soils for the flows
to the drains. Land drainage theory [47,48] shows that for steady-
state conditions with parallel drain lines, drainage equations are of
the form

$$\frac{q}{K} = f\left(\frac{H_m}{D}\right) \tag{30}$$

where q is the flux through the water table derived from a uniform
steady rainfall on the soil surface, hence given by the drain discharge rate per unit area of drained land, and $f(H_m/D)$ is a function of the ratio of the maximum water table height H_m midway between the drains to the half-drain spacing D (Fig. 16). The hydraulic conductivity K is thus given by

$$K = \frac{q}{f(H_m/D)} \tag{31}$$

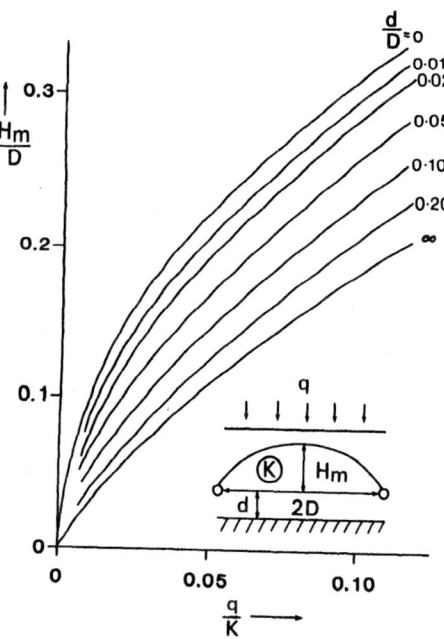

FIG. 16 Water flow to land drains: relationship between the maximum water table height H_m and the uniform rainfall rate q for various depths to the impermeable floor d.

so that from measurements of q, H_m, and D, and knowing the form of $f(H_m/D)$, K can be determined.

The difficulty in using this method of determining values of hydraulic conductivity from measurements on drained lands is to make a correct choice of drainage equation from the many available. These equations involve physical and mathematical assumptions in their derivation, and Lovell and Youngs [49] showed, in comparing ten commonly used equations, that these assumptions led often to large errors. However, one empirical equation that approximates well to the correct relationship when the drain is larger than the optimum size and so does not effect the water table height H_m midway between drains, is the power-law relationship

$$\frac{q}{K} = \left(\frac{H_m}{D}\right)^{\alpha} \tag{32}$$

where $\alpha = 2(d/D)^{d/D}$ for $0 < d/D < 0.35$ and $\alpha = 1.36$ for $d/D > 0.35$, and where d is the depth of an impermeable layer below the drains [50].

Equation 32 is particularly useful in analyzing moving water table situations and has been used in the characterization of hydrograph recessions of land drains [51]. However, analysis of drawdowns during non-steady-state water table recessions using the drain hydrograph also requires a knowledge of the specific yield if the hydraulic conductivity is to be obtained. Nevertheless, while it may not be possible to obtain hydraulic conductivity values directly from these drain hydrographs, a drain installation's characteristics, once determined from a recession, allow future drain performance to be predicted without the need of actual hydraulic conductivity values and, instead, using a parameter that involves the drain spacing and the soil's specific yield as well as the hydraulic conductivity [51].

The drainage inequality obtained from seepage analysis [41,52] can be used to interpret field results of drainage performance in terms of the depth-dependent hydraulic conductivity [29]. For parallel drains that lie on top of an impermeable layer, the depth-dependent hydraulic conductivity K(z) is given approximately by

$$K(z) = A\frac{d^2 q}{dH_m^2} \tag{33}$$

at $z = H_m$, where the factor A depends on the shape and dimensions of the drainage installation and for a parallel ideal drain system equals $D^2/2$. Thus the dependence of hydraulic conductivity with depth can be obtained by determining the relationship between the water-table height and drain discharge on a given drainage installation.

V. FIELD MEASUREMENTS IN THE ABSENCE OF A WATER TABLE

A. General Principles

The field methods described in the preceding section require the presence of a water table. On some occasions, however, values of hydraulic conductivity of saturated soils are required when there is no water table at the time measurements are to be made, to plan and design works for the future when the groundwater level is expected to rise. Techniques have been developed that allow measurements to be made in such circumstances. These measure the water uptake by the unsaturated soil from a saturated surface in a similar way to laboratory infiltration methods (see Sections III.F and III.G) and so rely for their interpretation on infiltration theory. Generally the measured flow depends not only on the hydraulic conductivity of the saturated soil but also on the capillary absorptive properties

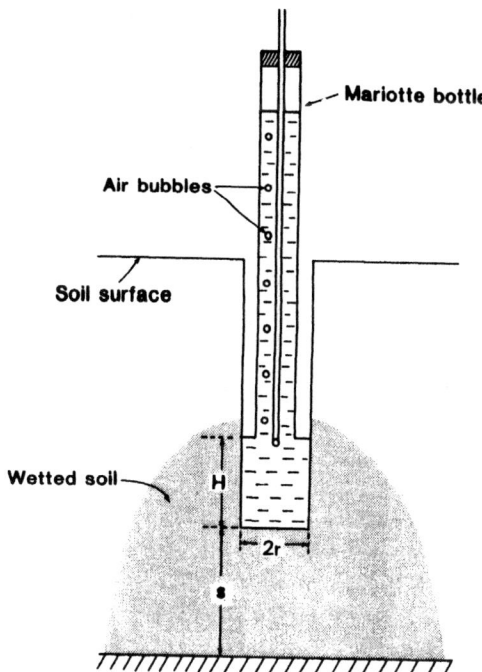

FIG. 17   Borehole permeameter.

of the unsaturated soil, often conveniently represented by the negative soil-water pressure head at the wetting front [21] or by the sorptivity [53]. Exact interpretation of experimental results to obtain hydraulic conductivities is difficult. Because many formulas for calculating values have been derived using very approximate theory, their reliability is often difficult to establish.

In the wetting-up process, entrapped bubbles of air may be left behind the advancing wetting front so that the soil is not completely saturated. These bubbles reduce the pore space for water conduction so that measurements of the hydraulic conductivity made in this way will generally be smaller, perhaps by as much as 50% [54], than those made with techniques that involve measurements below a water table. Thus caution should be exercised in using values obtained in this way for computing groundwater flows.

B. Borehole Permeameter

One of the oldest techniques for measuring the hydraulic conductivity of soils in the absence of a water table is the borehole permeameter, which uses water seeping into the soil from a vertical cylindrical
hole made in the unsaturated soil to the depth at which the measurement is required. Hydraulic conductivity values of the saturated
soil are obtained from the steady-state seepage from the borehole
that occurs after some time when the depth of water in the hole is
maintained at some constant level, often using a Mariotte bottle arrangement (see Fig. 17) [55-57]. The hydraulic conductivity is calculated from formulas that have been derived from an approximate
consideration of the physical situation. These formulas have been
cited in many reviews of the method (see, e.g., that by Stephens
and Neuman [58]).

For deep water tables Glover's formula [59] is commonly used,
giving K in the form

$$K = \frac{CQ}{2\pi H^2} \qquad\qquad (34)$$

with

$$C = \sinh^{-1}\left(\frac{H}{r}\right) = 1 \qquad \text{for} \quad H \gg r \qquad\qquad (35)$$

or more accurately, according to Reynolds et al. [56], by an expression that for $H \gg r$ reduces to

$$C = 2\left[\sinh^{-1}\left(\frac{H}{2r}\right) - 1\right] \qquad\qquad (36)$$

where Q is the steady seepage rate, H the depth of water in the
borehole, and r the radius of the borehole.

When an impermeable layer is at a relatively small depth s below
the borehole (s < 2H), K is given by [9,60]:

$$K = \frac{3Q}{\pi H(3H + 2s)} \ln\left(\frac{H}{r}\right) \qquad\qquad (37)$$

The formulas given above to obtain K are not derived from a
sound theoretical treatment but rely on approximations. These formulas overestimate values of hydraulic conductivity [61]; better values can be obtained using an extension of theory that takes into
account the effect of flow in the unsaturated soil [62]. Although
the borehole method has been considered to have great potential
for field measurements [56], theory shows that there is some doubt
concerning the utility of the method because of the difficulties in
the theoretical interpretation of the field data [63].

C. Auger-Hole Method

A variation of the borehole method uses an auger hole made to a given depth in the soil in the absence of a water table [10]. (This is sometimes referred to incorrectly as the "inverse" auger-hole method; the inverse of an auger hole would be a cylinder of soil!) Water is added to fill the hole to a given level and then the fall of of the water level is observed with time. The hydraulic conductivity is given approximately by

$$K = \frac{r \ln[(1 + 2H_0/r)/(1 + 2H/r)]}{2(t - t_0)} \tag{38}$$

where H_0 and H are the depths of water in the hole at time t_0 when measurements are begun and time t, respectively, and r is the radius of the hole.

In the derivation of Eq. 38, a unit hydraulic head gradient is assumed for the flow through the bottom and side of the hole. Because of this crude assumption, the method can be expected to give only a very approximate indication of the actual hydraulic conductivity.

D. Air-Entry Permeameter

With Bouwer's air-entry permeameter [9,64], a column of soil is contained within an infiltration cylinder driven into the soil. Water under a pressure head is infiltrated into the soil and the rate measured after the wetting front has penetrated some distance down the isolated column of soil. The hydraulic conductivity is obtained from this rate and the hydraulic head gradient causing the flow, deduced from measurements on the depth of the wetting front and the pressure head at this front.

The experimental arrangement is shown in Fig. 18. With a transparent lid sealed on the top of the infiltration cylinder, water is infiltrated at a large head from a reservoir with air escaping initially from an escape valve in the lid. The valve is closed when all air has escaped as observed through the transparent lid. After some time has elapsed for the wetting front to penetrate down the enclosed soil column to a depth of about 100 mm, no further water is added to the reservoir and the infiltration rate is measured by observing the rate of fall of the head. The supply valve is then closed, while a valve to a vacuum gauge that measures the pressure of the water standing on the soil surface is opened. This gauge shows when the water pressure reaches a minimum value before beginning to rise as air starts bubbling through the soil surface. The minimum reading is negative and is the air-entry valve p_a, which Bouwer [9,64] equates to $-2h_f$, where $-h_f$ is the pressure head of

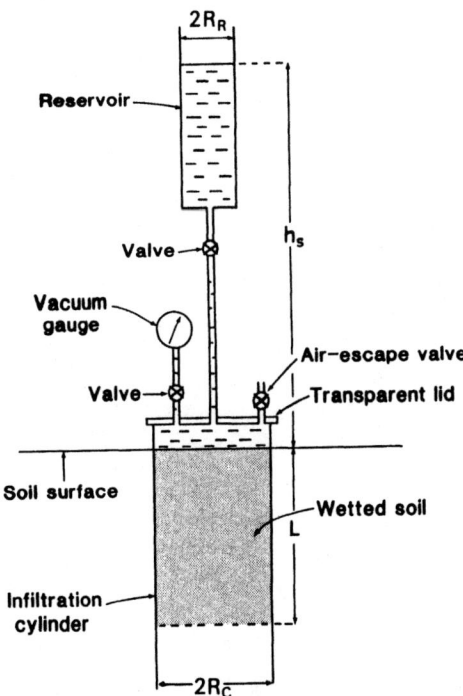

FIG. 18   Air-entry permeameter.

the water at the wetting front. The depth of wetting is then found
either by direct sampling as Bouwer suggested, or by employing a
tensiometer [65] to show the position of the wetting front at the
time of measurement. The hydraulic conductivity is given directly
by Darcy's law as in the Green and Ampt analysis

$$K = \frac{L(R_R/R_C)^2}{h_s + L - p_a/2} \frac{dh_s}{dt} \qquad (39)$$

where h_s is the head of water on the soil surface, L the depth of
the wetting front at the time of sampling, and dh_s/dt the rate of
fall of the water level. R_R is the radius of the reservoir containing the falling head and R_C that of the infiltration cylinder.

E.   Ring Infiltrometer Method

Since the infiltration capacity (i.e., the steady infiltration rate that
is approached at large times when water infiltrates over the whole

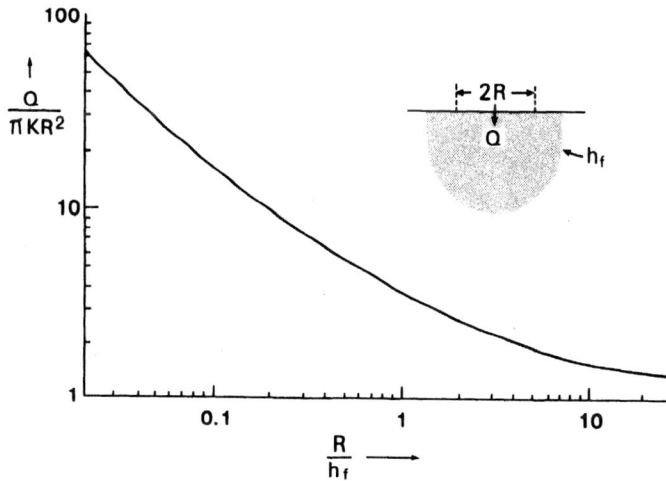

FIG. 19 Type curve of $Q/\pi KR^2$ plotted against R/h_f for steady-state flow from infiltrometer rings.

land surface) is identified with the hydraulic conductivity of the saturated soil, infiltration measurements into dry soil provide a means of obtaining hydraulic conductivity values. With ring infiltrometers there is a lateral component of flow due to capillarity as the water seeps into the soil, so that the infiltration rate at large times is not simply related to the hydraulic conductivity. However, the three-dimensional flow with infiltrometer rings quickly approaches a steady infiltration rate when water has wetted the soil to only a small depth. Thus the analysis of data from infiltrometer rings has the potential of being the means of obtaining the hydraulic properties of soils very close to the surface.

To obtain the hydraulic conductivity of saturated soils, the rings must be pressed into the soil to give a seal against leaks around the edge when a small head of water is maintained on the soil surface within the ring. The cumulative infiltration is measured with time, and the steady rate—which is approached, typically within 15 minutes for a fine sandy soil—is obtained from the results.

There are several ways of obtaining the hydraulic conductivity from the infiltration data. The type curve shown in Fig. 19 may be used [54]. This shows $Q/(\pi KR^2)$ plotted against R/h_f on log-log graph paper, where Q is the steady rate of water infiltrating into the soil after large times, R the radius of the ring, and h_f the negative pressure head at the wetting front of the saturated zone that is assumed to advance into the soil. By obtaining values of $Q/\pi R^2$ for rings with different radii R and plotting these on identical log-

log graph paper, the data can be superimposed on top of the type curve, keeping the axes on each parallel during matching. Values of K and h_f are the values of the coordinates $Q/\pi R^2$ and R, respectively, that superimpose unit values on the type curve.

Alternatively, the hydraulic conductivity can be obtained from infiltrometer results at early times using the equation

$$K = \frac{\rho g \eta R^4 (\theta_0 - \theta_i)^2}{\sigma^2 t^2} \left\{ -0.365 + \left[0.133 + \frac{I}{R^3(\theta_0 - \theta_i)} \right]^{1/2} \right\}^2 \quad (40)$$

where I is the total volume of infiltration up to time t, R the radius of the infiltration ring, θ_0 and θ_i the saturated and initial water contents of the soil, g the acceleration due to gravity, and ρ, η, and σ the density, viscosity, and surface tension, respectively, of water [66]. Equation 40 was obtained by curve fitting to laboratory experimental results that were scaled according to similar media theory [67], with a microscopic characteristic length defined in terms of the hydraulic conductivity of the porous material. This equation can be used only during the early stage of infiltration when $I < R^3(\theta_0 - \theta_i)$. If the unit of length is the centimeter and the unit of time is the second, $\rho g \eta / \sigma^2 = 0.00187$ cm^{-3} s to give the units of K in centimeters per second.

Another way of interpreting the infiltrometer data is to use Wooding's analysis [68] of the water flow from circular surface ponds (I. White, personal communication). This analysis assumes the hydraulic conductivity of the unsaturated soil to be an exponential function of the matric potential. When the initial hydraulic conductivity of the unsaturated soil is negligible in comparison with the saturated value, the steady infiltration rate Q is given by

$$Q = \pi R^2 K + \frac{4RK}{\alpha} \quad (41)$$

where α can be defined in terms of the sorptivity S by

$$\alpha = \frac{2K(\theta_0 - \theta_i)}{S^2} \quad (42)$$

so that K is given by

$$K = \frac{Q}{\pi R^2} - \frac{2S^2}{(\theta_0 - \theta_i)R} \quad (43)$$

The sorptivity S in Eq. 43 can be obtained from the infiltration at the beginning of the test from

$$S = \frac{dI}{d(\sqrt{t})} \quad \text{for} \quad t = 0 \tag{44}$$

As noted earlier, the infiltrometer method can give results that can be analyzed after only a short time of infiltration allowing hydraulic conductivity values to be measured near the soil surface. It thus provides a means to monitor structural changes of the soil. The method is very sensitive to worm and root holes as well as to structural fissures [6,66], and care must be taken to use rings large enough to sample a representative area.

The Bouwer double-ring method [69] is an infiltration method performed at the bottom of an auger hole. The rates of flow in a central ring and in a peripheral ring are measured when the heads feeding the water in each section are maintained at the same height, and also when no water is fed to maintain the head of the central ring, causing this head to fall. A lateral flow of water is thus induced between the inner and outer rings. The hydraulic conductivity is obtained from sets of graphs that have been obtained with an electrical analog. The method is sensitive to the hydraulic conductivity of the soil in the vicinity of the inner ring where soil disturbance is likely to occur during the installation of the rings, and thus results may not give the soil's undisturbed hydraulic conductivity.

VI. SUMMARY AND DISCUSSION

In making measurements of hydraulic conductivity of saturated soils and also in using measured values in calculations of the flow of groundwater, it is assumed implicitly that Darcy's law describes the flow of water in the vicinity of the measurement and in the flow region as a whole. This implies that the soil is "uniform" and that the same "uniformity" is "seen" by the measurement technique as is relevant for the calculation of the flow in the soil region under the given boundary conditions. A uniform soil is considered to be one that occupies enough volume to include several representative elementary volumes repeated throughout the space, and so in making a measurement of hydraulic conductivity the volume of soil involved must be large enough to contain several representative elementary volumes. When a system of inter-ped macropores is superimposed on the micropore space of the inter-particle channels, a larger representative elementary volume applies for flows through soil volumes containing macropores than for flows at the smaller scale inside the peds. Even larger representative elementary volumes might apply to flows at the field or regional scale. Thus methods of measuring hydraulic conductivity must ensure that techniques use flow regions of a volume relevant for the purpose for which measurements are required.

The concept of a representative elementary volume assumes a stochastic heterogeneity at some scale. By considering processes in volumes much larger than this, one can assume the soil to be uniform and apply continuum physics such as that embodied in Darcy's law. However, soil properties often change abruptly or vary as a trend. This deterministic heterogeneity requires the hydraulic conductivity to be known as a function of position, so that continuum physics can be applied taking into account this known variability. Techniques must allow this functional relationship, if present, to be measured.

When there is soil variability in the form of either stochastic or deterministic heterogeneity, methods of measuring hydraulic conductivity give the hydraulic conductance of the flow region between boundaries and for conditions imposed by the method employed. Values of "hydraulic conductivity" are usually calculated assuming that Laplace's equation in saturated soil, or Richards' equation in unsaturated soil, describes the hydraulic head as if the soil were uniform, and so these values are in fact equivalent values for the flow generated by imposing the boundary conditions of the method. The relevance of such values depends on how closely the flow patterns in the measurement technique conform to that of the flow in the region for which predictions are required. Values of "hydraulic conductivity" can be obtained from a set of experimental results from any method. Without replication using different geometries and scales, such values should not be regarded as meaningful for general predictive purposes.

For measurements of hydraulic conductivity in the laboratory, samples have to be contained within permeameters. Granular materials under study require to be packed uniformly, since hydraulic conductivity measurements accentuate any variability that occurs if the packing is nonuniform [11]. Undisturbed soil cores from the field for tests have to be collected carefully [12-17] to avoid smearing, which might seal off conducting channels, and also to avoid creating in the sample any cracks that would act as preferential channels for flow. Samples need to be large enough to contain a representative volume of any macropore structure present in the soil.

The standard hydraulic conductivity measurement on samples in the laboratory usually repeats Darcy's experiment [1], using a constant head to produce flow or making measurements of flow with a falling head. The oscillating permeameter [20] is a modification of the falling head permeameter, requiring a very small volume of water and so is useful for surface-active materials.

Besides these laboratory methods that use completely saturated samples, other methods wet up samples of dry soil from saturated surfaces, relying on infiltration theory to give hydraulic conductivity values of the saturated soils. Thus hydraulic conductivity values can be obtained by analyzing vertical one-dimensional infiltration [3,21] down columns of soil materials, although this method is not without its difficulties. Some situations characterized by a decrease of hydraulic conductivity with depth give results that appear to imply a uniform hydraulic conductivity down the column [22-24], so care is needed. Another method of utilizing water uptake by unsaturated soils is the varying moment permeameter [25], which can be particularly useful if only a short column of soil is available.

With the difficulties involved in obtaining an undistorbed representative sample of soil for hydraulic conductivity measurements, an in situ measurement in the field will in general be preferable for use in obtaining values for groundwater flow calculations. The most commonly used field methods involve sinking lined or unlined wells below the water table and observing the water flowing to the wells when the water level in the well is perturbed from the equilibrium level. The auger-hole method is the simplest and the most widely used of the methods, requiring little equipment [30-35], while the use of the piezometer method [36,37] is also common. The latter method is useful in the measurement of the variation of hydraulic conductivity with depth, for which purpose wells can be sunk successively in steps to greater depths, or horizontal piezometers can be inserted at different depths from inspection pits [38]. The latter technique allows measurements to be made from season to season to monitor long-term changes. The two-well method [26,39,40] and other multiple-well methods [42-44] are more time-consuming but sample much larger volumes, giving very reliable results in field surveys for land drainage [70]. The method of measuring the drawdown of a piezometric head around a pumped well [45] is used mainly in the measurement of aquifer transmissivities.

With the auger-hole and well methods, the induced flow is mainly horizontal, so that measurements reflect the horizontal component of hydraulic conductivity in an anisotropic soil. In contrast, the flow with piezometers with small cavity lengths is more vertical. The use of the two methods in conjunction allows values of horizontal and vertical components to be obtained. Childs et al. [40] used the two-well method [39] and Kirkham's piezometer method [36] to obtain the horizontal and vertical components of hydraulic conductivity in anisotropic soils, while Topp and Sattlecker [71] used the same well for auger-hole and piezometer tests, employing an inflatable gasket to turn the auger hole into a piezometer with an impermeable side wall.

For measurements of hydraulic conductivity on larger soil samples, use can be made of any system of sources and sinks that sample a larger flow region. For example, land drains provide an excellent means of obtaining the hydraulic conductivity of soils at a large scale, provided the appropriate drainage equation is known [47-52].

Measurements of the hydraulic conductivity of saturated soils in the absence of a water table rely on the interpretation of infiltration theory. These methods include the borehole permeameter [55-63], the auger-hole method [10], the air-entry method [64,65], and the ring infiltrometer method [54,66-68]. Many of the formulas used for calculating the hydraulic conductivity from experimental data with some of these methods are founded on approximate theory, preventing the easy assessment of the accuracy of values. Thus we cannot attribute the lack of agreement between different methods of measuring hydraulic conductivity, such as that found by Lee et al. [72] with greater values given by the air permeameter method than with the borehole permeameter, to the use of approximate formulas for calculating the hydraulic conductivity values or to some physical differences in the flow behavior (e.g., because of air entrapment). If air entrapment occurs during the wetting-up process [54], values obtained by methods using infiltration theory in the absence of a water table may be irrelevant for groundwater studies when the water table rises.

Measurements of hydraulic conductivity are required usually for analyzing groundwater flow situations: for example, the drainage of lands for agricultural and engineering purposes, the pumping of water from aquifers for water supplies, and the seepage loss from canals and storage reservoirs. They are also used to provide an indirect measure of the soil structure. Because of the difficulty of obtaining large undistorted samples of soils on which to make laboratory measurements, generally in situ field methods are preferable to laboratory methods for obtaining values for these field purposes. Laboratory measurements are mainly used in special research investigations concerning flow behavior in soils. In all cases, in selecting the method and considering the size of sample, attention has to be paid to any natural macropore development [73] and consideration has to be given to the possibility of heterogeneity, both stochastic at some scale or deterministic, to ensure that the value of hydraulic conductivity obtained in the measurement is as relevant as possible for the intended purpose.

Generally, the measurement of hydraulic conductivity is certainly not an easy task. However, the flow of water through porous materials in general, including that through soils in particular, is central to plant life and animal survival, and a quantitative description of the water flow by means of hydraulic conductivity values is a prerequisite to any basic study of life sciences. It is therefore essential that there be a proper appreciation of the concept of hydraulic conductivity and a good understanding of methods of making meaningful measurements of this basic transport property of soils and other porous materials.

ACKNOWLEDGMENT

The author thanks his colleague Guy Towner for helpful comments on the first draft of this chapter.

REFERENCES

1. Darcy, H., *Les Fontaines Publiques de la Ville de Dijon*, Dalmont, Paris, 1856.
2. Muskat, M., *The Flow of Homogeneous Fluids Through Porous Media*, McGraw-Hill, New York, 1937.
3. Childs, E. C., *An Introduction to the Physical Basis of Soil Water Phenomena*, Wiley, London, 1969.
4. Slichter, C. S., *Theoretical Investigation of the Motion of Ground Waters*, U.S. Geol. Surv. 19th Annu. Rep., Part 2, 1899, pp. 295-384.
5. Maasland, M., Soil anisotropy and land drainage, in *Drainage of Agricultural Lands* (J. N. Luthin, ed.), Am. Soc. Agron., Madison, WI, 1957, pp. 216-285.
6. Youngs, E. G., Soil physical theory and heterogeneity, *Agric. Water Manage.*, 6: 145-159 (1983).
7. Youngs, E. G., The analysis of groundwater flows in unconfined aquifers with nonuniform hydraulic conductivity, *Transport Porous Media*, 1: 399-417 (1986).
8. Reeve, R. C., and J. N. Luthin, Drainage investigation methods: I. Methods of measuring soil permeability, in *Drainage of Agricultural Lands* (J. N. Luthin, ed.), Am. Soc. Agron., Madison, WI, 1957, pp. 395-445.
9. Bouwer, H., and R. D. Jackson, Determining soil properties, in *Drainage for Agriculture* (J. van Schilfgaarde, ed.), Am. Soc. Agron., Madison, WI, 1974, pp. 611-672.
10. Kessler, J., and R. J. Oosterbaan, Determining hydraulic conductivity of soils, in *Drainage Principles and Applications: III. Surveys and Investigations*, Publ. No. 16, ILRI, Wageningen, The Netherlands, 1974, pp. 253-296.
11. Youngs, E. G., and S. M. Marei, The influence of air access on the water movement down soil profiles with impeding layers, in *Proc. Int. Conf. Infiltration Development and Application* (Y.-S. Fok, ed.), Honolulu, 1987, pp. 50-58.
12. Leeds-Harrison, P. B., and C. J. P. Shipway, Variations in hydraulic conductivity under different wetting regimes, in

Proc. ISSS Symp. Water and Solute Movement in Heavy Clay Soils (J. Bouma and P. A. C. Raats, eds.), Publ. No. 37, ILRA, Wageningen, The Netherlands, 1985, pp. 67-70.

13. Leeds-Harrison, P. B., C. J. P. Shipway, N. J. Jarvis, and E. G. Youngs, The influence of soil macroporosity on water retention, transmission and drainage in a clay soil, *Soil Use Manage.*, *2*: 47-50 (1986).

14. Bedford, R. K., Collection and evaluation of large soil monoliths for soil and crop studies, *J. Soil Sci.*, *30*: 363-373 (1979).

15. Bouma, J., *Soil Survey and the Study of Water in Unsaturated Soil*, Soil Survey Papers, No. 13, Netherlands Soil Survey Inst., Wageningen, 1977, pp. 1-107.

16. Bouma, J., and L. W. Dekker, A method for measuring the vertical and horizontal K_{sat} of clay soil with macropores, *Soil Sci. Soc. Am. J.*, *45*: 662-663 (1981).

17. Bouma, J., J. H. van Hoorn, and G. H. Stoffelsen, Measuring the hydraulic conductivity of soil adjacent to tile drains in a heavy clay soil in the Netherlands, *J. Hydrol.*, *50*: 371-381 (1982).

18. McNeal, L., and C. Roland, Elimination of boundary-flow errors in laboratory hydraulic conductivity measurements, *Soil Sci. Soc. Am. Proc.*, *28*: 713-714 (1964).

19. Youngs, E. G., The measurement of the variation with depth of the hydraulic conductivity of saturated soil monoliths, *J. Soil Sci.*, *33*: 3-12 (1982).

20. Childs, E. C., and A. Poulovassilis, An oscillating permeameter, *Soil Sci.*, *90*: 326-328 (1960).

21. Green, G. H., and G. A. Ampt, Studies on soil physics. 1. Flow of air and water through soils, *J. Agric. Sci., Camb.*, *4*: 1-24 (1911).

22. Childs, E. C., Soil moisture theory, *Adv. Hydrosci.*, *4*: 73-117 (1967).

23. Childs, E. C., and M. Bybordi, The vertical movement of water in stratified porous material. 1. Infiltration, *Water Resour. Res.*, *5*: 446-459 (1969).

24. Youngs, E. G., Soil physics and the water management of spatially variable soils, in *Proc. FAO/IAEA Symp. Isotope and Radiation Techniques in Soil Physics and Irrigation Studies*, Aix-en-Provence, 1983, pp. 3-22.

25. Youngs, E. G., An estimation of sorptivity for infiltration studies from moisture moment considerations, *Soil Sci.*, *106*: 157-163 (1968).

26. Childs, E. C., A. H. Cole, and D. H. Edwards, The measurement of the hydraulic permeability of saturated soil in situ. II, *Proc. R. Soc. London*, *A216*: 72-89 (1953).

27. Smiles, D. E., and E. G. Youngs, Hydraulic conductivity determinations by several field methods in a sand tank, *Soil Sci.*, *99*: 83-87 (1965).

28. Hoffman, G. J., and G. O. Schwab, Tile spacing prediction based on drain outflow, *Trans. Am. Soc. Agric. Eng.*, *7*: 444-447 (1964).

29. Youngs, E. G., Determination of the variation of hydraulic conductivity with depth in drained lands and the design of drainage installations, *Agric. Water Manage.*, *1*: 57-66 (1976).

30. Diserens, E., Beitrag zur Bestimmung der Durchlässigkeit des Bodens in natürlicher Bodenlagerung, *Schweiz. Landw. Monatsh.*, *12*: 188-198, 204-212 (1934).

31. Hooghoudt, S. B., Bijdragen tot de kennis van eenige natuurkunige grootheden van den grond: 4, *Versl. Landb. Ond.*, *42*(13)B: 449-541 (1936).

32. Ernst, L. F., *Een Niewe Formule voor de Berekening van de Doorlaatfactor Met de Boorgatenmethods*, Rap. Landbouwproefsta. Bodemkundig Inst. T.N.O., Groningen, The Netherlands (mimeo), 1950.

33. Kirkham, D., and C. H. M. van Bavel, Theory of seepage into auger holes, *Soil Sci. Soc. Am. Proc.*, *13*: 75-82 (1949).

34. Boast, C. W., and D. Kirkham, Auger hole seepage theory, *Soil Sci. Soc. Am. Proc.*, *35*: 365-374 (1971).

35. Boast, C. W., and R. G. Langebartel, Shape factors for seepage into pits, *Soil Sci. Soc. Am. J.*, *48*: 10-15 (1984).

36. Frevert, R. K., and D. Kirkham, A field method for measuring the permeability of soil below a water table, *Proc. Highw. Res. Board*, *28*: 433-442 (1948).

37. Youngs, E. G., Shape factors for Kirkham's piezometer method for determining the hydraulic conductivity of soil in situ for soils overlying an impermeable floor or infinitely permeable stratum, *Soil Sci.*, *106*: 235-237 (1968).

38. Goss, M. J., and E. G. Youngs, The use of horizontal piezometers for in situ measurements of hydraulic conductivity below the water table, *J. Soil Sci.*, *34*: 659-664 (1983).

39. Childs, E. C., The measurement of the hydraulic permeability of saturated soil in situ. I. Principles of a proposed method, *Proc. R. Soc. London*, A215: 525-535 (1952).

40. Childs, E. C., N. Collis-George, and J. W. Holmes, Permeability measurements in the field as an assessment of anisotropy and structure development, *J. Soil Sci.*, *8*: 27-41 (1957).

41. Youngs, E. G., The analysis of groundwater seepage in heterogeneous aquifers, *Hydrol. Sci. Bull.*, *25*: 155-165 (1980).

42. Smiles, D. E., and E. G. Youngs, A multiple-well method for determining the hydraulic conductivity of a soil in situ, *J. Hydrol.*, *1*: 279-287 (1963).

43. Kirkham, D., Measurement of the hydraulic conductivity of soil in place, in *Proc. Symp. Permeability of Soils*, ASTM Spec. Tech. Publ. No. 163, 1955, pp. 80-97.

44. Snell, A. W., and J. van Schilfgaarde, Four-well method of measuring hydraulic conductivity in saturated soils, *Trans. Am. Soc. Agric. Eng.*, 7: 83-87, 91 (1964).

45. Theis, C. V., The relation between the lowering of the piezometric surface and the rate of duration of discharge of a well using groundwater storage, *Trans. Am. Geophys. Union*, 519-524 (1935).

46. Abramowitz, M., and L. A. Stegun, *Handbook of Mathematical Functions, Applied Mathematics Series No. 55*, National Bureau of Standards, Washington, DC, 1972.

47. van Schilfgaarde, J., F. Engelund, D. Kirkham, D. F. Peterson, and M. Maasland, Theory of land drainage, in *Drainage of Agricultural Lands* (J. N. Luthin, ed.), Am. Soc. Agron., Madison, WI, 1957, pp. 79-285.

48. Youngs, E. G., The contribution of physics to land drainage, *J. Soil Sci.*, *34*: 1-21 (1983).

49. Lovell, C. J., and E. G. Youngs, A comparison of steady-state land-drainage equations, *Agric. Water Manage.*, *9*: 1-21 (1984).

50. Youngs, E. G., A simple drainage equation for predicting water-table drawdowns, *J. Agric. Eng. Res.*, *31*: 321-328 (1985).

51. Youngs, E. G., Characterization of hydrograph recessions of land drains, *J. Hydrol.*, *82*: 17-25 (1985).

52. Youngs, E. G., Horizontal seepage through unconfined aquifers with hydraulic conductivity varying with depth, *J. Hydrol.*, *3*: 283-296 (1965).

53. Philip, J. R., The theory of infiltration: 4. Sorptivity and algebraic infiltration equations, *Soil Sci.*, *84*: 257-264 (1957).

54. Youngs, E. G., Two- and three-dimensional infiltration: Seepage from irrigation channels and infiltrometer rings, *J. Hydrol.*, *15*: 301-315 (1972).

55. Talsma, T., and P. M. Hallam, Hydraulic conductivity measurement of forest catchments, *Aust. J. Soil Res.*, *18*: 139-148 (1980).

56. Reynolds, W. D., D. E. Elrick, and G. C. Topp, A reexamination of the constant head well permeameter method for measuring saturated hydraulic conductivity above the water table, *Soil Sci.*, *136*: 250-268 (1983).

57. Nash, D. M., S. T. Willatt, and N. C. Uren, The Talsma-Hallam well permeameter—Modifications, *Aust. J. Soil Res.*, *24*: 317-320 (1986).

58. Stephens, D. B., and S. P. Neuman, Vadose zone permeability tests, *J. Hydrol. Div. ASCE*, *108* [(HY5): Proc. Paper 17058]: 623-639 (1982).

59. Glover, R. E., Flow from a test-hole located above groundwater level, appendix in C. N. Zangar, *Theory and Problems of Water Percolation*, Eng. Monogr. No. 8, U.S. Bur. Reclam., Washington, DC, 1953, pp. 69-71.

60. Jones, C. W., *Comparison of Seepage Based on Well Permeameter and Ponding Tests*, Earth Mater. Lab. Rep. No. EM-264, U.S. Bur. Reclam., Denver, CO, 1951.

61. Reynolds, W. D., and D. E. Elrick, In situ measurement of field-saturated hydraulic conductivity, sorptivity, and the α-parameter using the Guelph permeameter, *Soil Sci.*, *140*: 292-302 (1985).

62. Reynolds, W. D., D. E. Elrick, and B. E. Clothier, The constant head well permeameter: effect of unsaturated flow, *Soil Sci.*, *139*: 172-180 (1985).

63. Philip, J. R., Approximate analysis of the borehole permeameter in unsaturated soil, *Water Resour. Res.*, *21*: 1025-1033 (1985).

64. Bouwer, H., Rapid field measurement of air-entry value and hydraulic conductivity of soil as significant parameters in flow system analysis, *Water Resour. Res.*, *2*: 729-723 (1966).

65. Topp, G. C., and M. R. Binns, Field measurement of hydraulic conductivity with a modified air-entry permeameter, *Can. J. Soil Sci.*, *56*: 139-147 (1976).

66. Youngs, E. G., Estimating hydraulic conductivity values from ring infiltrometer measurements, *J. Soil Sci.*, *38*: 623-632 (1987).

67. Miller, E. E., and Miller, R. D., Physical theory for capillary flow phenomena, *J. Appl. Phys.*, *27*: 324-332 (1956).

68. Wooding, R. A., Steady infiltration from shallow circular ponds, *Water Resour. Res.*, *4*: 1259-1273 (1968).

69. Bouwer, H., A double-tube method for measuring hydraulic conductivity of soil in situ above a water table, *Soil Sci. Soc. Am. Proc.*, *25*: 334-339 (1961).

70. Towner, G. D., and E. G. Youngs, Application of drainage theory in the field, *Soil Use Manage.*, *2*: 44-47 (1986).

71. Topp, G. C., and S. Sattlecker, A rapid measurement of horizontal and vertical components of saturated hydraulic conductivity, *Can. Agric. Eng.*, *25*: 193-197 (1983).

72. Lee, D. M., W. D. Reynolds, D. E. Elrick, and B. E. Clothier, A comparison of three field methods for measuring saturated hydraulic conductivity, *Can. J. Soil Sci.*, *65*: 563-573 (1985).

73. Bouma, J., Use of soil survey data to select measurement techniques for hydraulic conductivity, *Agric. Water Manage.*, *6*: 177-190 (1983).

5

Unsaturated Hydraulic Conductivity

CHRISTIAAN DIRKSEN *Wageningen Agricultural University, Wageningen, The Netherlands*

I. INTRODUCTION

The unsaturated zone plays an important role in the hydrological cycle. It forms the link between surface water and groundwater and has a dominant influence on the partition of water between them. The hydraulic properties of the unsaturated zone determine how much of the water that arrives at the soil surface will infiltrate into the soil and how much will flow off overland causing floods, erosion, etc. In many areas of the world, most of the water that infiltrates into the ground is transpired by plants or evaporated directly into the atmosphere, leaving only a little water to percolate deeper and join the groundwater. Surface runoff and deep percolation, moreover, may carry pollutants. Thus it is important to know how long it will take for this water to reach surface or groundwater reserves.

Besides providing water for plants to transpire, the unsaturated zone provides oxygen and nutrients to plant roots, thus having a dominant influence on the production of food, fiber, etc. Water content also determines soil strength, with many implications for such phenomena as anchoring of plants, root penetration, compaction by cattle and machinery, and tillage operations. To mention just one other role of the unsaturated zone, its water content has a great influence on the heat balance at the soil surface. This is well illustrated by the large diurnal temperature variations in deserts.

To understand and describe these and other processes, the hydraulic properties that govern water transport in the soil must be quantified. Of these, the unsaturated hydraulic conductivity is, if not the most important, certainly the most difficult to measure accurately. It varies over many orders of magnitude not only between different soils, but also for the same soil as a function of water content. Much has been published on the determination and/or measurement of the unsaturated hydraulic conductivity, including good reviews [1-7]. There is no single method that is suitable for all soils and circumstances. Methods that require taking "undisturbed" samples are not well suited for soils with many stones or with a highly developed, loose structure. It is better to select an in situ method for such soils. Hydraulic conductivity for relatively dry conditions cannot be measured in situ when the soil in its natural situation is always wet. It is then necessary to take samples and dry them first. The latter process presents problems if the soil shrinks excessively on drying. These and other factors that influence the choice between laboratory and field methods are discussed separately in Section IV.

Selection of the most suitable method for a given set of conditions is a major task. The literature is so exhaustive that it is neither necessary nor possible to give a complete review and evaluation of all available methods. Instead, I have focused on what I think should be the selection criteria (Section III) and have described the most familiar types of method (Sections VI-IX) with these criteria in mind. This includes some very recent work. The need for and selection of a standard method is discussed separately (Section V).

There are two soil water transport functions which, under restricting conditions, can be used instead of hydraulic conductivity, namely hydraulic diffusivity and matric flux potential. Diffusivity can be measured directly in a number of ways that are easier and faster than the methods available for hydraulic conductivity. Moreover, the latter can also be derived from the former. The same is true for yet another transport function, the sorptivity, which can also be measured more easily than the hydraulic conductivity. At the outset I have summarized the theory and transport coefficients used to describe water transport in the unsaturated zone (Section II). Theoretical concepts and equations associated with specific methods are given with the discussion of the individual methods. Readers who have little knowledge of the physical principles involved in unsaturated flow and its measurement can find these discussed at a more detailed and elementary level in soil physics textbooks [8-10] and would be advised to consult one of these before attempting this chapter.

Apparatus for determining unsaturated hydraulic conductivity is not usually commercially available as such. However, many of the methods involve the measurement of water content, hydraulic head, and/or the soil water characteristic, and methods and commercial supplies of equipment to determine these properties are given in Chaps. 1, 2, and 3, respectively. Requirements for specialized or specially constructed equipment are indicated in the discussions of individual methods.

In general, it is difficult if not impossible to measure the soil hydraulic transport functions quickly and/or accurately. Therefore, it is not surprising that attempts have been made to derive them indirectly. The derivation of the hydraulic transport properties from other, more easily measured soil properties is discussed in Section X and the inverse approach of parameter optimization in Section XI.

II. TRANSPORT COEFFICIENTS

A. Hydraulic Conductivity

In general, water transport in soil occurs as a result of gradients in the hydraulic potential [10]

$$H = h + z \tag{1}$$

where H is hydraulic head, h is pressure head, and z is gravitational head or height above a reference level. These symbols are generally reserved for potentials expressed on a weight basis, having the dimension $J \ N^{-1} = m$. Although h is called a pressure head, in unsaturated flow it will have a negative value with respect to atmospheric pressure and can be referred to as a suction or tension. In rigid soils there exists a relationship between water content [usually expressed as volume fraction, θ ($m^3 \ m^{-3}$)] and pressure head, called the soil water retention characteristic, $\theta[h]$ (see Chap. 3)]. Here, as well as throughout this chapter, square brackets are used to indicate that a variable is a function of the quantity within the brackets. The function $\theta[h]$ often depends on the history of wetting and drying; this phenomenon is called hysteresis. Water transport in soils obeys Darcy's law, which for one-dimensional, vertical flow in the z-direction, positive upward, can be written as

$$q = -k[\theta] \ \frac{dH}{dz} = -k[\theta] \ \frac{dh}{dz} - k[\theta] \tag{2}$$

where q is water flux density ($m^3 \ m^{-2} \ s = m \ s^{-1}$) and $k[\theta]$ is the hydraulic conductivity function ($m \ s^{-1}$); k is in the first place a function of θ, $k[\theta]$, since water content determines the fraction of the sample cross-sectional areas available for water transport. Indirectly, k is also a function of pressure head. $k[h]$ is hysteretic to the extent that $\theta[h]$ is hysteretic. Hysteresis in $k[\theta]$ is of second order and is generally negligible. Determinations of k usually consist of measuring corresponding values of flux density and hydraulic potential gradient, and calculating k with Eq. 2. This is straightforward and can be considered as a standard for other, indirect measurements.

B. Hydraulic Diffusivity

For homogeneous soils in which hysteresis can be neglected or in which only monotonically wetting or drying flow processes are considered, h[θ] is a single-valued function. Then, for horizontal flow in the x-direction, or when gravity can be neglected, Eq. 2 yields

$$q = -D[\theta] \frac{d\theta}{dx} \quad \text{for} \quad D[\theta] = k[\theta]\left(\frac{dh}{d\theta}\right)[\theta] \tag{3}$$

where D[θ] is the hydraulic diffusivity function (m s^{-2}). Thus, under the above-stated conditions, the water content gradient can be thought of as the driving force for water transport, analogous to a diffusion process. Of course, the real driving force remains the pressure head gradient. Therefore, D[θ] is different for wetting and drying. There are many methods to determine D[θ], some of which are described later. They usually require a special theoretical framework with simplifying assumptions. Once D[θ] and h[θ] are known, the hydraulic conductivity function can be calculated according to:

$$k[\theta] = D[\theta]\left(\frac{d\theta}{dh}\right)[\theta] \tag{4}$$

Because of hysteresis, one should combine only diffusivities and derivatives of the soil water retention characteristic that are both obtained either by wetting or by drying. Since k[θ] is basically non-hysteretic, the k[θ] functions obtained in the two ways should agree closely.

C. Matric Flux Potential

Water transport in soils in response to pressure (matric) potential gradients can also be described in terms of the matric flux potential [11,12]

$$\Phi = \int_{-\infty}^{h} k[h] \ dh = \int_{0}^{\theta} D[\theta] \ d\theta \tag{5}$$

Equation 3 then becomes

$$q = - \frac{d\Phi}{dz} \tag{6}$$

The matric flux potential integrates the transport coefficient and the driving force; it has the dimension square meters per seconds. In homogeneous soil without hysteresis, the horizontal water flow density is simply equal to the gradient of Φ. This formulation of the water transport process offers distinct advantages in certain situations, especially in the simulation of water transport under steep potential gradients [12-14]. It also allows one to obtain analytical solutions for steady-state, multidimensional flow problems, including gravity, when the hydraulic conductivity is expressed as an exponential function of pressure head [15,16]. Like k and D, Φ is a soil property that characterizes unsaturated water transport and is a direct function of θ and only indirectly of h. A method for measuring Φ directly [13] is described in Section VI.D.

D. Sorptivity

Sorptivity is an integral soil water property that contains information on the soil hydraulic properties $k[\theta]$ and $D[\theta]$, which can be derived from it mathematically. Generally, sorptivities can be measured more accurately and/or more easily than $k[\theta]$ and $D[\theta]$, so it is worth considering whether to determine the latter in this indirect way [17, 18]. One-dimensional absorption (gravity negligible), initiated at time t = 0 by a step-function increase of water content from θ_0 to θ_1 at the soil surface, x = 0, is described [17,19] by

$$i = S[\theta_1, \theta_0]\sqrt{t} \tag{7}$$

where i is cumulative absorbed amount (m) at any given time t, and sorptivity S (m $s^{-1/2}$) is a soil property that depends on the initial and final water content, usually saturation. Saturated sorptivity characterizes ponding infiltration at small times, as it is the first term in the infiltration equation of Philip [19] and equal to the amount of water absorbed during the first time unit. With the flux-controlled sorptivity method [17], the dependence of S on θ_1 at constant θ_0 is determined experimentally. From this $D[\theta]$ can be derived algebraically (see Eq. 27, below). The $\sqrt{t}$ relationship of Eq. 7 has also been used for scaling soils and estimating hydraulic conductivity [20] and diffusivity [21] of similar soils (Section X.B).

III. SELECTION FRAMEWORK

A. Types of Method

Many methods have been reported in the literature for the determination of soil water transport properties. There is no single method best suited to all circumstances. Therefore, it is necessary to select the method most suited to any given situation, and time spent on this selection is well used. Table 1 lists various methods that have been proposed and evaluates them according to the five gradations of the selection criteria listed in Table 2. These tables form the nucleus of this chapter. In subsequent sections the methods are reviewed in varying detail. In general, the theoretical framework and/or main working equations are described and other pertinent information is added to help substantiate the scores given for the criteria in Table 1. For the more familiar methods, mostly only evaluating remarks are made; some experimental details are given also for the less familiar and newer methods. The scores are a reflection of my own insight and experience and are not (and cannot be) based solely on the information provided. For additional information, the reader is advised to consult the listed references.

A major division is made between steady-state and transient measurements. In the first category, all parameters are constant in time. For this reason, steady-state measurementa are almost always more accurate than transient measurements, usually even with less sophisticated equipment. Their main disadvantage is that they take much more time, often a prohibitive amount. Therefore, the choice between these two categories usually involves balancing necessary costs, available time, and required accuracy. The methods are divided further into field and laboratory methods, the choice of which is discussed in Section IV. Methods for measuring soil water transport coefficients can also be divided into those that measure hydraulic conductivity directly and all other methods (column A of Table 1). From what follows it should become clear that one should measure hydraulic conductivity whenever possible. The distinction made between wetting and drying flow regimes (B) is important because the hysteretic character of soil water retention may affect any application that calls for hydraulic diffusivity or hydraulic conductivity as a function of pressure head.

B. Selection Criteria

The methods listed in Table 1 are evaluated on the basis of the criteria in Table 2 which include the following: the degree of exactness of the theoretical basis (C), the experimental control of the required initial and boundary conditions (D), the inherent accuracy of the measurements (E), the propagation of errors in the experimental data during the calculation of the final results (F); the range of

TABLE 1 Evaluation of Methods to Measure Soil Water Transport Properties According to Criteria and Gradations in Table 2

| | | | | | | Steady-state methods: criteria | | | | | | |
A	B	C	D	E	F	G	H	I	J	K	L	M
Laboratory methods												
Head-head	k	w/d	5	5	3(5)	5	3(4)	2(1)	3(2)	3(2)	4	4	4
Flux-head (infiltration)	k	w	5	5	3	5	3	2	3	3	4	4	4
Head-flux (evaporation)	k	d	3	3	3	3	3	2	3	3	5	5	4
Regulated evaporation	k	d	2	2	3	3	2	2	3	3	4	2	4
Long column infiltration	k	w	4	4	4	5	4	1	3	3	5	4	4
Matric flux potential	Φ	d	3	3	5	3	3	3	3	4	5	5	4
Field methods												
Sprinkling infiltrometer	k	w	5	4	2	5	3(4)	2(1)	1	2	1	1	3
Isolated column (crust)	k	w	4	3	3	2	2	3	3	3	2	2	3
Spherical cavity	k	w	4	3	3	3	4	2	4	2	3	4	3
Ponded disk/dripper	k	w	2	3	4	3	3	4	4	4	5	4	4

TABLE 1 (continued)

Transient methods: criteria

	A	B	C	D	E	F	G	H	I	J	K	L	M
Laboratory methods													
Instantaneous profile	k	d	5	5	2	2	3	2	2	2	2	2	2
Pressure plate outflow	D	d	2	4	5	3	2	2	3	4	3	4	3
One-step outflow	D	d	2	4	5	3	2	3	3	4	3	4	3
Boltzmann, fixed time	D	w	4	5	2	1	5	4	3	4	5	3	3
Boltzmann, fixed position	D	w	4	5	2	1	5	5	1	2	4	2	2
Hot air	D	d	4	1	2	1	5	4	4	4	4	3	2
Flux-controlled sorp-tivity	D	w	4	4	4	3	5	4	3(1)	4	3	2	4
Field methods													
Instantaneous profile	k	d	5	4	2(4)	2	3	2	2	2	2	2	2
Unit gradient, prescribed	D	d	2	3	2	2	3	3	4	2	2	4	2
Unit gradient, simple	k/D	d	1	1	4	2	3	2	4	3	3	4	2
Sprinkling infiltrometer	k	w	4	3	2	2	3	2	1	1	1	1	2

TABLE 2   Selection Criteria and Gradations for Methods to Measure Soil Water Transport Properties

---

A. Parameter measured

   k Hydraulic conductivity

   D Hydraulic diffusivity

   Φ Matric flux potential

B. Flow regime

   w Wetting

   d Drying

C. Theoretical basis

   5 Simple Darcy law or rigorously exact

   4 Exact, with minor simplifying assumptions

   3 Quasi-exact, with simplifying assumptions

   2 Major simplifying assumptions

   1 Minimal theoretical basis

D. Control of initial or boundary conditions

   5 Exact—no requirements

   4 Indirect and accurate

   3 Approximate

   2 Approximate part of the time

   1 Little control, if any

E. Accuracy of measurements

   5 Weight, (external) volume of water, and time

   4 Water contents, direct

   3 Pressure heads

   2 Indirect measurements and/or other sources of error

   1 Approximate measurements without calibration

F Error propagation in data analysis

   5 Simple quotient (Darcy law)

   4 Accurate algebraic operations with accurate data

   3 Inaccurate operations with accurate data

   2 Accurate algebraic operations with inaccurate data

   1 Inaccurate operations with inaccurate data

TABLE 2 (continued)

G. Range of application (pressure heads)

    5 Saturation to wilting point (0 to -160 m)

    4 Tensiometer range (0 to -8.5 m)

    3 Hydrological range (0 to -2.5 m)

    2 Dry range (-2.5 to -150 m)

    1 Wet range (0 to -0.5 m)

H. Duration of method

    5 1 hour

    4 1 day

    3 1 week

    2 1 month

    1 More than 1 month

I. Equipment

    5 Standard for soil laboratory

    4 General purpose, off the shelf

    3 Easily made in average machine shop

    2 Special purpose, off the shelf

    1 Special purpose, custom-made

J. Operator skill

    5 No special skill

    4 Some practice

    3 General measuring experience

    2 Special training of good experimentalist

    1 Highest degree of specialization

K. Operator time

    5 Simple and fast manipulations only at beginning and end

    4 Elaborate manipulations at beginning and/or end

    3 Simple and fast operations at regular time intervals

    2 Elaborate operations at regular time intervals

    1 Operator required during entire measuring period

L. Simultaneous measurements

    5 No limit

    4 Large number, at significant costs

TABLE 2 (continued)

---

3 Small number, at little costs

2 Small number, at substantial costs

1 No potential

M. Check on measurements in progress or afterward

5 Continuous monitoring of all parameters possible

4 Verification easy at any time

3 Each verification requires considerable effort

2 Single check is major effort

1 Check not possible

---

pressure heads over which the method can be used (G), the time (duration) required to obtain the particular transport coefficient function over the indicated pressure head range (H), the necessary investment in workshop time and/or money (I), the skill required by the operator (J), the operator time required while the measurements are in progress (K), the potential for measurements to be made simultaneously on many soil samples (L), and the possibility for checking during and/or after the measurements (M).

Depending on the particular situation, only a few or all of these criteria must be taken into account to make a proper choice. For example, accuracy will be a prime consideration for detailed studies of water transport processes at a particular site, whereas for a study of spatial variability, the ability to make a large number of measurements in a reasonably short time is mandatory. These measurements often do not have to be very accurate. If the absolute accuracy of a newly developed method must be established, the most accurate method already available should be selected, since there is no "standard" material with known properties available against which the method can be tested. The need for the selection of a "standard method," as alternative, is discussed separately (Section V). When facilities for routine measurements must be set up, the last four criteria of Table 2 are particularly pertinent. Finally, there may be particular (difficult) conditions under which one method is more suitable than others, and these conditions may dominate the choice of method. Such criteria are not covered by Table 1, but are mentioned with the description of individual methods when appropriate.

The five gradations used with the selection criteria (Table 2) are mostly self-explanatory and will become clearer with the discussion of the individual methods. At this stage only a few general remarks are made about accuracy (relating to criteria C-F) and the range of application (G) which, out of practical considerations, is associated with pressure heads. For examples, reference is made to methods that are described later in more detail.

C. Accuracy

Direct measurements of weight, volume of water, and time, made in connection with the determination of soil hydraulic properties, are simple and very accurate (maximum score 5). An exception is measuring very small volumes of water while maintaining a particular experimental setup, for example, a small hydraulic head gradient. Although the mass and water content of a soil sample usually can be measured accurately, the water content may not conform to the values corresponding to the theoretically assumed flow system. For example, for Boltzmann transform methods a water content profile must be determined after an exact time period of wetting or drying. It is not possible to do this instantaneously, and during sampling for gravimetric determinations, water contents will change due to redistribution and evaporation of water and due to manipulation of the soil. Indirect water content measurements can be made nondestructively and thus repeatedly during a flow process, but the accuracy of these measurements is normally not very good. Extensive calibration under identical conditions can improve the accuracy, but usually this is not possible or takes too much time.

Derivation of hydraulic properties from other measured parameters introduces two kinds of error. First, the theoretical basis of the method may not be exact, either because it involves simplifying assumptions or because the theoretical analysis of the water flow process yields only an approximation of the transport property. Second, errors in the primary experimental data are propagated in the calculations required to obtain the final results. Mathematical manipulations each have their own inherent inaccuracies, a good example being differentiation. Another common source of error is that the theoretically required initial and/or boundary conditions cannot be attained experimentally. For example, it is impossible to impose the step-function decrease of the hydraulic potential at the soil surface under isothermal conditions, as is assumed with the hot air method.

Hydraulic potential measurements are relatively difficult and can be very inaccurate. Water pressure inside tensiometers in equilibrium with the soil water around the porous cup can in principle be measured to any desired accuracy with pressure transducers, but temperature variations can render such measurements very inaccurate. Mercury manometers are probably the least sensitive to large errors, but their accuracy is limited to about ±2.5 cm (see Chap. 2).

In steady-state measurements near saturation, water manometers appear to be the most accurate. Beyond the tensiometer range, soil water potentials are mostly determined indirectly from soil water characteristics or by measuring the electrical conductivity, heat diffusivity, or other properties of probes in equilibrium with soil water, with all the inaccuracies associated with indirect measurements. Direct measurements can be made with psychrometers (which also measure the osmotic component of the soil water potential), but these can be used only by workers experienced with sophisticated equipment and are at best accurate to about ±500 cm. However, for many studies, such as that of the soil-water-plant-atmosphere continuum, such accuracies are acceptable, because hydraulic conductivities in this dry range are so low that hydraulic head gradients must be very large to obtain significant flux densities.

D. Range of Application

The range of application of a particular method depends to a large extent on whether and, if so, how soil water potentials are to be measured. Out of convenience and based on practical experience, therefore, the range of application is described in somewhat vague terms, which are identified further by approximate ranges of pressure head, even for methods in which only water contents or flux densities are measured. Tensiometers can theoretically be used down to pressure heads of about -8.5 m, but in practice air intrusion usually causes problems at much higher values. Fortunately, hydraulic transport properties need not be known in the drier range, except where water transport over small distances is concerned (e.g., evaporation at the soil surface, water transport to individual plant roots). Water transport over large distances occurs mostly in the saturated zone (or as surface water), for which the saturated hydraulic conductivity must be known. However, there are some exceptions, such as saline seeps caused by unsaturated water transport over large distances during many years. Although unsaturated water transport normally occurs over short distances, it plays a key role in hydrology, as mentioned in the introduction. The unsteady, mostly vertical water transport in soil profiles is only significant when the hydraulic conductivity is in the range from the maximum value at saturation down to values of about 0.1 mm per day, since precipitation, transpiration, and evaporation cannot generally be measured to that accuracy. This corresponds with a range in pressure head between 0 and -1.0 to -3.0 m, depending on the soil type.

The pressure head range over which hydraulic transport properties are wanted should be carefully considered; this should be a major consideration in the selection process. It makes no sense, for

instance, to determine hydraulic conductivities with the hot air
method (which yields very inaccurate results over the entire pressure head range) when the results will be used only in the
hydrological range, for which much better methods are available.
Conversely, it is dangerous to extrapolate to a dryer range the results of an attractive method suitable only in the wetter range. In
practice, the range of application of a particular method depends
also on the time required to attain appropriate measurement conditions. Criteria G and H are mutually dependent: the time needed
to measure the soil water property function often increases exponentially with increases in the pressure head range toward drier
conditions.

E. Alternative Approaches

Because measurements of the soil water transport properties leave
much to be desired in terms of accuracy, cost, applicability, and
time, it is not surprising that other ways to obtain these soil properties have been investigated. The most extreme of these approaches
is not to make any water transport measurements, but to derive the
water transport functions from other, more easily measured soil properties (e.g., particle size distribution or the soil water characteristic). These procedures are usually based on a theoretical model of
the relationship [5,6], but they can also be of a purely statistical
nature [22,23], in which case their application is limited to the range
of soils used to derive the relationship. An intermediate approach
is the so-called inverse approach, which has recently received renewed attention as the "parameter optimization technique" [7,24,25].
To be able to decide how the hydraulic transport functions can best
be determined in a given situation, the possibilities and limitations
of these alternative approaches should also be considered (Sections
X and XI).

IV. LABORATORY VERSUS FIELD METHODS

A. Working Conditions

A major division between available methods is that of laboratory versus field methods. Laboratory measurements have many advantages
over field measurements. In the laboratory all the usual facilities
(e.g., electricity, gas, water, vacuum) are available, and temperature variations are usually modest and can be controlled if necessary.
Standard equipment (e.g., balances and ovens) is also more readily
available than in the field. Expensive and delicate equipment often
cannot be used in the field because of weather conditions, theft,
vandalism, etc. One usually can save much time by working in the
laboratory—samples from many different locations can be collected

first, then measurements carried out consecutively or in series. Considering all these advantages, it would seem to be good practice to carry out measurements in the laboratory, unless there are overriding reasons to perform them in situ. For hydraulic conductivity measurements, such reasons normally exist only when the hydraulic properties of a strongly layered soil profile are needed as a whole, or when heterogeneity and instability of soil structure make it very difficult if not impossible to obtain large enough, undisturbed soil samples and transport them to the laboratory.

B. Sampling Techniques

Because the hydraulic conductivity of soil is very sensitive to changes in soil structure due to sampling and/or preparation procedures, these operations should be carried out with the utmost care. Fractures formed during sampling which are oriented in the direction of flow are disastrous for saturated hydraulic conductivity determinations but have very little influence on unsaturated hydraulic conductivities. Fractures perpendicular to the direction of flow have the very opposite effect on both types of measurement. Soil columns consisting of entire soil profiles can be obtained by driving a cylinder supplied with a sharp, hardened steel cutting edge into the soil with a hydraulic press. If the stroke of this press is smaller than the height of the sample, care should be taken to ensure that with each stroke the press is lined up in exactly the same way. We have been able to accomplish this easily and satisfactorily by pushing a sample holder hydraulically against a horizontal cross-bar anchored firmly by four widely spaced tie lines (Fig. 1). To reduce compaction of the soil inside the cylinder due to the friction between the cylinder wall and the soil, the diameter of the cylinder should be kept large and/or a sampling tool with a moving sleeve should be used [26]. Driving cylinders into the ground by repeated striking with a hammer should not be tolerated for quantitative work, not even for short samples, because of the lateral forces that are likely to be applied. A compromise between a hammer and a hydraulic press is a heavy metal cylinder that is dropped repeatedly onto a sampleholder while being constrained by a steady vertical rod attached to the sampleholder. For measurements of hydraulic conductivity of packed soil columns, it is essential that the packing be done systematically to attain the best possible reproducibility and uniformity. At the moment this appears to be more an art than a science.

C. Sample Representativeness

Other important aspects of soil sampling are the size and number of samples required to be representative in view of soil heterogeneity

FIG. 1 Hydraulic apparatus for obtaining short (left) and long
(right) "undisturbed" soil columns. The apparatus is stabilized by
a cross-bar and four widely anchored tie lines.

and spatial variability. The development and size of the natural
structural units (peds) dictate the size of the sample needed for a
particular measurement. If a soil property were measured repeatedly on samples of increasing size, the variance of the results normally would decrease until it reached a constant value, the variance of
the method alone. The smallest sample for which a constant variance
of a specific soil property is obtained is called the representative
elementary volume (REV) for that property [27]. Assuming that a
soil sample should contain at least 20 peds to be representative,
Verlinden and Bouma [28] estimated REVs for various combinations
of texture and structure. These varied from the commonly used
50 mm diameter (100 cm^3) samples to characterize the hydraulic
properties of field soils with little structure, to 10^5 cm^3 soil samples for heavy clays with very large peds or soils with strongly
developed layering. The desirable length of (homogeneous) soil
samples depends on the particular measurement method that is used.
     Considering the number of soil samples needed, Warrick and
Nielsen [29] listed the unsaturated hydraulic conductivity under the
category of soil properties with the highest coefficient of variation.
They reported that about 1300 independent samples from a normally
distributed population (field) were needed to estimate mean hydraulic
conductivity values with less than a 10% error at the 0.05 significance

level. The recently developed theory of regionalized variables or geostatistics [30] provides insight into the minimum number and spatial distribution of soil samples required to obtain results with a certain accuracy and probability. Of course, the same applies to the required number of locations of sites for in situ measurements.

V. STANDARD METHOD

A major problem associated with the determination of soil hydraulic transport properties is lack of unchanging, uniform soils or other porous materials with constant, known transport properties, which could serve as standard reference materials with which to establish the absolute accuracy of any method. It is impossible to pack granular material absolutely reproducibly, and consolidated porous materials (e.g., sandstone) are not suitable for most of the methods used on soil materials. Also, repeated wetting or drying of a soil sample to the same overall water content does not lead to the same water content distribution and hydraulic conductivity. Given these insuperable difficulties, hydraulic transport properties are almost always presented without any indication of their accuracy. Only the method used to determine them is described and sometimes, for good measure, a comparison between the results of two methods is given. Agreement between two methods is still not a guarantee that both are correct. Often the results of two methods are said to correspond well when in fact they differ by as much as an order of magnitude over part of the range. There is no way to decide which is the most accurate. The only recourse is to evaluate the available methods on their potential accuracy based on such criteria as theoretical exactness, inherent accuracy of the required measurements, possibility of experimentally attaining the theoretically required initial and boundary conditions, and error propagation in the required calculations. In this way, instead of a standard material with accurately known properties, a "standard reference method" would be chosen.

In searching for such a standard method, it should be realized that hydraulic conductivity is theoretically the most correct parameter for characterizing water transport in soils, since it is directly associated with the driving force for the movement of water, the hydraulic potential gradient. Moreover, it can be measured more directly and probably more accurately than any of the other parameters characterizing water transport, especially when measured during steady-state conditions. From this it follows that steady-state measurements of hydraulic conductivity in vertical soil columns between two porous plates, in which purely graviational flow (no

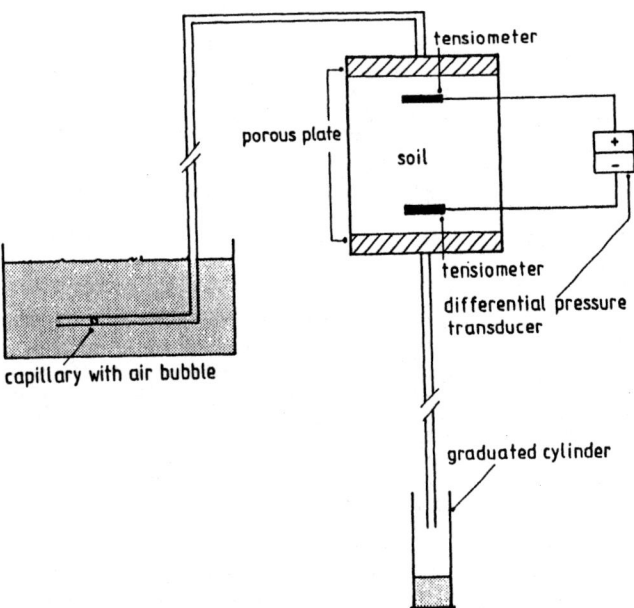

FIG. 2   The "standard reference method" (head-head).

pressure head gradient) is established (Fig. 2), approach most
closely the requirements for a "standard method."   Since the pressure head is everywhere the same, the water content and thus the
hydraulic conductivity are uniform throughout the column.   Therefore, there is no question (error) regarding the water content and/
or pressure head with which the obtained hydraulic conductivity
should be associated.   Because the contact resistances between the
soil column and the porous plates are often too large and unpredictable to permit reliance on measurement of the externally applied
hydraulic gradient, the hydraulic head gradient should be measured
within the soil column with accurate tensiometer equipment.   To as-
sign the status "standard" to this method, both the influx and out-
flux should be measured until they have become equal.   These
fluxes can be measured accurately down to very low values by observing the movement of air bubbles in thin glass capillaries.
    Once this experimental setup has been assembled, it can be
used at various pressure heads.   The range of pressure heads is
theoretically limited to that of tensiometers, approximately 0 to -8.5
m of water.   Another limitation of the two-plate method is the time
needed to reach a steady state.   This can become prohibitively long,
either due to practical considerations or because long-term effects

(e.g., microbial activity, loss of water through tubing walls) reduce
the overall accuracy to an unacceptable level. Therefore, the practical range probably does not extend much below a pressure head of
-3.0 m. This is sufficient for characterization of water transport
over relatively large distances. However, for analyses of water
transport to plant roots, and of evaporation near the soil surface,
etc., hydraulic conductivities for much lower pressure heads and
water contents are needed. These can be determined only with
other, usually indirect methods. Selection of a standard method
for this higher tension range does not yet seem to be possible.
For field measurements, steady infiltration over a large surface area
(with tensiometer measurements in the center) with a sprinkling infiltrometer approaches most closely the requirements for a "standard
method." Further comments about these methods follow in Section
VI.

VI.   STEADY-STATE LABORATORY METHODS

A.   Head-Controlled (Head-Head)

The method featured in most soil physics textbooks involves steady-
state measurements on a soil column in which the pressure head is
controlled at both ends (usually by two porous plates) such that it
is uniform over the entire length (Fig. 2). Principles, apparatus,
procedures, required calculations, and general comments are given
in great detail by Klute and Dirksen [3]. The preceding section
identified this method as most suitable for use as a "standard method." This is reflected in the maximum scores in Table 1 for theoretical basis (C), control of initial and boundary conditions (D), and
error propagation in data analysis (F). Tensiometric measurements
generally are tedious and error-prone, but they can be very accurate when made carefully with good equipment (this is indicated by
the additional score within parentheses in column E). Also, the
ease with which fluxes can be measured accurately decreases with
their magnitude. The installation of the tensiometers and the porous plates in good contact with the soil column may take considerable time. The time required to reach steady state at unit hydraulic
gradient (i.e., gravitational flow) increases rapidly with decreasing
hydraulic conductivity. Therefore, while theoretically the entire
tensiometer range can be covered, this method will in practice probably not be used at pressure heads below -2.0 to -3.0 m. If the
hydraulic conductivity is to be measured over an extensive range
of water contents (as is warranted when the method is used as a
standard to establish the accuracy of another method), the measurements will take much longer than 1 month (parentheses for criteria
G and H).

Near saturation, one such measurement takes little time for all but the least permeable soils. For this reason, and because of the inherent accuracy of the measurements, I use this method to obtain the one hydraulic conductivity value (at about h = -0.1 m) normally used to correct hydraulic conductivities derived theoretically from other data (e.g., the soil water characteristic; see Section X.A). Most often, the saturated hydraulic conductivity is used as such a correction (matching) factor. This is often the worst possible choice. Saturated hydraulic conductivities of different samples of the same soil can vary tremendously because of such anomalies as imperfections in the sampling procedure, worm and root channels, and structural cracks and fissures. If present, these large pores are filled with water at saturation and completely dominate water transport through the soil sample; yet they have little if any relation to the properties of the soil matrix from which the hydraulic conductivity function is derived. However, even at small suctions, all these large spaces are empty, and the then-prevailing hydraulic conductivity is a truer reflection of the soil matrix.

B. Flux-Controlled (Flux-Head, Head-Flux, Regulated Evaporation)

Hydraulic conductivities can also be measured under steady-state conditions by controlling the flux density rather than the hydraulic head at one end of a vertical soil column [3]. If the water flows toward a water table at the bottom ("flux-head"), the range of pressure heads that can be covered is limited to the height above that water table. The range can be extended by maintaining a controlled suction at the bottom of the soil column with either a porous plate or another soil column with a water table at some depth. Steady-state conditions can also be attained when the water flows upward from a water table or a water supply at constant negative pressure head and is evaporated at the soil surface at a constant rate ("head-flux"). In the latter case, it is no longer possible to have a measuring zone with uniform pressure head and water content. As the soil becomes drier, the hydraulic gradient will become larger and more difficult to measure accurately. The derived hydraulic conductivity then will be for some kind of average of a range of water contents, and the correct water content to which it should be assigned will be uncertain.

A slightly different experimental arrangement was used by Gardner and Miklich [31]. Their soil column was closed at one end, which made it theoretically impossible ever to reach a steady state. Nevertheless, they claimed that various constant fluxes could be attained by regulating evaporation from the other end of the column according to the size and number of perforations in a cover plate ("regulated evaporation"). This would seem to require a lot of manipulation. The rates of water loss were determined by weighing the entire column. The hydraulic gradient was measured with two tensiometers, and for each evaporation rate, k and θ were assumed to be constant between the tensiometers. The hydraulic conductivity is then approximated by

$$k = \frac{(x_1^2 - x_2^2)q}{2L(h_1 - h_2)} \tag{8}$$

where x_1, x_2 are the positions of the tensiometers and L is the length of the soil column. These rather severe assumptions limit the applicability of the method, and it has not been frequently used.

C. Long-Column Infiltration

When a constant water flux density of water is applied to a long dry vertical soil column, the flow system can reach a "quasi" steady state [32,33]. True steady-state conditions, of course, will never be attained, because although the potentials on both ends of the flow system are constant, the distance between these ends keeps increasing with time. As a result, the pressure head gradient keeps diminishing with time. Eventually, it may become small enough to be negligible with respect to the constant, unit gravitational potential gradient. Then, a "quasi" steady state is attained. If the soil column is long enough to permit a zone to develop at the top of the column in which the hydraulic gradient can be assumed to be unity, the hydraulic conductivity there is equal to the externally imposed known flux density. Thus, tensiometers are not needed and, if the hydraulic conductivities are assigned to measured water contents, the pressure head range of the method can theoretically extend beyond the tensiometer range. While this method does not present problems with contact resistances between soil and porous plates, it does require a device to deliver small fluxes uniformly over the soil surface (see, e.g., Refs. 34 and 35).

D. Matric Flux Potential

The configuration of a controlled evaporative flux from a short soil column in which the pressure head at the other end is controlled (Section VI.B) was used by Ten Berge et al. [13] in a steady-state method for measuring the matric flux potential as a function of water content. These investigators assumed that the matric flux potential function has the form

$$\Phi[\theta] = -\frac{A}{x + B} \qquad \text{for} \quad x = 1 - \frac{\theta}{\theta_0} \tag{9}$$

where A is a scale factor (m^2 s^{-1}) and B is a dimensionless shape factor, both typical for a given soil, and θ_0 is a reference water content, experimentally controlled at the bottom of the soil column. Whereas Ten Berge et al. use the earlier [36] proposed diffusivity function

$$D[\theta] = a(b - \theta)^{-2} \tag{10}$$

where a and b are constants, the method can be used with any set of two-parameter functions of $\Phi[\theta]$ and $D[\theta]$.

After a small soil column has been brought to a uniform water content (pressure head) and weighed, it is exposed to artificially enhanced evaporation at the top, while the bottom is kept at the original condition with a Mariotte-type water supply. When the flow process has reached a steady state, the flux density is measured, as well as the wet and oven-dry weight of the soil column. From these simple, accurate experimental data, the parameters A and B, and thus $\Phi[\theta]$ and $D[\theta]$, can be evaluated by assuming that gravity can be neglected. In this case the matric flux potential in the steady state decreases linearly with height, so that this method does not suffer from any ambiguity (generally associated with upward flow) in the assignment of appropriate values of water content and pressure head to the calculated values of the water transport parameter.

It is better not to start from saturation, but at a small negative pressure head, to reduce the influence of gravity and be able to meet the theoretically required upper boundary condition ($\theta = 0$). The method is rather slow and covers a limited range of θ and h, but the measurements require little attention while in progress. The major source of errors appears to be that the theoretically prescribed initial and boundary conditions are hard to obtain experimentally. Furthermore, the theoretical basis involves a number of assumptions. However, direct measurement of $\Phi[\theta]$ is likely to be more accurate than methods involving separate measurements of $D[\theta]$ and $h[\theta]$ for flow processes involving steep gradients, thin, brittle soil layers, etc. For an analysis of the propagation of errors, see Ten Berge et al. [13].

VII. STEADY-STATE FIELD METHODS

A. Sprinkling Infiltrometer

Analogous to the long-column measurements in the laboratory (Section VI.C), hydraulic conductivities can be measured directly in the field under quasi-steady-state conditions with a sprinkling infiltrometer [4, 37]. This is the closest counterpart to the two-plate laboratory method as a "standard reference method" for the field. In such applications it is warranted to use very elaborate sprinkling equipment, which normally must be attended whenever it is in operation. This may extend over days or even weeks, depending on the range of water contents to be covered. This range is technically limited by the ability to reduce the sprinkling rate while retaining uniformity. This can be done best by intercepting an increasing proportion of the artificial rain, rather than reducing the discharge from a nozzle [35,38,39]. Green et al. [4] give 1 mm h^{-1} as a practical lower limit for the flux density. To prevent hysteresis, the flux density of the applied water should be increased monotonically with time. Because soil profiles are frequently inhomogeneous, and because of the possibility of lateral flow, the hydraulic gradient cannot be assumed to be unity and it should be measured when a high accuracy is required. Sprinkling infiltrometers are used frequently for soil erodability studies. In such applications, the impact energy of the water drops emitted by the sprinkling infiltrometer should be as nearly equal to that of natural raindrops as possible [40], since changes of the physical soil properties due to structural breakdown of the soil (e.g., crust formation) have a great effect on the erosion process [41,42]. For hydraulic conductivity measurements, in contrast, the soil surface generally should be protected against crust formation as much as possible (e.g., by covering the soil surface with straw).

Field measurements of hydraulic conductivity with a sprinkling infiltrometer may take a long time, during which large temperature variations may occur. Temperature changes and gradients may have a significant influence on the water transport process, especially for small water flux densities and/or hydraulic head gradients near the soil surface. Therefore, it is good practice to ensure that all field measurements minimize temperature changes as much as possible— for example, by shielding the soil surface from direct sunlight.

B. Isolated Soil Column (Crust Method)

This is analogous to the long-column method. A soil column is isolated in situ by carefully excavating the surrounding soil. Although not strictly necessary for unsaturated conditions, usually a plaster of Paris jacket is cast around the soil column-cylinder assembly for protection, transportation, and/or subsequent saturated conductivity measurements. Use of such a truly undisturbed soil column is especially suitable for soils with a well-developed structure, since large-scale "undisturbed" samples, which are easily damaged during transport, would otherwise be required.

Usually, the pressure head, rather than the flux, has been controlled, for example, with a crust [43,44]. After the soil column surface has been smoothed at the desired depth, a close-fitting cylinder is pushed into the top of the column. A crust of uniform

FIG. 3 Isolated soil column method. Water supply via hypodermic
needles is regulated by stroke and frequency of pulsating pump.
Tensiometers are hydraulically switched to pressure transducer with
digital voltmeter.

thickness and composition (usually a mixture of hydraulic cement
and sand) is applied inside the cylinder. After the crust has cured,
normally 24 hours, the cylinder is sealed off and water is applied to
the soil column via the crust at constant head with a Mariotte device.
Supposedly, the crust soon causes the flow density to attain a steady
state at unit hydraulic gradient, after which time the hydraulic conductivity is equal to the prevailing flux density. Measurement of the
pressure head in the soil just below the crust with a single tensiometer provides the pressure head corresponding to this value of hydraulic conductivity. However, because the assumption of unit hydraulic gradient is often invalid, the hydraulic gradient should be
measured with at least two tensiometers. By using different values
of the controlled pressure head and/or crust resistance, a number

of points on the hydraulic conductivity function can be obtained. In doing this, one should proceed from dry to progressively wetter conditions (by replacing more resistant crusts with progressively less resistant ones), since the wetter wetting fronts will quickly overtake each other. Letting the soil dry before applying a smaller flux density takes much time and introduces hysteresis into the measurements. The minimum pressure head that can be attained with crusts appears to be, practically, not much lower than -50 cm.

In comparison with ponding infiltration, the claim that crusts enhance the attainment of a steady state is correct. The hydraulic head loss across the relatively less permeable crust decreases the pressure head difference between the ends of the extending zone of wetted soil. Thus the pressure head gradient will become negligible with respect to the constant, unit gravitational potential gradient more quickly with a crust. I suspect, however, that often the final measurements with the "crust method" are made before a "quasi" steady state has been reached. The crust does not add to the speed of attaining a steady state in comparison with the application of a non-saturating, constant water flux to a soil column (the preceding method). On the contrary, it may well be slower, and it introduces other experimental problems. Crust resistances have proved to be quite unpredictable, often nonuniform and unstable in time. Making and replacing good crusts is tedious work, and curing of the crusts takes time and may add to the soil solution chemicals that alter the hydraulic conductivity. I advocate, therefore, that the "crust method" in its present form no longer be used.

The isolation of a soil column is an attractive feature that can be retained, but the water should be applied uniformly over the soil surface at easily changed, constant rates that can be verified. We have been exploring application of water from a reservoir with hypodermic needles (Fig. 3) suspended just above the isolated soil column. When the water is applied with a pulsating pump, each needle can be made to release just one water drop per pulse, down to fairly low average flux densities of about 2 mm d^{-1}. The uniformity of water supply can be determined easily by placing a rack of reaction tubes in the same pattern under the needles. Additional study is needed to see whether flux density can be reduced further by decreasing the pulse frequency and/or the needle density without unduly affecting the flow process by the inhomogeneous water application. When electricity is not available, a constant head water supply (Mariotte bottle) can be used, but the water application becomes nonuniform at flux densities less than about 10 cm d^{-1}. This variant of the isolated soil column method appears to be a very attractive, much simplified version of the sprinkling infiltrometer.

C. Spherical Cavity

The preceding discussions make it clear that in one-dimensional flow, a steady state can be achieved only when there are two controlled, steady boundaries, either potentials or flux densities. Both features are inconvenient under field conditions, particularly when measurements must be repeated many times. It is not too difficult to force the flow to be one-dimensional by isolating a soil column, either as practiced with the "crust method" or by making vertical trenches, covering the vertical walls with plastic sheet and refilling the trenches with soil. However, a major experimental effort is required to impose a steady boundary condition at the bottom of a flow system in the field. The practical solution is usually to perform measurements in a deep uniform soil profile in the center of a larger area wetted by a sprinkling infiltrometer, allowing the "quasi" steady state of a constant-shape wetting front moving downward at constant velocity. This is then due to the action of gravity. Without gravity (i.e., in a horizontal direction or when the pressure head gradient is large enough to permit the effect of gravity to be neglected), the wetting front advances as a function of the square root of time, as long as water is applied at the soil surface. This process is often referred to as adsorption.

In contrast, three-dimensional infiltration from a point source reaches a "large-time steady state" with and without the influence of gravity [19]. The influence of gravity is much smaller in three-dimensional than in one- or two-dimensional flow. Without gravity, three-dimensional infiltration from a point source is spherically symmetric. Raats and Gardner [11] showed that the hydraulic conductivity can be derived from a series of such steady flows. This presents a very attractive set of conditions for measuring hydraulic conductivity, especially in situ because (1) only one controlled boundary is required, (2) the influence of gravity, which must be neglected, is especially small, and (3) steady-state measurements are inherently accurate. For these reasons, I have explored the possibilities of this "spherical cavity" method and have analyzed the influence of gravity [45]. Water is supplied to the soil (which needs to be initially at uniform pressure head) through the porous walls of a spherical cavity maintained at a constant pressure head until both the flux F and the pressure head h_a at the radial distance r = a from the center of the spherical cavity, have become constant. This is repeated for progressively larger (less negative) controlled pressure heads in the cavity. Hydraulic conductivity can then be calculated according to

$$k[h_a] = \frac{1}{a}\frac{dF}{dh_a} \tag{11}$$

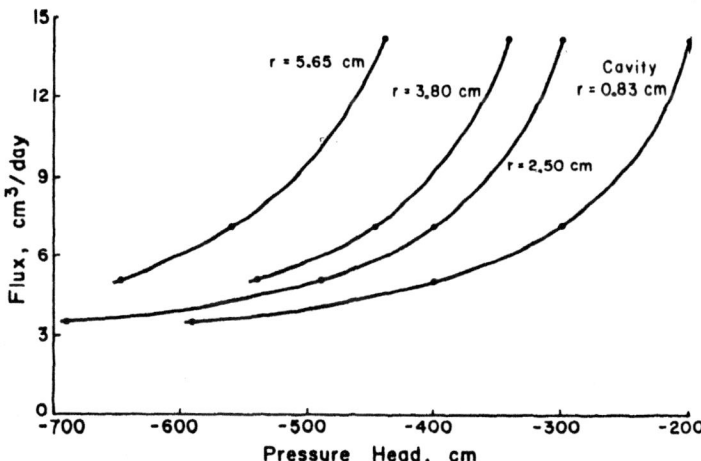

FIG. 4 Steady fluxes from a spherical cavity versus steady pressure heads in cavity and in three tensiometers at indicated radial distances. (From Ref. 45.)

which is simply the slope of the graphs in Fig. 4 at any desired pressure head, divided by the radial distance of the particular measuring point. In this way hydraulic conductivities down to h = -700 cm were obtained in about 2 weeks, with each tensiometer and the cavity yielding its own result. This overlap provides an internal check. Note that the pressure head range can be expanded downward easily by increasing the radial distance of the measuring point. Of course, the time required to reach a steady state increases then also. It is possible to use the regulated pressure head in the cavity as the only "tensiometer" data. This reduces the experimental operations to a minimum. The resistance between the water supply and the soil (porous walls and soil-ceramic interface) must then be negligible. The effect of gravity is minimized when tensiometers, if used, are placed directly below the cavity. The method has been demonstrated only in the laboratory, although there have been some exploratory measurements in the field. Because of its very attractive features, especially as an in situ method, the approach is worthy of further investigation. If tensiometer measurements can be omitted, placement of the spherical cavity without undue contact resistance with and disturbance of the soil presents the only great experimental challenge.

D. Ponded Disk

After a complicated mathematical analysis, wherein he assumed

$$k = k_s \exp \alpha h \tag{12}$$

where k_s is saturated hydraulic conductivity and α is a constant characterizing different soils, Wooding [46] obtained a simple, linear equation for the steady infiltration of water from a shallow circular pond

$$q = \alpha \Phi_s + \frac{4\Phi_s}{\pi r} \tag{13}$$

or

$$q = k_s + \frac{4k_s}{\pi \alpha r} \tag{14}$$

where Φ_s is the matric flux potential. The first term is the contribution of gravity, the second that of the matric potential gradient. Scotter et al. [47] used this result to determine k_s, Φ_s, and sorptivity S, the latter for soils with a delta-function diffusivity.
When (average) steady infiltration flux densities, q, are measured with shallow rings of two different radii, r, then

$$k_s = \frac{q_1 r_1 - q_2 r_2}{r_1 - r_2} \tag{15}$$

$$\Phi_s = \frac{\pi}{4} (q_1 - q_2) \left(\frac{1}{r_1} - \frac{1}{r_2} \right) \tag{16}$$

and

$$S = [2\Phi_s (\theta_s - \theta_n)]^{1/2} \tag{17}$$

From the same results, the parameter α in the exponential hydraulic conductivity function can also be derived:

$$\alpha = \frac{4(q_1 r_1 - q_2 r_2)}{\pi (r_1 r_2)(q_1 - q_2)} \tag{18}$$

Strictly speaking, these are saturated measurements and belong in Chap. 4. However, because of the preassumed functional relationships, they yield hydraulic properties of unsaturated soil. It seems appropriate, therefore, to review a few details of the experimental aspects. The measurements are clearly simple enough to be carried out in great number. Apart from the flux measurements, only volumetric water contents before and immediately after each infiltration run must be determined.

Scotter et al. [47] presented equations for the standard deviations of k_s and S, whether normally or log-normally distributed. They performed sufficient measurements (4-25 per ring) to investigate the spatial variability of k_s and S. The rings, with radii ranging from 25 to 204 mm ($r_1 > 2r_2$), were gently pushed into the soil only about 10 mm, keeping disturbance to a minimum and making the method suitable for a wide range of soils. The ponding depth, also about 10 mm, was maintained with a Mariotte device or by hand. Measurements were continued for an hour after a steady state appeared to have been reached, which occurred after elapsed time periods ranging from 5 to 100 minutes (in soils ranging from sandy loam to silt loam). However, Scotter et al. warned that much more time may be necessary and cited a case that took 14 hours. They also suggested plotting q versus log t rather than t to judge whether a steady state has been reached.

E. Dripper

Shani et al. [48] used the theoretical basis of the preceding method for estimating the hydraulic conductivity function. Instead of confining the saturated zone at the soil surface with rings and waiting until the flux had become steady, they used commercially available drippers, used for drip irrigation, to apply water at different steady discharge rates and waited until the diameter of the ponded area at the soil surface had become steady. They stated that this usually occurred within 15 minutes. They dubbed this the "dripper" method. Also, rather than substituting average values of q in Eq. 14, they estimated first k_s from the intercept of a linear regression of q versus $1/r$ and then determined α from the slope of the linear regression equation, b, according to:

$$\alpha = \frac{4k_s}{b\pi} \tag{19}$$

These saturated measurements yield unsaturated results owing to the preassumed functional relationships. Therefore, the results cannot be better than the degree to which these relationships hold. It should also be realized that these functions are based on measurements in the wet range. They can easily be extrapolated to lower pressure heads, but there is no guarantee that this is valid.

Shani et al. [48] used the same data also to determine the parameters of the Brooks and Corey [49] relationship for hydraulic
conductivity.

$$k = k_s \left(\frac{h_w}{h} \right)^\mu \tag{20}$$

Because of the interrelationship between the equations of Brooks
and Corey, this also yields the soil water characteristic. Equation
20 contains two soil parameters: μ, which is related to a pore size
distribution index, and the air-entry or bubbling head h_w. Both
can be determined from the dripper measurements if, again, the
sorptivity is also measured. Shani et al. did this by measuring the
horizontal wetting front advance from the steady ponded-zone perimeter at the soil surface as a function of time. They checked their
results by, among other things, measuring the air-entry head directly [50], but this is not unambiguous, especially in structured soils.
The determinations of the pore size distribution index and residual
saturation, required for the Brooks and Corey equations, are also
not always straightforward. Brooks, Corey, and their coworkers
invariably tested these equations with the hydrocarbon fluid "Soltrol,"
which has soil wetting properties altogether different from water.
There is, therefore, some doubt whether these equations are valid
for soil-water systems. Van Schaik [51] found large internal discrepancies, even for studies that have been claimed to yield the
best results for the Brooks and Corey equations. For these reasons, I caution against the use of these equations.

VIII. TRANSIENT LABORATORY METHODS

A. Instantaneous Profile

In contrast to the steady-state methods, most transient laboratory
methods yield in the first place hydraulic diffusivities. Then $k[\theta]$
must be derived from $D[\theta]$ with the soil water characteristic (see
Section II.B). The one major exception is the instantaneous profile
method. In its many variants it is probably the most used method
to determine nondestructively the hydraulic conductivity of laboratory columns in which other water transport processes are studied
and for which $k[\theta]$ must be known. Often, there is already available quite sophisticated equipment, such as automated gamma attenuation scanners and multiple tensiometer apparatus [52], which
allows more complete and/or accurate determination of $k[\theta]$ than is
normally the case. This is reflected in the scores for the various
criteria for this method as a laboratory method, in comparison with
the scores as a field method. Since this method is especially suited
for use in situ, it is discussed in more detail in the next section.

B. Pressure Plate Outflow

Gardner [53] proposed the pressure plate outflow method. A soil sample at hydraulic equilibrium on a porous plate is subjected to a step decrease in the pressure head in the porous plate (e.g., a hanging water column) or a step increase in the air pressure. The resulting outflow of water is measured with time. The step decrease or increase must be so small that the hydraulic conductivity can be assumed constant and that the water content is a linear function of pressure head. The experimental water outflow as a function of time is matched with a theoretical solution, yielding after many approximations

$$\ln(Q_0 - Q) = \ln\left(\frac{8Q_0}{\pi^2}\right) - \left(\frac{\pi}{2L}\right)^2 Dt \tag{21}$$

where Q is the cumulative outflow at time t, Q_0 is the total outflow, and L is the length of the soil sample. The diffusivity for the mean pressure head can be derived from the slope of a plot of $\ln(Q_0 - Q)$ versus t. This is repeated for other step increases in pressure, which must be initiated only after a new state of hydraulic equilibrium has been reached. The pressure increments must be small enough for the assumptions to be valid, but large enough to allow accurate measurement of water outflow, while the more steps there are, the more time it takes to cover the desired range of water content. This method was initially widely used but generally failed to yield satisfactory results. Much effort was spent to improve it, especially with respect to the correction for the resistance of the porous plate or membrane, but without too much success.

C. One-Step Outflow

Doering [54] proposed the one-step variant of the pressure plate outflow method, which is much faster and not very sensitive to the resistance of the plate or membrane. If uniform water content in the soil column is assumed at every instant, diffusivities can be calculated from instantaneous rates of outflow and average water content

$$D[\theta] = \frac{-4L^2}{\pi^2(\theta - \theta_f)} \frac{d\theta}{dt} \tag{22}$$

where L is the length of the soil sample, θ is the average water content when the outflow rate is $d\theta/dt$, and θ_f is the final water content. These can be determined by measuring the cumulative outflow and the final weight. Doering found the results to be as reliable as those obtained with the original version (Section VIII.B), and there were large time savings.

Gupta et al. [55] showed that the analysis of one-step outflow data according to Gardner [56] and used by Doering can be in error by a factor of 3. They improved the analysis by first estimating a weighted mean diffusivity. This does not require the assumption of a constant diffusivity over the pressure increment, nor over the length of the soil sample, and it also reduces the effect of membrane impedance. Passioura [57] obtained about the same improvement in accuracy with a much less complicated calculation procedure (with detailed stepwise instructions) by assuming that the rate of change of water content at any time is uniform throughout the entire soil sample. He also estimated that a soil sample 60 mm long will take about 5 weeks to run and a 30 mm sample about 1 week. Measurements have been automated recently for up to 16 samples [58].

The one-step outflow method is attractive for its experimental simplicity; the theoretical analysis of the data remains its weakest point. Since this limitation does not apply to the simulation of the flow process, it is not surprising that recently the same measurements were selected as basis for the parameter optimization approach (Section XI).

D. Boltzmann Transform

There are three variants of the transient, so-called Boltzmann transform methods. The theoretical framework on which these methods are based is well known and can be found in soil physics textbooks [10,59]. If gravity is neglected (e.g., using horizontal columns), the flow equation can be expressed in the diffusivity form of Eq. 3. For a step-function increase or decrease of the water content at the adsorption or desorption interface of an effectively semi-infinite uniform soil column, this partial differential equation can be transformed into an ordinary differential equation using the Boltzmann variable $\tau = x/\sqrt{t}$, where x is the distance from the sample surface and t is time. Integration of this equation for the also transformed initial and boundary conditions yields the diffusivity as

$$D[\theta'] = \frac{1}{2}\left(\frac{d\tau}{d\theta}\right)_{\theta'} \int_{\theta'}^{\theta_1} \tau[\theta] \ d\theta \qquad (23)$$

where θ_1 is the final water content at the adsorption or desorption interface, θ' is the water content at which D is evaluated, and θ is the water content as a function of x and t.

Thus the diffusivity at any water content is equal to half the product of the slope and area indicated in Fig. 5. The function $\tau[\theta]$ can be determined experimentally by measuring either the water content distribution in a soil column at a fixed time [60] or the change of water content with time at a fixed position [61]. The

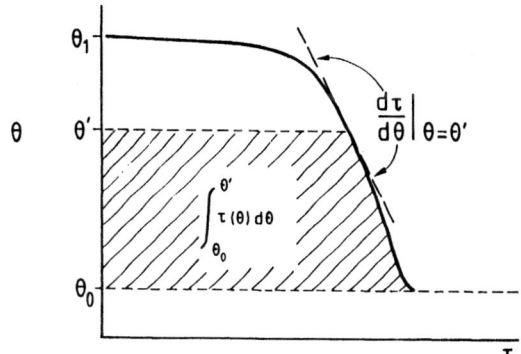

FIG. 5   Graphical solution of the Boltzmann transform equation (Eq. 23).

first measurement is often done gravimetrically; the latter needs to be done nondestructively with specialized equipment (e.g., gamma attenuation, capacitance sensors). Gravimetric measurements must be done very quickly to minimize redistribution and evaporation of water during sampling. The main drawback of the fixed-time method is the sensitivity of the calculated diffusivities to irregularities in the bulk density and water content in the soil column and the consequent propagation of errors from errors in the water contents. At first thought, the fixed-position method would seem to eliminate most of these problems. However, indirect, nondestructive water content measurements are inherently less accurate and the propagation of errors is therefore similar in both cases. A comparative study of the two variants [62] yielded similar errors.

Derivation of a $D[\theta]$ function from experimental $\tau[\theta]$ data according to Eq. 23 involves differentiating experimental data with scatter, which is inherently inaccurate and yields poor results, especially near saturation, where the water content profile is quite flat [63,64]. Clothier et al. [64] showed that it is much better to find a value for a parameter p by fitting the experimental $\tau[\theta]$ data to the function

$$\tau[\theta] = \varepsilon(1 - \Theta)^p \qquad \text{for } p > 0 \tag{24}$$

where ε is a parameter that can be derived from p and the sorptivity, Θ is the dimensionless soil water content

$$\Theta = \frac{\theta - \theta_0}{\theta_1 - \theta_0} \tag{25}$$

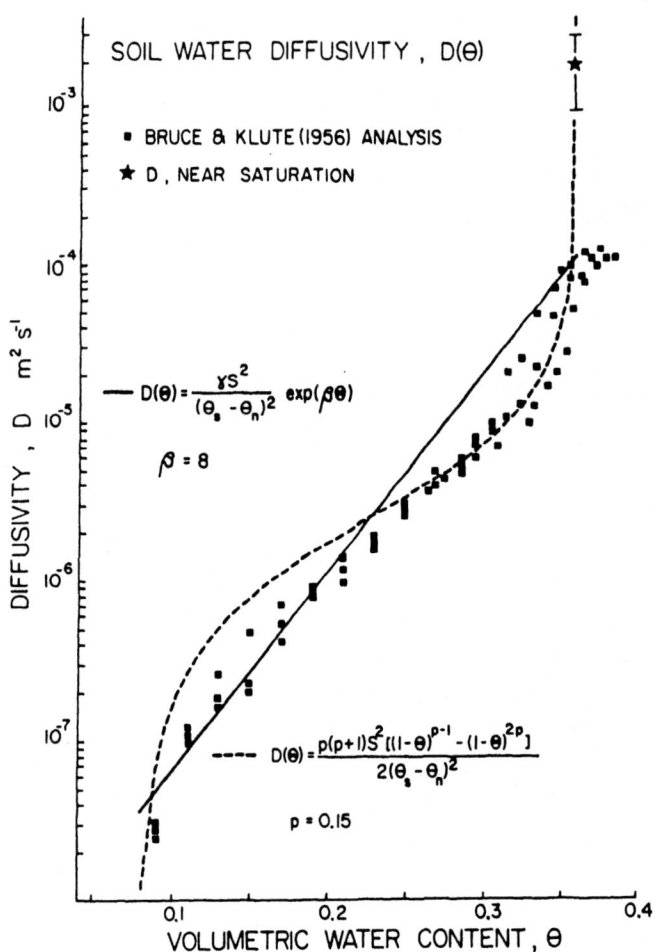

FIG. 6 Diffusivity function derived graphically according to Fig. 5 and derived from fit to Eq. 26, for p = 0.15, and diffusivity measured near saturation. [From Ref. 64; for Bruce and Klute (1956), see Ref. 60.]

where θ_1 is the final water content at the adsorption-desorption interface and θ_0 is the initial water content. The corresponding equation for the diffusivity is then

$$D[\theta] = p(p + 1)S^2 \frac{(1 - \theta)^{p-1} - (1 - \theta)^{2p}}{2(\theta_s - \theta_0)^2} \qquad (26)$$

This analysis of the experimental data ensures correct integral properties of the obtained $D[\theta]$ function, because it is fitted to the primary data set $\tau[\theta]$ and the measured value of the sorptivity. Moreover, it never leads to physically nonsensical $D[\theta]$ functions, which decrease with increasing θ, as least-squares fitting of $\tau[\theta]$ can do. Instead, it yields S-shaped diffusivity curves with infinite diffusivity at saturation (Fig. 6), as observed for many soils [65]. More details on this recently proposed improved form of data analysis can be found in the original publication [64].

E. Hot Air

A third variant of the Boltzmann transform method was reported by Arya et al. [66]. As the "hot air" method, this variant has become quite popular in some areas, undoubtedly due to the simplicity and speed of the required measurements, and the large range of θ over which $D[\theta]$ values are obtained. It is the drying counterpart of the Bruce and Klute variant. It has not only all the disadvantages of this variant, but also many others. Whereas the required boundary condition of a step-function change in potential (water content) can be attained easily in the case of wetting, a drying step function is nearly impossible experimentally. It is imposed by a stream of hot air directed at the soil surface, while the rest of the soil column (usually 10 cm long and 5 cm diameter) is shielded from it as much as possible. Air temperatures of up to 240°C have been required for sandy soils. Even then it takes normally a few minutes to dry the soil surface, while the total evaporation period normally lasts from 10 to 15 minutes. Whereas temperatures in excess of 90°C have been measured in the soil [67], the data can be analyzed only by assuming isothermal conditions. The effects of temperature on variables (viscosity, surface tension, etc.) and of any water transport due to the thermal gradient are significant but must be ignored. Because the soil is hot, there is significant water loss due to evaporation during sampling. Finally, the measurements are usually performed on initially saturated, vertically oriented soil columns. This introduces errors due to gravity during a run and water loss at the wet end due to compaction during sampling. This can be reduced by equilibrating the soil column at a moderate negative pressure head (around -50 cm).

Without arbitrary manipulation of the water content profile of the sample, the data often yield diffusivities decreasing with water content. This is physical nonsense. To prevent this, computer programs have been devised [68] that keep the analysis within the theoretically acceptable framework; the results, however, are still based on very dubious experimental measurements. When the method appears to yield useful results, this may be accidental; several

sources of errors appear to cancel each other [67]. I feel, therefore, that the hot air method should be abandoned. It may be possible to find a way to impose the boundary condition by using hy-
groscopic agents, eliminating the temperature effects, but in view
of all the other objections the approach does not seem worth the
effort. In this connection, it should be pointed out that it is not
necessary to dry the soil instantaneously at the surface; only a constant water content or pressure head must be imposed. This does
not need to go beyond the range over which the diffusivity or conductivity function is required.

F. Flux-Controlled Sorptivity

The sorptivity method is related to the Boltzmann transform methods
in that the same transformation is used in the derivation of the working equation [17,69]

$$D[\theta_1] = \frac{\pi S^2}{4(\theta_1 - \theta_0)^2}\left[\frac{\theta_1 - \theta_0}{(1 + \tau)\log e}\frac{d}{d\theta_1}\ (\log\ S^2[\theta_1,\theta_0 = \text{const}]) \right.$$

$$\left. - \frac{1 - \tau}{1 + \tau}\right] \tag{27}$$

where τ is a constant that can be varied between 0.50 and 0.67
without significant effect [70]. Detailed information on required experimental apparatus and a step-by-step description of the experimental procedure of the sorptivity method can be found in Klute
and Dirksen [3].

Experimentally, the method entails the determination of $S[\theta_1,$
$\theta_0 = $ constant], the sorptivity as function of the water content at
the adsorption interface, θ_1, for constant initial water content, θ_0
(see Eq. 7). This can be accomplished by means of a series of
one-dimensional absorption runs, each yielding one set of (S,θ_1)
values. Rather than regulating θ_1 via h_1, each sorptivity is controlled by mechanically controlling the supply of water to the adsorption interface according to the $\sqrt{t}$ relationship of Eq. 7. Then,
after each run a single soil sample is required for gravimetric determination of θ_1. This takes only about 10 seconds, which virtually eliminates errors due to evaporation and redistribution during
sampling. Moreover, near the soil surface θ changes neither with
time ("pseudo" steady state) nor with position. With proper functioning of a somewhat complex apparatus, experimental errors are
thus limited to a minimum, and thus any propagation of errors in
the calculation of $D[\theta]$ according to Eq. 27 is also minimized. The
required differentiation is performed algebraically on a polynomial
regression of log S^2 in terms of θ_1. Depending on the desired

accuracy, a diffusivity function can be obtained from one to three soil samples of 10 cm length. These samples are first dried to guarantee the required uniform initial water content, and a large water content (pressure head) range can be covered. For each run, a new dry soil surface must be carefully prepared.

The effect of nonuniformity of soil samples on the final results still requires further investigation. The theoretical basis of Eq. 27, although not rigorously exact, appears to be accurate [18,69,71]. Although water is applied through porous plates, diffusivities well beyond the "tensiometer" range have been obtained. This is possible, because the individual runs need to be continued for only a few minutes near saturation to a maximum of 1 hour when the final water content is very low. A complete diffusivity function can be determined in a day.

During sorptivity measurements in the wetter range, h_1 was measured by an isolated small tensiometer, slightly protruding in the center of the porous plate, and a pressure transducer that needed virtually no water displacement for a full-scale measurement (zero-balance principle). Later tests yielded the best pressure transducer response with tensiometers of only 1.5 mm diameter. Such simultaneous pressure head measurements allow immediate determination of $k[\theta]$, which is convenient because wetting $h[\theta]$ functions are not normally available. The line in Fig. 7 labeled "Sorptivity method" was in the wetter region obtained with such simultaneous measurements. Only seven sorptivity runs, each lasting 6–12 minutes, yielded $k[\theta]$ values for water contents less than $\theta = 0.10$. The results with the instantaneous profile method, obtained on the same packed soil before the samples for the sorptivity measurements were taken, required several weeks and still yielded $k[\theta]$ values only for water contents larger than 0.20. The experimental results presented here and in Ref. 17 were all obtained with apparatus fabricated in our own machine shop. More versatile apparatus is commercially available [3], as indicated in Table 1 in parentheses.

F. Other Methods

Several other methods have been proposed in the literature, which fall in the category of transient, laboratory methods. A few of these are mentioned, without being exhaustive, and without evaluating them in Table 1.

Wind [72] proposed a modified instantaneous profile method to measure simultaneously the water retention characteristic and the hydraulic conductivity of the same soil sample. An initially saturated and homogeneous sample is allowed to evaporate at the top. The total weight and the pressure heads at at least two depths are recorded. From these data one can calculate the water retention

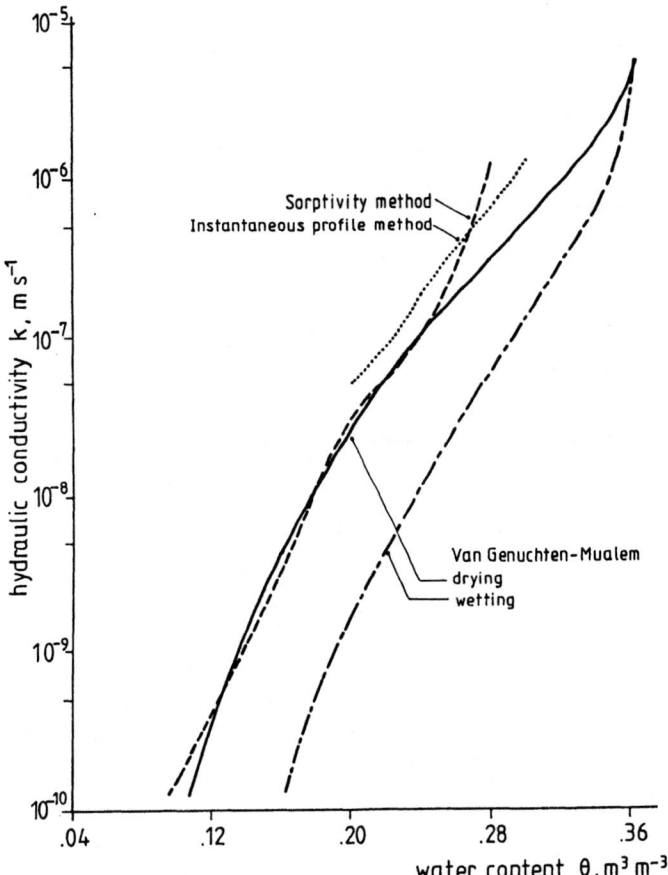

FIG. 7 Hydraulic conductivity functions of Pachappa sandy loam measured with flux-controlled sorptivity method (and simultaneously measured pressure heads) and with instantaneous profile method. The Van Genuchten-Mualem functions (Eq. 39) are based on the fitted soil water retention characteristics in Fig. 10.

characteristic with an iterative method. Knowing this, one can determine the flux densities at the bottom (zero), at the top (measured evaporation rate), and in between the depths were the pressure heads are recorded. From then onward, the calculation is the same as for the instantaneous profile method (Section IX.A). Boels et al. [73] designed an automatic recording system for these measurements on many soil samples. They also proposed a direct calculation method by approximating the soil water retention characteristic by a polygon. The data of these experiments can also be used for the inverse parameter optimization approach (Section XI), allowing a comparison between the two approaches [74]. All this has been improved and automated further to the point that it is now the major method at their institute (Halbertsma, Staring Center, Wageningen, personal communication).

Ahuja and El-Swaify [75] determined the soil hydraulic properties by measuring one-step cumulative inflow or outflow from short soil cores through high-resistance plates at one end and measuring the pressure head at the other end. They obtained good results for pressure heads down to -150 cm. Scotter and Clothier [76] claimed, without referring to the previous authors, that it is better to analyze the results of a series of small pressure head changes than of one large change, because the former approach does not involve the difficult task of measuring small flow rates. The accuracy relies mainly on the time delay of the outflow, not on the shape of the outflow curve.

IX. TRANSIENT FIELD METHODS

A. Instantaneous Profile

The relative merits of laboratory and field measurements were discussed in Section IV. Especially for layered soils or soils with a well-developed structure, the unsaturated hydraulic conductivity function can best be determined in situ. For drying conditions, this is done most frequently with the instantaneous profile method, also called the unsteady drainage flux method [1,4,52,77,78]. Water contents and hydraulic potentials are measured as functions of time and depth during drainage of an initially saturated, bare soil profile. When the water flux density q is known for all time at one depth z_0, the flux density can be calculated for any depth and time from the water contents:

$$q[z,t] = q[z_0,t] - \int_{z_0}^{z} \frac{\delta\theta}{\delta t} [z,t] \, dz \tag{28}$$

This equation assumes vertical transport only, without root uptake. The boundary condition $q[z_0,t]$ is usually set as a zero flux at the soil surface obtained by covering the surface to prevent evaporation. Hydraulic conductivities can then be obtained from calculated flux densities and measured hydraulic potentials found for a set of times and depths (if needed after smoothing and interpolation) from Eq. 29.

$$k[\theta,z] = \frac{q[z,t]}{(\delta H/\delta z)[z,t]} \tag{29}$$

Unless the draining surface area is very small, water contents can be determined gravimetrically by taking soil samples with an auger. This is accurate and does not take a great deal of time. Often, however, water contents are measured indirectly and nondestructively (e.g., by neutron scattering, gamma attenuation, or capacitance sensors). Hydraulic potentials should be measured directly with tensiometers, using mercury manometers or pressure transducers. Hydraulic conductivities can thus be obtained for any layer between two tensiometers. Within the range of the experimental data, a soil water characteristic can also be constructed for each distinct soil layer from the values of θ and h already measured.

The range of water contents that can be covered is limited at the wet end by the degree of saturation that can be attained by ponding the water on the soil surface. This is often no more than 90% of the available pore volume because air tends to be entrapped by the wetting front. At the drier end, the water content range is limited by the drainage characteristics of the particular soil in its hydrological setting. At first, near saturation, θ and H should be measured as frequently as possible, because they vary so quickly that it is hard to obtain accurate results without automated data collection. After the first few days, further accurately measurable differences in water contents will take days or weeks, and even then will yield k values only for pressure heads that usually do not go below -200 cm. This is the main disadvantage of the method, namely the rather limited range of θ and h over which $k[\theta]$ can be determined. This is reflected in the concept of field capacity, which still appears to be useful in practice in spite of theoretical misgivings.

An analysis of the error propagation of this method [79] is not very encouraging; especially toward the dry end, errors can be very large (Fig. 8). At small times tensiometer errors predominate, while later water content measurements introduce the largest errors. To reduce errors in fine-textured soils, water content measurements should be intensified; in coarse-textured soils it is better to increase the number and/or frequency of tensiometer measurements. Contrary to the usual laboratory conditions, which allow only nondestructive indirect soil water content measurements in soil columns, it is often quite possible to make repeated direct (gravimetric) soil water content determinations in instantaneous profile measurements in the field. Since this will improve the accuracy of the final results, if enough soil samples are taken, such data are indicated in parentheses in Table 1. The h range can be expanded by allowing evaporation from the soil surface and determining the zero-flux

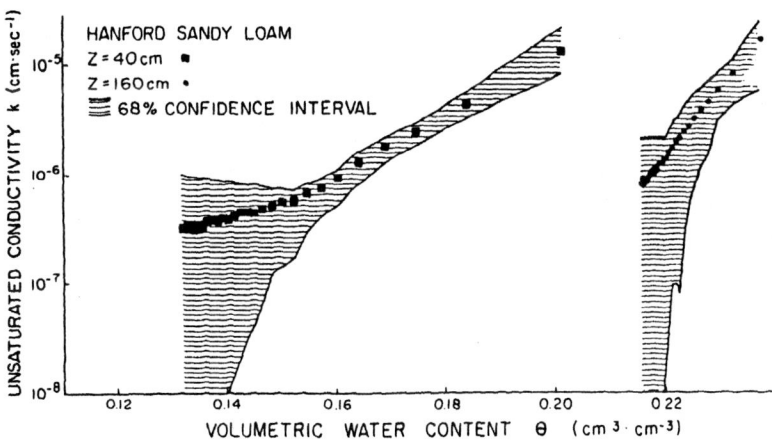

FIG. 8 Confidence intervals (68%) for hydraulic conductivity at two depths due to error propagation in instant profile calculations. (From Ref. 79.)

plane from the tensiometer data [80]. However, the overall results will be even less accurate. The same is true if only either water contents or hydraulic potentials are measured and the others are derived from an independently determined soil water characteristic.

B. Unit Gradient with Prescribed k Function

With the present emphasis on studying the spatial variability of soil hydraulic properties, there is a need for simple in situ measurements. Tensiometric measurements are much less convenient for this purpose than water content measurements, especially when the latter are performed with neutron probes. A simplified version of the instantaneous profile method involving only water content measurements was recently used by Jones and Wagenet [81]. They installed 100 neutron access tubes in a 50 m × 100 m fallow field and wetted the soil around them by ponding water in rings of 37 cm diameter, inserted 15 cm into the soil. When water contents were steady down to 120 cm, the access tube sites were covered and redistribution was followed for 10 days. At the end, gravimetric samples were taken to back up the neutron measurements. The results were analyzed in five somewhat different ways, all assuming the hydraulic gradient to be unity at all times and exponential hydraulic conductivity functions

$$k[\theta] = k_0 \exp[\beta(\theta - \theta_0)] \tag{30}$$

where k_0 and θ_0 are values measured during steady ponded infiltration, sometimes called "satiation." All five analyses yielded values of the constants k_0 and β, with their mean and variance, for selected depths. The difference between the analyses mostly concerned further assumptions on the water content distributions. For instance, in one analysis, already proposed by Libardi et al. [82], the average water content θ^* to depth z is assumed to be a linear function of the water content θ at depth z.

$$\theta^* = a\theta + b \tag{31}$$

For larger times, this leads to

$$\theta - \theta_0 = \frac{1}{\beta} \ln t + \frac{1}{\beta} \ln\left(\frac{\beta k_0}{za}\right) \tag{32}$$

Thus, for each depth, a plot of $\theta - \theta_0$ versus $\ln t$ yields β as the reciprocal slope, and the intercept, given a, yields k_0.

Jones and Wagenet concluded that the five approximate analyses will be most useful in developing relatively rapid, preliminary estimates of soil water properties over large areas but not as useful when k_0 and β at a particular location need to be known precisely.

C. Simple Unit Gradient

In an even more simplified version, uniform water content and pressure head (and thus unit hydraulic gradient) are assumed throughout the draining profile [4]. This implies that the increase of k with depth, which is needed to accommodate the increasing flux density with depth, is assumed to occur with a negligible increase of θ. The hydraulic conductivity is then

$$k[\bar{\theta}] = L\left(\frac{d\bar{\theta}}{dt}\right) \tag{33}$$

where $\bar{\theta}$ is the average water content of the profile above depth L. With a single tensiometer at depth L and making the same assumptions, the diffusivity can be determined analogously [83] as follows.

$$D[h] = L\left(\frac{dh}{dt}\right) \tag{34}$$

Unless the soil profile is highly uniform, it is doubtful that these versions can yield results better than an educated guess.

D. Sprinkling Infiltrometer

If hydraulic properties must be known for wetting conditions, the instantaneous profile analysis may be used on transient data ob-

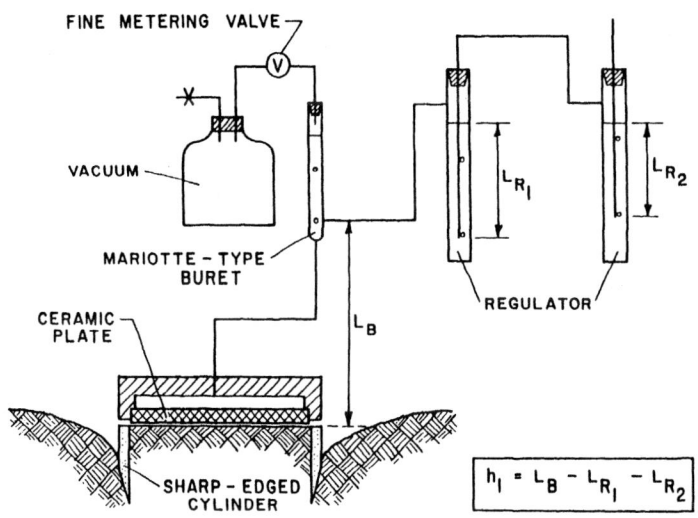

FIG. 9 Apparatus for in situ unsaturated sorptivity measurements. (From Ref. 69.)

tained with a sprinkling infiltrometer. However, this equipment is much more elaborate (see Section VII.A) than that needed simply to saturate a soil profile, and it normally must be attended whenever it is in operation.

E. Sorptivity Measurements

Sorptivity is the first term in the Philip infiltration equation [19] (see Section II.D) and is a function of initial and final water content. This function contains composite information on the other soil hydraulic transport properties [18,71], which can be obtained mathematically. At saturation, sorptivity is measured easily in the field [84]. To prevent macropores from dominating saturated sorptivity measurements, Clothier and White [85] measured "saturated" sorptivity under very small negative pressure heads. Dirksen [69] used the apparatus in Fig. 9 to measure sorptivities in situ over a large range of pressure heads. This was used by Russo and Bresler [50], with other measured soil parameters (saturated hydraulic conductivity air-entry value of pressure head, and residual water content) to determine the probability density functions of $k[\theta]$ and $h[\theta]$ for statistical analysis.

X. DERIVATION FROM OTHER SOIL PROPERTIES

A. Soil Water Retention Characteristic

Physical measurements of soil hydraulic conductivities and other
transport parameters are time-consuming and tedious, and therefore expensive. Moreover, despite considerable effort, the accuracy most often is very poor. With the tremendous variability of
these soil properties, both in space and in time, the practical value
of such measurements is difficult to estimate. It is worthwhole,
therefore, to consider the possibility of deriving these properties
from more easily measured soil properties. The soil water retention
characteristic is most often used for this purpose because, at least
in the range of water contents where the capillary binding of water
is predominant, it reflects the geometry of the pores and this geometry, in turn, determines to a large extent the hydraulic transport properties. The pressure head difference across an air-water
interface is [10]

$$h = \frac{2\sigma}{\rho g R} \tag{35}$$

where σ is the surface tension of the air-water interface (N m^{-1}),
ρ is the density of water (kg m^{-3}), g is the gravitational constant
(N kg^{-1}), and R is the equivalent radius of the interface (m). If
the soil material is perfectly hydrophyllic (i.e., zero angle of contact), then R is equal to the (equivalent) radius of the pore at the
interface, and the soil water retention characteristic can be converted into an equivalent pore size distribution, since the water
content at any given pressure head is equal to the porosity con-
tributed by the pores that are smaller than the equivalent diameter
corresponding to that pressure head (measured with respect to atmospheric pressure), as given by Eq. 35.

There are two approaches to calculating soil hydraulic conductivities from soil water retention characteristics. One was originated
by Childs and Collis-George [32] and later modified [86,87]. The
other, based on the generalized Kozeny equation, had its origin in
the oil industry and was introduced into the soil literature by
Brooks and Corey [49]. This approach is not discussed here further. For a good summary of the theory and final working equations,
see Laliberte et al. [88], and the comments in Section VII.E.

Childs and Collis-George assumed that the soil consists of randomly distributed pores of various sizes, which can be divided into
a number of size classes. If two imaginary cross-sections of a soil
were to be brought into contact with each other, the hydraulic conductivity of the assembly would depend on the number and sizes of
pores on each side that connect with each other. The chance of

pores of two sizes connecting is proportional to the product of the relative contributions of their respective pore size classes to the total cross-sectional area. Childs and Collis-George assumed further that since according to Poiseuille's law the flow of water through a pore is proportional to the square of its diameter, the flow through two matching pores is determined by the smaller of the two. By dividing the soil water retention characteristic into a number of pore size classes, based on Eq. 35, they finally obtained

$$k = F \frac{\rho g}{\eta} \sum_{\Gamma=0}^{\Gamma=R} \sum_{\delta=0}^{\delta=R} \delta^2 f(\Gamma) \, dr \, f(\delta) \, dr \qquad (36)$$

where F is a correction (matching) factor to match the calculated hydraulic conductivity at a single water content to a measured value at the same water content, η is the viscosity of water (Pa·s), and $f(\Gamma)$ dr and $f(\delta)$ dr are the partial areas occupied by pores of radii Γ to Γ + dr and δ to δ + dr, respectively.

With this equation, the hydraulic conductivity for a selected water content can be obtained by carrying out the calculations up to the value of r for which the pores are still just water-filled. Jackson [89] reviewed and summarized the various versions of this equation and, since the calculations were quite cumbersome, proposed a simpler procedure without making basic changes. For a complete example of the required calculations according to Jackson, see Hillel [8, p. 223]. Many experimental verifications of this approach have been reported (e.g., Refs. 89-92). In all these, the matching factor F (based on measured saturated hydraulic conductivities) was unpredictable and varied between 2.0 and 0.004. Often, the shapes of the theoretical and experimentally determined curves for $k[\theta]$ also differed substantially.

Mualem [93] introduced a few basic changes to the theory of Childs and Collis-George [32]. For instance, he calculated the contribution to the hydraulic conductivity of a larger pore (radius r_1) following a smaller one (radius r_2). Assuming that the length of a pore is equal to its diameter, Mualem was able to define an equivalent radius of the two pores as $\sqrt{(r_1 r_2)}$. Combining his theory with elements of the model of Brooks and Corey [49] for the soil water characteristic and of Burdine [94] for the relative hydraulic conductivity, he found that, based on a comparison with experimental results of 45 soils, the relative hydraulic conductivity was described best by

$$k_r[\theta] = \theta^{1/2} \left[\frac{\int_0^\theta (1/h) \, d\theta}{\int_0^1 (1/h) \, d\theta} \right]^2 \qquad (37)$$

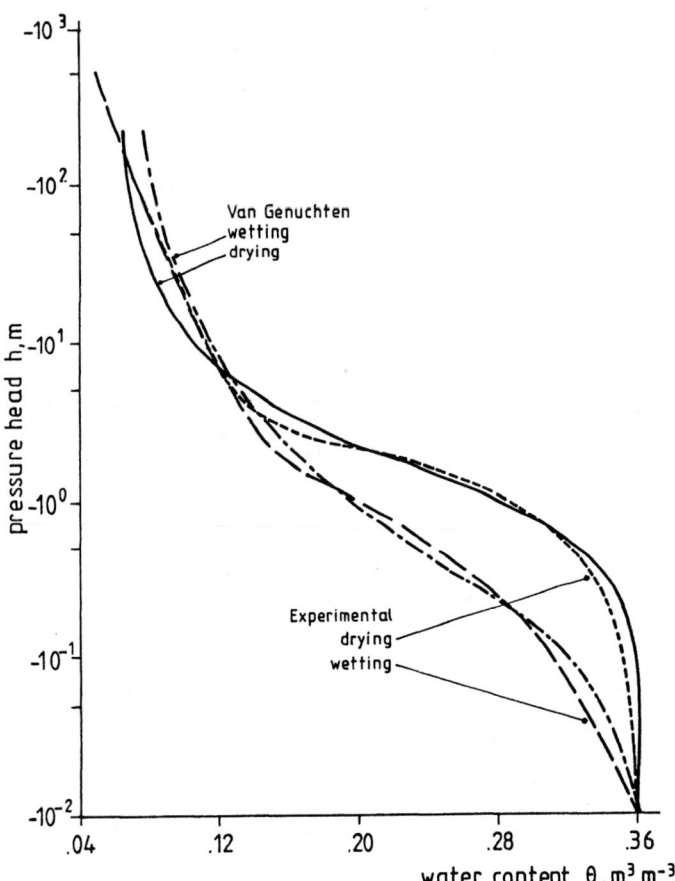

FIG. 10   Soil water retention characteristics of Pachappa sandy loam
composed of various experimental data, and the fits of these to Eq.
38.   The corresponding hydraulic conductivity functions according
to the Van Genuchten-Mualem model are shown in Fig. 7.

where $k_r = k/k_s$ is the relative hydraulic conductivity, Θ is a dimensionless water content (see Eq. 25), and θ_r is the residual water
content, which is the water content at which the hydraulic conductivity becomes negligibly small.

Van Genuchten [95] proposed as approximation for the soil
water retention characteristic

$$\Theta = [1 + (-\alpha h)^n]^{-m} \tag{38}$$

where α, n, and m are fitting constants. He then combined Eq. 38 with the model of Mualem (Eq. 37).

$$k_r[\theta] = \theta^{1/2}[1 - (1 - \theta^{1/m})^m]^2 \quad \text{for} \quad m = 1 - \frac{1}{n} \tag{39}$$

By substituting into Eq. 39 the parameter values obtained in fitting Eq. 38 to a soil water retention characteristic, a relative hydraulic conductivity function is obtained without additional measurements. For absolute hydraulic conductivities, the hydraulic conductivity must be determined for one water content. Figure 10 shows the fits of Eq. 38 to experimental wetting and drying soil water retention characteristics of Pachappa fine sandy loam. The corresponding absolute hydraulic conductivity functions according to Eq. 39 were given in Fig. 7. The absolute values were obtained with an independently determined hydraulic conductivity at "satiation": $\theta = 0.36$. The comparison with the experimental hydraulic conductivity data is very good for drying, especially in the drier range, but very poor for wetting. The reason for this is not clear, nor is it known whether this result can be expected generally. For a more extensive review of this and other models for calculating hydraulic conductivities, see Van Genuchten and Nielsen [6].

It is common practice to use measured saturated (or "satiated") hydraulic conductivities to match calculated and measured values. In general, this is about the worst choice one can make. The standard deviation of such measurements is normally very large, since they can be totally dominated by wormholes, old root channels, fractures resulting from poor sampling procedures, etc. More important, such features have no relation with the pore size distribution of the soil matrix. At small negative pressure heads, all large spaces not associated with the soil matrix are empty and do not conduct water. Therefore, I recommend that hydraulic conductivities measured at small negative pressure heads be used in the calculation procedure outlined above. These can be measured accurately and fast with the "head-head" technique (Section VI.A).

The determination of θ_r, especially, is problematic. Van Genuchten developed a procedure to determine the parameters θ_r, α, n, and m simultaneously with a least-squares curve-fitting algorithm of the soil water characteristic. This approach is used by many investigators and already has earned a certain reputation. More recently, Van Genuchten has developed a program with which up to seven parameters (the four mentioned above, plus θ_s, k_s, and the exponent of θ, which in Eq. 39 has the value 1/2) can be optimized, based on differently weighted experimental data of $h[\theta]$ as well as $k[\theta]$. If desired, even the relationship between n and m, given with Eq. 39, can be left out.

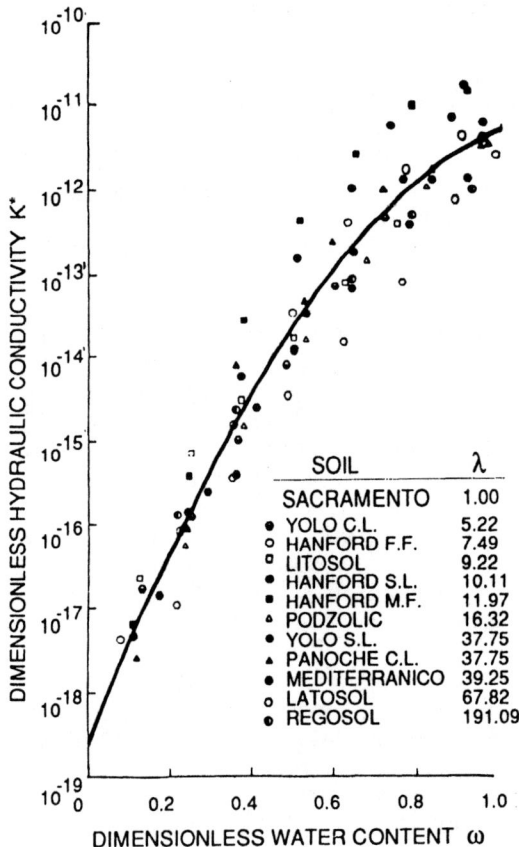

FIG. 11 Hydraulic conductivities of 12 soils scaled according to λ^2 (or m^4) versus dimensionless water content. [Reproduced, with permission, from K. Reichardt, P. L. Libardi, and D. R. Nielsen, Unsaturated hydraulic conductivity determination by a scaling technique, *Soil Science*, 120: 165-168 (1975), © by Williams and Wilkins, Baltimore, MD.]

B. Scaling

If the scaling relationships of Miller and Miller [96,97] are assumed, soil hydraulic properties often can be determined with much less work than otherwise is required. For example, Reichardt et al. [98] measured hydraulic diffusivities of 12 different soils with the fixed-time Boltzmann method [60] and converted these to hydraulic conductivities according to Eq. 4. When these hydraulic conductivities were scaled according to the square of a characteristic microscopic length λ, the data coalesced nicely into one relationship (Fig. 11). For k in centimeters per second, the solid line in Fig. 11 can be described by [20]:

$$k[\theta] = 1.942 \times 10^{-12} m^4 \exp(-12.235 \; \theta^2 + 28.061\theta) \tag{40}$$

λ was assumed proportional to the square of the slope m of the linear relationship between the advance of wetting front and the square root of time during horizontal infiltration (see Eq. 7) and is listed for each soil in Fig. 11 as a ratio to the value for the standard soil. If a soil belongs to the group for which this assumed scaling relationship is valid (which normally will not be known beforehand and must be verified), the hydraulic conductivity function can be obtained with Eq. 40 and just one simple, short infiltration run to measure m, θ_1, and θ_0.

Miller and Bresler [21] showed that the experimental data of Reichardt et al. [98], on which Eq. 40 is based, can be transformed to what they suggest to be a "universal" equation for the diffusivity

$$D[\theta] = \alpha m^2 \exp[\beta\theta] \tag{41}$$

with $\alpha = 10^{-3}$ and $\beta = 8$.

Bresler et al. [99] derived a relationship for the hydraulic conductivity from the same experimental data:

$$k[\theta] = 0.27 \; m^4 \theta^{7.2} \tag{42}$$

C. Texture

Hydraulic conductivities have also been correlated with soil textural data. These are more abundantly available than soil water retention characteristics and, therefore, are attractive. However, the results do not have a physical basis and the observed relationships can be used as statistics only. They must be verified by measurements on a large number of soils, while it remains uncertain whether they can be extrapolated to soils outside the group used to obtain this relationship. If such correlations are shown to be reliable predictors, a lot of work could be saved.

Bloemen [22] defined a particle size distribution index

$$f = \sum_{i=1}^{n} [(p_{i+1} - p_i)\log(p_{i+1}/p_i)/\log(S_{i+1}/S_i)] / \sum_{i=1}^{n} (p_{i+1} - p_i) \tag{43}$$

where p_i is the cumulative weight percentage and S_i is the corresponding particle size class boundary.

Based on data for a large number of Dutch soils, Bloemen found

$$k_s = 0.02M_d^{1.93}f^{-0.74} \qquad (cm\ d^{-1}) \qquad\qquad (44)$$

$$h_a = 2914M_d^{-0.96}f^{0.79} \qquad (cm) \qquad\qquad (45)$$

$$n = 1.4 + 4.536(e^{0.3f} - 1) - 0.75f^{1.6}\ \log\ OM \qquad\qquad (46)$$

$$k[h] = k_s\left(\frac{h_a}{h}\right)^n \qquad\qquad (47)$$

where M_d is the median particle size, h_a is the pressure head at air entry (cm), n is an empirical coefficient, and OM is the organic matter weight percentage.

It is doubtful that these results can be extrapolated to soils in other parts of the world. Schuh and Bauder [23] did a similar study on a number of soils in the United States. They found particularly good correlations between n and the ratio of sand to silt.

XI. PARAMETER OPTIMIZATION

Recently, the so-called inverse approach has received renewed attention in the form of a parameter optimization technique. First proposed around 1970 [100,101], the inverse approach calls for the performance of a relatively simple experiment with inherently accurate measurements. Subsequently, assuming algebraic forms of the hydraulic property functions, the water transport process is simulated on a computer, starting with guessed values of the parameters in the transport functions and then repeated with the newly estimated values until the simulated results agree with the experimental results to within the desired degree of accuracy. Thus the problem is reduced to optimizing the parameters in the transport functions. Optimization is a specialized mathematical process for which computer programs are available [102]. Mathematical details are not discussed at this point. The technique appears to have been improved recently so that it has become attractive for solving soil water flow problems. To be able to decide how the hydraulic transport functions can best be determined in a given situation, the merits of this inverse approach should be appreciated. Only a few aspects of it are discussed here. Further details can be found in the references. An up-to-date review is given in Kool et al. [7].

Whereas in principle many flow systems with different initial and/ or boundary conditions can be used for the parameter optimization, the one-step outflow method is especially suitable [25,103]. It requires no more than inherently accurate measurements of cumulated (external) outflow as a function of time from an initially saturated short soil column as a result of a step increase of the air pressure in a pressure plate apparatus. It allows a large water content range to be covered in a reasonably short time. The influence of the resistance of the porous plate on the outflow, which complicates the traditional analysis of the experimental results, is easily accounted for in the simulation. A draining soil column in which water content profiles must be measured at different times [24,104,105] is less attractive experimentally and can cover a much smaller water content range. Sir et al. [106] used one-dimensional infiltration as the flow process for optimization. The remarks in the following paragraphs specifically apply to optimization of the parameters in the Van Genuchten-Mualem functions (Eqs. 38 and 39), based on the experimental one-step outflow data of Parker et al. [103]. The same authors were also able to evaluate hysteresis in the hydraulic functions by solving the inverse problem consecutively for outflow and inflow on the same soil column [107].

A major aspect of the inverse approach is *convergence*. The first guess of the parameter values may be so far from the actual values that the optimization procedure cannot yield the correct values or can do this only after a prohibitively long computing time. As a first guess for medium-textured soils, the "average" values $\alpha = 2.50$ m^{-1}, n = 1.75, and $\theta_r = 0.150$ may be taken, with suitable adjustments for differently textured soils. Convergence also may be a problem when the information contained in the input data is too scanty. Therefore, the input data should cover as large a range of water contents, time, and other variables as is practical. To prevent undue use of computer time, a maximum number of function evaluations may be set. If the solution fails to converge within this number, a new solution can be started with different initial parameter values.

Another aspect of the inverse approach is *uniqueness*: there may be more solutions to the problem as stated, and the solution obtained may not be the correct one. This is not expected to be a serious problem with the one-step outflow measurements, if the pressure step and the time period are kept relatively large. However, the solutions obtained should be verified and again, in case of doubt, the optimization process should be repeated with different initial estimates of the parameters.

The *accuracy* of the optimized parameters depends on the accuracy of the experimental data used as input in the optimization procedure. The sensitivity for this source of errors is different for each combination of flow process and parametric functions and deserves further study. Of course, if the preselected algebraic functions are incapable of describing the actual soil hydraulic properties accurately, even a perfect optimization process will not yield an accurate result.

XII. SUMMARY AND CONCLUSIONS

Water transport in soils that are not fully saturated with water plays an important role in hydrology, water uptake by plant roots, irrigation management, transport of pollutants through the environment, and other areas. This transport is to a large extent characterized by the dependence on the volume fraction of water θ, of hydraulic conductivity k, diffusivity D, matric flux potential Φ, and sorptivity S. For a given soil, these soil water transport functions ($k[\theta]$, $D[\theta]$, etc.) vary over several orders of magnitude and can differ by orders of magnitude between soils. Measuring these functions is a difficult task, which continues to absorb much time and effort. Many methods have been proposed, but no single approach is suitable for all conditions and/or purposes. Most methods lack accuracy, take a prohibitively long time, and/or are costly. In general, steady-state methods are more accurate than transient methods, but they take a lot more time and are therefore more expensive. One also must choose between laboratory and field measurements. The former may have many advantages, which were spelled out above, but they require the acquisition of undisturbed soil samples and the transport of these to the laboratory.

The absolute accuracy of any given method cannot be established by using it on a "standard" porous medium with very accurately known hydraulic properties. As a result, it is standard practice to compare the results obtained by two (or more) different methods, without knowing the accuracy of either of them separately. It is necessary, therefore, to evaluate the available methods on their inherent features and potential accuracy. Methods of various types were described and evaluated in Table 1 with respect to a number of criteria and gradations, given in Table 2. Where the highest accuracy is required, methods should be selected according to soundness of theoretical basis (criterion C), control of initial and boundary conditions (D), inherent accuracy of the required measurements (E), and error propagation (F). On these criteria, "head-head" measurements on undisturbed soil cores between two porous plates score the highest. It is proposed, therefore, in view of the lack of a "standard" material, to elevate this method to the status of "standard method," against which other available methods could and should be evaluated. A disadvantage of this method is that it can be used conveniently only over a pressure head range from saturation down to about -2.5 m (G). This is normally more than sufficient for hydrological studies. With special effort (parentheses in Table 1), a larger pressure head range can be covered at the expense of more time (H) and better equipment (I). This is justified when a "standard" measurement is needed. Of the other laboratory methods, the "flux-head" variant, long-column infiltration, and flux-controlled sorptivity methods score the highest for criteria C-G.

As for field methods, the instantaneous profile method might seem to have only one big disadvantage, namely the very limited pressure head range over which it can yield results, even after rather long time periods. Unfortunately, the error analysis of Flühler et al. [79] shows that even with directly measured pressure heads and using only Darcy's law, the accuracy of the final results can be very poor. Use of the sprinkling infiltrometer under steady-state conditions at least eliminates large errors introduced when fluxes are calculated from indirectly measured water contents. Therefore, the sprinkling infiltrometer appears to be the strongest candidate for "standard field method." Operation of this equipment is very cumbersome and time-consuming. However, if accuracy is of overriding importance, criteria of required time (H), investments (I), skill (J), and operator time (K) should play a secondary role.

When accuracy is not as important as speed and minimizing cost, criteria H-K, as well as the potential for simultaneous measurements (L), become dominant. When many simultaneous measurements are made, it is also important (especially when these are carried out by unskilled workers) to provide for some check on the quality of the work (M). The recently proposed matric flux potential and ponded disk/dripper methods score quite high on these criteria. Also the hot air method is very attractive with respect to these criteria. However, the theoretical basis, control of boundary conditions, error propagation, and limitations on measurement accuracy are, in my opinion, so totally unacceptable that the hot air method should no longer be used.

The other Boltzmann-type methods do not have the disadvantage of poor boundary control and nonisothermal conditions, but the inaccuracy of the measurements and the unreliability of the analysis thereof are serious disadvantages. The spherical cavity method has a number of attractive features that appear to deserve further investigation. The pressure plate outflow method in its one-step variant is not good as a direct method, due to the approximate nature of the analysis of the experimental data. As a basis for the inverse approach of parameter optimization, however, the simple, accurate measurements involved make this method very attractive.

Given the unpredictability and nonuniformity of the conductivity of the crusts, as they are presently being made for the "crust method," the potential accuracy of this approach is questionable. Moreover, the pressure head range is very small. The crust method is too cumbersome and too time-consuming to be suitable for routine measurements at many sites. The use of hypodermic needles with a pulsating pump as a substitute for the crust promises to eliminate most of these limiting factors and give improved performance. This makes it a small, much simplified version of the sprinkling infiltrometer, which may well prove to be very useful.

Derivation of the water transport functions from other soil properties may be a good alternative to direct measurements, particularly when absolute accuracy is not of primary importance but many results are required (e.g., in studies of spatial or temporal variability as such). Often, the required input data are already available. The Van Genuchten-Mualem model appears to have an edge on other alternatives. It has an adequate theoretical basis, is generally available in user-friendly PC programs (and is, therefore, widely used), and has given good results for many studies. The same model is also used for the parameter optimization technique. This "inverse" approach seeks the values of the parameters of the model that give the best agreement between measured and numerically simulated quantities. It would seem that as the mathematical procedure is further improved in terms of convergence, uniqueness, and accuracy, this approach should be used more and more. This will be true, particularly, if the selected experimental flow system can be tailored to the actual situation and conditions in which the results will be used.

REFERENCES

1. Klute, A., The determination of the hydraulic conductivity and diffusivity of unsaturated soils, *Soil Sci.*, 113: 264-276 (1972).
2. Bouwer, H., and R. D. Jackson, Determining soil properties, in *Drainage for Agriculture* (J. van Schilfgaarde, Ed.), Am. Soc. Agron., Madison, WI, 1974, pp. 611-672.
3. Klute, A., and C. Dirksen, Hydraulic conductivity and diffusivity: Laboratory methods, in *Methods of Soil Analysis, Part I, Physical and Mineralogical Methods*, 2nd ed. (A. Klute, Ed.), Am. Soc. Agron., Madison, WI, 1986, pp. 687-734.
4. Green, R. E., L. R. Ahuja, and S. K. Chong, Hydraulic conductivity, diffusivity, and sorptivity of unsaturated soils: Field methods, in *Methods of Soil Analysis, Part I. Physical and Mineralogical Methods*, 2nd ed. (A. Klute, Ed.), Am. Soc. Agron., Madison, WI, 1986, pp. 771-798.
5. Mualem, Y., Hydraulic conductivity of unsaturated soils: Prediction and formulas, in *Methods of Soil Analysis, Part I. Physical and Mineralogical Methods*, 2nd ed. (A. Klute, Ed.), Am. Soc. Agron., Madison, WI, 1986, pp. 789-823.
6. Van Genuchten, M. T., and D. R. Nielsen, On describing and predicting the hydraulic properties of unsaturated soils, *Ann. Geophys.*, 3: 615-628 (1985).
7. Kool, J. B., J. C. Parker, and M. T. Van Genuchten, Parameter estimation for unsaturated flow and transport models—A review, *J. Hydrol.*, 91: 255-293 (1987).

8. Hillel, D., *Fundamentals of Soil Physics*, Academic Press, New York, 1980.

9. Hillel, D., *Applications of Soil Physics*, Academic Press, New York, 1980.

10. Koorevaar, P., G. Menelik, and C. Dirksen, *Elements of Soil Physics*, Elsevier, Amsterdam, 1983.

11. Raats, P. A. C., and W. R. Gardner, Comparison of empirical relationships between pressure head and hydraulic conductivity and some observations on radially symmetric flow, *Water Resour. Res.*, 7: 921-928 (1971).

12. Shaykewich, C. F., and L. Stroosnijder, The concept of matric flux potential applied to simulation of evaporation from soil, *Neth. J. Agric. Sci.*, 25: 63-82 (1977).

13. Ten Berge, H. F. M., K. Metselaar, and L. Stroosnijder, Measurement of matric flux potential: A simple procedure for the hydraulic characterisation of soils, *Neth. J. Agric. Sci.*, 35: 371-384 (1987).

14. Vauclin, M. R., R. Haverkamp, and G. Vachaud, *Résolution numerique d'une équation de diffusion non-linéaire*, Presse Universitaire de Grenoble, France, 1979.

15. Warrick, A. W., Time-dependent linearized infiltration: I. Point sources, *Soil Sci. Soc. Am. J.*, 38: 383-386 (1974).

16. Raats, P. A. C., Laterally confined, steady flows of water from sources and to sinks in unsaturated soils, *Soil Sci. Soc. Am. J.*, 41: 294-304 (1977).

17. Dirksen, C., Flux-controlled sorptivity measurements to determine soil hydraulic property functions, *Soil Sci. Soc. Am. J.*, 43: 827-834 (1979).

18. White, I., and K. M. Perroux, Use of sorptivity to determine field soil hydraulic properties, *Soil Sci. Soc. Am. J.*, 51: 1093-1101 (1987).

19. Philip, J. R., Theory of infiltration, *Hydroscience*, 5: 215-296 (1969).

20. Reichardt, K., P. L. Libardi, and D. R. Nielsen, Unsaturated hydraulic conductivity determination by a scaling technique, *Soil Sci.*, 120: 165-168 (1975).

21. Miller, R. D., and E. Bresler, A quick method for estimating soil water diffusivity functions, *Soil Sci. Soc. Am. J.*, 41: 1020-1022 (1977).

22. Bloemen, G. W., Calculation of hydraulic conductivities of soils from texture and organic matter content, *Z. Pflanzenernaehr. Bodenk.*, 143: 581-605 (1980).

23. Schuh, W. M., and J. W. Bauder, Effect of soil properties on hydraulic conductivity-moisture relationships, *Soil Sci. Soc. Am. J.*, 50: 848-854 (1986).

24. Dane, J. H., and S. Hruska, In-situ determination of soil hydraulic properties during drainage, *Soil Sci. Soc. Am. J.*, 47: 619-624 (1985).

25. Kool, J. B., J. C. Parker, and M. T. Van Genuchten, Determining soil hydraulic properties from one-step outflow experiments by parameter estimation: I. Theory and numerical studies, *Soil Sci. Soc. Am. J.*, 49: 1348-1354 (1985).

26. Begemann, H. K. S. P., The 66 mm continuous sampling apparatus, Delft Soil Mechanics Lab., Delft, The Netherlands, 1988.

27. Peck, A. J., Field variability of soil physical properties, *Adv. Irrig.*, 2: 189-221 (1980).

28. Verlinden, H. L., and J. Bouma, *Fysische Bodemonderzoekmethoden voor de Onverzadigde zone*, VROM-Rapport BO 22, The Netherlands, 1983.

29. Warrick, A. W., and D. R. Nielsen, Spatial variability of soil physical properties in the field, in *Applications of Soil Physics* D. Hillel, Ed.), Academic Press, New York, 1980, pp. 319-344.

30. Journel, A., and C. Huibregts, *Mining Geostatistics*, Academic Press, New York, 1978.

31. Gardner, W. R., and F. J. Miklich, Unsaturated conductivity and diffusivity measurements by a constant flux method, *Soil Sci.*, 93: 271-274 (1962).

32. Childs, E. C., and N. Collis-George, The permeability of porous materials, *Proc. R. Soc. Aust.*, 201: 392-405 (1950).

33. Youngs, E. G., An infiltration method of measuring the hydraulic conductivity of unsaturated porous materials, *Soil Sci.*, 97: 307-322 (1964).

34. Wesseling, J., and K. E. Wit, An infiltration method for the determination of the capillary conductivity of undisturbed soil cores [Wageningen Symp. Water in the Unsaturated Zone], *Int. Assoc. Sci. Hydrol., Proc.*: 223-234 (1966).

35. Kleijn, W. B., J. D. Oster, and N. Cook, A rainfall simulator with nonrepetitious movement of drop outlets, *Soil Sci. Soc. Am. J.*, 43: 1248-1251 (1979).

36. Knight, J. H., and J. R. Philip, Exact solutions of non-linear diffusion, *J. Eng. Math.*, 8: 219-227 (1974).

37. Hillel, D., and Y. Benyamini, Experimental comparison of infiltration and drainage methods for determining unsaturated hydraulic conductivity of a soil profile in situ, in *Isotope and Radiation Techniques in Soil Physics and Irrigation Studies*, IAEA, Vienna, 1974, pp. 271-275.

38. Amerman, C. R., D. Hillel, and A. E. Petersen, A variable-intensity sprinkling infiltrometer, *Soil Sci. Soc. Am. Proc.*, 34: 830-832 (1970).

39. Rawitz, E., M. Margolin, and D. Hillel, An improved variable intensity sprinkling infiltrometer, *Soil Sci. Soc. Am. Proc.*, 36: 533–535 (1972).

40. Petersen, A. E., and G. D. Bubenzer, Intake rate: Sprinkler infiltrometer, in *Methods of Soil Analysis, Part I. Physical and Mineralogical Methods*, 2nd ed. (A. Klute, Ed.), Am. Soc. Agron., Madison, WI, 1986, pp. 845–870.

41. Baver, L. D., W. H. Gardner, and W. R. Gardner, *Soil Physics*, 4th ed., Wiley, New York, 1972.

42. Lal, R., and D. J. Greenland, *Soil Physical Properties and Crop Production in the Tropics, Part 8, Soil and Water Conservation*, Wiley, New York, 1979.

43. Bouma, J., D. Hillel, F. D. Hole, and C. R. Amerman, Field measurement of unsaturated hydraulic conductivity by infiltration through artificial crusts, *Soil Sci. Soc. Am. Proc.*, 35: 362–364 (1971).

44. Bouma, J., and J. L. Denning, Field measurement of unsaturated hydraulic conductivity by infiltration through gypsum crusts, *Soil Sci. Soc. Am. Proc.*, 36: 846–847 (1972).

45. Dirksen, C., Measurement of hydraulic conductivity by means of steady, spherically symmetric flows, *Soil Sci. Soc. Am. Proc.*, 38: 3–8 (1974).

46. Wooding, R. A., Steady infiltration from a shallow circular pond, *Water Resour. Res.*, 4: 1259–1273 (1968).

47. Scotter, D. R., B. E. Clothier, and E. R. Harper, Measuring saturated hydraulic conductivity and sorptivity using twin rings, *Aust. J. Soil Res.*, 20: 295–304 (1982).

48. Shani, U., R. J. Hanks, E. Bresler, and C. A. S. Oliveira, Field method for estimating hydraulic conductivity and matric potential-water content relations, *Soil Sci. Soc. Am. J.*, 51: 298–302 (1987).

49. Brooks, R. H., and A. T. Corey, Hydraulic properties of porous media, Hydrology Paper No. 3, Colorado State University, 1964.

50. Russo, D., and E. Bresler, Field determination of soil hydraulic properties for statistical analysis, *Soil Sci. Soc. Am. J.*, 44: 697–702 (1980).

51. Schaik, J. C. Van, Soil hydraulic properties determined with water and with a hydrocarbon liquid, *Can. J. Soil Sci.*, 50: 79–84 (1970).

52. Dirksen, C., Relationship between root uptake-weighted mean soil water salinity and total leaf water potentials of alfalfa, *Irrig. Sci.*, 6: 39–50 (1985).

53. Gardner, W. R., Calculation of capillary conductivity from pressure plate outflow data, *Soil Sci. Soc. Am. Proc.*, 20: 317–320 (1956).

54. Doering, E. J., Soil-water diffusivity by the one-step method, *Soil Sci.*, 99: 322-326 (1965).

55. Gupta, S. C., D. A. Farrell, and W. E. Larson, Determining effective soil water diffusivities from one-step outflow experiments, *Soil Sci. Soc. Am. Proc.*, 38: 710-716 (1974).

56. Gardner, W. R., Note on the separation and solution of diffusion type equations, *Soil Sci. Soc. Am. Proc.*, 26: 404 (1962).

57. Passioura, J. B., Determining soil water diffusivities from one-step outflow experiments, *Aust. J. Soil Res.*, 15: 1-8 (1976).

58. Chung, C. L., S. H. Anderson, C. J. Ganzer, and Z. Haque, Automated one-step outflow method for measurement of unsaturated hydraulic conductivity, *1988 Agron. Abstr.*: 181 (1988).

59. Kirkham, D., and W. L. Powers, *Advanced Soil Physics*, Wiley, New York, 1972.

60. Bruce, R. R., and A. Klute, The measurement of soil-moisture diffusivity, *Soil Sci. Soc. Am. Proc.*, 20: 458-462 (1956).

61. Whisler, F. D., A. Klute, and D. B. Peters, Soil water diffusivity from horizontal infiltration, *Soil Sci. Soc. Am. Proc.*, 32: 6-11 (1968).

62. Selim, H. M., D. Kirkham, and M. Amemiya, A comparison of two methods for determining soil water diffusivity, *Soil Sci. Soc. Am. Proc.*, 34: 14-18 (1970).

63. Jackson, R. D., Porosity and soil-water diffusivity relations, *Soil Sci. Soc. Am. Proc.*, 27: 123-126 (1963).

64. Clothier, B. E., D. R. Scotter, and A. E. Green, Diffusivity and one-dimensional absorption experiments, *Soil Sci. Soc. Am. J.*, 47: 641-644 (1983).

65. Reichardt, K., and P. L. Libardi, A new equation for the estimation of soil water diffusivity, in *Isotope and Radiation Techniques in Studies of Soil Physica, Irrigation and Drainage in Relation to Crop Production*, IAEA, Vienna, 1974, pp. 45-51.

66. Arya, L. M., D. A. Farrell, and G. R. Blake, A field study of soil water depletion patterns in presence of growing soybean roots: I. Determination of hydraulic properties in the soil, *Soil Sci. Soc. Am. J.*, 39: 424-430 (1975).

67. Van Grinsven, J. J. M., C. Dirksen, and W. Bouten, Evaluation of the hot air method for measuring soil water diffusivity, *Soil Sci. Soc. Am. J.*, 49: 1093-1099 (1985).

68. Van den Berg, J. A., and T. Louters, An algorithm for computing the relationship between diffusivity and soil moisture content from the hot air method, *J. Hydrol.*, 83: 149-159 (1986).

69. Dirksen, C., Determination of soil water diffusivity by sorptivity measurements, *Soil Sci. Soc. Am. Proc.*, 39: 22-27 (1975).

70. Dirksen, C., Determination of soil water diffusivity by sorptivity measurements: Reply to Dr. Parlange's letter, *Soil Sci. Soc. Am. Proc.*, 39: 1012-1013 (1975).

71. Brutsaert, W. H., The concise formulation of diffusive sorption of water in a dry soil, *Water Resour. Res.*, 12: 1118-1124 (1976).

72. Wind, G. P., Capillary conductivity data estimated by a simple method, in *Proc. UNESCO/IASH Symp. Water in the Unsaturated Zone*, Wageningen, The Netherlands, 1966, pp. 181-191.

73. Boels, D., J. B. H. M. Van Gils, G. J. Veerman, and K. E. Wit, Theory and system of automatic determination of soil moisture characteristics and unsaturated hydraulic conductivities, *Soil Sci.*, 126: 191-199 (1978).

74. Feddes, R. A., P. Kabat, P. J. T. van Bakel, J. J. B. Bronswijk, and J. Halbertsma, Modelling soil water dynamics in the unsaturated zone—State of the art, *J. Hydrol.*, 100: 69-111 (1988).

75. Ahuja, L. R., and S. A. El-Swaify, Determining both water characteristics and hydraulic conductivity of a soil core at high water contents from a transient flow experiment, *Soil Sci.*, 121: 198-204 (1976).

76. Scotter, D. R., and B. E. Clothier, A transient method for measuring soil water diffusivity and unsaturated hydraulic conductivity, *Soil Sci. Soc. Am. J.*, 47: 1069-1072 (1983).

77. Watson, K. K., An instantaneous profile method for determining the hydraulic conductivity of unsaturated porous materials, *Water Resour. Res.*, 2: 709-715 (1966).

78. Hillel, D., V. D. Krentos, and Y. Stylianou, Procedure and test of an internal drainage method for measuring soil hydraulic characteristics in situ, *Soil Sci.*, 114: 395-400 (1972).

79. Flühler, H., M. S. Ardakani, and L. H. Stolzy, Error propagation in determining hydraulic conductivities from successive water content and pressure head profiles, *Soil Sci. Soc. Am. J.*, 40: 830-836 (1976).

80. Richards, L. A., W. R. Gardner, and G. Ogata, Physical processes determining water loss from soil, *Soil Sci. Soc. Am. Proc.*, 20: 310-314 (1956).

81. Jones, A. J., and R. J. Wagenet, In situ estimation of hydraulic conductivity using simplified methods, *Water Resour. Res.*, 20: 1620-1626 (1984).

82. Libardi, P. L., K. Reichardt, D. R. Nielsen, and J. W. Biggar, Simple field methods for estimating soil hydraulic conductivity, *Soil Sci. Soc. Am. J.*, 44: 3-7 (1980).

83. Gardner, W. R., Field measurement of soil water diffusivity, *Soil Sci. Soc. Am. Proc.*, 34: 832-833 (1970).

84. Talsma, T., In situ measurement of sorptivity, *Aust. J. Soil Res.*, 7: 269-276 (1969).

85. Clothier, B. E., and I. White, Measurement of sorptivity and soil water diffusivity in the field, *Soil Sci. Soc. Am. J.*, 45: 241-245 (1981).

86. Marshall, T. J., A relation between permeability and size distribution of pores, *J. Soil Sci.*, 9: 1-8 (1958).
87. Millington, R. J., and J. P. Quirk, Permeability of porous solids, *Trans. Faraday Soc.*, 57: 1200-1207 (1961).
88. Laliberte, G. E., R. H. Brooks, and A. T. Corey, Permeability calculated from desaturated data, *J. Irrig. Drainage Div. ASCE*, 94: 57-71 (1968).
89. Jackson, R. D., On the calculation of hydraulic conductivity, *Soil Sci. Soc. Am. Proc.*, 36: 380-382 (1972).
90. Jackson, R. D., R. J. Reginato, and C. H. M. Van Bavel, Comparison of measured and calculated hydraulic conductivities of unsaturated soils, *Water Resour. Res.*, 1: 375-380 (1965).
91. Brust, K. J., C. H. M. Van Bavel, and G. B. Stirk, Hydraulic properties of a clay loam and the field measurement of water uptake by roots: III. Comparison of field and laboratory data on retention and of measured and calculated conductivities, *Soil Sci. Soc. Am. Proc.*, 32: 322-326 (1968).
92. Green, R. E., and J. C. Corey, Calculation of hydraulic conductivity: A further evaluation of some predictive methods, *Soil Sci. Soc. Am. Proc.*, 35: 3-8 (1971).
93. Mualem, Y., A new model for predicting the hydraulic conductivity of unsaturated porous media, *Water Resour. Res.*, 12: 513-522 (1976).
94. Burdine, N. T., Relative permeability calculations from pore-size distribution data, *Trans. AIME*, 198: 71-78 (1953).
95. Van Genuchten, M. T., A closed-form equation for predicting the hydraulic conductivity of unsaturated soils, *Soil Sci. Soc. Am. J.*, 44: 892-898 (1980).
96. Miller, E. E., and R. D. Miller, Physical theory for capillary flow phenomena, *J. Appl. Phys.*, 27: 324-332 (1956).
97. Miller, E. E., Similitude and scaling of soil water phenomena, in *Applications of Soil Physics* (D. Hillel, Ed.), Academic Press, New York, 1980, pp. 300-318.
98. Reichardt, K., D. R. Nielsen, and J. W. Biggar, Scaling of horizontal infiltration into homogeneous soils, *Soil Sci. Soc. Am. Proc.*, 36: 241-245 (1972).
99. Bresler, E., D. Russo, and R. D. Miller, Rapid estimate of hydraulic conductivity function, *Soil Sci. Soc. Am. J.*, 42: 170-172 (1978).
100. Whisler, F. D., and K. K. Watson, One-dimensional gravity drainage of uniform columns of porous materials, *J. Hydrol.*, 6: 277-296 (1968).
101. Skaggs, W. R., E. J. Monk, and L. F. Huggins, An approximate method for defining the hydraulic conductivity-pressure potential relationship for soils, *Trans. ASAE, 14*: 130-133 (1971).

102. Kool, J. B., J. C. Parker, and M. T. Van Genuchten, *ONE-STEP: A Nonlinear Parameter Estimation Program for Evaluating Soil Hydraulic Properties from One-step Outflow Experiments*, Virginia Agric. Exp. Stn. Bull. No. 85-3, 1985.

103. Parker, J. C., J. B. Kool, and M. T. Van Genuchten, Determining soil hydraulic properties from one-step outflow experiments by parameter estimation: II. Experimental studies, *Soil Sci. Soc. Am. J.*, 49: 1354-1359 (1985).

104. Zachmann, D. W., P. C. Du Chateau, and A. Klute, The calibration of the Richards flow equation for a draining column by parameter identification, *Soil Sci. Soc. Am. J.*, 45: 1012-1015 (1981).

105. Zachmann, D. W., P. C. Du Chateau, and A. Klute, Simultaneous approximation of water capacity and soil hydraulic conductivity by parameter identification, *Soil Sci.*, 134: 157-163 (1982).

106. Sir, M., M. Kutilek, V. Kuraz, M. Krejca, and F. Kubik, Field estimation of the soil hydraulic characteristics, *Soil Technol.*, 1: 63-75 (1988).

107. Kool, J. B., J. C. Parker, and M. T. Van Genuchten, The inverse problem for hysteretic unsaturated flow, Proc. 6th Int. Conf. on Finite Elements in Water Resources, Lisbon, Portugal, June 1986, Springer Verlag, pp. 337-346.

6
Particle Size Analysis

PETER J. LOVELAND *Soil Survey and Land Research Centre, Cranfield Rural Institute, Silsoe, Bedfordshire, England*

W. RICHARD WHALLEY *AFRC Institute of Engineering Research, Silsoe, Bedfordshire, England*

I. FOREWORD

This chapter is not a laboratory manual. It is more concerned with the principles underlying the concepts of particle, size, and distribution, the relationships between them, and the methods by which they may be measured. Soil scientists have amassed a large body of measurements of particle size distribution using simple methods and equipment, principally sieving, gravitational settling, the pipet, and the hydrometer. There is also a large body of experience in interpreting these data. There is still a surprising lack of uniformity in these simple procedures, and for that reason we consider them in some detail. However, it is also true that soil science can make little further headway in the interpretation of particle size distribution at the very small size end of the range, because the simple methods are incapable of further resolution. For that reason we have reviewed a number of less common or more recent instrumental techniques, which are capable of extending our understanding of the distribution of particles, in particular at the submicrometer level. Such methods are not without their problems, but if progress is to be made in improving our understanding of soil behavior, especially in terms of the influence of the very smallest particles, this seems to be the way forward. We hope this review will encourage others to try newer techniques so that a body of knowledge and experience will accrue which will ultimately lead to improvements in our ability to understand, hence predict, soil behavior, with greater confidence.

II. INTRODUCTION

The fractionation of crushed ores by sieving has been known since
Neolithic times [1].   Similarly, the sorting of cereal grains by sieving and winnowing, respectively, is depicted in Egyptian wall paint-
ings of approximately 1900 B.C. [1] and early Chinese texts [2].
The influence of particle size on the properties of single substances,
notably pigments, was also appreciated in ancient times [3].   Mech-
anized sieving was well known by the sixteenth century [4], while
the rise of industrial chemistry in the nineteenth century brought
growing realization of the importance of particle size in the behavior
of materials [5].

The classification of soils in terms of particle size stems essentially from the work of Atterberg [6], further developments being
summarized in Oden [7].   The International Society of Soil Science
adopted, in 1927, proposals to standardize the method for the "mechanical analysis" of soils by a combination of sieving and pipeting,
and equally important, resolved to analyze (at least for agricultural
soils) only the fraction passing a round-hole 2 mm sieve—the so-
called fine earth [8].   This simplification of a complex story does
not end there.   There have been many revisions of the particle size
classes promulgated in 1927, many other disciplines need to determine the size of particles and, particularly germane to soil science,
sedimentologists especially have continued to question the concepts
of particle, size, and shape.   It is widely recognized that there are
particular problems at the very small end of the particle size range,
where the methods have traditionally relied on Stokes' law with its
implications of spherical particles.   There are now some 400 reported
techniques for the determination of particle size [9].

Relatively few have been applied to soils, clays, or other sediments.   These are commonly analyzed by a combination of sieving
and sedimentation, and a large number of standard methods for such
size analysis based on these principles are available (see, e.g., Refs.
10-15).   This chapter examines some traditional techniques and some
newer ones, but first there follows a brief section on what is meant
by "particle" and "size."

III. BASIC CONCEPTS

A. Particles

A particle can be regarded as any coherent body bounded by a
clearly recognizable surface.   Particles may consist of one kind of
material with uniform properties, or of smaller particles bonded together in some way, the properties of each being, possibly, very
different.   It should be remembered that a soil is formed under
particular conditions, and the particles are to a greater or lesser

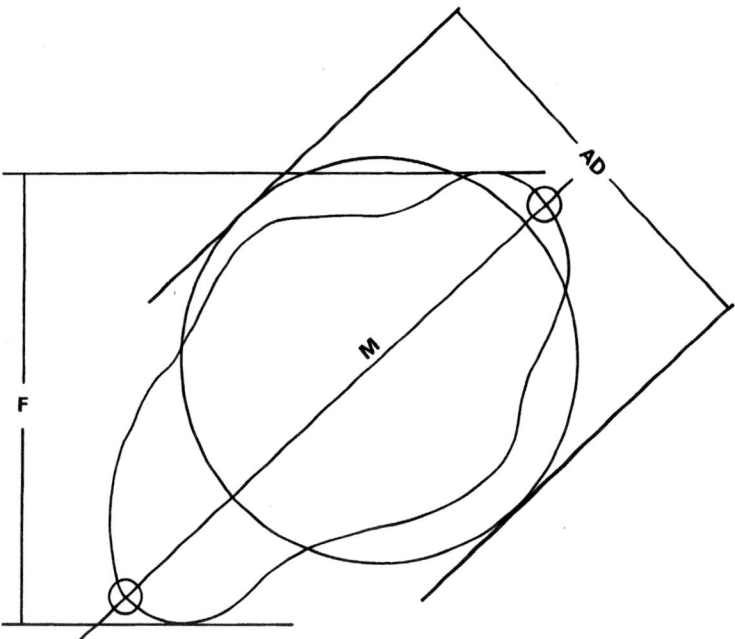

FIG. 1 Some two-dimensional measures of particle size: F (Feret's diameter) is the distance between two parallel tangents drawn on opposite sides of the particle. A particle can have several Feret diameters, one of which may be the maximum particle dimension; M (Martin's diameter) is the length of a line (O-O) that bisects the particle; the projected area diameter (AD) is the area of a circle having the same area as the particle projected onto a plane surface.

extent products of those conditions. If the soil is disturbed, the particles may change: for example, salts and cements can dissolve, organic remains can be fragile, bonding ions can hydrolyze, and bonds thus be weakened. Not all these changes may be desirable if the original material is to be fully and properly characterized.

Few natural particles are spheres, and often the smaller they are, the greater is the departure from sphericity. Allen [16] makes the points that one method of size analysis may not be enough, and the methods chosen should reflect the information desired. There may be little point in characterizing as spheres particles that are plates. Sedimentologists and powder technologists have given much attention to the relationship between particle dimensions and particle shape. Allen [16] lists 13 measures of particle size applicable to powders. In soil analysis the commonest by far is the volume diameter, which is generally equated with Stokes' diameter. Recently,

Murphy et al. [17] and Ringrose-Voase and Bullock [18] used Feret's diameter (Fig. 1) to characterize irregular soil pores, which have many geometric properties in common with irregular particles. Sedimentologists have made considerable efforts to characterize irregular particles in terms of "sphericity" or, more usually, an index to indicate departure from sphericity (see, e.g., Cui and Komar [19], Ehrlich and Weinberg [20], and especially Griffiths [21]). Many of these methods were developed to differentiate depositional environments, something that ought to concern at least some soil scientists. All the methods, however, tend to involve an inordinate amount of labor to acquire enough measurements on enough grains to obtain statistically valid data [21].

The recent introduction of image-analyzing computers has made the task of size analysis much easier and has extended the techniques beyond the range of the optical microscope [e.g., 22-24]. In an interesting development, Gutteridge has shown that good particle size data can be obtained from acetate peels [25]. Kaye [26] has applied fractal geometry, in conjunction with image analysis, to the characterization of carbon flocs, a technique that seems to be capable of extension to irregular soil particles.

B. Size and Related Matters

Soils may contain particles ranging from more than one meter in a maximum dimension to fractions of a micrometer (i.e., a size ratio of 1,000,000:1 or more). It is hardly surprising that it is difficult to conceive of one measurement technique that covers this range adequately. For the larger particles, which can be viewed easily by the naked eye, a fairly crude measure of size is often the maximum dimension from one point on the particle to another (Fig. 1). In many cases it is necessary only to give some scale to the coarse material present—for example, as a guide to the practicalities of plowing a piece of land or to assess the portion of the soil mass occupied by particles above a certain approximate size. It is the smaller particles, however, on which most interest focuses, as these tend to have a disproportionate influence on the physical and chemical behavior of soils.

Size and shape are indissolubly linked when considering particle size determination. The only particle whose dimensions can be specified by one number (viz., its diameter) is the sphere. Other particle shapes can be related to a sphere by means of their volume. For example, a cube of side 1 cm has the same volume as a sphere of 1.24 cm diameter. This is the concept of equivalent sphere (or spherical) diameter, often written esd. It is clearly relatively easy to characterize the behavior of spheres of differing diameters, and to equate particles of similar behavior to those spheres in terms of their esd.

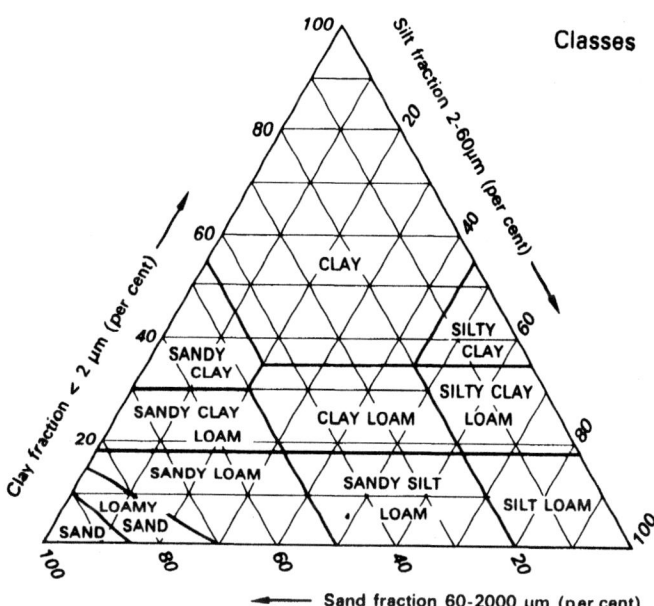

FIG. 2 Triangular diagram relating proportions of sand, silt, and clay to particle size classes as defined in England and Wales. [After Ref. 28, courtesy of Soil Survey and Land Research Centre, Silsoe, U.K.)

The limitations of the equivalent sphere diameter concept are illustrated by the following:

A sphere of diameter 2 µm has a volume of approximately 4×10^{-12} cm^3; the same volume is occupied by a particle 100 nm $\times$ 2 µm $\times$ 20 µm.

Whichever method of particle size determination is used, most soil scientists are interested in the amount (usually a weight-percent) of particles within any given size class, as defined by an upper and lower limit (e.g., 60-200 µm). Size classes usually are identified by name, such as clay, silt, or boulder, and each class corresponds to a grade. The rationalization of grade names in relation to specified size ranges owes much to the work of Wentworth [27]. It is common, particularly among sedimentologists, to describe a deposit in terms of its principal particle size class, for example, of being "sand grade." Soil scientists use a similar system when using the proportions of material in different size fractions to construct so-called texture triangles (Fig. 2) or particle size class

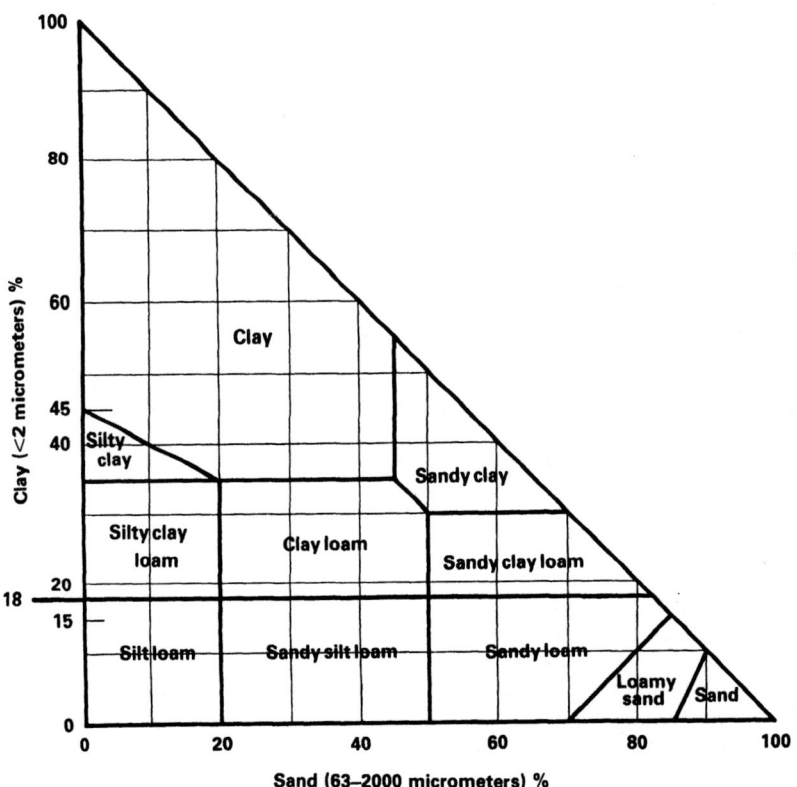

FIG. 3 Particle size classes from Fig. 2 redrawn as an orthogonal graph using only clay and sand fractions.

triangles. Orthogonal axis figures are equally valid and easier to use but have not been widely adopted (Fig. 3). There is, however, considerable variation among countries as to the limits of the different particle size classes, as discussed in detail by Hodgson [28]. Many countries have official standards (e.g., Refs. 10 and 29) in which these matters are thoroughly specified.

The distribution of particles in the different size classes can be used to construct particle size distribution curves, the commonest of which is the cumulative curve, although there are others [16,21, 30]. It should be stressed, however, that interpolation of intermediate values of particle size from such curves should be undertaken with care. The curves are only as good as the method used to obtain the data and the number of points used to construct them. Serious errors can arise if the latter are inadequate [31].

TABLE 1   Minimum Quantities of Soils for Sieve Analysis

Maximum size of particle forming more than 10% of soil (mm)	Minimum mass of soil for sieve analysis (kg)
63	50
50	35
37.5	15
28	5
20	2
<20[a]	1

[a]It is recommended that the minimum sample mass be 1 kg, however small the particles.

Source: Modified from Refs. 12 and 96.

Alderliesten [32] presenta a scheme for characterizing particle size distributions in terms of a mean diameter function.

C.   Sampling and Treatment of Data

Sampling and treatment of data, essentially statistical matters, are discussed exhaustively by many authors [11,16,21,33]. The cardinal principle is that the sample must be representative of the soil under study; otherwise, the resulting data will be inadequate or misleading, and no amount of statistical massaging will compensate for this.   The immensely costly failure of the post-World War II "groundnuts scheme" for agricultural development in what was then British East Africa was due partly to inadequate soil data [34].

Head [12] gives recommended minimum quantities of soil to be taken for analysis based on the maximum size of particle forming more than 10% of the soil (Table 1), the general rule being that the amount of material taken should be about 100 times greater than the mass of the largest particle present.   It is clear that as particle size increases, the problems of representative sampling become formidable.

Ideally, laboratory subsamples should be taken from a moving stream of the bulk material [16]. A rotary sampler or chute spitter is the best tool for obtaining relatively small samples of soil of less than 2 mm size from a larger bulk sample [35], while riffling can be used up to about 10 cm, and the only practicable method thereafter is coning and quartering [29].

The *accuracy* of particle size analysis methods for soils is difficult to establish in the absence of any precisely sizable natural soils (i.e., soils made up of perfectly spherical particles) for use as standards and for comparisons among different methods. In practical terms, if a properly subsampled reference soil, or soils, is introduced into each analytical batch, the accumulated results from this exercise can be used to assess the *precision* (reproducibility) of the method used, rather than its absolute accuracy.

There is a surprisingly widespread view, in our experience, that a few percent error either way in the particle size determination of a specific size class is not very important. This idea seems to stem from the beliefs that soils are inherently variable and that in most cases the analytical data are used only to place a soil in a particle size class. However, size classes have numerical boundaries, and major decisions can flow from which class a soil is placed in. Therefore, the class should be decided on the best possible data that can be obtained.

IV. PARTICLE SIZING TECHNIQUES

A. Introduction

Methods for determining particle size can be divided into the following broad groups:

Direct measurement (ruler, caliper, microscope, etc.)
Sieving
Elutriation
Sedimentation (gravity, centrifugation)
Interaction with radiation (light, laser light, X-rays, neutrons)
Electrical properties
Optical properties
Gas adsorption
Permeability

Some methods make use of combinations of these methods. This chapter cannot do more than touch on some of the techniques available, especially insofar as so many seem not to have been applied to soils. We make no attempt to provide a recipe book; rather, we aim to discuss the principles, origins, and limitations of some of the standard methods and to point to newer methods that may provide more and/or better information as to how particles in soils may be characterized, hence how soil behavior may be better predicted. It is becoming clear, for example, that the material of less than 2 µm esd in soil can have very different size distributions, and this must have profound effects on soil properties. Table 2 gives commercial sources of some of the instrumentation.

TABLE 2 Suppliers of Equipment[a]

1. General equipment (samplers, sieves, shakers, splitters, crushers, etc.)

   Alpine Process Technology, Rivington Road, Whitehouse, Runcorn, Cheshire WA7 3DS, U.K.
   Telex: 628051 ALPINE G

   Eijkelkamp Agrisearch Equipment, P.O. Box 4, 6987 ZG Giesbeek, The Netherlands
   Telex: 35416 EYKEL NL

   ELE International (Agronomics), Eastman Way, Hemel Hempstead, Hertfordshire, HP2 7HG, U.K.
   Telex: 825239 ELELTO G

   Endecotts Limited, 9 Lombard Road, Morden Factory Estate, London SW19 3BR, U.K.
   Telex: 929395 ENDSIV G

   Fritsch Laborgerätebau GmbH, D-6580 Idar-Oberstein 1, Federal Republic of Germany
   Telex: 426203 FRITS D

   The Giddings Machine Company, P.O. Drawer 2024, Fort Collins, 80521 Colorado

   Gilson Screen Company, P.O. Box 99, Malinta, 43535 Ohio
   Telex: 128285 WHOM FOR NYK: ATTGSC

   Glen Creston Ltd., 16 Dalston Gardens, London NW9 OHL, U.K.
   Telex: 925792 GLENCR G

   Humax Bohr- und Grundwassersonden, Postfach 376, CH-1211 Geneva 4, Switzerland
   Telex: 423683

   Ladal (Scientific Equipment) Ltd., Warlings, Warley Edge, Halifax, Yorkshire HX2 7RL, U.K.

   Pascal Engineering Co. Ltd., Gatwick Road, Crawley, Sussex RH10 2RD, U.K.
   Telex: 878386 PASCO G

   Seishin Enterprise Co. Ltd., Nippon Brunswick Buildings, 5-27-7 Sendagaya, Shibuya-ku, Tokyo, Japan
   Telex: 232-4242 SEISIN J

   Vi-cor Technologies, Inc., 400-601 West Cordova Street, Vancouver, British Columbia V6B 1G1, Canada
   Telex: 04-55465

   Wykeham Farrance Engineering Ltd., Weston Road Trading Estate, Slough, Berkshire SL1 4HW, U.K.
   Telex: 847301 WKENG G

TABLE 2 (continued)

2. Centrifugal analyzers

   Brookhaven Instruments Corp., 200 Thirteenth Avenue,
   Ronkonkoma, NY 11779
   Telex: WUI 6852252 BRKHUN INSTR

   Fritsch Laborgerätebau GmbH, D-6580 Idar-Oberstein 1,
   Federal Republic of Germany
   Telex: 426203 FRITS D

   Horiba Ltd., Miyanohigashi, Kisshaoin, Kyoto, Japan
   Telex: 5422130 HORIBA J

   Joyce-Loebl, Marquis Way, Team Valley, Gateshead NE11 0QW,
   U.K.
   Telex: 537270 JLOEBL G

3. Digital Density Meters

   Anton Paar K.G., Postfach 58, A-8054 Graz, Austria
   Telex: 03-1805

4. Electrical sensing zone devices

   Coulter Electronics Ltd., Northwell Drive, Luton,
   Bedfordshire LU3 3RH, U.K.
   Telex: 820574 COULTER G

   Particle Data, Inc., Elmhurst, Illinois 60126
   Telex: 910 254 0180 PARDATA ELM H

5. Field-flow chromatography

   Du Pont Ltd., Wedgwood Way, Stevenage, Hertfordshire
   SG1 4QN, U.K.
   Telex: 825591 DUPONT G

6. Light-scattering devices

   Brookhaven Instruments Corp., 200 Thirteenth Avenue,
   Ronkonkoma, NY 11779
   Telex: WUI 6852252 BRKHUN INSTR

   Cilas Alcatel, Route de Nozay, B.P. 27, 91460 Marcoussis, France
   Telex: 601862 F

   Coulter Electronics Ltd., Northwell Drive, Luton,
   Bedfordshire LU3 3RH, U.K.
   Telex: 825074 COULTER G

   Delcita Limited, Ver House, London Road, Markyate,
   Hertfordshire AL3 8JT, U.K.

   High Accuracy Products Corp., 141 Spring Street, Claremont,
   CA 91711

TABLE 2 (continued)

Kane-May Ltd., Northey International Division, Nortec House, Chaul End Lane, Luton, Bedfordshire LU4 8EZ, U.K.
Telex: 827504 NORINT G

Leeds and Northrup Co., 4907 Stenton Avenue, Philadelphia, PA 19144

Malvern Instruments Ltd., Spring Lane, Malvern, Worcestershire WR14 1AL, U.K.
Telex: 339679 MALINS G

Polytech GmbH, 7517 Waldbronn, Karlsruhe, Federal Republic of Germany
Telex: 0782859

Seishin Enterprise Co. Ltd., Nippon Brunswick Buildings, 5-27-7 Sendagaya, Shibuya-ku, Tokyo, Japan
Telex: 232-4242 SEISIN J

7. X-ray sedimentation (Sedigraph)

Micromeritics Instrument Corp., One Micromeritics Drive, Norcross, GA 30093-1877
Telex: 6827018

---

[a]We list manufacturers/suppliers only of equipment specific to particle size analysis, general laboratory supplies being widely available. For multinational companies, or those with a large dealer network, we generally give the headquarters address. The mention of any company or product is not intended as a recommendation or warranty of any kind, but is given merely for information.

B. Direct Measurement

Soil scientists generally concentrate on the soil fraction passing a 2 mm aperture sieve; the so-called fine earth. However, many soil classification systems categorize soils (as skeletal, fragmental, etc.) if they contain a certain percentage of particles greater than a given size [28]. Engineers faced with moving much soil may find the complete grading of a soil to be absolutely vital [29]. Although even very large particles may be sized by sieving, the task can be extremely laborious. It is often more practical to resort to direct measurement in situ. The very largest particles can be measured with a tape, whereas those up to some tens of centimeters in size can be measured by means of wooden or light alloy templates into which are cut holes of differing shapes and dimensions [36]. Caroni and Maraga [37] used an adjustable caliper connected to a tape-punch so that the results could be fed directly to a computer when field investigators returned to the laboratory. An obvious extension of this idea is to use an electronic caliper in conjunction with a data-logger. Laxton [38] has used a photographic technique for estimating the grading of the boulder- and cobble-grade material in exposed working faces of quarries. Smith and Thomasson [39] give a method by which the *volume* of particles above a particular sieve size may be estimated by means of plastic balls.

For particles between about 10 cm and 1 mm, there is little *practical* alternative to sieving (Section IV.C), as the particles are usually too numerous to lend themselves to the methods outlined above. Between 1 mm and about 20 µm, optical microscopic determination of particle size has long been a popular technique with sedimentologists [21,30], while for the smaller end of the size spectrum electron microscopy can be used. The great advantage of microscopic methods is that they allow full consideration of shape factors as well as size [16,21].

A potential disadvantage of microscopic methods is that they are slow, requiring the measurement of many individual particles to obtain statistically valid results [21]. Kiss and Pease [40] have discussed methods of determining the optimum sampling strategy for microscopic methods. The use of automatic image analysis can also speed matters [23,24].

A requirement for all microscopic techniques, but especially those for very small particles, is good dispersion of the material This usually means destruction of organic matter, possible solvation with a particular cation, commonly sodium, with subsequent removal of excess salt, and possibly dissolution of cementing agents [41-43].

The basic techniques for sizing by microscopy are reviewed by Allen [16] and Gahm [44]. British Standards give specific procedures for optical microscopy [45]. Smart and Tovey [41] discuss electron microscopy techniques in detail, while Bradley [46], Comer and Turley [47], and Nixon and Weir [48] treat the practicalities of measuring the "thickness" of very small particles and clay mineral platelets by shadowing.

Where particles are roughly equidimensional, microscopy can yield useful particle size data in terms of a single or average dimension, which can be relatively easily checked against the dimensions of accurately sized graticules [45] or spheres [49]. However, the smaller particles in soils (i.e., < ~5 µm) are usually far from equidimensional [50], and the sizes measured along different particle axes may differ enormously [51]. In such cases, it may be more useful to express size in terms of particle thickness or equal volume diameter, together with the *aspect ratio*—that is, the distance between parallel crystallographic faces divided by thickness, itself often the distance between two other crystallographically related surfaces such as cleavage planes [52].

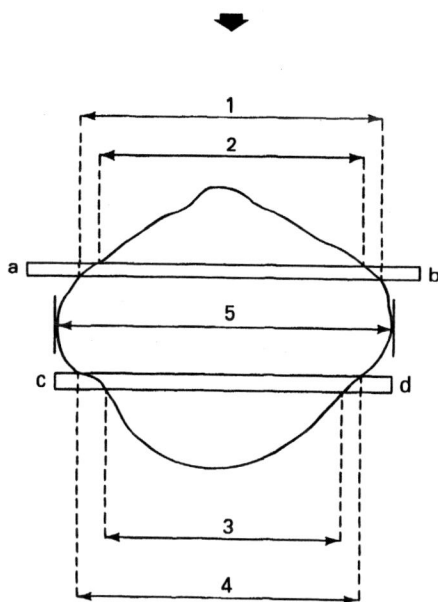

FIG. 4 Side view of two sections a-b and c-d through a particle showing how the dimensions measured can differ depending on the plane in which the measurement is made.

With nonspherical, platy, or angular particles, there is also the problem of overestimating size because the surface whose dimensions are measured rarely corresponds exactly in geometric terms with the surface resting on the support (Fig. 4). Where the particles are very thin and the dimensions measured are very large in relation to the vertical dimension, the error is small. When the vertical dimension increases greatly in relation to dimensions in the horizontal plane, however, the error can be much greater [16]. One possible solution to these problems is to section the particles as for normal petrological work. The particles are then much thinner. If the sectioning is done carefully, the dimensions in the plane of the section can be used to calculate the particle size from a probabilistic approach [53]. However, there will always be uncertainty as to how well the plane of section represents a random pass through the "true" dimensions of the particles. In optical microscopy, because of the difficulty of locating the particle edges because of diffraction effects, there can be substantial errors when working with very small particles near to the limits of resolution of the optical system [54]. For this reason, BS 3406 [45] recommends that optical microscopy not be used for particles smaller than 0.8 μm and qualifies the accuracy obtainable below 2.3 μm.

C. Sieving

Sieves are available with apertures ranging from 125 mm to 5 μm, either in round-hole or square-hole forms, depending on aperture size, and made from woven wire or punched plate, or by an electro-forming process. Round-hole sieves are, however, becoming less common. The ratio between aperture sizes was originally based on the square root of 2 (i.e., 1.414), by von Rittinger [55], but it now may be based on other roots, most commonly the fourth root, depending on which national or international standard sieves are specified (Table 3). The availability of any particular aperture as round-below 1 mm aperture, square-hole, woven-wire sieves are usual, while electroformed square-hole sieves are increasingly popular below about 37 μm. For fibrous materials (e.g., peats), it may be necessary to consider the use of special slotted-aperture sieves. Round-hole sieves size material by one dimension only, whereas square-hole sieves size particles by two dimensions [16] (the distances between two parallel faces and the diagonal between corners, respectively). Tanner and Bourget [70] comment on the dislocation that a mixture of round-hole and square-hole sieves can cause in constructing particle size distribution curves of soils.

Some form of mechanical sieve shaking is commonly used in preference to hand sieving. With careful control, mechanical siev-ing can give very precise results, most errors arising from worn or damaged sieve screens, variation in sieve loading, variation in shak-ing time, and poor fit between sieves, lids, and receivers. Metz [71] has shown that significant errors can arise if sieves are not kept horizontal when using the "Ro-Tap" type of sieve shaker. Kennedy et al. [72] have commented on the sorting and sizing of particles during sieving, according to their shape.

Generally speaking, sieving becomes increasingly laborious below an aperture size of approximately 30 μm, because the area of hole drops sharply as a percentage of total sieve area (Fig. 5), and dry sieving is not to be recommended in this range. If such sieving is attempted, the air-jet technique is both quicker and more reproduci-ble than conventional sieving [73,74]. Again, for finer materials that may "ball" (aggregate), even under air jetting, wet-sieving equipment is available [75].

It is well known that sieve apertures tend to become blocked, and the offending material is usually brushed out. It is probably this operation that does most damage to sieves, especially those of smaller aperture, both by stretching and by breaking the weave. To reduce mesh wear, cleaning by means of an ultrasonic bath filled with propan-2-ol can be used, although the frequency of oscillation must be chosen with care to avoid cavitation, hence mesh weakening [16,69].

TABLE 3 Sieve Ranges (Max-Min, mm) According to Standards[a,b]

Origin of standard	Preformed plate		Woven wire	Electroformed[c]
	Round hole	Square hole		
International	125-1.0[A] [56]	125-4.0[A] [56]	125[A]-0.02[A,B] [56]	0.5-0.005[A,B] [56]
Federal Reblic of Germany	125-0.63[C] [60]	125-2.0[C] [61]	16.0-0.025[C] [62]	
France	125-0.5[C] [57,58]	125-3.15[C] [57,58]	125-0.02[C] [58]	0.200-0.005[D] [59]
Japan	125-5.0[E] [63]	125-5.0[E] [63]	125-0.02[F] [63]	
United Kingdom	125-1.0[F] [64]	125-4.0[F] [64]	16.0-0.032[F] [64]	
United States	125-1.0[F] [65,66]	125-3.35[F] [65,67]	125-0.038[F] [68]	0.150-0.005[F] [69]

[a]Numbers in brackets indicate listed references.

[b]Key: A = $\sqrt{2}$ basis; B = $\sqrt[3]{2}$ from 0.032 to 0.005 mm; C = $\sqrt[3]{2}$ basis; D = in 0.005 mm increments to 0.03 mm, then 0.01 mm increments to 0.20 mm; E = ~ $\sqrt[6]{2}$ basis (not given in standard but calculated from stated sieve apertures); F = $\sqrt[4]{2}$ basis.

[c]Most standards accept electroformed sieves as long as they conform to given tolerances.

One still finds sieves described by their *mesh number*, a practice that is, in our view, to be deplored. The mesh number of a sieve is the number of wires per linear inch, which will (in theory) be one more than the number of holes over the same distance. However, without a knowledge of wire diameter, one cannot derive the sieve aperture from the mesh number. While it is perfectly possible to memorize a table of mesh numbers and apertures, there seems to be little point to this exercise when the aperture itself can be stated so simply.

It is often forgotten that sieve apertures are manufactured to tolerances, not to absolute values; that is, the stated aperture may vary between given limits. For example, the nominal 2 mm aperture of a wire-woven sieve may have an average variation of ±3% (1.94-2.06 mm), with no one aperture being more than 12% larger than the nominal aperture (i.e., 2.24 mm) [64]. Note also that it is very

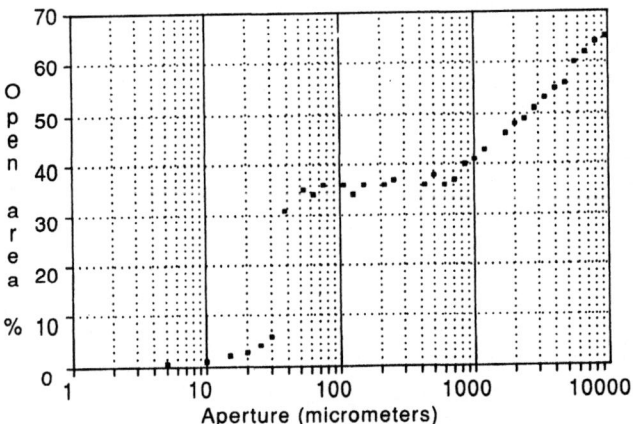

FIG. 5   Relationship between open area of sieve and sieve aperture
(square-hole sieves).

common to round off sieve sizes when reporting results; for example,
53 μm will be given as 50 μm, 63 μm as 60 μm, 106 μm as 100 μm, and
212 μm as 200 μm.   This practice is very widespread but the reason
for it is obscure: *There are no such things as 50, 60, 100, and 200
μm aperture sieves.*

D.   Sedimentation

1.   Introduction

Methods of particle size determination using a combination of sieving
and sedimentation are undoubtedly the commonest in soil science.
"Sedimentation" means the settling of particles in a fluid under the
influence of gravity or centrifugation.   The amount of material above
or below a specified size is determined either by abstraction of an
aliquot of suspension that is then dried and the residue weighed,
by measuring the change in the density of the suspension, or by
measuring the amount of sediment that has settled in a suitable vessel after a certain time.

Whichever method of measurement is chosen, *all* assume that the
particles in suspension behave according to the Stokes' equation [76],
which can be written as follows:

$$t = \frac{18\eta h}{(\rho - \rho_0)gd^2} \tag{1}$$

where t is the time in seconds for a particle to fall h cm once terminal velocity has been attained, ρ is the particle density (g cm^{-3}), ρ_0 is the density of the suspending medium (g cm^{-3}), g is the acceleration due to gravity (cm s^{-2}), d is the equivalent sphere particle diameter (cm), and η is the viscosity of the suspending medium (poise, where 1 poise = 0.1 Pa·s). Because this is not an empirical equation, it is equally valid if SI units are used throughout.

This equation is modified in a centrifugal field [77] to

$$t = \frac{18\eta}{(\rho - \rho_0)\omega^2 d^2} \ln\left(\frac{R}{S}\right) \qquad (2)$$

where ω is angular velocity of the centrifuge (i.e., the number of revolutions per second × 2Π), S is the distance (cm) of particles from the axis of rotation of the centrifuge at the start of analysis and is measured from the surface of the suspension, and R (cm) is the distance the particle has reached in time t (s).

Stokes' equation for spheres is applicable when the following criteria are met: (1) particles are rigid and smooth, (2) particles settle independently of each other, (3) there is no interaction between fluid and particle, (4) there is no "slip" or shear between the particle surface and the fluid, (5) the diameter of the column of suspending fluid is large compared to the diameter of the particle, (6) the particle has reached its terminal velocity, and (7) the settling velocity is small.

Stokes' law refers to an equation that describes the drag force on a particle of any shape. Thus it is valid for nonspherical particles if (and *only* if) the concept of equivalent sphere diameter is used.

Allen [16] points out that Stokes' equation is valid only under conditions of laminar flow when Reynolds number is 0.2 or less [78] and that the critical value of the Stokes' diameter (d), which sets an upper limit to the use of Stokes' law, is

$$d = \frac{3.6\eta^2}{(\rho - \rho_0)\rho_0 g} \qquad (3)$$

For quartz particles settling in water, for example, Allen shows that Stokesian behavior for spherical particles holds only for those less than about 61 μm in diameter.

Allen [16] has considered each of the criteria listed above in considerable detail. For soils and clays his findings may be summarized:

1. Flat, thin plates will settle more slowly than their equivalent spheres; hence the amount of such material may be overestimated.

This slowing of the fall rate is partly because the plates trace out a zig-zag path as they settle.

2. Below about 1 μm equivalent sphere diameter, Brownian motion can displace a settling particle in any direction by an amount equal to or greater than the settling induced by gravitation. This sets a lower limit beyond which gravitational sedimentation becomes increasingly unreliable.

3. Electrical interactions between a dilute electrolyte and soil particles have a negligible effect on settling, as does the time taken for particles to reach terminal velocity.

The question of particle-particle interaction is more difficult to deal with, as the number of particles in a soil suspension will vary enormously, depending on the very particle size distribution one is trying to measure. Robinson [79] suggested that the maximum concentration of suspended material should be no more than 1% by volume, or about 2.5% by weight, and general experience suggests that this is a sensible upper limit. Note, however, that suspensions of bentonitic soils may exhibit thixotropic behavior at smaller concentrations of suspended solids. Dilution of the suspension will usually overcome this problem but may also introduce greater possibility of error because of the difficulty of determining very small residue weights accurately. It is axiomatic that the soil should be well dispersed in an electrolyte, usually following the destruction of organic matter. The latter may be accomplished by hydrogen peroxide [42], hypobromite [43] or hypochlorite solution [80], or by low-temperature ashing [81]. Dispersion is almost always carried out using an alkaline solution, most commonly sodium hexametaphosphate buffered to about pH 9.5 with sodium carbonate or ammonia solution [42], although there are many others in common use [13]. Soils treated with hypobromite or hypochlorite solution may not need further dispersion. Dispersion may be aided by ultrasonic treatment [81,82], particularly in volcanic ash soils, for which dispersion in alkaline media is inappropriate due to their often large content of positively charged material. For these soils an acid dispersion routine should be followed [83].

The persistent use of the word "Calgon" in discussing soil dispersants is to be deplored. Calgon is the trade name for a phosphate-based water softener for domestic and industrial use, and it is of variable composition. At least one soil laboratory has gone badly astray in using this product uncritically [84]. There is little excuse for not using proper reagents of known composition in a soil laboratory. The savings in cost of not doing so are trivial, but the consequences could be serious.

In summary, the two greatest problems associated with gravity sedimentation methods are the uncertainty over behavior of the

particles, particularly in the smaller size range, and the variation
in particle density due to differences in mineralogy and to the presence (or absence) of adhering particles of iron "oxides." The latter
may disperse to a greater or lesser degree and settle as individual
particles. Alternatively, if these materials are present in amounts
or forms that prevent proper soil dispersion, or affect grossly the
sedimentation rate of the soil particles, they may be removed by,
for example, sodium dithionite reagent [42].

2. The Pipet Method

The pipet method was developed by Robinson [79] and refined to
its present form by Andreasen [85]. For the size fractions smaller
than 63 µm obtained after sieve analysis, it is the officially preferred method in the United Kingdom [13,86], the Federal Republic
of Germany [14], and France [59], and is the method of choice of
the U.S. Soil Conservation Service [15] and Agriculture Canada [87].
The referenced publications describe the method in considerable
detail.

Gee and Bauder [88] have discussed the basic pipet methodology
for routine soil analysis. A common complaint about the method is
that it is rather tedious, especially with respect to the time that has
to be allowed for sedimentation of the fraction smaller than 2 µm.
Kilmer and Mullins [89] describe a method of sampling large numbers
of sedimentation tubes, while Coventry and Fett [90] have shown how
the efficiency of pipet analysis may be greatly improved by attention
to time-saving details at every step of the process. In the Soil Survey laboratory we have greatly improved the turn-around time of
the pipet method by developing a programmable automatic sampling
device for taking the silt-plus-clay and clay aliquots (Fig. 6).
Miller and Miller [91] have described recently a micropipet method
that seems to be adequate for placing soils in the appropriate texture class. Given sufficient care in dispersion and sampling, the
pipet method is capable of great precision [88]. However, the relatively large spread of values found during an interlaboratory comparison of the pipet method when each laboratory was asked to follow a detailed worksheet [92] shows that there is still room for improvement [93]. Computer processing of the calculations involved in
conventional particle size analysis can give large savings in operator
time if many samples are involved [94].

3. Density Methods

The density of a suspension is proportional to the amount of solid
present and to the difference between the densities of the suspending liquid and the suspended solid. The density of the liquid is
usually fixed by controlling its temperature and electrolyte content

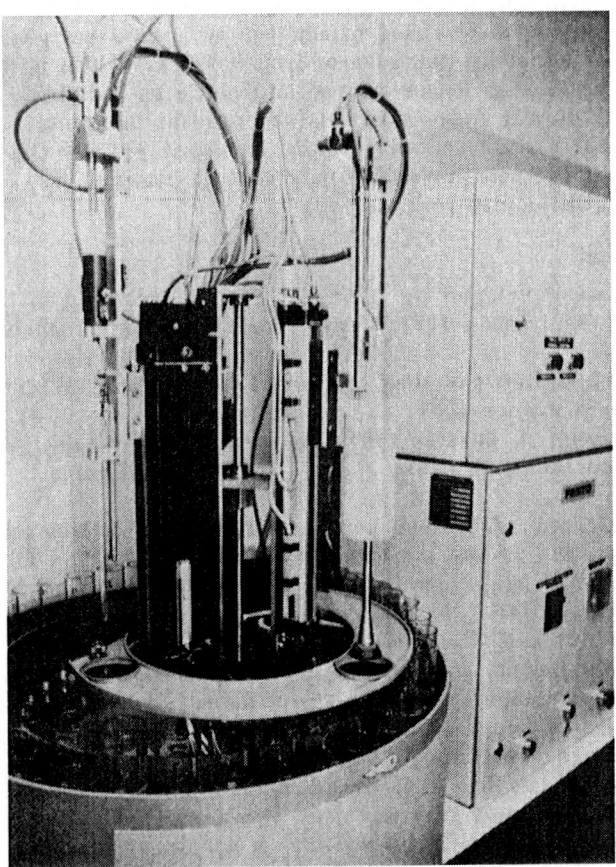

FIG. 6   General view of semiautomatic pipetting device used by the
Soil Survey and Land Research Centre.

(e.g., a dispersing agent), while the solid is usually assigned some
constant value, commonly 2.65 g cm^{-3} (Mg m^{-3}) for soils and clays.
If the soils under study contain considerable amounts of soluble
salts, these can greatly affect the principles on which a routine
method of density measurement of particle size analysis is based.

  If the density of a suspension is measured at known depths and
time intervals following agitation, it is relatively easy to relate this
to the mass of material above or below the Stokes' diameter calculated
from sampling time and depth. By far the most widespread procedure
is that based on the hydrometer developed by Bouyoucos [95]. It is
still the officially preferred method in the United States [96], and is

a subsidiary, but accepted, procedure in the United Kingdom [86]. A detailed procedure suitable for agricultural soils is given by Gee and Bauder [88]. Head [12] gives a detailed account of the precautions necessary for the proper use and calibration of soil hydrometers. Gee and Bauder [97] present an alternative, simplified version of their procedure and make the important point that the greatest source of error in hydrometer methods is the reading of the hydrometer scale. It becomes almost impossible to carry out such readings accurately if there is a layer of undecomposed organic matter on the surface of the suspension, and even after suitable oxidation treatment or with purely mineral soils, frothing following agitation can be a problem. The latter may be controlled by adding a drop or two of a surfactant such as octan-2-ol *after* the suspension has been stirred. [*Warning*: Some authors recommend the use of pentan-n-ol, n = 1 or 2 (amyl alcohol or isoamyl alcohol) to control frothing. This is effective, but amyl alcohols can become addictive. Octan-2-ol is equally effective but has an unpleasant smell and is less likely to encourage addiction.]

A further difficulty that may arise with the hydrometer method is related to the density of the suspension. For accurate determination, this should be significantly different from that of the suspending fluid. Gee and Bauder [88] recommend 40 g of soil per liter of suspension. This should ensure that even where the soil contains only a few percent of clay or silt, there should be enough to give an accurately measurable increase in the suspension density. Should *all* the soil be of clay or silt size, the suspension may contain so many particles that hindered settling occurs, whereupon the determinations may need to be made with less soil [98]; also, bentonitic clays will gel at this concentration. Allen [16] cautions against the use of hydrometers in suspensions that are not reasonably continuous distributions of sizes, as the sharp boundaries produced in such settling suspensions can affect hydrometers in unpredictable ways. This effect arises because of the relatively large length of the hydrometer bulb, which may straddle one or more such boundaries between differently sized particles. The hydrometer may thus give an average density for two or more zones of very different density, and when plotted as a particle size distribution, the effect is to smooth out any sharp changes in the grading that actually occur.

For accurate work, the particle size to which a hydrometer measurement relates can be determined only after measurement, because the depth to which the hydrometer will settle cannot be known in advance. Attempts to overcome this problem have been made by Puri and Puri [99] with the so-called Chaino-hydrometer, by Arulanandan and Smith [100] with a device for maintaining the hydrometer at a fixed depth, and by Bascomb and Pritchard [101], who devised a smaller, Cartesian-type hydrometer.

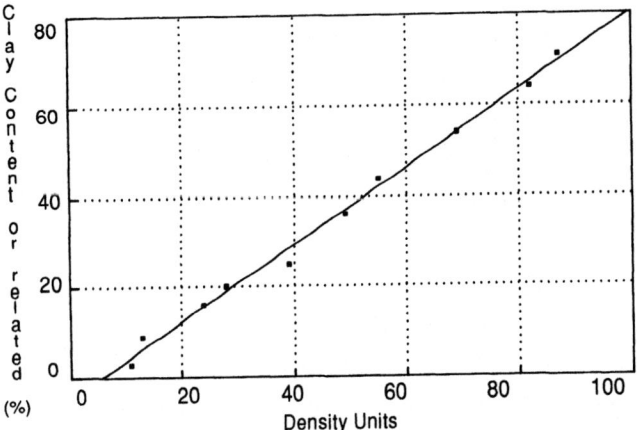

FIG. 7 Relationship between clay content by the pipet method and
density units measured by a digital density meter. [Density unit
is calculated from: (density of suspension minus density of electrolyte) × 10^4.]

Stabinger et al. [102] have measured the density of suspensions
by an ultrasonic technique. The equipment requires only a small
volume of suspension, which can be abstracted from a larger volume
automatically and with little disturbance. The ultrasonic signal can
be processed digitally, hence offers the prospect of automation.
Work done in the Soil Survey laboratory indicates a reasonable relationship between measured suspension density and clay (< 2 μm esd)
content determined by the pipet method (Fig. 7).
    There have been numerous comparisons between the pipet and
hydrometer methods, and it is generally agreed that the former is
more precise (see Refs. 88 and 97 for relevant references), but the
hydrometer method does have the attraction of simplicity.
    Marshall [103] proposed the use of a plummet balance to measure
the density of soil suspensions. Slager and Koenigs [104], Sheldrick
and McKeague [105], and Nagaraj and Silvapullaiah [106] found good
agreement between the plummet balance and the pipet methods, the
first authors claiming better precision for the former. Bannister et
al. [107] describe a plummet balance method that combines speed
with accuracy.
    Moum [108] has used the rate of fall of a drop of the soil sus-
pension in an immiscible organic liquid to determine the density of
the drop, hence its particle concentration. However, no direct comparisons with other methods of size analysis are given in the paper.

4. Other Methods

Bascomb [109] has described a device in which a sensitive pressure transducer measures the changes in hydrostatic pressure exerted by a column of suspension during sedimentation. The hydrostatic pressure decreases in proportion to the decreasing mass of suspended solids above the transducer and can be related to Stokes' diameters. Bascomb thought the device useful over the size range 20-200 μm esd.

Sedimentation balances that record the mass of material accumulating at the base of a column of suspension are described by Allen [16]. It is our experience that such devices require considerable experimental practice before consistently reliable operation can be achieved.

Syvitski and Swinbanks [110] have developed a volume size analysis (VSA) technique that relates the increase in volume of sediment in a tube, with time, to the particle size distribution. The technique is said to be valid for particles from 0.45 to 61 μm esd and the size distributions obtained not to differ significantly from those obtained by the pipet method.

Just [111] describes a field method for estimating the amounts of silt and clay from their settling volumes in which test sand was used to establish the cutoff between silt and very fine sand. It is not clear how silt and clay are distinguished, and the method seems unlikely to be more precise than carefully controlled hand-texturing [112].

E. Centrifugation

1. Introduction

Centrifugation is an extension of sedimentation under gravity, and it offers a means of determining the amounts of particles smaller than 1 μm esd in suspension (i.e., those whose settling under gravity is seriously affected by Brownian motion). Tanner and Jackson [113] published comprehensive nomograms for the settling times of particles of different Stokes' diameters under centrifugation. This approach has been adopted by Jackson [42], by the U.S. Soil Conservation Service [114], and by Avery and Bascomb [115] for the determination of particles smaller than 0.2 μm esd (the so-called fine clay). With laboratory centrifuges the volumes of suspension involved are usually large, and the design of standard laboratory centrifuges is not well suited to controlled sedimentation. This problem arises because the cylindrical sedimentation vessels are usually long compared with the centrifuge radius, thus the radial sedimentation of particles in centrifugal fields results in the particles colliding with the vessels' walls. The designs of modern centrifugal particle size analyzers reflect attempts to overcome this problem and can be divided into two groups, which are defined by the dimensions R and S, the radius of the measurement zone and the radius to the inner surface of the sedimenting column, respectively. In the most common type of centrifugal particle size analyzers S/R tends to zero, and radial sedimentation occurs in a hollow rotating disk. Hence they are known as disk centrifuges. Typically such disks are no more than a few centimeters thick and perhaps 20 cm in diameter. In the second type, which are often called long-arm centrifuges, S is large, S/R tends to 1, and the sedimentation paths of particles are assumed to be parallel. The two types of centrifugal particle size analyzer can be distinguished by observing whether concentration of an initially homogeneous suspension is reduced at the sampling point immediately after start-up. In the case of a disk centrifuge this is due to the dilution effect of radial sedimentation, whereas in long-arm centrifuges the suspension concentration remains constant until the larger size fractions sediment out of the measurement zone.

The upper limit of Stokes' diameters that can be determined in centrifugal particle size analysis depends on the reliability of measurements taken during the first minute or two and the limit to laminar flow for the larger particles. With water as a suspension medium the upper limit is about 7 µm esd, but the range can be extended by the use of more viscous liquids. The lower limit is still controlled by Brownian motion and it is thought to lie between 10 and 50 nm esd [116].

Centrifugal particle size analyzers are operated in one of two modes. Either the sedimentation vessel is filled with a homogeneous suspension at the start of analysis or the vessel is filled with a clear carrier liquid onto which the suspension is floated. These two techniques are known as the *homogeneous-start* and *line-start* techniques, respectively. Pipet sampling is not recommended for use with the line-start technique because the suspension concentration involved is usually very low [16]. Examples of common types of centrifugal particle size analysis are discussed in the following sections.

2. Pipet Sampling Centrifuges

The modern version of the pipet disk centrifuge was first developed by Slater and Cohen [117], who used a pipet to abstract aliquots of suspension at intervals from an initially homogeneous suspension centrifuged in a horizontally mounted hollow disk. The instrument was used by Vaughan et al. [118] to measure the size distribution of industrial clays between 0.16 and 5 µm. The disadvantage of this particular instrument was its large disk volume (~ 2.5 dm^3), which was necessary to permit investigators to ignore the change in initial suspension level (i.e., at the radius S) as a consequence of the withdrawal of 40 cm^3 aliquots to measure suspension concentration.

This led to the development of a much smaller disk centrifuge [119], with a disk volume of 150 cm^3 (requiring samples of 5 g at most) and the facility to withdraw 10 cm^3 aliquots, in parallel with the theory required to compensate for changes in initial suspension level.

When disk centrifuges are used with the homogeneous-start technique, as is the case with pipet sampling, the reduction in suspension concentration at the sampling point can be attributed to two factors. The first is a result of the sedimentation of various size fractions. Second, the diverging radial sedimentation paths of particles give rise to additional dilution. To calculate particle size distributions, this radial dilution effect must be corrected. Allen [16], Kamack [120], and Slater and Cohen [117] describe the necessary theory, whose exact solution is complex. However, provided sampling is modified such that successive values of Stokes' diameter occur in a ratio of 1:$\sqrt{2}$, a much simpler approximate solution can be applied [16,117,120]. The use of this approximation may lead to some error when the sample under analysis has a bimodal particle size distribution. It has been suggested that in some cases improved results can be obtained by fitting experimental data to a curve defined by a mean and a standard deviation or other assumed functions. It seems paradoxical that a technique based on accurate measurements of suspension concentration should produce approximate results. If exact results are required, the exact solution to the theory of Kamack [120] must be applied to experimental data. A complete mathematical analysis of this theory is presented by Svarovsky and Svarovska [121], along with a design of an analog computer that can perform the necessary calculations.

The Allen and Svarovsky centrifuge [119] is essentially similar to the instrument marketed by Fritsch of Germany as the "Analysette 21 Pipette Centrifuge." Allen [16] claims that the data obtained from this instrument are "in line" with those obtained by the conventional pipet technique, but offers no examples. The technical literature of Fritsch shows excellent agreement between the size distribution of BCR Reference Quartz particles by both conventional pipet and disk centrifuge methods (Fig. 8). Judging from the lack of published work quoting its use, the pipet centrifuge is little known among soil scientists and would seem to be worthy of further investigation as a tool for exploring the small (< 2 μm) size fraction (see Table 2).

3. X-Ray and Photosedimentation Centrifuges

The centrifugal X-ray and photosedimentation techniques continually monitor the sedimenting suspension by measuring the transmission of radiation (either visible or X-ray) in a well-defined measurement zone. Thus continuous particle size distribution data can be obtained, as opposed to the discrete sampling that is provided by pipet analysis.

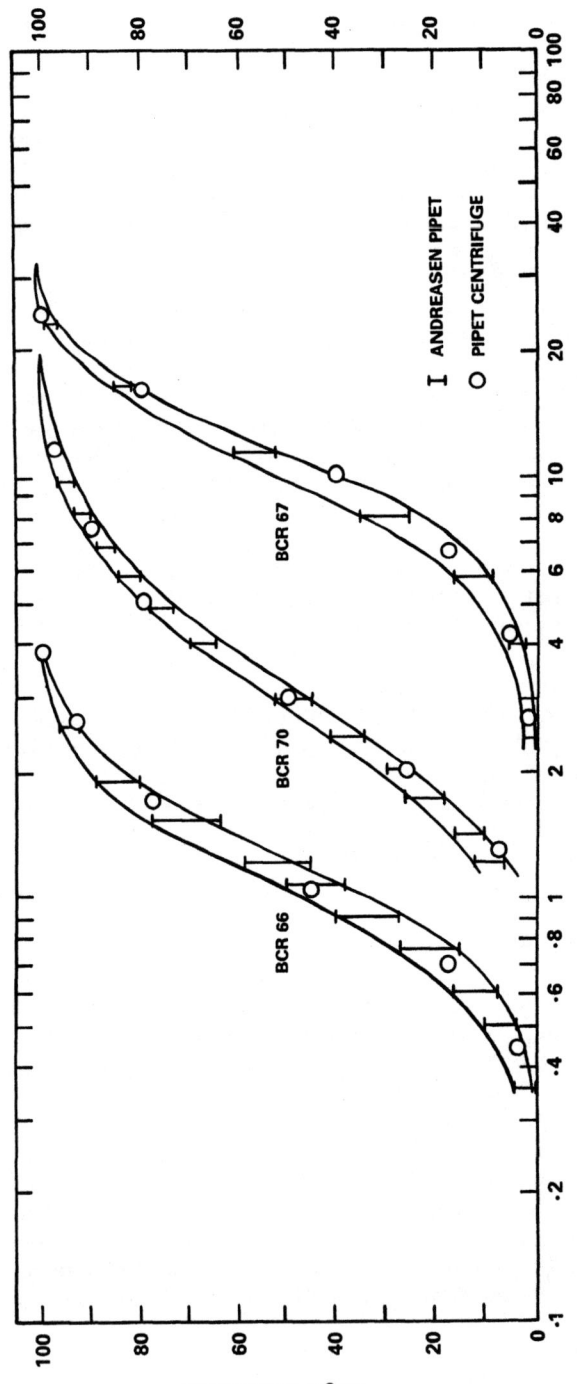

FIG. 8 Particle size distribution of BCR Reference Quartz by pipet analysis and by Fritsch pipet-centrifuge. (Courtesy of Fritsch Laborgerätebau, Idar-Oberstein, Federal Republic of Germany.)

A centrifugal disk X-ray particle size analyzer operates on principles essentially similar to those of gravitational X-ray sedimentation described below (Section IV.G.2). However, since a homogeneous start is used with a disk centrifuge, the data analysis must follow the theory of Kamack [120] and Svarovsky and Svarovska [121] to compensate for the radial dilution effect.

Centrifugal photosedimentation (i.e., using visible radiation) has been widely used for particle size analysis (see, e.g., Refs. 122 and 125). The use of light is better suited for soils than X-rays, because, as explained below (Section IV.G.2), quartz and clay minerals can be translucent to X-rays. However, since clay contains particle sizes both greater and smaller than the wavelength of light, photosedimentation data must be corrected for the large variation in light scattering that occurs with particles in this situation. The theory and techniques of this correction are described by Coll and Searles [122]. Analysis may be performed with either the line-start or the homogeneous-start techniques, and examples of both modes of use will be discussed.

Homogeneous-start sedimentation produces a relationship between turbidity (the absorption coefficient of the suspension) and Stokes' diameter that increases monotonically with Stokes' diameter. The initial suspension concentration has to be adjusted to ensure that the turbidity data obtained from the start of the analysis are within the region in which the Beer-Lambert law is valid (i.e., suspension concentration is proportional to turbidity). When analyzing clays or other very small particles, it is preferable to split the whole analysis into a series of overlapping or contiguous runs (e.g., 20 nm–0.1 µm, 0.1–2 µm, and 1–10 µm) [123]. This is necessary because the smaller particles scatter very little light compared to larger particles, so to obtain measurable turbidity values, higher suspension concentrations are required for smaller particles. Typically, initial suspension concentrations of 10 g dm^{-3} are required for the 20 nm to 0.1 µm size range to obtain reliable turbidity data, while the suspension concentrations in the 1–10 µm size range may have to be as low as 0.3 g dm^{-3} to ensure compliance with the Beer-Lambert law [123].

After completion of the photosedimentation, the turbidity data can be normalized to a single suspension concentration to give a continuous curve that covers the overlapping runs. After correction for the variation in light scattering with particle size, the results from a long-arm centrifuge (i.e., neglecting radial dilution effects) represent a particle size distribution by area. Some assumption about particle shape is necessary to convert it into a particle size distribution by mass [123], and suitable theories and methods for correcting for both light scattering and absorption effects in clays are given by Whalley [123].

In line-start centrifugal photosedimentation, the dispersed sample is floated on top of the already spinning disk of liquid and the sedimentation of the particles out of their narrow start zone is monitored at some fixed distance in the disk fluid by light transmission. It is usual for the disk liquid to be slightly denser (e.g., 10% glycerol/90% water) than the suspension to prevent irregular streaming of the sample from the narrow start zone. Once the relationship between turbidity and Stokes' diameter obtained from line-start photosedimentation analysis has been corrected for the variation in light scattering with particle size, it represents a particle size distribution by mass, in contrast to the distribution by area initially given by homogeneous-start photosedimentation [122]. Correction of disk centrifuge data for light-scattering effects is described by Oppenheimer [124]. The Joyce-Loebl centrifugal photosedimentometer was used in the line-start mode by Kane [125], whose work was drawn on by Churchman and Tate in an investigation of allophanic soils in New Zealand [126]. Kane [125] found, from the sedimentation of clays in slowly setting agar, that in high centrifugal fields platelike clay particles sediment with their minimum dimension in the direction of motion. This was also deduced by Whalley [123] from observing the turbidity of the same sample at various centrifuge speeds. This phenomenon is in accordance with hydrodynamic theory [127], and excessive force fields should therefore be avoided in all types of centrifugal particle size analysis. Coll and Searles [122] claim that line-start analysis is preferable to the homogeneous-start variety because the former gives a higher resolution of data.

The main criticism of all photosedimentation analysis, particularly with fine clays, is that large corrections to the experimentally obtained data are required to compensate for light-scattering effects. However, the particle size distributions obtained are continuous and have a very good size resolution. The use of centrifugal pipet analysis in conjunction with photosedimentation would form a very powerful approach toward particle size analysis of clays, since it would combine accuracy with a high degree of size resolution in the data. The study of the effect of saturating cation on aggregate (tactoid) size in dilute bentonite suspensions by Whalley [123] provides a good example of the advantages obtained from the high size resolution of photosedimentation.

4. Other Centrifuges

McCormick [128] has described the use of the ultracentrifuge (rotor speeds up to 60,000 rpm) for the determination of latex particles ranging from 0.02 to 0.5 μm in diameter and found excellent agreement with diameters measured by electron microscopy. Note, however, that the latex particles were *spheres*.

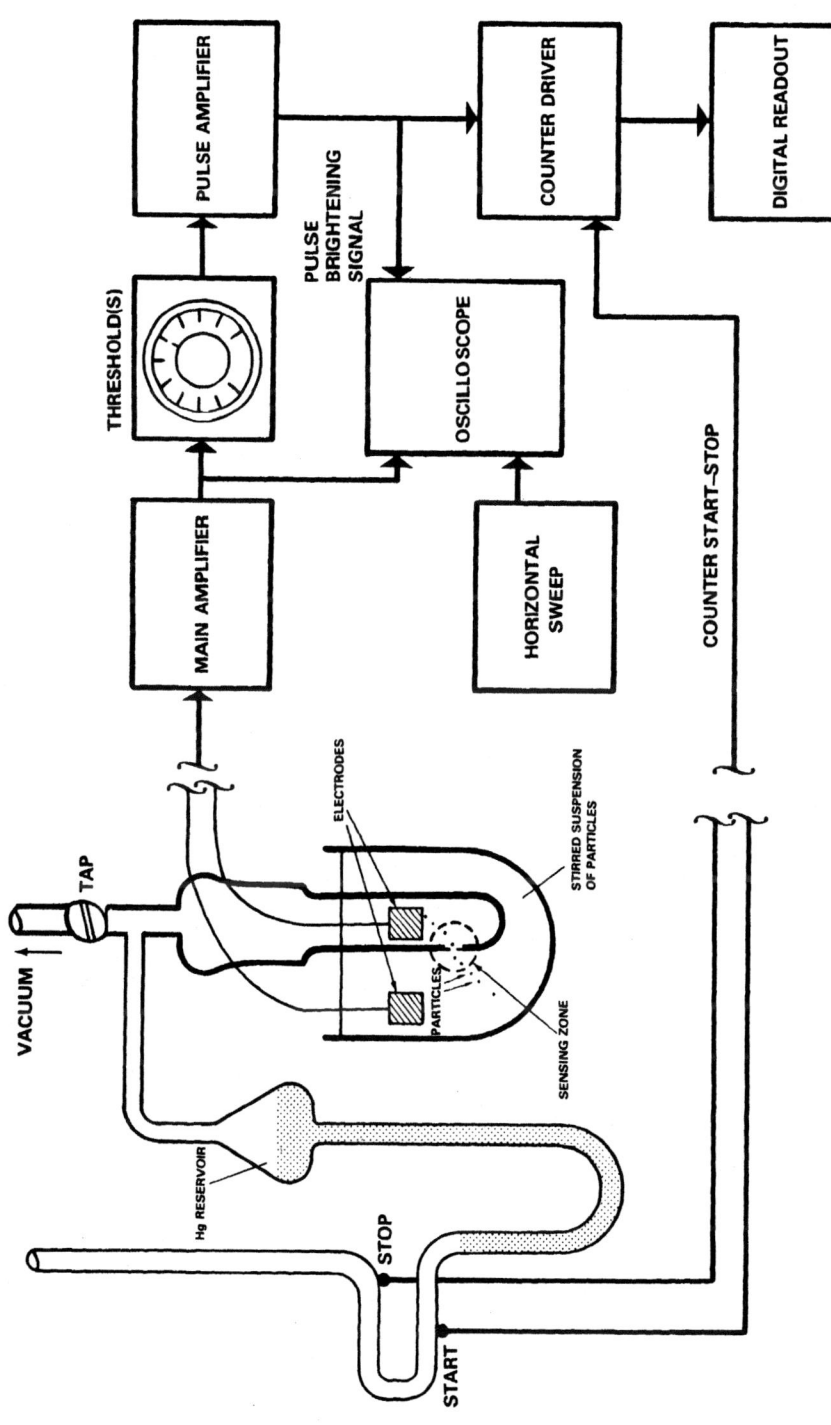

FIG. 9 Schematic representation of the principal components of the Coulter Counter. (Courtesy of Coulter Electronics Ltd., Luton, U.K.)

Allen [16] reports the use of a Sharples supercentrifuge for size splitting of particles in large volumes of fluid (tens of liters), and work at Rothamsted Experimental Station has shown that this device can yield particles within very narrow size limits [Weir, personal communication]. A similar machine is the hydrocyclone, but its use in soil science seems to be restricted largely to fairly crude beneficiation [129].

F. The Electrical Sensing Zone Method

The electrical sensing zone method is commonly known as the Coulter principle approach, from its discoverer, and commercially available instruments, although not all made by the Coulter Corporation, are generally called Coulter counters. Coulter's discovery was that the resistance measured between two electrodes in an electrolyte, and separated by an aperture, changes in proportion to the volume of a particle passing through the aperture. These changes in resistance can be scaled and counted at the rate of several thousand per second.

Allen [16] gives a general account of the electrical sensing zone method (esz), and the basic principles are illustrated in Fig. 9. A measured volume of suspension is drawn through the aperture by automatic operation of the manometer. As the particles stream through the aperture, the change in resistance between the electrodes caused by the passage of each particle is detected as a voltage pulse, which is scaled, amplified, and assigned electronically to a particular size class or channel. There may be up to 256 such channels to cover the range of the particular aperture in use. With the aid of microprocessors, the instrument output can be expressed directly as "percent oversize," as a cumulative distribution curve, and so on. It is important to remember, however, that the output is a number size distribution, in which the total volume of the particles is deduced (with some assumptions) from the size class itself. The mathematics of conversion to a weight basis are considered by Batch [130], who also gives a detailed mathematical analysis of the principle and shows that for a rod-shaped particle, the response of the electrode system is

$$\Delta R = \frac{(1 - \rho_0/\rho)\rho_0 V}{A^2[1 - (\rho - \rho_0)a/A]} \tag{4}$$

where ΔR is the change in resistance between the electrodes, ρ_0 is the resistivity of the electrolyte, ρ is the resistivity of the particle, V is particle volume, A is the cross-sectional area of the aperture across which the resistance between the electrodes is measured, and a is the cross-sectional area of the particle.

It is usually assumed that the resistivity of the particles is extremely high, due either to very stable electrical double layers or to the presence of oxide films, although this may not be true for some of the iron and iron-titanium minerals found in sediments [131]. The crucial parameter is the relationship between particle and aperture cross-sectional areas. Lines [132] has shown that oversizing by between 3.6 and 6.2% occurs (depending on particle shape) when the particle occupies more than 40% of the aperture area and recommends this ratio be adopted for routine analysis. Harfield and Knight [133], working with spheres and rigid disks 30-400 μm in diameter, suggest that an 80% particle/aperture ratio is possible, and 60% is acceptable, although at the higher ratios aperture blockage becomes a real possibility [16]. Lloyd [134] investigated the response of the aperture to nonspherical particles using a model system and found no serious deviations. However, the nature of his experimental system makes it difficult to relate these results to natural particles.

Two kinds of coincidence counting can occur. In one (primary coincidence) two particles give rise to overlapping pulses; that is, they pass through the aperture so closely together that the instrument counts them as one. In the other (secondary coincidence) two small particles, which are normally below the detection or "threshold" voltage measurement limit, give rise to a combined signal that is above the limit. The answer is to use extremely dilute suspensions, as it then becomes possible to meaningfully calculate primary coincidence, which follows a Poisson distribution and yields the following relationship [16]:

$$N = n + \frac{V_z N^2}{2V_m} \qquad (5)$$

where N is the true count, n is the observed count, V_m is the volume monitored for each count, and V_z is the sensing zone volume.

Sheldon and Parsons [135] give a more direct expression:

$$N = n + p\left(\frac{n}{1000}\right)^2 \qquad (6)$$

where p, the coincidence factor, is 2.5 $(D/100)^3$ $(500/V_m)$ and D is the aperture diameter (μm). Secondary coincidence is generally regarded as the much less severe problem of the two.

It is clear from this discussion of coincidence that if precise results are to be obtained from the esz method, particles must be effectively dispersed to ensure that they are counted singly and separately.

The size range that can be studied with this technique is from
1.5 mm to 0.5 μm, the largest aperture available being 2 mm diameter, with the lower size limit being set by electronic noise. To
cover the entire range of size, several apertures may thus be necessary [16]. Large particles cannot be kept suspended adequately
in water, and McCave and Jarvis [136] successfully used 10:90 sa-
line/glycerol solution to suspend quartz particles up to 1 mm in
diameter.

Hunt and Woolf [137] and Schrag and Corn [138] have compared
the size distribution obtained by esz and optical microscopy using
calibrated glass beads. In both it was found that the electrical
method slightly oversized the beads (which accords with Lines' calculations [132]) but that the oversizing was small. Hunt and Woolf
express no serious reservations about the accuracy of esz over the
size range 5-30 μm, but Schrag and Corn consider the oversizing at
the coarse end to be too large and would restrict the method to the
1-10 μm size range.

Johnston and Swanson [139] used a fine test dust as a standard
during an interinstrument comparison and concluded that the esz
method performed well over the 0.3-80 μm range. This lower limit
is much below that thought advisable for routine work by most
authors. Ormsby and Odom [140] carried out a similar interlaboratory test using standard whiteware clays. They found that the esz
technique compared very unfavorably with other methods, but offered no explanation. It is noteworthy, however, that the geometric mean diameters of the test samples, obtained by more conventional techniques, were generally less than 2 μm esd and in many
cases less than 1 μm. Seville et al. [141] used esz as a reference
method over the 0.5-5 μm range in a comparison of light-scattering
particle size analyzers. Griffiths and Rood [142] found comparable
size distributions of BCR Reference Quartz by both esz and image
analysis, the latter giving the volume of the particle calculated from
its projected area diameter.

The esz technique is the subject of at least two national standards [143,144], which consider the problem of standardizing the
equipment. This has also been examined by Harfield [145], who
used plastic spheres as test material in a "round-robin" of several
laboratories, and by Atkinson and Wilson [146], who advocate the
"mass integration" technique. The latter is the only practicable
means of dealing with materials such as soils in which the shapes
of the particles are known only approximately. Full details of the
method are given in the British Standard [143], but essentially a
sample of the unknown of a very narrow size range is used to derive an aperture constant [i.e., a figure that reflects the response
of the aperture in use to the material being analyzed and, in particular, will make (at least some) allowance for variations in particle

shape]. This aperture constant can then be used in resizing some other standard material (e.g., latex spheres), which can then be used to check the instrument at intervals or derive constants for other apertures.

The most thorough report on the use of the esz technique for soils is that of Walker and Hutka [131]. Although the equipment they used is now outmoded, many of their findings are still relevant. As well as considering the principles of esz equipment, they give particular attention to dispersion and the need to prevent flocculation in the electrolyte, and the establishment of a procedure for the routine analysis of soils. Their broad conclusions about use of the esz method for particle size analysis of soils are:

1. The satisfactory size range is 2-100 µm using apertures of 50, 100, and 200 µm.

2. It is necessary to split soil suspensions at 31.5 µm to avoid blockage of the 50 and 100 µm apertures.

3. Careful attention needs to be given to a choice of electrolyte to ensure that flocculation does not occur. The electrolyte may need to be different for different apertures.

4. The clay fraction (< 2 µm esd) can be determined with reasonable accuracy by a difference technique based on the measurement of the 0-31.5 µm and the 2-31.5 µm fractions.

5. Clear relationships exist between esz size fraction percentages and sieve weight percentages in the 37.2-88.5 µm range. However, conversion of one to the other requires a different factor for each size fraction.

6. Materials of low resistivity (e.g., magnetite, haematite, ilmenite) are probably not sized properly. (But then, neither are they in conventional sedimentation because of their large specific gravities.)

7. The technique is especially useful where only very small amounts of sample are available or for already existing very dilute suspensions (e.g., river and marine waters).

8. The electrical sensing zone technique compares well with conventional sieving and sedimentation in terms of reproducibility and efficiency for *detailed* size analysis. However, the needs to change apertures and electrolytes and to perform considerable mathematical analysis of the data to achieve results on a mass basis make the technique difficult to use for rapid routine use. The use of a multiaperture instrument, with all the apertures in operation in the same suspension at the same time, coupled with computerized data processing, could overcome many of these difficulties. As far as we know, such an instrument has never been built.

Sheldon and Parsons [135] give detailed instructions for the application of the electrical sensing zone method in marine science, and their examples show that it is possible to size material at concentrations as low as about 1.5 mg kg^{-1}. Davidson and Keen [147] found the esz method useful over the 2-64 μm range in the analysis of turbidity current sediments, using pollen grains for calibration.

Walker et al. [148] applied the method to the analysis of very small deposits such as laminae and to suspended sediment in freshwater streams. The question of flocculation has been further considered by Ozturgut and Lavelle [149], who used diluted salt water suspensions to preclude flocculation. Conversely, Gibbs [150] studied floc stability during esz analysis and found that flocs of 78-12 μm disintegrated during their passage through the instrument aperture, irrespective of the latter's size. Santero and Stotzky [151] reported that clays flocculate under the influence of microbial metabolites, and this effect could be detected by esz analysis. Sodium metaphosphate was said to prevent such flocculation.

Dudley [152] applied the esz method to the discrimination of soils for forensic purposes and found the reproducibility over the 2-60 μm range to be extremely good. Duke et al. [153] also found the method to be highly reproducible for lunar soil between 1 and 125 μm esd, using 200 and 50 μm apertures. Good agreement was found over the same sieve and esz equivalent ranges. Sapetti [154] considered esz to be superior to the Andreasen pipet and to agree well with results from a sedimentation balance, the latter finding being in accord with Walker and Hutka [131]. The esz method and the hydrometer technique have been shown to diverge at small sizes [155]. Shideler [156] compared the esz method with pipet sedimentation and claimed that the former gave significantly coarser size distributions in the 63-0.63 μm range. However, Behrens [157] was critical of Shideler's technique and presentation of data, and showed that Shideler's omission from his data of the fraction smaller than 0.63 μm was the principal cause of the coarsening effect.

Rybina [158] and Pennington and Lewis [159] have also compared esz and pipet techniques. Some of Rybina's results, presented graphically in Fig. 10, show that the esz method oversizes the finer material relative to the pipet. Furthermore, the esz method generally *undersizes* the 10-50 μm fraction, which Walker and Hutka [131] also reported to be the case for the 44-53 μm fraction of their soils. Pennington and Lewis [159] found a reasonable linear relationship between silt content (53-2 μm) by both esz and pipet methods using 43 soils of different particle size classes and mineralogies:

$$\text{silt}_{\text{pipet}} = 0.91 \text{silt (esz)} + 5.538 \qquad (r^2 = 0.80, \ p < 0.1) \qquad (7)$$

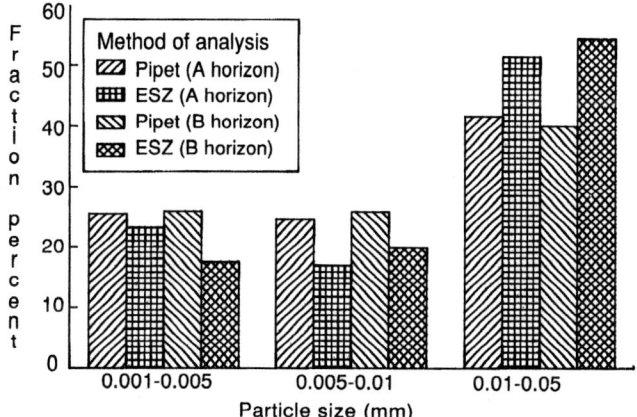

FIG. 10 Comparison of particle size distribution obtained by an esz instrument and pipet analysis for a Chestnut soil. (Derived from Ref. 158, courtesy of Scripta Technica, Inc., Silver Spring, Md.)

For the clay fraction ($< 2 \ \mu m$), the relationship was:

$$\text{clay}_{\text{pipet}} = 0.842 \ \text{clay (esz)} + 2.542 \qquad (r^2 = 0.61, \ p < 0.1) \quad (8)$$

However, inspection of their data suggests that the latter relationship was curvilinear. These authors also note that background "noise" in esz systems can be greatly reduced if all water and electrolytes are filtered at 0.45 and 0.22 μm before use. Lewis et al. [160] used an esz instrument to identify loess by constructing very detailed particle size distribution curves between 2 and 50 μm esd.

In summary, the esz method is probably best used to obtain very detailed particle size distributions over a narrow range of equivalent sphere diameters. The lengthy calculations involved can now be done by microprocessor. There is little doubt, however, that esz instruments do not always "see" particles in the same way as more conventional methods, such as sedimentation. This, however, is true of all methods and does *not* mean that the electrical sensing zone approach is, thereby, invalidated. One drawback to the esz method is the need to work with more than one aperture to cover a range exceeding 50 μm esd.

G. Interaction with Electromagnetic Radiation

1. Introduction

A particle may absorb, scatter, refract, diffract, or reradiate incident electromagnetic radiation. Such interactions may be used to estimate the mass of material encountered by a beam of radiation, or they may be used directly to yield information about the size of the particles encountered. Generally speaking, modern instruments utilizing these principles fall into two groups, radiation absorbers and radiation scatterers. Full discussion of these two principles is given by Kerker [161], and applications to particle size analysis are discussed by Barth [162] and Mully and Frock [163].

2. Absorption

The simplest case is represented by total light extinction, in which each particle intercepts a collimated beam of light. The amount of light obscured from the beam is determined electronically. Because the sample cell is designed to cause turbulent flow, the particles present a constantly changing cross-section to the beam as they pass through, and it is the *maximum* cross-sectional area that is recorded. This principle has been incorporated in the HIAC instrument, which (in theory at least) can cover the range from about 1 to 9000 µm esd [162]. In many ways the instrument is the optical analog of the electrical sensing zone concept.

West [164] considered that the system tends to oversize slightly with respect to microscopy, while Gibbs [165] found that floc breakage was a severe problem as material passed through the sensor. Northey International (Kane-May) market an instrument based on similar principles but claim to have greatly improved instrument performance by redesign of the optical system.

Turbidimetric measurement of particle size distribution in soils has been attempted by Lamp and Schroeder [166], who found the effective range of the technique to be 1-300 µm. Over this range, good agreement was found with the pipet method. Zaneveld et al. [167] used optical attenuation in conjunction with sedimentation (i.e., photosedimentation) and found good agreement with the electrical sensing zone and gravitational settling tube techniques. Eaton et al. [168] report the use of continuous-flow instruments to measure suspended sediment in estuarine waters. Jordan et al. [169] described a photoextinction technique, which has been reappraised recently by Coates and Hulse [170]. The latter authors found that, despite good precision, this so-called hydrophotometer gave results very different from those yielded by the pipet and hydrometer methods. Melik and Fogier [171] have recently examined both the theory and practice of turbidimetric particle size analysis and conclude that for particles with regular shapes, the method is reliable between ~ 0.1 and 3 µm.

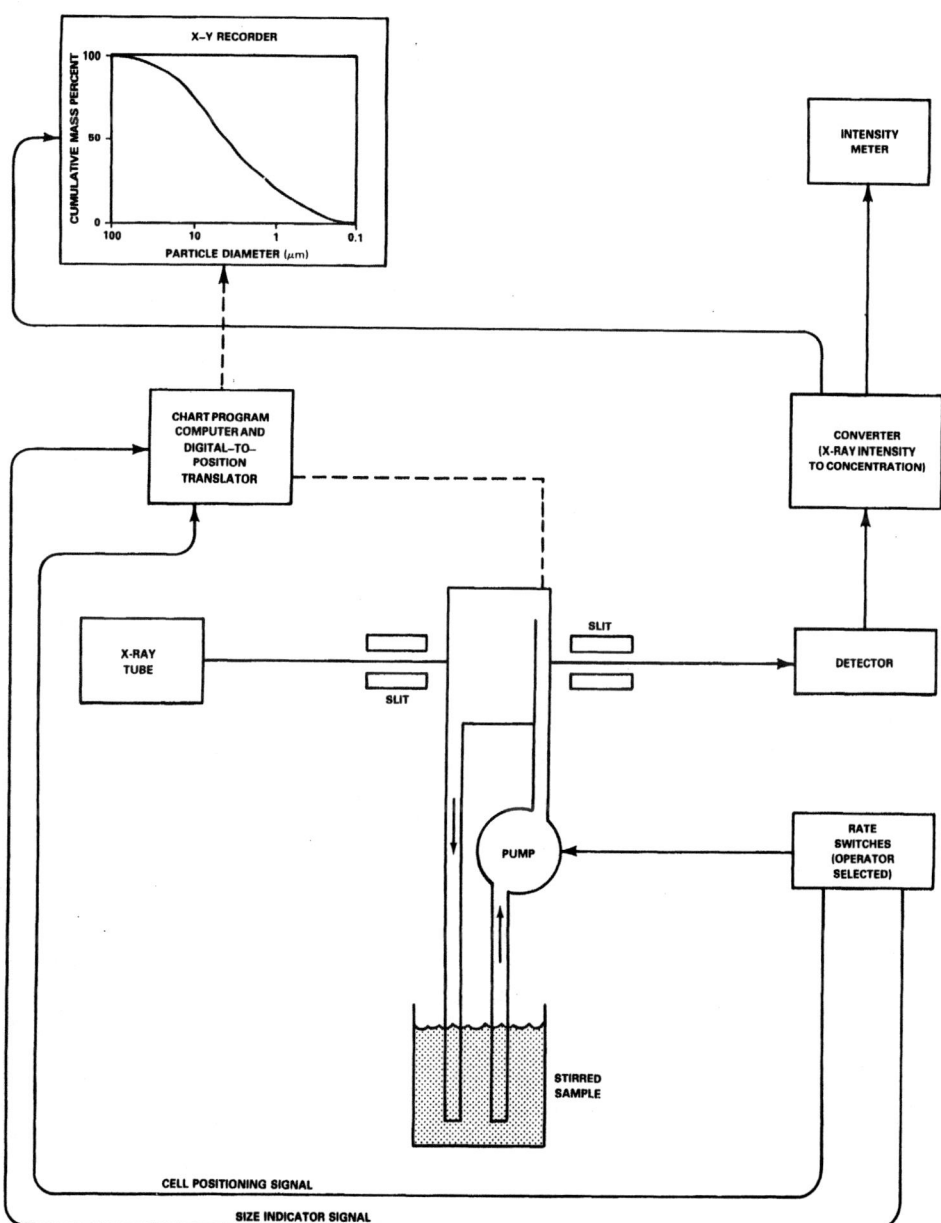

FIG. 11 Schematic representation of the Sedigraph X-ray sedimen-
tometer. (Courtesy of the Micromeritics Instrument Corp., Norcross,
GA.)

Conlin et al. [172] describe an apparatus in which the mass of material in suspension at a fixed depth in a cell is determined from the attenuation of a beam of X-rays. Carr-Brion and Mitchell [173] applied the principle to the measurement of size in a stream of moving particles. The same principle is used in the Micromeritics Corporation Sedigraph, first described by Hendrix and Orr [174]. This instrument consists of an X-ray source (Tungsten L-line, wavelength 14.76 nm), a cell (~ 1.25 cm wide, 3.5 cm high, and 0.35 cm thick; volume ~ 1.65 cm^3) through which the finely collimated X-ray beam passes, an X-ray detector and signal processor, a pump, and a chart recorder (Fig. 11). The chart is set at 100% with the pump in operation; that is, the suspension thoroughly agitated. Once the pump is switched off, a "run" begins and the particles sediment. The unique feature of the Sedigraph is that the cell is slowly lowered relative to the X-ray beam during measurement, thus greatly reducing the effective settling time. The manufacturers state that the suspension density is measured every 1.88 μm throughout the cell length (a total of more than 13,000 measurements). The instrument is programmed to solve automatically the Stokes equation as modified by the movement of the cell, and it plots the cumulative mass percentage versus equivalent sphere diameter on an integral chart recorder.

Olivier et al. [175] discuss instrument performance and show that as long as the area irradiated by the beam is small (achieved by the fine collimation), the errors from irradiation of the cell wall are negligible, and attenuation of the beam is dependent on the mass absorption coefficients of the suspending liquid and the particles in suspension. This raises two problems:

1. The absorption of X-rays becomes increasingly poor for elements below atomic number 14. This includes aluminum and silicon.
2. The mass absorption coefficients of soil materials cover a range of values, and average values have to be assumed. However, it is unlikely that these values will remain constant over the whole size range being examined in polymineralic mixtures such as soils [176].

The appropriate suspension concentration has been considered by several authors. Bernhardt [177] recommended a minimum concentration of about 50 g dm^{-3} (2% v/v) for silicate analysis, while Ullrich et al. [176] found that concentrations below 1% v/v were necessary for bentonite clays if thixotropy/hindered settling were to be avoided. Stein [178] presents good experimental evidence to show that the suspension concentration should be less than 2% v/v to achieve reproducible results using smaller than 63 μm fractions, but even at such a relatively low concentration, samples with more than 50% montmorillonite in the same size fraction (i.e., < 63 µm) gave unreliable results due to thixotropic effects. Berezin and Voronin [179] compared the Sedigraph with the pipet method and found that, in general, the former gave higher percentages of fractions of less than 5 µm esd but smaller percentages of the 50-5 µm fraction. Above 50 µm neither method showed a clear pattern. Welch et al. [180] found excellent agreement between the pipet and Sedigraph methods for particles of less than 31 µm esd using a sample concentration of 50 g dm^{-3} but suggested that a concentration of half this value (i.e., ~ 1% v/v) would be adequate. Coates and Hulse [170] found that the Sedigraph was of inherently great precision when used as a research tool, but for routine use less rigorous experimental procedures were necessary if a reasonable throughput of samples was to be maintained. Under these conditions the precision of the instrument was inferior to that of pipet analysis. They also found that the Sedigraph gave good agreement with the pipet between 16 and 2 µm esd, but agreement was poor at greater sizes. Vitturi and Rabitti [181] reported that relatively high concentrations of organic matter did not affect instrument performance.

Many of the authors quoted above have utilized the Sedigraph most successfully between 63 and 2 µm. Since with a cell volume of 1.65 cm^3 and, say, 50 g of < 100 µm soil in the suspension the cell will contain less than 0.1 g of material, this may simply yield too few particles to give reliable values for the larger particles. Because of Brownian motion (see Section IV.D.1), the determination of the proportion of particles below about 1 µm esd is unreliable by gravitational sedimentation. The Sedigraph takes about 20 minutes to determine the size distribution between 63 and 2 µm.

Given these constraints, and the need to bear in mind the mineralogy of the sample, the Sedigraph offers a rapid method of determining the size distribution of soil material between about 60 and 2 µm. The smaller (< 2 µm) fraction may need to be determined by difference.

Burr [182] has described a combined pipet and X-ray attenuation technique for the analysis of commercial kaolin suspensions, using a ^{238}Pu X-ray source. Karsten and Kotze [183] have used a gamma-ray technique in conjunction with a very shallow sedimentation depth, which gave results for clay contents comparable to those obtained with a hydrometer. Barthakur [184] describes a system based on a beta-ray gauge, which also gave results that compared well with hydrometer results.

In all cases, one of the great advantages is claimed to be the *speed* with which measurement can be made, compared with more conventional techniques.

3. Scattering

Developments in modern electronics, signal processing, and micro-computing ensure that scattering is the most rapidly developing area of particle size measurement. Two problems are, however, inherent in *all* light-scattering devices:

1. The theories on which they are based, and that can readily be evaluated, are available *only* for spheres.
2. There are considerable theoretical and technical problems in obtaining meaningful information for particles whose size is of the order of (or smaller than) the wavelength of the incident radiation.

Size information about smaller particles is yielded by large-angle scattering (commonly 90° to the plane of the incident light), and for larger particles by so-called forward scattering. The former is dealt with by the Mie theory, the latter usually by Fraunhofer diffraction theory [161,185]. By careful instrument design, the smaller particle region can be considered to cover the range from about 0.04 to 3 μm, and the larger particle region from about 1 to 2000 μm, or more [162]. Some modern instruments attempt to combine the two techniques.

The submicrometer range can be dealt with by photon (or auto-) correlation spectroscopy. This relies on the fluctuations in light intensity with time caused by Brownian movement of particles. Although the theory is well understood for monodisperse systems of spheres, this is not the case for polydisperse systems of particles of differing shapes and refractive indices. Thomas [186] describes a mathematical technique for obtaining improved size distribution data from photon correlation spectroscopy. So far photon correlation spectroscopy seems not to have been applied to mineral systems, but it may have applications for monomineralic samples with particles of similar shape.

A related device, which also depends on the fluctuation of light intensity is the fiber-optic Döppler anemometer (FODA). In this case laser light is passed down a fiber into a suspension and particles passing the end of the fiber reflect light back to a detector. There is a Döppler shift in the wavelength of the reflected light due to the Brownian motion of the particle, which is related to the particle size [187]. Kosmas et al. [188] used this method to obtain size distribution information for synthetic iron oxides, but no comparison is given with more conventional methods. Since in fiber-optic anemometry there is no sample cell, the fiber can be dipped into a vessel. With this advantage, it becomes possible, in theory, to follow the change in particle size inside a reaction vessel.

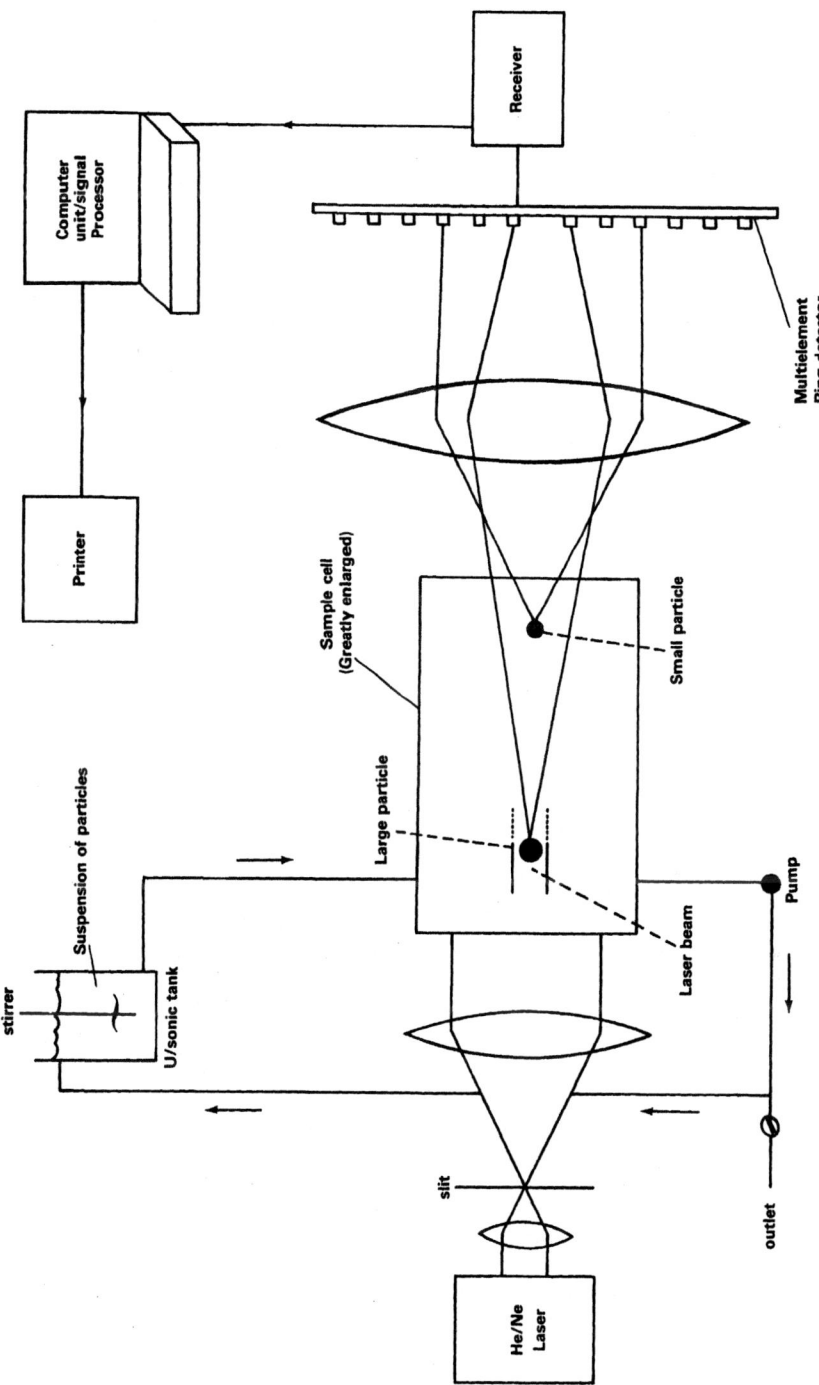

FIG. 12 Schematic representation of a light-scattering particle size analyzer. (Courtesy of Malvern Instruments Ltd., Malvern, U.K.)

There are several Fraunhofer-based light-scattering devices on the market (e.g., Microtrac, Cilas, Malvern Instruments; see Table 2). All use low-power helium-neon lasers as light sources; the generalized layout is shown in Fig. 12. There is considerable variation in the manner in which the signal is detected, and detailed consideration of the physical principles is given by Plantz [189], Cornillault [190], and Swithenbank et al. [191], for the Microtrac, Cilas, and Malvern Instruments machines, respectively. All work over a fixed number of preset size ranges within an overall range (e.g., from 1.9 to 176 µm in steps of 1.9-3.9, 3.9-7.8, 7.8-16, etc.), the overall range being expandable in some of the instruments by changing the optical system. This is currently an area of intensive instrument development, and improvements are taking place very rapidly. Those considering the use or purchase of such instruments should realize that some cheaper machines are designed to give only a median particle size and the statistics associated with that function. These machines are really intended for quality control operations in industrial processes, not for the measurement of particle size distribution as conceived in this review.

The application of light-scattering instruments to sedimentological analysis has, to date, been very limited. Cooper et al. [192] have reported the use of the Microtrac in soil particle size analysis, over the size range 1.9-176 µm, in comparison with the pipet method. They presented their data as a statistical comparison of the percentage of each size fraction found by each method, with and without removal of organic matter and soluble salts. Their findings were as follows:

1. Removal of organic matter improved the agreement between each particle size range.
2. Agreement was better between separate size fractions than between the complete range studied.
3. Agreement was best between the ranges 62-31 and 31-16 µm ($r^2 = 0.92$ in both cases).
4. Statistical agreement for all size ranges improved when the 176-62 µm sieve data were omitted.
5. The greatest differences between the pipet and Microtrac methods were found in the 3.9-1.9 µm fractions.

Differences were also found on the basis of mineralogy. Samples containing a greater proportion of platy minerals such as mica and kaolin, and expansible clays, gave higher contents for the finer fractions than did samples in which such minerals were less abundant. In general, there was no very clear pattern of agreement between the methods for any given sample.

Mohnot [193] found the Cilas instrument to be useful as a rapid means of checking flocculation phenomena in drilling muds, but reported no details of his comparisons with other methods. He also appears, like everyone else, to have ignored the possible role of the instrument pump in floc breakage, as was found for the esz method [150].

McCave et al. [194] evaluated the Malvern Instruments 3600E laser particle size analyzer using both 63 and 100 mm focal length lenses. They also compared the instrument with data obtained from the same samples by an electrical sensing zone machine. Their principal finding was that the laser-based instrument seemed to be severely affected by light scattered by particles under 2 μm, which showed as modes in the cumulative particle size curves irrespective of sample type and treatment and did not occur in the curves obtained from esz measurements. The effect was most pronounced with the 63 mm lens, but it also occurred with the 100 mm lens, varied in magnitude and, to some extent, with clay content. The effect is most marked in samples with clay contents (< 2 μm esd) of 35% or more.

These findings reflect uncertainties found by others in the use of laser light particle size analyzers for very small particles. Dodge [195] reported discrepancies during calibration of such instruments, while Evers [196] found that the Malvern and Microtrac instruments gave very different results for the same material.

In summary, light-scattering instruments offer the possibility of measuring particle size very rapidly with very small samples of material. However, the theories on which they are based are known rigorously only for spheres, and the instruments clearly have problems in dealing with variation in this factor and with systems of particles of differing refractive index. For the soil scientist they represent very much a "black box" approach and should be used with some caution and a good grasp of physics if sensible progress is to be made in their application.

4. Other Electromagnetic Methods

Ballinger and Jennings [197] reported the possibility of using acoustic birefringence for the rapid determination of the size of submicrometer particles. However, Darlington [198] questioned the usefulness of the method for materials such as bentonite. Oakley and Jennings [199] reported on the technique of transient electric birefringence and found excellent agreement between sizes measured by this method and by transmission electron microscopy.

A very recent technique, for which commercial instrumentation is only just becoming available, is field-flow chromatography. The particles are injected as a suspension into a channel across which is applied an electrical field. The rate at which the particles are eluted along the channel depends on their size and the degree to which they are affected by the field. So far the principle seems not to have been applied to complex systems such as very fine sediments, but it may be capable of development in this direction. The technique is discussed at length in Barth [162], while more general aspects of particle sizing by chromatography are treated by Janca [200].

H. Organic Soils

Organic soils behave quite differently from mineral soils, usually being much more fragile. The commonest technique for sizing organic materials is sieving [201-203], possibly following ultrasonic dispersion [204]. There is a considerable body of opinion that regards the *rubbed fiber* content of an organic soil as a more meaningful guide to its degree of humification, and this measurement figures largely in the classification of such soils [205-207].

ACKNOWLEDGMENTS

We are greatly indebted to the staff of the Library of Rothamsted Experimental Station, Harpenden, Hertshire, U.K., especially Mrs. Joan Johnson and Mrs. Elizabeth Allsopp, for their help with the bibliography, and also to Mrs. Jan Shuttleworth and especially Mrs. Carole Scott for word-processing the manuscript.

REFERENCES

NOTE: Readers requiring a general overview of the subject should consult Allen [16]. Modern instrumental methods are dealt with by Barth [162], Dahneke [185], and Janca [200], while the theory of applications of electromagnetic radiation is covered by Bayvel and Jones [208]. The *Proceedings* of the "Particle Size Analysis" conferences (usually denoted by the year) are an excellent source for new developments and applications (see Ref. 146 for an example of this series). From time to time reviews of recent developments are published as supplements in *Analytical Chemistry*; Ref. 9 is an example of such a review.

1.   Singer, C., E. J. Holmyard, and A. R. Hall, *A History of Technology*, Vol. I, *From Early Times to Fall of Ancient Empires*, Clarendon Press, Oxford, 1956, pp. 274-279 and pp. 566 et seq.

2. Bray, F., *Science and Civilisation in China*, Vol. 6, *Biology and Biological Technology*, Part II, *Agriculture*, Cambridge University Press, Cambridge, 1984, pp. 363-378.

3. Partington, J. R., *Origins and Development of Applied Chemistry*, Macmillans, London, 1935.

4. Ramelli, A., *La Diverse et Artificiose Machine*, Paris, 1588.

5. Mellor, J. W., *A Comprehensive Treatise on Inorganic and Theoretical Chemistry*, Vol. 1, Longmans, London, 1922, pp. 1-119 and 506-578.

6. Atterberg, A., Die Klassifikation der humusfreien und der humusarmen Mineralboden Schwedens nach den Konsistenz verhaltnissen derselben, *Int. Mitt. Bodenkd.*, 6: 27-37 (1916).

7. Oden, S., The size distribution of particles in soils and the experimental methods of obtaining them, *Soil Sci.*, 19: 1-36 (1925).

8. Commission I, The study of soil mechanics and physics, in *Proc. 1st Int. Congr. Soil Sci.*, Part II, Washington, DC, 1928, pp. 359-404.

9. Barth, H. G., and S. T. Sun, Particle size analysis, *Anal. Chem.*, 57: 151R-175R (1985).

10. American Society for Testing and Materials, *Annual Book of ASTM Standards*, Part 19, *Natural Building Stones, Soil and Rock, Peats, Mosses and Humus*, ASTM, Philadelphia, 1974.

11. Klute, A. (Ed.), *Methods of Soil Analysis*, Part I, *Physical and Mineralogical Methods*, 2nd ed., Am. Soc. Agron., Madison, WI, 1986.

12. Head, K. H., *Manual of Soil Laboratory Testing*, Vol. I, *Soil Classification and Compaction Tests*, Pentech Press, London, 1980, pp. 152-155.

13. British Standards Institution, *Methods for the Determination of Particle Size Distribution*, BS 3406, Part 2: *Recommendations for Gravitational Liquid Sedimentation Methods for Powders and Suspensions*, British Standards Institution, London, 1984.

14. Deutsches Institut für Normen, *Partikelgrossenanalyse; Sedimentationanalyse in Schwerefeld; Pipette-Verfahren*, DIN 66115, Deutsches Institut für Normen, Berlin, 1983.

15. Soil Conservation Service, *Procedures for Collecting Soil Samples and Methods of Analysis for Soil Survey*, Soil Survey Investigations Report No. 1, U.S. Dept. Agric., Washington, DC, 1984, pp. 15-19.

16. Allen, T., *Particle Size Measurement*, 3rd ed., Chapman and Hall, London, 1981.

17. Murphy, C. P., P. Bullock, and R. H. Turner, The measurement and characterisation of voids in soil thin sections by image analysis: I. Principles and Techniques, *J. Soil Sci.*, 28: 498-508 (1977).

18. Ringrose-Voase, A., and P. Bullock, The automatic recognition and measurement of soil pore types by image analysis and computer programs, *J. Soil Sci.*, 35: 673-684 (1984).

19. Cui, B., and P. D. Komar, Size measures and the ellipsoidal form of clastic sediment particles, *J. Sedimentol. Petrol.*, 54: 783-797 (1984).

20. Ehrlich, R., and B. Weinberg, An exact method for characterization of grainshape, *J. Sedimentol. Petrol.*, 40: 205-212 (1970).

21. Griffiths, J. C., *Scientific Method in Analysis of Sediments*, McGraw-Hill, New York, 1967.

22. Ralph, B., Application of image analysis to the measurement of particle size and shape, *Anal. Proc.*, 21: 506-508 (1984).

23. Mazzullo, J., and S. K. Kennedy, Automated measurement of the nominal sectional diameters of individual sedimentary particles, *J. Sedimentol. Petrol.*, 55: 593-595 (1985).

24. Raeymaekers, B., P. van Espen, and F. Adams, The morphological characterization of particles by automated scanning electron microscopy, *Mikrochem. Acta*, 2: 437-454 (1984).

25. Gutteridge, P., Grain-size measurement from acetate peels, *J. Sedimentol. Petrol.*, 55: 595-596 (1985).

26. Kaye, B. H., Characterization of the surface area of a fine-particle profile by its fractal dimension, in *Particle Size Analysis* (M. J. Groves, Ed.), Heyden, London, 1978, pp. 250-259.

27. Wentworth, C. K., A scale of grade and class terms for clastic sediments, *J. Geol.*, 30: 377-392 (1922).

28. Hodgson, J. M., *Soil Sampling and Soil Description*, Oxford University Press, London, 1978.

29. British Standards Institution, *Code of Practice for Site Investigation*, BS 5930, British Standards Institution, London, 1981.

30. Carver, R. G. (Ed.), *Procedures in Sedimentary Petrology*, Wiley-Interscience, New York, 1971.

31. Walton, E. K., W. E. Stephens, and M. S. Shawa, Reading segmented grain-size curves, *Geol. Mag.*, 117: 517-524 (1980).

32. Alderliesten, M., A nomenclature for mean particle diameters, *Anal. Proc.*, 21: 167-173 (1984).

33. Webster, R., *Quantitative and Numerical Methods in Soil Classification and Soil Survey*, Oxford University Press, London, 1977.

34. Wood, A., *The Groundnut Affair*, Bodley Head, London, 1950.

35. Mullins, C. E., and B. J. Hutchinson, The variability introduced by various subsampling techniques, *J. Soil Sci.*, 33: 547-561 (1982).

36. Billi, P., Quick field measurement of gravel particle size, *J. Sedimentol. Petrol.*, 54: 658-660 (1984).

37. Caroni, E., and F. Maraga, Misure granulometriche in alvei naturali con un compasso registratore, *Geol. Appl. Idrogeol.*, 18: 19-31 (1983).

38. Laxton, J. L., A method for estimating the grading of boulder and cobble grade material, in *IGS Short Communications*, Report Inst. Geol. Sci. No. 80/1, HMSO, London, 1980, pp. 31-35.

39. Smith, P. D., and A. J. Thomasson, Density and water-release characteristics, in *Soil Survey Laboratory Methods* (B. W. Avery and C. L. Bascomb, Eds.), Tech. Monogr. No. 6, Soil Survey, Harpenden, U.K., 1974, pp. 42-44.

40. Kiss, K., and R. N. Pease, Quantitative analysis of particle sizes: Estimation of the most efficient sampling scheme, *J. Microsc.*, 126: 173-178 (1982).

41. Smart, P., and N. K. Tovey, *Electron Microscopy of Soils and Sediments: Techniques*, Clarendon Press, Oxford, 1982.

42. Jackson, M. L., *Soil Chemical Analysis—Advanced Course*, 2nd ed., published by the author, Madison, WI, 1975.

43. Fischer, W. R., Oxidation of soil organic matter by potassium hypobromite for particle size determination, *Commun. Soil Sci. Plant Anal.*, 15: 1281-1284 (1984).

44. Gahm, J., Mikroskopische Kornungsanalyse: III, *Verfahrenstechnologie*, 16: 869-871 (1982).

45. British Standards Institution, *Methods for the Determination of Particle Size Distribution*, BS 3406, Part 4, Optical Microscope Method, British Standards Institution, London, 1985.

46. Bradley, D. E., Evaporated carbon films for use in electron microscopy, *Br. J. Appl. Phys.*, 5: 65-66 (1954).

47. Comer, J. J., and J. W. Turley, Replica studies of bulk clays, *J. Appl. Phys.*, 26: 346-350 (1955).

48. Nixon, H. L., and A. H. Weir, The morphology of the Unter-Rupsroth montmorillonite, *Min. Mag.*, 31: 413-416 (1957).

49. Hunt, C. M., and A. R. Woolf, Comparison of some different methods for measuring particle size using microscopically calibrated glass beads, *Powder Technol.*, 3: 9-23 (1969).

50. Gard, J. A. (Ed.), *The Electron-Optical Investigation of Clays*, Mineralogical Society, London, 1971.

51. Cases, J. M., P. Cunin, Y. Grillet, C. Poinsignon, and J. Yvon, Methods of analysing morphology of kaolinites: Relations between crystallographic and morphological properties, *Clay Mineral.*, 21: 55-68 (1986).

52. Nadeau, P. H., M. J. Wilson, W. J. McHardy, and J. M. Tait, Inter-particle diffraction: A new concept for interstratification of clay minerals, *Clay Mineral.*, 19: 757-770 (1984).

53. Kellerhals, R., J. Shaw, and V. K. Arora, On grain size from thin sections, *J. Geol.*, 83: 79-96 (1975).

54. Born, M., and E. Wolf, *Principles of Optics*, 5th ed., Pergamon Press, Oxford, 1975, pp. 459-490.

55. Rittinger, P. Ritter, von, *Lehrbuch der Aufbereitungskunde*, Vol. II, *Das Separiren: 2. Separation durch Classieren und nachheriges Sortiren*, Ernst und Korn, Berlin, 1867, pp. 222-232.

56. International Organisation for Standardisation, *Test Sieves—Woven Metal Wire Cloth, Perforated Plate and Electro-formed Sheet—Nominal Sizes of Openings*, ISO-565-1983(E), ISO, Geneva, 1983.

57. Association Française de Normalisation, *Ouvertures et entraxes des tôles et feuilles perforées à trous ronds et à trous carrés*, NF-E-81-061, Association Française de Normalisation (AFNOR), Paris, 1986.

58. Association Française de Normalisation, *Analyse granulométrique: Tamisage de contrôle*, NF-X-11-507, Association Française de Normalisation (AFNOR), Paris, 1970.

59. Association Française de Normalisation, *Tamisage en milieu liquide des poudres de granulométrie inférieure à 200 micromètres*, NF-X-11-642, Association Française de Normalisation (AFNOR), Paris, 1982.

60. Deutsches Institut für Normen, *Rundloch-platten, Masse*, DIN 24 041, Deutsches Institut für Normen, Berlin, 1981.

61. Deutsches Institut für Normen, *Quadratloch-platten, Masse*, DIN 24 042, Deutsches Institut für Normen, Berlin, 1981.

62. Deutsches Institut für Normen, *Siebboden: Drahtgewebe aus Stahl—Nichtrostendem Stahl - und NE - Metalldraht, Masse*, DIN 4189, Teil 1, Deutsches Institut für Normen, Berlin, 1968.

63. Japanese Standards Association, *Test Sieves*, Japanese Industrial Standard JIS-Z-8801 (English transl.), Japanese Standards Association, Tokyo, 1982.

64. British Standards Institution, *Specification for Test Sieves*, BS 410, British Standards Institution, London, 1986.

65. American Society for Testing and Materials, *Standard Specification for Perforated-Plate Sieves for Testing Purposes*, in *Annual Book of Standards*, Vol. 14.02, Designation E323-80, ASTM, Philadelphia, 1986, pp. 321-325.

66. American Society for Testing and Materials, *Standard Specification for Industrial Perforated Plate and Screens (Round Opening Series)*, in *Annual Book of Standards*, Vol. 14.02, Designation E674-80, ASTM, Philadelphia, 1986, pp. 518-536.

67. American Society for Testing and Materials, *Standard Specification for Industrial Perforated Plate and Screens (Square Opening Series)*, in *Annual Book of Standards*, Vol. 14.02, Designation E454-80, ASTM, Philadelphia, 1986, pp. 408-423.

68. American Society for Testing and Materials, *Standard Specification for Wire-Cloth Sieves for Testing Purposes*, in *Annual Book of Standards*, Vol. 14.02, Designation E11-81, ASTM, Philadelphia, 1986, pp. 5-10.

69. American Society for Testing and Materials, *Standard Specification for Precision Electroformed Sieves*, in *Annual Book of Standards*, Vol. 14.02, Designation E161-70, ASTM, Philadelphia, 1986, pp. 112-116.

70. Tanner, C. W., and S. J. Bourget, Particle shape discrimination of round and square holed sieves, *Soil Sci. Soc. Am. Proc.*, 16: 88 (1952).

71. Metz, R., The importance of maintaining horizontal sieve screens when using a Ro-Tap, *Sedimentology*, 32: 613-614 (1985).

72. Kennedy, S. K., T. P. Meloy, and T. E. Durney, Sieve data—Size and shape information, *J. Sedimentol. Petrol.*, 55: 356-360 (1985).

73. Jones, T. M., The Alpine air-jet sieve, *Proc. Soc. Anal. Chem.*, 7: 159-163 (1970).

74. Association Française de Normalisation, *Analyse granulometrique des poudres fines sur tamiseuse à depression d'air*, NF-X-11-640, Association Française de Normalisation (AFNOR), Paris, 1979.

75. British Standards Institution, *Method for Test Sieving*, BS 1796, British Standards Institution, London, 1985.

76. Stokes, G. G., On the theories of the internal friction of fluids in motion, and of the equilibrium and motion of elastic solids, *Camb. Phil. Trans.*, 8: 287-319 (1849).

77. Dewell, P., A centrifugal sedimentation method for particle size analysis, in *Particle Size Analysis*, Society for Analytical Chemistry, London, 1967, pp. 268-280.

78. Thewlis, J. (Ed.), *Encyclopaedic Dictionary of Physics*, Vol. 5, Pergamon Press, Oxford, 1962, pp. 313-316.

79. Robinson, G. W., A new method for the mechanical analysis of soils and other dispersions, *J. Agric. Sci. Camb.*, 12: 306-321 (1922).

80. Anderson, J. U., An improved pretreatment for mineralogical analysis of samples containing organic matter, *Clays Clay Mineral.*, 10: 380-388 (1963).

81. Pritchard, D. T., A method for particle-size analysis using ultrasonic disaggregation, *J. Soil Sci.*, 25: 34-40 (1974).

82. Genrich, D. A., and J. M. Bremner, A reevaluation of the ultrasonic-vibration method of dispersing soils, *Soil Sci. Soc. Am. Proc.*, 36: 944-947 (1972).

83. Maeda, T., H. Takenaka, and B. H. Warkentin, Physical properties of allophane soils, *Adv. Agron.*, 29: 229-264 (1977).

84. Yaalon, D. H., "Calgon" no longer suitable, *Soil Sci. Soc. Am. J.*, 40: 333 (1976).

85. Andreasen, A. H. M., Zur Kenntnis des Mahlgutes, *Kolloid-chem. Beihefte*, 27: 349-458 (1928).

86. British Standards Institution, *Methods of Test for Soils for Civil Engineering Purposes*, BS 1377, British Standards Institution, London, 1975, pp. 30-46.

87. McKeague, J. A. (Ed.), *Manual on Soil Sampling and Methods of Analysis*, Canada Soil Survey Committee, Ottawa, 1976, pp. 4-32.

88. Gee, G. W., and J. W. Bauder, Particle size analysis, in *Methods of Soil Analysis*, Part I, 2nd ed. (A. Klute, Ed.), Am. Soc. Agron., Madison, WI, 1986, pp. 383-411.

89. Kilmer, V. J., and J. F. Mullins, Improved stirring and pipetting apparatus for mechanical analysis of soils, *Soil Sci.*, 77: 437-441 (1954).

90. Coventry, R. J., and D. E. R. Fett, *A Pipette and Sieve Method of Particle-Size Analysis and Some Observations on Its Efficiency*, Div. Soils Divisional Rep. No. 38, CSIRO, Queensland, Australia, 1979.

91. Miller, W. P., and D. M. Miller, A micro-pipette method for soil mechanical analysis, *Commun. Soil Sci. Plant Anal.*, 18: 1-15 (1988).

92. van Reeuwijk, L. P. (Ed.), *Procedures for Soil Analysis*, Tech. Paper No. 9, International Soil Reference and Information Centre, Wageningen, The Netherlands, 1986, pp. 4-13.

93. Pleijsier, L. K., *The Laboratory Methods and Data Exchange Programme: Interim Report on the Exchange Round 86-1*, Working Paper and Preprint No. 86/4, International Soil Information and Reference Centre, Wageningen, The Netherlands, 1986.

94. MacKenzie, R. L., and J. A. Westgate, A microcomputer program for the ASTM method of grain size analysis, *Can. J. Earth Sci.*, 23: 737-739 (1986).

95. Bouyoucos, G. J., The hydrometer as a new and rapid method for determining the colloidal content of soils, *Soil Sci.*, 23: 319-331 (1927).

96. American Society for Testing and Materials, *Standard Method for Particle-Size Analysis of Soils*, in *Annual Book of Standards*, Vol. 04.08, Designation D422-63, ASTM, Philadelphia, 1986, pp. 116-126.

97. Gee, G. W., and J. W. Bauder, Particle size analysis by hydrometer: A simplified method for routine textural analysis and a sensitivity test of measurement parameters, *Soil Sci. Soc. Am. J.*, 43: 1004-1007 (1979).

98. Scarlett, B., and S. K. Cowlam, Concentration effects in particle size analysis by sedimentation, in *Particle Size Analysis* (M. J. Groves, Ed.), Heyden, London, 1978, pp. 430-440.

99. Puri, A. N., and B. R. Puri, Physical characteristics of soils: IV. Density gradients in sedimenting columns and a chaino-hydrometer for mechanical analysis of soils, *Soil Sci.*, 48: 149-160 (1939).

100. Arulanandan, K., and S. S. Smith, A new hydrometer, *Soil Sci.*, 112: 206-212 (1971).

101. Bascomb, C. L., and D. T. Pritchard, A Cartesian hydrometer and its application to sedimentation analysis, *J. Sci. Instrum.*, 40: 30-31 (1963).

102. Stabinger, H., H. Leopold, and O. Kratky, Eine neue Präzisionsmethode zur Bestimmung der Dichte von Flüssigkeiten, *Monatsh. Chem.*, 98: 436-438 (1967).

103. Marshall, T. J., A plummet balance for measuring the size distribution of soil particles, *Aust. J. Appl. Res.*, 7: 142-147 (1956).

104. Slager, S., and F. F. R. Koenigs, Particle-size analysis of soils by means of a torsion balance and a plummet, *Sedimentology*, 3: 240-252 (1964).

105. Sheldrick, B. H., and J. A. McKeague, Test of the plummet balance and a comparison of some procedures for particle-size analysis, *Can. J. Soil Sci.*, 56: 55-57 (1976).

106. Nagaraj, T. S., and P. V. Sivapullaiah, Plummet balance—A potential tool for subsieve particle size analysis, *Geotech. Test. J. (ASTM)*, 4: 36-40 (1981).

107. Bannister, P., P. R. Tomlinson, and P. H. T. Beckett, A simple procedure for soil mechanical analysis, *Plant Soil*, 40: 211-215 (1974).

108. Moum, J., Falling drop used for grain-size analysis of fine-grained materials, *Sedimentology*, 5: 343-347 (1965).

109. Bascomb, C. L., A new apparatus for recording particle size distribution, *J. Sedimentol. Petrol.*, 38: 878-884 (1968).

110. Syvitski, J. P. M., and D. D. Swinbanks, V.S.A.: A new fast size analysis technique for low sample weight based on Stokes' settling velocity, *Can. Geotech. J.*, 27: 304-312 (1980).

111. Just, G., Eine Feldmethode zur Ansprache des Schluff- und Tongehaltes von Bodenproben—Kurzmitteilung, *Arch. Acker Pflanz. Bodenkd.*, 17: 255-258 (1973).

112. Hodgson, J. M., J. M. Hollis, R. J. A. Jones, and R. C. Palmer, A comparison of field estimates and laboratory analyses of the silt and clay contents of some West Midland soils, *J. Soil Sci.*, 27: 411-419 (1976).

113. Tanner, C. B., and J. L. Jackson, Nomographs of sedimentation times for soil particles and gravity or centrifugal acceleration, *Soil Sci. Soc. Am. Proc.*, 12: 60-65 (1947).

114. U.S. Department of Agriculture, *Soil Survey Laboratory Methods and Procedures for Collecting Soil Samples*, Soil Survey Investigations Rep. No. 1, Soil Conservation Service, Washington, DC, 1972, pp. 11-12.

115. Avery, B. W., and C. L. Bascomb (Eds.), *Soil Survey Laboratory Methods*, Tech. Monogr. No. 6, Soil Survey, Harpenden, U.K., 1982, pp. 18-19.

116. British Standards Institution, *Methods for the Determination of Particle Size Distribution, BS 3406, Part 6, Recommendations for Centrifugal Liquid Sedimentation Methods for Powders and Suspensions*, British Standards Institution, London, 1985.

117. Slater, C., and L. Cohen, A centrifugal particle size analyzer, *J. Sci. Instrum.*, 39: 614-617 (1962).

118. Vaughan, G. N., R. W. Ford, and H. W. West, The application of the Simcar centrifugal particle size analyser to clays, *Proc. Br. Ceram. Soc.*, 13: 47-56 (1969).

119. Allen, T., and L. Svarovsky, A new high speed disc centrifuge with pipette withdrawal, *Dechem Monogram. (Nürnburg) Part 1589-1615*: 279-292 (1976).

120. Kamack, H. J., Particle size determination by centrifugal pipet sedimentation, *Anal. Chem.*, 23: 844-850 (1951).

121. Svarovsky, L., and J. Svarovska, Centrifutal sedimentation data analysis by analogue deconvolution, *J. Phys. D: Appl. Phys.*, 8: 181-190 (1975).

122. Coll, H., and C. G. Searles, Particle size analysis with the Joyce-Loebl disk centrifuge: A comparison of the line-start with the homogeneous-start method, *J. Colloid Interface Sci.*, 115: 121-129 (1987).

123. Whalley, W. R., Theory and use of centrifugal photosedimentation for particle size analysis of clays, Ph.D. thesis, University of Aberdeen, U.K., 1988.

124. Oppenheimer, L., Interpretation of disk centrifuge data, *J. Colloid Interface Sci.*, 92: 350-357 (1983).

125. Kane, P., The nature of the translocated unit and structure of the related cutanic forms in certain loess and loess-like soils, Ph.D. thesis, University of Reading, U.K., 1981.

126. Churchman, G. J., and K. R. Tate, Stability of aggregates of different size grades in allophanic soils from volcanic ash in New Zealand, *J. Soil Sci.*, 38: 19-28 (1987).

127. Davis, C. N., Sedimentation of small suspended particles, *Trans. Ind. Chem. Eng.*, S25: 25-39 (1947).

128. McCormick, H. W., Determination of latex particle size distribution by analytical centrifugation, *J. Colloid Sci.*, 19: 173-184 (1964).

129. Bain, J. A., and D. J. Morgan, Laboratory separation of clays by hydrocycloning, *Clay Mineral.*, 18: 33-47 (1983).

130. Batch, B. A., The application of an electronic particle counter to size analysis of pulverized coal and fly-ash, *J. Inst. Fuel*, 37: 455-461 (1964).

131. Walker, P. H., and J. Hutka, *Use of the Coulter Counter (Model B) for Particle-Size Analysis of Soils*, Div. of Soils Tech. Paper No. 1, CSIRO, Melbourne, Australia, 1971.

132. Lines, R. W., Particle counting by Coulter counter, *Anal. Proc.*, 18: 514-519 (1981).

133. Harfield, J. G., and P. Knight, Experimental evidence of the linear response of the Coulter counter, in *Particle Size Analysis 1981* (N. G. Stanley-Wood and T. Allen, Eds.), Wiley-Interscience, New York, 1982, pp. 151-164.

134. Lloyd, P. J., Response of the electrical sensing zone method to nonspherical particles, in *Particle Size Analysis 1981* (N. G. Stanely-Wood and T. Allen Eds.), Wiley-Interscience, New York, 1982, pp. 199-208.

135. Sheldon, R. W., and T. R. Parsons, *A Practical Manual on the Use of the Coulter Counter in Marine Science*, Coulter Electronics of Canada Ltd., Ontario, 1967.

136. McCave, I. N., and J. Jarvis, Use of the Coulter counter in size analysis of fine to coarse sand, *Sedimentology*, 20: 305-315 (1973).

137. Hunt, C. M., and A. R. Woolf, Comparison of some different methods for measuring particle size using microscopically calibrated glass beads, *Powder Technol.*, 3: 1-8 (1969).

138. Schrag, K. R., and M. Corn, Comparison of particle size determined with the Coulter counter and by optical microscopy, *Am. Ind. Hyg. Assoc. J.*, 31: 446-453 (1976).

139. Johnston, P. R., and R. Swanson, A correlation between the results of different instruments used to determine the particle size distribution in AC fine test dust, *Powder Technol.*, 32: 119-124 (1982).

140. Ormsby, W. C., and P. R. Odom, Particle size analysis of whiteware clays: Interlaboratory comparison of methods, *J. Mater. JMLSA*, 5: 486-601 (1970).

141. Seville, J. P. K., J. R. Coury, M. Ghadiri, and R. Clift, Comparison of techniques for measuring the size of fine nonspherical particles, *Part. Charact.*, 1: 45-52 (1984).

142. Griffiths, W. D., and A. P. Rood, Comparison of the size distribution of the "BCR" reference quartz by ESZ and image analysis techniques, *Anal. Proc.*, 21: 165-167 (1984).

143. British Standards Institution, *Methods for Determination of Particle Size Distribution*, BS 3406, Part 5, *Recommendations for Electrical Sensing Zone Method (the Coulter Principle)*, British Standards Institution, London, 1983.

144. American Society for Testing and Materials, *Tentative Method of Test for Concentration and Particle Size Distribution of Airborne Particulates Collected in Liquid Media Using an Electronic Counter*, Designation D3365-74T, ASTM, Philadelphia, 1974.

145. Harfield, J. G., Standardization of the Coulter counter for sizing and counting, *Anal. Proc.*, 21: 162-165 (1984).

146. Atkinson, C. M. L., and R. Wilson, The mass integration method for the calibration of the electrical sensing zone technique used for the sizing and counting of fine particles, in *Particle Size Analysis 1981* (N. G. Stanley-Wood and T. Allen, Eds.), Wiley-Interscience, New York, 1982, pp. 185-197.

147. Davidson, C., and M. J. Keen, Size analysis of turbidity current sediment, *Nature* (London), 197: 372-373 (1963).

148. Walker, P. H., K. D. Woodyer, and J. Hutka, Particle-size measurements by Coulter counter of very small deposits and low suspended sediment concentrations in streams, *J. Sedimentol. Petrol.*, 44: 673-679 (1974).

149. Ozturgut, O., and J. W. Lavelle, Settling analysis of fine sediment in salt-water at concentrations low enough to preclude flocculation, *Mar. Geol.*, 69: 353-362 (1986).

150. Gibbs, R. J., Floc stability during Coulter size analysis, *J. Sedimentol. Petrol.*, 52: 657-660 (1982).

151. Santero, T., and G. Stotzky, Influence of cations on flocculation of clay minerals by microbial metabolites as determined by the electrical sensing zone particle analyzer, *Soil Sci. Soc. Am. Proc.*, 31: 761-765 (1967).

152. Dudley, R. J., The particle size analysis of soils and its use in forensic science—The determination of particle size distribution within the silt and sand fractions, *J. Forens. Sci. Soc.*, 16: 219-229 (1976).

153. Duke, M. B., C. C. Woo, M. L. Bird, G. A. Sellers, and R. B. Finkelman, Lunar soil: Size distribution and mineralogical constituents, *Science*, 167: 648-650 (1970).

154. Sapetti, C., Misure granulometriche sul terreno, *Ann. Sper. Agrar.*, 17: 583-615 (1963).

155. Muller, R. N., and G. T. Tisne, Preparative-scale size fractionisation of soils and sediments and an application to studies of plutonium geochemistry, *Soil Sci.*, 124: 191-198 (1977).

156. Shideler, G. L., A comparison of electronic particle counting and pipet techniques in routine mud analysis, *J. Sedimentol. Petrol.*, 46: 1017-1025 (1976).

157. Behrens, E. W., Further comparisons of grain size distributions determined by electronic particle counting and pipet techniques, *J. Sedimentol. Petrol.*, 48: 1213-1218 (1978).

158. Rybina, V. V., Use of conductimetry for the determination of the particle-size composition of soils, *Sov. Soil Sci.*, 11: 482-486 (1979).

159. Pennington, K. L., and G. C. Lewis, A comparison of electronic and pipet methods for mechanical analysis of soils, *Soil Sci.*, 128: 280-284 (1979).

160. Lewis, G. C., M. A. Fosberg, A. L. Falen, and B. J. Miller, Identification of loess by particle size distribution using the Coulter counter TAII, *Soil Sci.*, 137: 172-176 (1984).

161. Kerker, M., *The Scattering of Light and Other Electromagnetic Radiation*, Academic Press, New York, 1969.

162. Barth, H. G. (Ed.), *Modern Methods of Particle Size Analysis*, Wiley, New York, 1984.

163. Mully, E. C., and H. N. Frock, Submicrometer to millimeter particle size measurement using light scattering, in *Particle Size Analysis 1981* (N. G. Stanley-Wood and T. Allen, Eds.), Wiley-Interscience, New York, 1982, pp. 385-391.

164. West, G. C., The application of the light obscuration principle to the particle size analysis of powdered materials, in *Particle Size Analysis 1977* (M. J. Groves, Ed.), Heyden, London, 1978, pp. 347-355.

165. Gibbs, R. J., Floc breakage during HIAC light blocking analysis, *Environ. Sci. Technol.*, 16: 298-299 (1982).

166. Lamp, J., and D. Schroeder, Zur Anwendung der photometrischen Kornungs-analyse, *Z. Pflanzenernaehr. Bodenkd.*, 125: 240-243 (1970).

167. Zaneveld, J. R. V., R. W. Spinrad, and R. Bartz, An optical settling tube for the determination of particle-size distributions, *Mar. Geol.*, 49: 357-376 (1982).

168. Eaton, A., V. Grant, O. Bricker, and D. Wells, On the use of the nephelometer in estuarine waters, *Estuaries*, 4: 379-384 (1981).

169. Jordan, J. R., G. E. Fryer, and E. H. Hemmen, Size analysis of silt and clay by hydrophotometer, *J. Sedimentol. Petrol.*, 41: 489-496 (1971).

170. Coates, G. F., and C. A. Hulse, A comparison of four methods of size analysis of fine-grained sediments, *N.Z. J. Geol. Geophys.*, 28: 369-380 (1985).

171. Melik, D. H., and H. S. Fogier, Turbidimetric determination of particle size distributions of colloidal systems, *J. Colloid Interface Sci.*, 92: 161-181 (1983).

172. Conlin, S. G., W. J. Levene, and W. F. Volume, An instrument for size analysis of very fine powders by X-ray absorption, *J. Sci. Instrum.*, 44: 606-610 (1967).

173. Carr-Brion, K. G., and P. J. Mitchell, An "on-stream" particle-size sensor, *J. Sci. Instrum.*, 44: 611-614 (1967).

174. Hendrix, W. P., and C. Orr, Automatic sedimentation size analysis instrument, in *Particle Size Analysis 1970* (M. J. Groves and J. L. Wyatt-Sargent, Eds.), Society for Analytical Chemistry, London, 1971, pp. 133-146.

175. Olivier, J. P., G. K. Hickin, and C. Orr, Jr., Rapid, automatic particle size analysis in the sub-sieve range, *Powder Technol.*, 4: 257-263 (1971).

176. Ullrich, B., D. Schmidt, M. Storr, and E. M. Arndt, Erfahrungen mit einer veranderten Messzelle für das Röntgensedimentometer "Sedigraph 5000-D," *Silikattechnik*, 35: 312-314 (1984).

177. Bernhardt, C., Zu einigen Problemen bei der Kornungsanalyse von Kaolin mit dem Sedigraph 5000, *Silikattechnik*, 33: 102-104 (1982).

178. Stein, R., Rapid grain-size analysis of clay and silt fraction by Sedigraph 5000-D: Comparison with Coulter counter and Atterberg methods, *J. Sedimentol. Petrol.*, 55: 590-593 (1985).

179. Berezin, P. N., and A. D. Voronin, Use of a Sedigraph for the particle-size analysis of soils, *Sov. Soil Sci.*, 13: 101-109 (1981).

180. Welch, N. H., P. B. Allen, and D. J. Galdino, Particle-size analysis by pipet and Sedigraph, *J. Environ. Qual.*, 8: 543-546 (1979).

181. Vitturi, L. M., and S. Rabitti, Automatic particle-size analysis of sediment fine fraction by Sedigraph 5000-D, *Geol. Appl. Idrogeol.*, 15: 101-108 (1980).

182. Burr, K. J., Improved methods of particle size analysis, *Anal. Proc.*, 21: 133-134 (1984).

183. Karsten, J. H. M., and W. A. G. Kotze, Soil particle size analysis with the gamma attenuation technique, *Commun. Soil Sci. Plant Anal.*, 15: 731-739 (1984).

184. Barthakur, N. N., Clay fraction determinations with a beta-ray gauge, *Commun. Soil Sci. Plant Anal.*, 17: 533-545 (1986).

185. Dahneke, B. E. (Ed.), *Measurement of Suspended Particles by Quasi-Elastic Light Scattering*, Wiley, New York, 1982.

186. Thomas, J. C., The determination of log normal particle size distributions by dynamic light scattering, *J. Colloid Interface Sci.*, 117: 187-192 (1987).

187. Ross, D. A., H. S. Dhadwal, and R. B. Dyott, The determination of the mean and standard deviation of the size distribution of a colloidal suspension of submicron particles using the fiber-optic Döppler anemometer (FODA), *J. Colloid Interface Sci.*, 64: 533-542 (1978).

188. Kosmas, C. S., D. P. Franzmeier, and D. G. Schulze, Relationship among derivative spectroscopy, color, crystallite dimensions, and Al-substitution of synthetic goethites and hematites, *Clays Clay Mineral.*, 34: 625-634 (1986).

189. Plantz, P. E., Particle size measurements from 0.1 to 1000 μm, based on light scattering and diffraction, in *Modern Methods of Particle Size Analysis* (H. G. Barth, Ed.), Wiley, New York, 1984, pp. 173-209.

190. Cornillault, J., HR850 Granulometer, *Spectra 2000*, 111: 27-29 (1986).

191. Swithenbank, J., J. M. Beer, D. S. Taylor, and G. C. McCreath, A laser diagnostic technique for the measurement of droplet and particle size distribution, *Progr. Astron. Aeronaut.*, 53: 421-447 (1977).

192. Cooper, L. R., R. L. Haverland, D. M. Hendricks, and W. G. Knisel, Microtrac particle-size analyzer: An alternative particle-size determination method for sediment and soils, *Soil Sci.*, 138: 138-146 (1984).

193. Mohnot, S. M., Characterisation and control of fine particles involved in drilling, *J. Petrol. Technol.*, 37: 1622-1632 (1985).

194. McCave, I. N., R. J. Bryant, H. F. Cook, and C. A. Coughanowr, Evaluation of a laser-diffractions size analyzer for use with natural sediments, *J. Sedimentol. Petrol.*, 56: 561-564 (1986).

195. Dodge, L. G., Calibration of the Malvern particle sizer, *Appl. Opt.*, 23: 2415-2419 (1984).

196. Evers, A. D. Methods for particle size analysis of flour: A collaborative test, *Lab. Practice*, 31: 215-219 (1982).

197. Ballinger, K. W. A., and B. R. Jennings, Rapid size measurement of submicron particles, *Nature* (London), 282: 699-701. (1979).

198. Darlington, M. W., The acoustic double refraction of pure liquids and macromolecular solutions, Ph.D. thesis, University of Southampton, U.K., 1967.

199. Oakley, D. M., and B. R. Jennings, Swift particle size characterization by transient electric birefringence, *J. Colloid Interface Sci.*, 91: 188-198 (1983).

200. Janca, J., *Field Flow Fractionation: Analysis of Macromolecules and Particles*, Dekker, New York, 1988.

201. American Society for Testing and Materials, *Standard Test Method for Particle Size Range of Peat Materials for Horticultural Purposes*, in *Annual Book of Standards*, Vol. 04.08, Designation D2977-71, ASTM, Philadelphia, 1986.

202. Williams, B. L., The nitrogen content of particle size fractions separated from peat and its rate of mineralization during incubation, *J. Soil Sci.*, 34: 113-125 (1983).

203. Boelter, D. H., Physical properties of peats as related to degree of decomposition, *Soil Sci. Soc. Am. Proc.*, 33: 606-609 (1969).

204. Watson, J. R., and J. W. Parsons, Studies of soil organomineral fractions: I. Isolation by ultrasonic dispersion, *J. Soil Sci.*, 25: 1-8 (1974).

205. Lynn, W. C., W. E. McKinzie, and R. B. Grossman, Field laboratory tests for characterization of Histosols, in *Histosols: Their Characteristics, Classification and Use* (A. R. Aandahl, S. W. Buol, D. E. Hill, and H. H. Bailey, Eds.), Soil Sci. Soc. of Am. Spec. Publ. No. 6, Madison, WI, 1974.

206. Levesque, M., and H. Dinel, Fiber content, particle-size distribution and some related properties of four peat materials in eastern Canada, *Can. J. Soil Sci.*, 57: 187-195 (1977).

207. Day, J. H., P. J. Rennie, W. Stanek, and G. P. Raymond (Eds.), *Peat Testing Manual*, Associate Committee on Geotechnical Research, National Research Council of Canada Tech. Memor. No. 125, 1979, pp. 35-46.

208. Bayvel, L. P., and A. R. Jones, *Electromagnetic Scattering and Its Applications*, Applied Science Publishers, London, 1981.

7

Bulk Density

DONALD J. CAMPBELL and J. KENNETH HENSHALL *Scottish Centre of Agricultural Engineering, Penicuik, Midlothian, Scotland*

I. INTRODUCTION

The wet bulk density of a soil is its mass, including any water present, per unit volume, and its dry bulk density is the mass of oven-dry soil per unit volume of moist soil. The dry bulk density ρ_s and the wet bulk density ρ are related to the soil gravimetric water content W as follows:

$$\rho_s = 100 \; \frac{\rho}{100 + W} \tag{1}$$

where gravimetric water content is the mass of water expressed as a percentage of the mass of dry soil. When a soil undergoes compaction, its bulk density increases. While dry bulk density per se is a satisfactory measure of the state of compaction of a soil, a knowledge of the soil particle density allows the porosity and void ratio to be calculated. The latter properties are often of more interest in studies of crop growth and the consolidation of soils, respectively.

The methods available for the measurement of soil bulk density fall into two groups. In the first group are the long-established methods, which involve measurement of the sample mass and volume. Such methods have been used by both agricultural soil scientists [1] and civil engineers [2], and many of these direct methods reduce essentially to the problem of the accurate determination of the sample volume. These methods are not always entirely effective,

however, and therefore a second group of methods has evolved in
which the attenuation or scattering of nuclear radiation by soil is
used, in conjunction with a calibration relationship, to give an indirect measurement of bulk density. Radiation methods are capable
of measurements that are generally more accurate and precise than
those involving measurement of sample mass and volume, but they
are not without limitations of their own.

Thus, there is no single measurement method suitable for all
circumstances, and it is important to consider the advantages and
limitations of each method in relation to the reason for making the
measurement. Sometimes a very crude but quick measurement is all
that is required to characterize soil conditions. On other occasions—
when, for example, comparisons of the compaction produced by
wheels or the loosening effects of cultivation implements are required—it may well be appropriate to use a slow method involving
expensive equipment if the detailed differences between treatments
could not otherwise be detected.

II. THEORY

A. Direct Measurement of Sample Mass and Volume

In direct determinations of dry bulk density, the mass M_s of the
oven-dry sample is obtained by weighing and the total volume V of
the soil including air and water is obtained by measurement or indirect estimation. The dry bulk density ρ_s is then given by

$$\rho_s = \frac{M_s}{V} \qquad\qquad (2)$$

B. Radiation Methods

Radiation methods depend on the measurement of the effect of soil
on radiation. Empirical calibration relationships are used to relate
the magnitude of such effects to soil bulk density. The most commonly used radiation methods involve the attenuation or scattering
of gamma radiation by the soil, both of which increase with density.

Gamma-ray photons are emitted by radioactive nuclei as they
decay to form more stable nuclei of lower excitation. These photons
are emitted with an energy characteristic of the decay transition.
A specific source will therefore emit gamma photons with one or more
characteristic energies. In passing through any medium, the probability that these photons will interact with the atoms of the medium
is dependent on the density of the medium, as well as other factors
such as the energy of the photon and the chemical composition of
the medium. These interactions take the form of either complete
absorption or scattering of the photon. During scatter, the photon

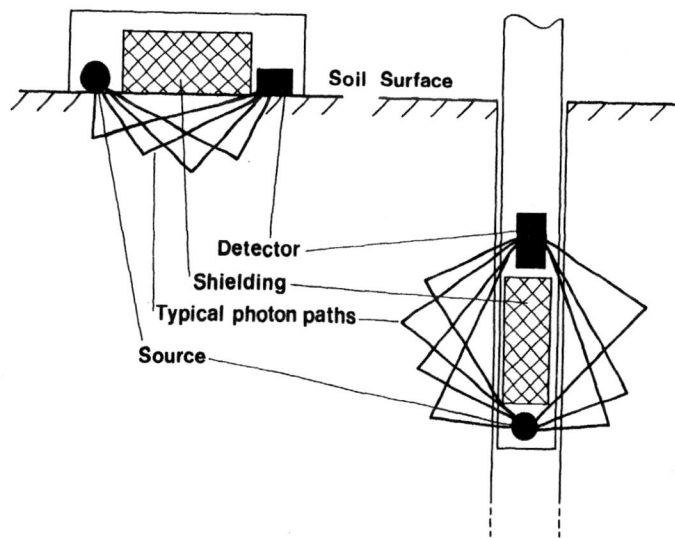

FIG. 1 Schematic diagrams of backscatter gamma-ray gauges: left, source-detector assembly lies on the soil surface; right, source-detector is lowered into an access hole in the soil.

is deflected by the electrons within the medium with a consequent loss of energy related to the angle of deflection. Since the photons interact principally with the electrons, the extent of the interaction depends on the electron density, which is related to the bulk density of the medium.

Basically, two types of gamma-ray density equipment exist: backscatter gauges, which are designed to detect only scattered photons, and transmission gauges, which detect only unscattered photons. However, some simpler transmission systems, which employ detectors without energy discrimination, will also detect some scattered photons.

1. Backscatter Gauges

In backscatter gauges, the gamma-ray source and detector are fixed relative to, and shielded from, each other in an assembly designed to prevent measurement of directly transmitted photons. This assembly either rests on the soil surface or, in some designs, is lowered into an access hole in the soil (Fig. 1). In either case, any photons incident upon the detector must have been deflected by one or more scattering interactions in the medium. Since there is only a low probability that a photon that has traveled an appreciable

distance from the source will reach the detector, it follows that
only a restricted volume of the medium close to the source-detector
axis will influence the detected photon count rate. In practice [3],
with a probe that is used in an access hole, it is found that the
zone of influence does not extend more than about 75 mm from the
source-detector axis and that 50% of the count results from material
within about 25 mm of this axis.

The relation between count rate and bulk density is complex,
since the degree of scattering increases with density, thereby increasing the count rate, but absorption of both scattered and unscattered photons also increases with density and so reduces count
rate. Thus, theoretical calibrations of backscatter gauges are im-
practical and empirical calibrations must be made. Backscatter
gauges are also very sensitive to design features. For example,
the maximum depth of the zone of influence of surface gauges can
be slightly increased by raising the source into a more shielded
position.

Surface backscatter gauges are almost entirely nondestructive,
requiring only that the surface of the soil be made perfectly level,
to exclude air gaps; but as a result, these instruments yield little
information, merely indicating the average density of the top 50-75
mm of the soil profile. Their main use is in civil engineering applications, where bulk densities that are generally uniform with depth
are to be measured. A typical level of accuracy for these gauges
is ±0.16 Mg m^{-3} [4].

Single-probe backscatter gauges are normally lowered into lined
access holes in a manner similar to neutron moisture probes (Chap.
1) and are available in combination with such probes. The major
failing of these gauges results from their bias toward gamma photons scattered by material close to the source-detector axis, which
causes the clearance gap around the probe and the material of the
liner tube to influence measurements unduly. The gauge will also
be very susceptible to any changes in bulk density of the soil surrounding the liner tube that are inadvertently caused during installation of the liner tube.

2. Transmission Gauges

In transmission gauges (Fig. 2), the sample to be tested is located
between the source and the detector of the gauge and, ideally, only
photons that have passed unattenuated directly from source to detector are counted. In this ideal case, where none of the photons
has been degraded, the detected photon count rate I obeys Beer's
law

$$I = I_0 \exp[-\mu\rho x] \tag{3}$$

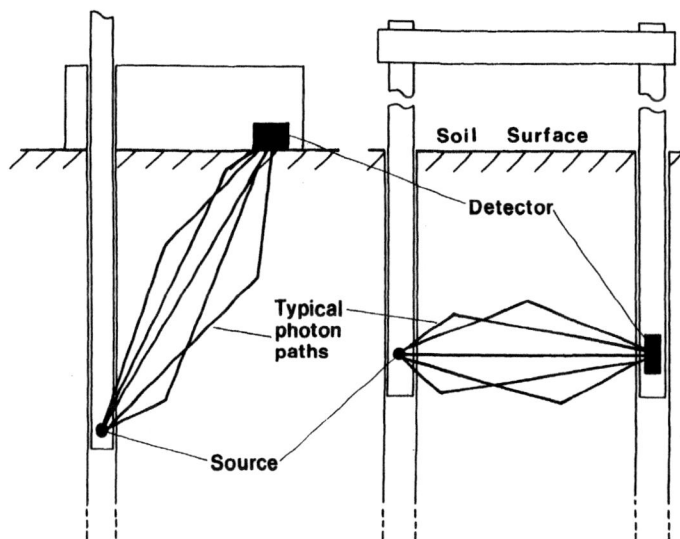

FIG. 2 Schematic diagrams of transmission gamma-ray gauges: left, detector remains on the soil surface and the source is lowered into an access hole in the soil; right, both source and detector are lowered into separate access holes.

where I_0 is the photon count rate in the absence of a sample, μ is the mass attenuation coefficient for the photon energy and sample material concerned, ρ is the wet bulk density of the sample, and x is the sample length. Theoretically the bulk density of the sample can then be calculated as

$$\rho = \frac{-1}{\mu x} \ln\left(\frac{I}{I_0}\right) \tag{4}$$

if values are available for μ, x, and I_0.

In practice, several factors make such a theoretical calculation of density impracticable. The most important of these are:

1. The presence of both scattered and unattenuated photons
2. The determination of a single mass attenuation coefficient for soils of variable composition
3. The estimation of the photon count rate in the absence of a sample

These difficulties and the way they are overcome by gauge design or experimental procedure are considered next.

Scattered Photons. With the exception of laboratory equipment in which a high degree of both collimation and energy discrimination is possible, scattered photons will always be included to some extent in the detected count rate. The presence of scattered photons affects the linearity of the relationship between ρ and $\ln I/I_0$. The reduced energy of these scattered photons also increases the dependence of the detected count rate on the chemical composition of the soil sample, as discussed later. A further effect of scattered photons is to reduce the spatial resolution of the gauge by increasing the volume of soil, which influences the count rate.

While it is possible to reduce the number of scattered photons by collimation, space is usually limited in field gauges. An alternative is to use an energy-discriminating detector, which can be set to exclude photons with energies lower than the emission energy of the source. Gauges that have this facility generally use a scintillation detector such as a sodium iodide crystal linked to a photomultiplier tube and a pulse height analyzer. It is essential that such equipment can be stabilized against temperature changes.

Most commonly available transmission gauges use Geiger-Müller detectors, which are not capable of energy discrimination, and hence are susceptible to the effects of scattered photons. In effect, these gauges operate in both the transmission and backscatter modes simultaneously. Provided such a gauge is calibrated in an empirical way, the only major disadvantage, other than a slight dependence on the chemical composition of the soil, is the unusual effect of the scattered photons when measuring close to discontinuities such as the soil surface or at the upper limit of a plow pan. While the count rate due to transmitted photons decreases exponentially with increasing density, the scattered component increases in a less well-defined manner. Henshall and Campbell [5], using a typical Geiger-Müller based gauge, found that the error in indicated bulk density exceeded 5% at a distance of 90 mm below an air-water surface and that the error exceeded 35% at a depth of 20 mm.

Thus, it can be seen that a gauge employing an energy-discriminating detector most closely resembles the ideal transmission system. However, since energy discrimination does not always exclude all scattered photons, such gauges will require empirical calibration in the same way as gauges that do not employ energy discrimination.

Soil Composition. As used in Eq. 3, the mass attenuation coefficient μ is an overall value for the bulk material examined. A theoretical value of μ would be the mean of the individual mass attenuation coefficients for each of the constituent elements, weighted according to the mass fraction of each element in the sample. Differences in the chemical composition of the soil can therefore affect the overall mass attenuation coefficient.

The mass attenuation coefficient of a chemical element varies with the atomic number of the element Z and the incident photon energy. Coppola and Reiniger [6] showed that μ increased with increasing photon energy but that for photon energies above about 0.3 MeV, there was little dependence of μ on Z below Z = 30, except for hydrogen, as discussed later. Cesium-137, which emits monoenergetic photons of 0.662 MeV, is the radioactive source most commonly employed in soil bulk density gauges. At this photon energy, calculations based on theoretical values of mass attenuation coefficient for nine different soils show that the error in estimated density due to the effect of composition is of the order of 0.5% in the most extreme case [7]. An energy discriminating system set to exclude photons of energy lower than the ^{137}Cs emission energy would therefore not be expected to show a significant dependence on chemical composition of the soil. In contrast, detectors such as Geiger-Müller detectors, which do not employ energy discrimination, are sensitive to photon energies as low as 0.04 MeV [8]; consequently a significant proportion of the detected count rate will be from photons with energies below 0.3 MeV, and hence susceptible to composition effects. Nevertheless, only a small proportion of the photons will have been scattered through angles large enough to result in such low energies, so that the effect of composition on count rate is unlikely to be serious except for backscatter gauges, where only low-energy scattered photons are counted. Generally, transmission gauges, especially those with energy discrimination, are not susceptible to soil composition effects except in soils that have a large proportion of heavy elements, such as iron [9].

Photon Count Rate in the Absence of a Sample. To apply Beer's theoretical equation (Eq. 3), it is necessary to know the photon intensity I_0 in the absence of a sample. This can be difficult to measure since, for safety reasons, the source should always be shielded either by the test material or by the shield in its transport case. Except for systems with the highest degree of collimation or energy discrimination, the presence of any such shielding material will affect the measured count rate.

This problem is usually overcome by making a reference measurement in a material of fixed density such as a steel plate. The reference count rate I_r can be written as

$$I_r = I_0 \exp[\mu_r \rho_r x] \tag{5}$$

where ρ_r is a mean density, over the sample length, of the reference plate and air gap, and μ_r is the corresponding mass attenuation coefficient. This, combined with Eq. 3, gives

$$\frac{I}{I_r} = \exp[-x(\mu\rho - \mu_r\rho_r)] \tag{6}$$

thereby eliminating I_0. By relating both test and empirical calibration count rates to the same reference plate, we can eliminate ρ_r. Calibration of gauges in this way has the further dadvantage of accounting for the gradual decrease with time in the activity of the source and the decreasing efficiency of the detection system.

3. Calibration

Since gauge design and practical considerations restrict the application of the theoretical relation between density and count rate, some form of empirical calibration is necessary even for transmission gauges. When calibrated relative to a standard reference plate, the calibration is of the form of Eq. 6, which can be rearranged to give bulk density

$$\rho = \frac{1}{\mu x}\left[\ln\left(\frac{I_r}{I}\right) - \mu_r\rho_r x\right] \tag{7}$$

or

$$\rho = A\,\ln\left(\frac{I_r}{I}\right) + B \tag{8}$$

where A and B are empirically determined constants.

In soil, the bulk density measured is the wet bulk density. In most cases, dry bulk density is a more useful parameter, and its determination requires an independent measurement of gravimetric water content, usually by oven-drying of augered samples. Then the dry bulk density ρ_S is calculated from Eq. 1.

As mentioned earlier, hydrogen, which in soil is most abundant in the water, does not conform with other elements in its attenuation of gamma photons. This is because hydrogen possesses only one nucleon per electron, whereas other atoms typically possess approximately two. Since the gamma-ray attenuation system effectively measures the number of electrons per unit volume whereas bulk density is related to the number of more massive nucleons per unit volume, the mass attenuation coefficient of hydrogen exceeds that of other atoms by a factor of approximately 2. Consequently, if the greater attenuation coefficient of hydrogen were not corrected for, the bulk density of soil samples with gravimetric water contents of 10, 25, and 100% would theoretically be overestimated by 1, 2, and 5%, respectively. In many applications, this level of accuracy may be acceptable, but, if required, the error can be corrected for during calibration. When the effects of water and soil have been separated, Eq. 3 becomes

$$I = I_0 \exp\{-x[\mu_s \rho_s + \mu_w \rho_w]\} \tag{9}$$

where ρ_w is the mass of water per unit total sample volume, and μ_s and μ_w are the mass attenuation coefficients for soil and water, respectively. Expressing ρ_w as (ρ_s W/100) and incorporating a reference standard as in Eq. 6, we have

$$\frac{I}{I_r} = \exp\left\{-x\left[\rho_s\left(\mu_s + \mu_w \frac{W}{100}\right) - \mu_r \rho_r\right]\right\} \tag{10}$$

which leads to

$$\rho_s = \frac{\ln(I_r/I) + \mu_r \rho_r x}{x(\mu_s + \mu_w W/100)} \tag{11}$$

which again can be simplified to

$$\rho_s = \frac{A \ln(I_r/I) + B}{100 + CW} \tag{12}$$

where constants A, B, and C are determined empirically.

4. Gauge Design

Radioactive Source. The primary requirements of a radioactive source for a soil density gauge are as follows: it should have a single energy peak at an energy that is high enough to reduce composition effects, the emitted photons should have a suitable penetration range into the soil sample, and the half-life should be so long (preferably exceeding the expected life of the gauge) that it does not affect any series of experimental measurements. Most radioactive sources have emission energy spectra that are too complex or in unsatisfactory energy ranges, or they have inadequately short half-lives. Cesium-137, with a monoenergetic peak of 0.662 MeV and a half-life of 30 years, is the source most suited to these requirements. The optimum soil sample length for gamma photons of this energy has been suggested as 100-250 mm [10].

There is a level of uncertainty in any measurement of count rate, since the rate of emission of gamma photons from a radioactive source is not perfectly constant but subject to random fluctuations about a mean value. The resulting fractional error in any measurement of count rate can be shown to be inversely proportional to the square root of the total number of photons counted [10]. Theoretically, therefore, it is preferable to count as many photons as possible to achieve the highest level of precision. This can be achieved either by counting for long periods of time, as discussed later, or

by using the highest possible activity of source. In practice, the
rate of photon emission, or activity, of the source to be used is determined mainly by the manner in which the probe is to be used.
For portable field gauges, the practical limit of activity is set by
safety considerations. The maximum activity of source that can be
shielded to give the statutory levels of safety without the gauges
becoming unacceptably heavy for field use is of the order of 0.4
GBq (10 mCi). This level of activity could be exceeded, but safety
requirements would be more stringent, and the practicality of the
equipment would be reduced. In laboratory gauges, the source can
be permanently housed in a much larger shield with the sample remotely manipulated, enabling much larger sources to be used. In
this case the upper limit to source activity is determined by the
dead time of the detection system. This results from the inability
of the detector to respond to an incident photon sufficiently quickly
after a previous response, thereby restricting the detectable count
rate to a fixed maximum. With gauges based on NaI(Tl) detectors,
this limits source activity to about 7 GBq (200 mCi), although it
has been suggested [11] that this could be increased to 70 GBq
(2000 mCi) where plastic scintillators are used.

   Probe Design. Portable field transmission gauges are of either
single- or twin-probe design (Fig. 2). Single-probe gauges contain
the radioactive source within the single probe, which can be lowered
through the body of the gauge into a preformed access hole, normally to a depth of about 300 mm. The detector, which is generally of
the nondiscriminating type, is located in the body of the gauge at a
fixed distance from the source probe axis such that it is in contact
with the surface of the soil. Count rates are recorded with the
source at each of several depths in the soil. Each measurement
then relates to the average bulk density between the source depth
and the surface. This is the simplest form of transmission gauge;
it avoids some operational problems common to twin-probe gauges
but suffers from an inability to examine soil layers and also requires
separate calibrations for each measurement depth. Commercial gauges
are normally supplied with factory calibrations, but users generally
find that recalibration is necessary [9].
   Twin-probe gauges consist of separate parallel probes for the
source and detector assemblies (Fig. 2). The probes are normally
clamped rigidly at a fixed separation of between 140 and 300 mm so
that, after they have been lowered to any desired depth in the soil,
horizontal layers of soil can be examined. These gauges are more
suited to the study of soils in the context of agriculture, forestry,
and the natural environment, where considerable variation in bulk
density with depth is usually found. Conversely, in civil engineering applications, the soil is likely to be more uniform with depth,

FIG. 3 Prototype gamma-ray transmission gauge developed at the Scottish Centre of Agricultural Engineering, complete with scaler and a transport box, which incorporates material for making a reference measurement.

since only subsoils are of concern, either in situ or excavated and subsequently compacted as fill material. In such applications, single-probe gauges have proved to be more popular.

Because of the fixed probe separation in twin-probe gauges, a single calibration relationship is applicable to all depths, but it is essential that the parallelism of the access holes be maintained or that any deviation be corrected for. Most popular commercial gauges

incorporate nondiscriminating detectors and are therefore susceptible to problems of lack of resolution close to either air-soil interfaces
or abrupt soil density changes with depth. However, detectors that
employ energy discrimination are becoming more readily available
(Fig. 3).

5. Soil Water Content Determination

Accurate water content information is required from depths corresponding to the density measurements to permit the conversion of
the estimated values of wet bulk density to dry bulk density. While
water content data are normally obtained from samples that have
been extracted by auger and oven-dried at 105°C, some gauges incorporate a facility that allows water content to be estimated by nucleonic methods. This is achieved in some single-probe gamma transmission gauges by the incorporation of a neutron backscatter apparatus, either in the base of the gauge or in the probe. In conditions
of uniform water content, such systems would give an adequate overall estimate, but where water content varies with depth, the neutron
backscatter apparatus does not have sufficient resolution to allow
correction of individual density measurements, since it has a typical
sphere of influence of radius 250 mm.

A much more sophisticated method of simultaneously measuring
bulk density and water content is the double-energy gamma transmission gauge. By employing a low energy source, usually ^{241}Am
with an energy peak of 0.06 MeV, together with a ^{137}Cs source
(0.662 MeV), this technique makes use of the effect of chemical
composition, especially hydrogen content, on the attenuation of low-
energy photons. By using energy discrimination to separate the
effects of soil and water, as in Eq. 9, and calibrating for both energies, it is possible to solve the resulting simultaneous equations
for both dry bulk density and water content. The major drawback
to this method is that the dependence of the low-energy calibration
on chemical composition may necessitate different calibrations for
different soils or possibly even for different depths in the same soil.
This limitation effectively restricts the usefulness of this method to
repeated laboratory tests on a single soil, where only a single calibration would be needed. In addition, compensation is required for
contributions to the low-energy count from scattered high-energy
photons if truly simultaneous determinations of bulk density and
water content are required.

Where the monitoring of changes in water content is the prime
objective, a single-energy gamma-ray gauge can be used, provided
the dry bulk density remains constant. Since the gauge measures
wet bulk density, monitoring changes in this measurement at a fixed
dry bulk density will give a direct indication of changes in water
content. If absolute measurements of water content are required

rather than changes, independent measurements of water content must be made concurrently with one set of gamma-ray measurements to allow the dry bulk density to be accounted for in the calculation of water content. Such independent, destructive measures of water content would normally be made on the last measurement occasion.

III. METHODS OF MEASURING BULK DENSITY

A. Direct Measurement of Sample Mass and Volume

1. Core Sampling

In this widely used method a cylindrical sampler is hammered or pressed into the soil. Since the volume of the cylinder is known, trimming of the soil core flush with the ends of the cylinder allows the bulk density to be calculated [12,13]. The method works best in soft, cohesive soils sampled at water contents in the region of field capacity. Sands and gravels cannot be sampled satisfactorily.

A possible source of error in the method, which is difficult to quantify, is the disturbance of the soil core, especially by compression, during insertion of the sampler. Baver et al. [14] have suggested that insertion by hammering may cause shattering, while steady pressure may produce compression. In an extensive survey of core sampling for civil engineering purposes referred to by Frietag [1], Hvorslev considered sample distortion to be at a minimum when the sampler was pressed steadily rather than hammered into the soil. Hvorslev also built a core sampler in which a piston was used to reduce the air pressure acting on the upper surface of the sample in the cylinder. The diameter of the sample also influences the risk of compression, with small diameter samples being more susceptible. Baver et al. [14] proposed a diameter of 75-100 mm as a satisfactory compromise for most work, while Freitag [1] suggested that the diameter be selected to give a sample of adequate size and that the length be not more than about three times the diameter. Generally, the cylinder wall should be as thin as possible consistent with maintaining rigidity [2]. Further aids to easy insertion of the sampler include relieving both the inner and outer diameter immediately behind the cutting edge [15] and lightly greasing the inside of the sample cylinder [16].

To extend the range of soils from which core samples can be taken, rotary core samplers have been introduced for hard, brittle soils, which may shatter during conventional core sampling. Such rotary samplers comprise a conventional cylindrical sampler with an auger attached to the outside, which excavates an annulus of soil from around the sample as the sampling cylinder is inserted. It is important with this method that the auger cut rather than "corkscrew" its way into the soil. Otherwise, there is a risk of the sample undergoing torsional shear inside the cylinder. Modifications intended to overcome this risk include a double-walled sampler, in which the cylinder to which the auger is attached rotates outside a nonrotating inner cylinder, which contains the sample [1]. Alternatively, the auger may be engine-powered to ensure that the annulus of soil is removed ahead of the leading edge of the sample cylinder [15].

A brief review of core sampling equipment, including some tractor-mounted samplers, has been made by Erbach [17].

2. Rubber Balloon

In the rubber balloon method a hole is excavated in the soil to the bottom of the layer being tested, the soil removed is weighed, and its water content determined. The volume of the sample is determined by inserting a thin rubber balloon into the excavated hole and filling it with a measured volume of water. For accurate results to be obtained, the excavated hole should have a regular shape so that the balloon can reasonably be expected to fill any irregularities that arise [1,2,18]. To this end, apparatus has been developed in which the balloon is clamped to the base of a calibrated water container, which includes a pump to force the water into the balloon. Clearly, care must be taken not to distort the excavated hole, and pressures of 21-48 kPa have been suggested as appropriate [1]. Alternatively, water may be forced into the balloon until the apparatus is raised about 20 mm off the ground. An air-cock in the apparatus is then closed, and the operator places both feet on the base plate of the equipment to force the balloon into any irregularities in the hole [2].

Generally, the method is considered to give unreliable results, errors being associated with entrapped air and with the dependence of the fit of the balloon in the hole on the air pressure.

3. Sand Replacement

In the sand replacement method, the sample is excavated, weighed, and its water content determined as in the rubber balloon method. The hole produced is usually about 100 mm in diameter. A metal cylinder containing dry sand, usually referred to as a "sand bottle" (Fig. 4), is placed over the hole, and a tap in the base of the cylinder is opened to allow the sand to fill the hole. The difference in weight of the cylinder, before and after filling the hole, is recorded. The bulk density of the sand is obtained from a calibration test in which sand from the bottle is used to fill a can of known volume, and this allows the volume of the excavated hole to be calculated [2,18]. Allowance must be made for the sand between the tap and the soil surface level. This is done by opening the tap

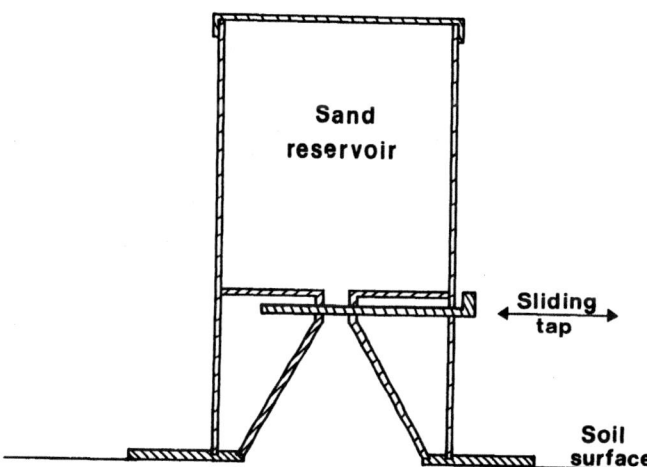

FIG. 4 Schematic section through a typical sand bottle used in the sand replacement method showing the sliding tap in the closed position.

while the equipment rests on a flat metal plate. In a new variation of the method, which does not involve determination of the bulk density of the sand, a container for the sand is calibrated in terms of volume, as in a measuring cylinder, and the difference in volume before and after filling the hole gives the volume of the hole. The new method is claimed to give smaller errors than the conventional sand replacement method [19].

Several aspects of the test procedure must be carefully controlled if reliable results are to be obtained. The volume of the calibration can should be of a volume similar to that of the hole from which the sample is to be excavated, since a 25 mm decrease in the depth of the can produces a decrease of about 1% in sand bulk density. A similar decrease in density is produced by a 50 mm reduction in the initial level of the sand in the cylinder [2]. The sand should be closely graded (typically, 0.2-2.0 mm material is used) to prevent segregation, and hence variation in sand bulk density occurring; indeed, grading is considered to be more important than the actual size range used.

The greatest care should be taken to ensure that the sand remains dry and uncontaminated by soil when it is recovered from the hole at the end of a test. Frequent checks on the calibration are the best way of checking whether sample integrity has been maintained [1]. Although the sand replacement method is relatively slow, with a typical test time of 30 minutes, it has the advantage that it can be used on all soil types [1].

Freitag [1] describes a technique in which oil is used instead of sand to fill the hole. The volume of oil used is not sensitive to handling techniques, and irregularities in the hole are readily filled. However, it is usually necessary to spray the inside of the hole with a lacquer or plastic to avoid penetration of the soil pore space by the oil. Other problems include the possible effect of temperature on oil bulk density and the need to excavate the hole from an initially horizontal surface.

4. Clod

In this method a clod is weighed and its volume is determined by coating it in paraffin wax and immersing it in a volumenometer. The volume of water displaced corresponds to that of the clod plus wax [2]. Alternatively, the waxed clod may be weighed in air and in water. In both versions of the method the wax coating must subsequently be removed and weighed. The wax coating is applied by suspending the clod from a fine wire and dipping the clod in paraffin wax at a temperature just above its melting point. Although the method gives satisfactory results, it is limited to cohesive soils and is a rather slow method when wax is used as the coating material.

A commercially available resin, Saran F-220, dissolved in methyl ethyl ketone, has been used as a substitute for wax [20]. It is available from the Dow Chemical Company (Suite 500/Tower 2, 1701 West Golf Road, Rolling Meadows, IL 60008). It is claimed that this product is flexible, does not melt during oven-drying at 105°C, and is permeable to water vapor but not to water. It can therefore be used to study the drying and shrinkage characteristics of a clod. Rubber solution has also been used as the coating material, with claims of improved accuracy and convenience over the paraffin wax method [21]. A flotation technique has also been used in which the clods were sprayed with a resin solution and then immersed sequentially in liquids of differing relative density. The relative densities of the two liquids in which the clods just sank and just floated provide an upper and a lower limit to the clod bulk density. Neither clod mass nor clod volume is determined, and the technique was shown to be 10 times as rapid as the wax coating method [22].

It is possible to avoid coating the clod at all if the immersion fluid will not penetrate the soil pores. Although various viscous oils and mercury have been used, the technique is probably restricted to soils with very small pores. Thus, one successful application has been in a study of the density of puddled soils [23]. Further techniques for clod bulk density measurement include the use of X-rays [24], elutriation in a vertical airstream [25], and immersion in a bed of glass beads [26]. A useful summary of many of these techniques is given by Russell and Balcerek [27].

B. Radiation Methods

Several users have designed and built gamma-ray gauges to suit specific purposes. Table 1 lists a selection of commercially available backscatter and transmission gauges.

1. Sample Preparation

For any type of nuclear density gauge it is important that the sample be presented to the gauge in a consistent manner. In laboratory transmission gauges, each sample is placed in turn in a container located between the source and detector. In field transmission gauges, either a single access hole or two parallel access holes must be made in the soil; equipment for this purpose is shown in Fig. 5. The most common means of forming access holes is by hammering solid spikes through an alignment jig lying on the soil surface [28]. Although a certain amount of disturbance takes place during the operation, this can be compensated for by providing access holes in calibration samples in exactly the same way, provided the soil is not fractured during spiking.

The provision of access holes by augering minimizes soil disturbance, but probe access can be more difficult, particularly where two parallel holes are required. Augering has several other advantages, however: the removed soil can be used for water content determination; calibration samples can be smaller; and it is easier to install liner tubes in the access holes where they are required [29]. In loose soil conditions, it is wise, while augering, to insert liners progressively; this prevents topsoil from entering the access hole.

Surface backscatter gauges require only that the soil surface be made perfectly flat so that no air gaps occur between the base of the gauge and the soil surface.

2. Calibration

Except for laboratory gauges with high levels of collimation, for which it is possible to use theoretical values for mass attenuation coefficients, some form of empirical calibration is required. Some gauge manufacturers supply specimen calibrations with gauges, but most workers involved with agricultural soils have found it desirable to recalibrate their gauges. Some manufacturers also supply standard density blocks for calibration, which can be useful for periodic checks on calibration stability but are unlikely to be suitable for a full calibration because these blocks can differ from the soil both in the mode of probe access and in composition.

Calibrations with field soils can either be made in situ by comparison with a direct method, normally core sampling, or by repacking field soils into bins and determining their density independently from measurements of sample mass and volume [5,28]. Both types

TABLE 1 Details of Some Commercially Available Gamma-Ray Gauges

Supplier	Model	Configuration	Detector
ELE Ltd., Eastman Way, Hemel Hempstead, Hertfordshire HP2 7HB U.K.	CPN Corp. MC-3 Porta-probe	Transmission (surface detector, single probe) OR Backscatter (surface source and detector)	Geiger-Müller
	CPN Corp. Strata gauge	Transmission (twin probe at ~300 mm separation)	Geiger-Müller
	CPN Corp. 501 B depth probe	Backscatter (source and detector in single probe)	Geiger-Müller
Wykeham-Farrance, Weston Road, Slough Berkshire SL1 4HW U.K.	Humboldt Mfg. Co. HS 5001	Transmission (surface detector, single probe) OR Backscatter (surface source and detector)	Geiger-Müller
Ronly Scientific Ltd., P.O. Box 168, Rishon-Lezivvon, 75101, Israel	SDG-39 Density gauge	Transmission (twin probe at 210 mm separation)	Geiger-Müller
	DDG-37 Density gauge	Backscatter (surface source and detector)	Geiger-Müller
Troxler Electronic Laboratories Inc., P.O. Box 12057 Research Triangle Park, NC 27709	2376 Density gauge	Transmission (twin probe at 300 mm separation)	Energy discrimination
Scottish Centre of Agricultural Engineering, Bush Estate, Penicuik, Midlothian, EH26 0PH U.K.	SCAE Density gauge	Transmission (twin probe at 220 mm separation)	Energy discrimination

Source and strength	Maximum measurement depth (m)	Data recording microprocessor	Approx. 1987 price ($)	Comments
^{137}Cs, 10 mCi (370 MBq)	0.2 or 0.3	Yes	9,300	Incorporates neutron backscatter gauge with source at surface
^{137}Cs, 10 mCi (370 MBq)	0.6	Yes	13,300	Incorporates neutron backscatter gauge with source in probe
^{137}Cs, 10 mCi (370 MBq)	10.0	Yes	10,900	Incorporates neutron backscatter gauge with source in probe
^{137}Cs, 10 mCi (370 MBq)	0.2 or 0.3	Yes	7,700	Incorporates neutron backscatter gauge
^{137}Cs, 1 mCi (37 MBq)	0.45	No		3 mCi optional
^{137}Cs, 5 mCi (185 MBq)	3.0	No		
^{137}Cs, 5 mCi (185 MBq)	2.0	No		
^{137}Cs, 5 mCi (185 MBq)	0.6	No	13,200	Detailed specification to order

FIG. 5 Equipment used to provide two parallel access holes for transmission gamma-ray gauges: left, by hammering spikes through an alignment jig; right, by augering. A liner tube has been inserted in the right-hand augered hole.

of calibration are slow, and each has its advantages and disadvantages. Comparison with core sampling has the advantage that soils of field structure are used, but core sampling, especially at depth, is time-consuming and unreliable. Such comparisons assume, without justification, that core sampling results are accurate. Unless minimal disturbance is ensured in the gauge method by using auger

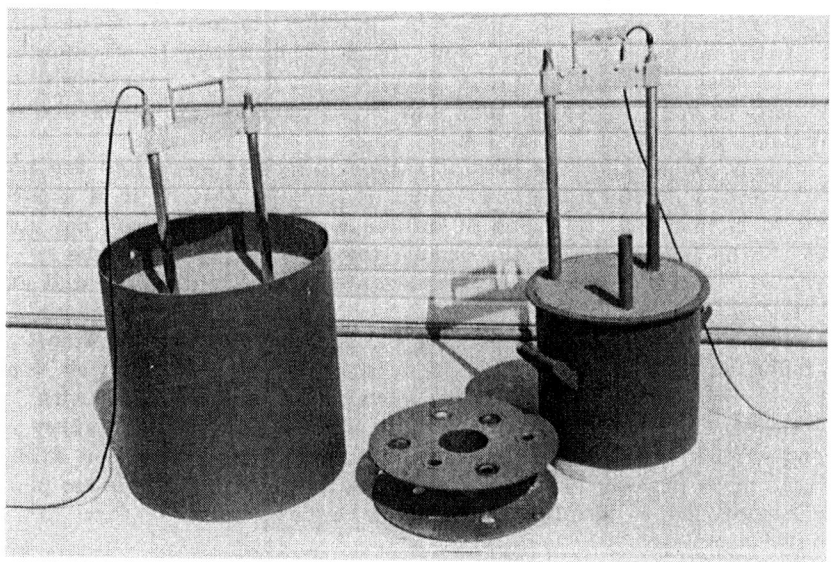

FIG. 6 Calibration samples for gamma-ray gauges in which access is provided by spiked holes (left) and by augered and lined holes (right). The alignment jig for the augered holes is also shown.

access, sampling at different positions for the two methods is required, with the resulting complication of having to account for the variability of field soils.

Calibration with remolded field samples packed in bins simplifies the direct measurement of bulk density [5,30]. Where gauge access is by spiking, however, samples must be large enough to ensure that the walls of the bin do not influence the soil disturbance that occurs during spiking, and tests may be restricted to a single access position to avoid interaction between multiple spikings. Where insertion is by augering and only unattenuated photons are counted, samples that are only marginally larger than the probe spacing can be used and multiple access positions will compensate for inconsistencies in the packing of the sample, which must otherwise be carefully prepared in thin layers to achieve uniform packing (Fig. 6).

In calibration, the measurements made with both the gauge and direct methods should offer similar precision. For example, when comparing gauge measurements with core samples, if the soil variability is high, there is no advantage in making long precise measurements of count rate with the gauge. Instead, shorter, less precise measurements should be made with the gauge and the time saved spent in further sampling with both methods.

Conversely, where samples can be expected to be uniform or where virtually the whole volume of remolded soil can be tested, it is justifiable to make longer, more precise measurements of count rate in more positions on fewer samples. In general, test counts normally comprise between 2000 and 10,000 counts, giving levels of precision of between 2.5 and 1%.

When considering the size of sample to be prepared and the positions and depth increments at which it is to be tested, it is important to ensure that the zone of influence of the gauge does not approach the boundaries of the sample too closely. It should be remembered that the zone of influence extends horizontally as well as vertically.

If the calibration is to include a correction for water content, a detailed water content profile will need to be obtained for each sample. Where there is a significant variation in water content with depth, it will be necessary to include the data for each measurement separately in the calibration. If there is little variation with depth, it is possible to use mean values for each profile or sample. Such a procedure simplifies laboratory calibrations on uniform remolded samples.

Standard reference counts should be made for each calibration sample, using the same reference plate as for test measurements. Since the reference count is related to all measurements in a sample, and any errors in it could have a significant effect on the calibration, the count period is usually longer than that for the test counts.

Finally, it should be stressed that it is essential for calibration samples to be tested in exactly the same manner as the experimental samples to which the calibration is applied. This is particularly important with respect to the method of providing probe access.

3. Experimental Considerations

As with calibration, the decision between making a few highly detailed measurements or more replication in less detail is based on sample variability. Since field soils tend to display large random variations in soil properties, it is generally more worthwhile to replicate measurements than to make very precise measurements in one sampling position. Typically, more than 5000 counts per measurement cannot be justified, and between 2000 and 3000 counts is adequate [8]. In replicated field experiments, the number of measurement positions used per treatment is typically two or three; this gives coefficients of variation of about 10% and is probably a good compromise [28]. However, measurements of soil properties in sampling positions that are close together generally tend to be more similar than those made farther apart [31]. When such spatial dependence is allowed for, the number of measurements required for a given level of precision can sometimes be reduced [32].

Stones contribute to the inherent variability of a soil and, in addition, they may present difficulties either by preventing the provision of access holes to the full depth desired or by deflecting the probes of a twin-probe system and so altering the source-detector separation.   Where access holes cannot be made, a new sampling position must be tested instead, with the result that the mean bulk density may be biased in favor of the samples for which stones occur between, rather than at, the positions of the two probes.   Thus, the bulk density of stony soil may be overestimated.   The effect, which will depend on both the number and size distribution of stones, appears not to have been investigated.   The problem of possible probe deflection by stones can be overcome only by measuring, and allowing for, the actual source-detector separation at each depth [29]. The statistical problems arising from soil variability and from stones have been examined in relation to the measurement of soil cone resistance; some of this information is relevant to the measurement of bulk density [33].

With the exception of the single-probe transmission gauge, for which the depths of measurement are restricted to those for which calibrations are available, measurements can be made at any depth. However, even where the maximum detailed information on bulk density variations within a soil profile is sought, it is not worthwhile choosing depth increments that are substantially smaller than the spatial resolution of the gauge.

4. Operational Safety

All nuclear density gauges incorporate sufficient quantities of radioactive material to make them a potential health hazard.   In the United Kingdom, it is a legal requirement for such devices to be registered with the Health and Safety Executive [34].   A similar situation exists in the United States.   In the United Kingdom, a "system of work" that describes an approved safe operating procedure for the gauge is normally incorporated in the registration.   Most manufacturers supply an example of such a document with the gauge but, for nonstandard gauges or procedures, a system that ensures minimum exposure to the operator and safe design of the probe must be devised, documented, and approved.

For field gauges, a safe operating procedure would ensure that the source is exposed for the minimum time possible.   This can be achieved by lowering the probes through the base of the gauge so that the source is always shielded by either its shield or the soil, and by ensuring that when not in use, the source is securely located in its shield.   For laboratory gauges, interlocking devices on the shield would probably be required to prevent accidental exposure, since much larger sources are generally used than in the field.

C. Comparison of Methods

The difficulty in extracting soil samples from the field without disturbance to the sample and to the wall of the remaining hole means that none of the direct methods of measuring bulk density can be relied on to be totally accurate. Erbach [35] describes the sand replacement method as "good for use in gravelly soil," but for most soils the core sampling method is generally taken to be the standard method. As mentioned earlier, the core method is susceptible to many forms of error. Raper and Erbach [36] stated that "it is disturbing that a method with this many inherent errors is referred to as a standard." It is therefore unfortunate that many workers, when finding that density measurements recorded by gamma-ray gauges do not agree with direct measurements have been inclined to dismiss the gamma gauge as inaccurate or unsatisfactory.

Several comparisons between direct and gamma-ray measurements have found general agreement between the two methods [9,18,28, 37-39], with discrepancies in some soil types that are normally attributed to inaccuracies in the gamma gauge. King and Parsons [37] found reasonable agreement (±3%) between a single-probe gamma gauge and the sand replacement method in sandy and clay soils but unacceptably large differences of 11% in gravelly soils. Several explanations of the discrepancy were given, such as variation in gamma-ray absorption according to particle size, but the more probable dependence of the sand replacement test on particle size was not considered [2].

Gameda et al. [9] compared single- and twin-probe gamma gauges with the core sampling method on three soils to a depth of 0.6 m. They found a good correlation between the gamma and core measurements on sandy and clay soils but not on loamy soil. The poor correlation in loamy soil was attributed to the presence of stones in the soil and its high iron content. The data as presented would support the view that the loam was very variable, perhaps due to stones; but since a pooled correlation for all three soils yielded a correlation coefficient better than for any individual soil, it appears unlikely that the iron content of the soil provided a significant bias in the results for the loamy soil. Although a good correlation was found between core density and the density values indicated by the factory calibration for the gamma gauges, the test values for the gauges were significantly different from each other, confirming the need for calibration of gamma gauges in field soils. Twin-probe gauge measurements were found to correlate better with the core measurements than with single-probe gauge measurements, which unexpectedly correlated better with the density at the depth of insertion than with the average density between the surface and that depth. Readings from neutron backscatter moisture meters incorporated in these gauges were not found to correlate well with field measurements.

Some comparisons between gamma-ray gauges and direct methods have been carried out under conditions such that reliable conclusions are not possible. Minaei et al. [38] attempted to compare a twin-probe gamma gauge with core sampling in an experiment on the compaction by wheels of a sandy clay loam. Since a linear regression through the data points did not yield a 1:1 relationship between gamma and core measurements, it was concluded that the gauge was not a viable tool in field conditions. The scatter of data points, however, was so large that the assumption of a 1:1 relationship would scarcely have had any effect on the errors of prediction. Unless the gauge was malfunctioning or the core sampling techniques were at fault, it can only be assumed that the pairs of measurements did not relate to unique samples and that a very high variability in soil density was the cause of such scattered data.

Soane et al. [28] found that on three contrasting mineral soils, density measurements from a twin-probe gamma gauge agreed with corresponding core sample measurements within 3% but that there was a discrepancy of 0.06 Mg m^{-3} on low density (0.28 Mg m^{-3}) organic peat samples. The coefficients of variation for both methods were found to be similar for a given soil. The gamma gauge was found to be faster in operation by a factor of 2 or 3; it also had the advantage that measurements could be made at close intervals down a soil profile with little disturbance. A single calibration relationship was applicable to all the soils and water contents tested. In a review of gamma-ray transmission systems, Soane [8] reported that laboratory measurements can be accurate to ±0.002 Mg m^{-3} and cited examples of field gauges with accuracies of ±1.2 to ±3%.

A useful indication of the potential accuracy of gamma gauges was carried out by Schafer et al. [39]. Over a 5-year period, core samples were removed from the field and tested in an empirically calibrated laboratory gamma gauge after direct measurement of their bulk density. For 80% of the 236 cores tested, the discrepancy was less than 1% and the results for only two samples disagreed by more than 2%.

The gamma-ray transmission method is therefore potentially at least equal in accuracy to any of the direct methods of density determination and is simpler and quicker to use, especially where measurements at depth are required. The twin-probe gamma gauge is more accurate than the single-probe version, allowing much more detailed information of soil layers to be acquired, provided the parallel access holes are carefully prepared or nonparallelism is allowed for. Calibration of gauges will be required for agricultural soils, since factory calibrations have generally been found to be unsatisfactory, and it is essential that the access method used in calibration be identical to that used in the field. Separate calibrations for different soil types are unlikely to be necessary unless exceptionally high

proportions of heavy elements are present or very high levels of
accuracy are required.

The high cost of gamma-ray gauges compared with equipment
for direct measurement and the requirement for compliance with radiation safety regulations [34] will offset the advantages of the gamma gauges where few measurements are required. In such cases,
the core sampling method has proved to be the most popular alternative except in gravelly soils or where looseness of the soil prevents
its retention within the core, in which case the sand replacement
method is the best option.

Some comparisons have been made of the various direct methods
available, both in terms of their practical advantages and disadvantages and in terms of the errors associated with them [2,19]. The
clod method might be expected to give bulk densities greater than
other measures of bulk density that include interclod spaces. Generally, however, core sampling and the clod method give similar results, while the sand replacement values are about 2% lower [2].
The rubber balloon method has proved to be relatively unreliable,
with systematic errors of 4.85% being found, in comparison with
2.95% for the sand replacement method and 0.53% when sand volume
rather than mass is measured [19].

All methods of bulk density measurement may be hindered by
the presence of stones, mainly because they can cause sample disturbance. As with most soil physical and mechanical measurements,
however, stones may also create complications in the interpretation
of treatment means from field experiments [33]. A further complication arises in the case of gamma-ray measurements, however, since
the presence of stones cannot be allowed for by their retrieval, as
is the case during sample destruction when direct methods are used.
Keisling and Smittle [40] made measurements of the bulk density at
which root growth was inhibited in a soil with between 5.8 and 11%
of stones of 3-13 mm size. They found that bulk densities were
between 0.097 and 0.12 Mg m^{-3} lower when the presence of stones
was allowed for and that the corrected values corresponded to the
limiting values for root growth in stone-free soil. The corrected
densities were therefore considered to be more appropriate to root
growth studies.

IV. APPLICATIONS OF BULK DENSITY
MEASUREMENTS

Many of the direct methods of bulk density measurement have been
widely used for civil engineering purposes. Since civil engineering
works generally involve excavation, remolding, replacement, and uniform compaction of subsoil material, the problem of bulk density

measurement in the completed earthworks probably will be little affected by soil organic matter, soil structure, and the variation of bulk density within any sample. Thus, the direct methods can be entirely appropriate. However, the limitations of all methods other than transmission methods employing energy discrimination can be very important in the context of agricultural soils, in which large variations in bulk density can occur over very short horizontal and, especially, vertical distances as a result of the localized effects of tillage and traffic. Thus, thin layers of soil of high bulk density, which may be very important in relation to such matters as root penetration or water infiltration, may pass undetected in the mean bulk density value derived from such methods.

The sample disturbance associated with the direct methods may be inappropriate for structured soils. With gamma gauges, however, such damage is minimized by the use of augered access holes, and even where spikes are used to provide access, disturbance is allowed for by preparing calibration samples in the same way.

Despite their limitations, the direct methods and the gamma-ray transmission methods that do not employ energy discrimination may be a good compromise in some circumstances. However, transmission radiation methods that employ energy discrimination are usually more appropriate to measurements in agricultural soils, provided the equipment is robust enough for field use and has adequate spatial resolution. Some examples of the use of such methods to detect the bulk density changes during soil compaction by wheels and loosening by tillage implements are considered next.

A. Soil Compaction by Wheels

Soil compaction by a wheel may be assessed by measuring bulk density at regular depth increments below the soil surface before the wheel runs over the soil and making similar measurements under the centerline of the wheel rut produced. The measurements may then be graphed as the variation of dry bulk density with depth both before and after the passage of the wheel. Figure 7 shows the results of such measurements made in a sandy loam in an indoor soil tank after the passage of an unladen tractor. Measurements were made with gamma-ray transmission equipment both with and without energy discrimination, and the data confirm that different results are produced by the two methods [41]. The depth interval between measurements can be varied so that measurements are more intensive in the region of any feature of interest such as the top of a plow pan, but an interval of about 30 mm has been found to be an appropriate compromise for general purposes [42-44].

In replicated field experiments, as discussed earlier, an adequate number of counts at each measurement depth in each measurement

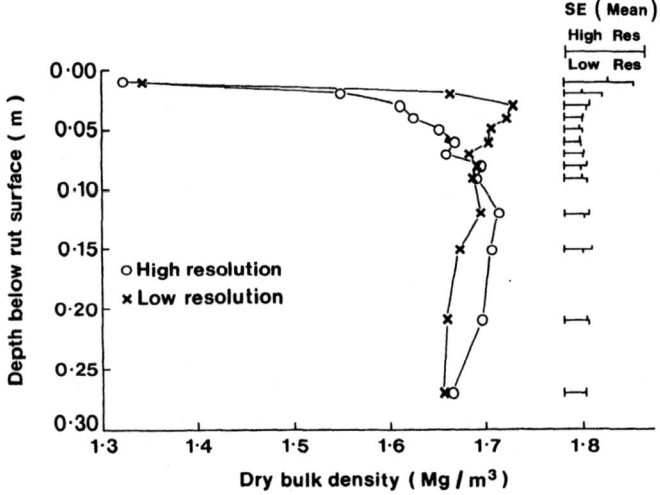

FIG. 7 Variation of dry bulk density with depth below a wheel rut produced in a sandy loam by an unladen tractor. Measurements were made with gamma-ray transmission equipment both with (high resolution) and without energy discrimination. (Based on data from Ref. 5.)

position is probably between 2000 and 3000. Two or three measurement positions per plot probably is an adequate number.

Presentation of data at fixed depths in relation to the undisturbed soil surface as shown in Fig. 7 is satisfactory for many purposes, but difficulties can arise when the effects of two or more vehicles are compared, especially when wheel ruts of different depths are produced. Henshall and Smith [45] developed a procedure in which bulk density measurements are used to trace vertical soil movement due to compaction. Consequently, comparisons between treatments can be made on soil elements that originated from the same depth in the undisturbed soil profile, irrespective of their depths in the compacted profiles (Fig. 8).

A further limitation to the value of the information provided by Fig. 7 is that very often the lateral distribution of compaction on either side of the centerline of the wheel rut is of interest. This is particularly the case when soil compaction is being studied in relation to crop growth. Such additional information can be obtained by making a series of measurements in a transect at right angles to the wheel rut, the gamma-ray transmission gauge being aligned parallel to the rut. With such an arrangement, sampling positions can usually be no closer than about 100 mm before probe access disturbs

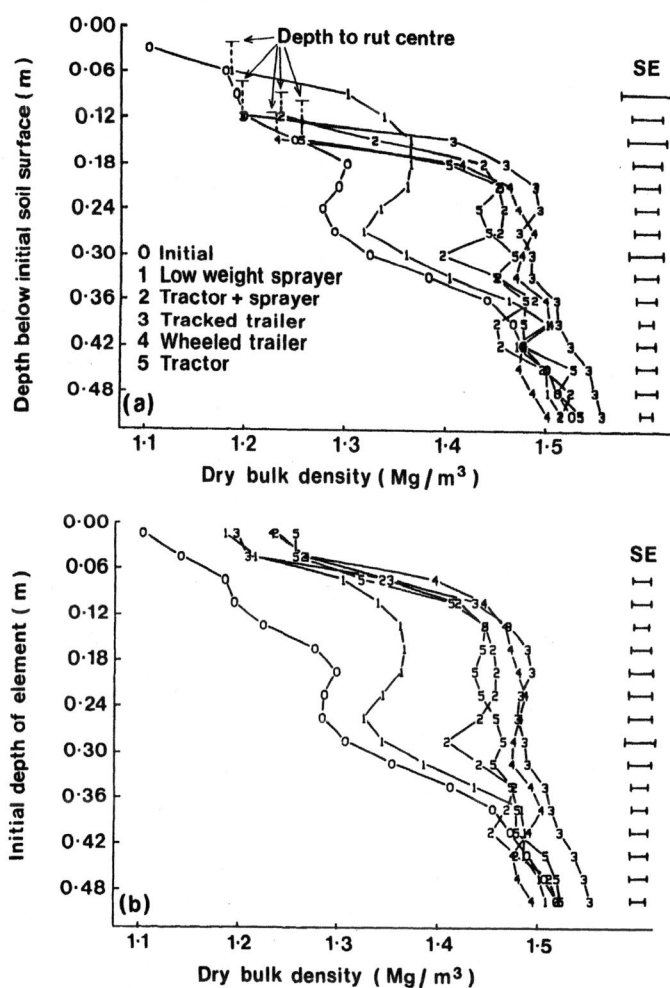

FIG. 8 Variation, for five treatments, of dry bulk density with (a) depth below the initial soil surface and (b) initial depth of each soil element. (Based on data from Ref. 45.)

adjacent positions [46]. The latter difficulty can be overcome by employing a two-dimensional scanning gamma-ray system to make measurements on a regular vertical grid on a rectangular block of soil that is at right angles to the wheel rut. Probe access necessitates the provision of a trench on each side of the sample, and sample preparation can be extremely slow (Fig. 9). Nevertheless, the method can provide a detailed description of both the vertical and

FIG. 9 Gamma-ray transmission system, designed and constructed at Scottish Centre of Agricultural Engineering, which provides a two-dimensional scan of an undisturbed block of soil at right angles to a wheel rut.

horizontal variation in bulk density at right angles to a wheel track (Fig. 10). Soane [47] used a version of the method that employed energy discrimination; the source and detector probes were mounted on an electrically-powered carriage, the movement of which was controlled automatically such that 825 bulk density readings were made on a 20 mm × 20 mm grid. The test sample was 1.4 m long at right angles to the wheel track and 0.3 m deep. Sample thickness was a nominal 0.3 m, and it was felt necessary to check the sample thickness with long-nosed calipers on a 40 mm × 40 mm grid on completion of the scan. Using this technique on a simulated seedbed in a sandy loam, it was possible to compare the distribution of compacted soil produced by a conventional tractor, the same tractor with the addition of cage wheels, and a crawler tractor. In particular, it was shown that although the fitting of cage wheels resulted in rather less compaction under the centerline of the adjacent tire, additional compaction occurred under the cage wheel. Similarly, the crawler tractor resulted in a reduction in the maximum bulk density produced in comparison with the conventional tractor, but it also compacted a greater volume of soil.

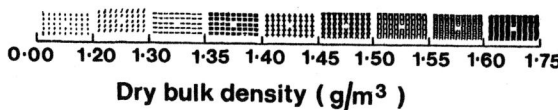

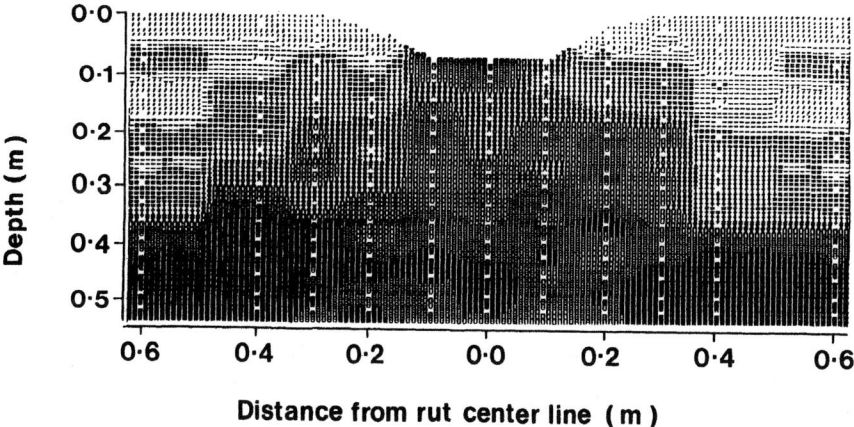

Distance from rut center line (m)

FIG. 10 The variation in bulk density produced in a sandy loam by a tire with an inflation pressure of 84 kPa and a load of 2.47 t as measured with a scanning gamma-ray transmission system that employed energy discrimination. (D. J. Campbell and J. K. Henshall, unpublished data.)

B. Soil Tillage

There have been many attempts to determine the limiting bulk density for root growth for a variety of crops in a range of soils [48-50]. Although good relationships have been found in the laboratory, such relationships are always much poorer in the field because of soil variability. Veihmeyer and Hendrickson [50], who found that the limiting bulk density for the growth of sunflower roots in the laboratory ranged from 1.46 to 1.90 g cm^{-3} depending on soil texture, demonstrated that the restriction to root growth was high bulk density and small pore size. Their conclusion was consistent with that of Wiersum [51], who proved that the tip of a growing root will enter a pore only if that pore is larger than the root tip diameter. Wiersum [51] also concluded that for satisfactory root growth, the pore structure must not be too rigid, implying that both soil bulk density and soil strength are important in this context. Thus, it is easily seen that with the inherent variability of soils in the field, any effect of bulk density on root growth will interact with the effects of soil strength, water status, aeration, and structure.

Many researchers have felt it worthwhile to measure soil bulk density in tillage experiments. Usually, these measurements are made in conjunction with measurements of soil water content so that air-filled porosities may be derived. In some experiments that require measurements of water release characteristics or permeability to air or water, the soil cores needed for such measurements are also used for bulk density determination [52]. Typically, two or three cores per plot at each depth are considered to be sufficient in replicated experiments.

Bulk density measurements by the gamma-ray method are often used to measure the degree of loosening provided by tillage treatments or the extent of compaction following direct drilling [53,54]. However, because it is the change in soil bulk density that is of interest in such circumstances, the effect of high soil variability both before and after the treatments can result in the need for large numbers of measurements if treatment differences are to be detected. Soane [55] has used the scanning gamma-ray method in unreplicated measurements to illustrate the distribution of compacted soil in moldboard-plowed land and in potato ridges and furrows. The two-dimensional scan possible with a cone penetrometer [56] is more useful for detecting compacted soil in such circumstances than is a scan of bulk density, because of the vastly greater speed of the cone penetrometer test, which in turn allows the replicated measurements required to overcome problems of soil variability.

Hand-held gamma-ray transmission equipment has been used successfully in tillage experiments. In a long-term experiment to compare three alternative plowing treatments with direct drilling [54,57-59], such measurements were made in two positions per plot in each of the four replications of the four treatments (Fig. 11). It was possible to conclude from a series of these measurements that the direct-drilled soil reached a bulk density that was in equilibrium with the applied traffic after 3 years. Moreover, this equilibrium bulk density differed for the two soil textures involved in the experiment. It was shown that most of the soil that was loosened by the three plowing treatments had compacted to its original bulk density by the end of the growing season. Although each soil was compacted to a different bulk density in response to traffic, measurements of cone resistance showed no difference between soils. Thus, although cone resistance depended on tillage and traffic only, bulk density was also influenced by soil compactibility, and hence texture and water status. These results emphasize the potential dangers of assessing soil compaction in terms of changes in only one soil physical property. In this instance, measurements of cone resistance in isolation would not have detected the difference in response to the tillage treatments of the two soil textures [54].

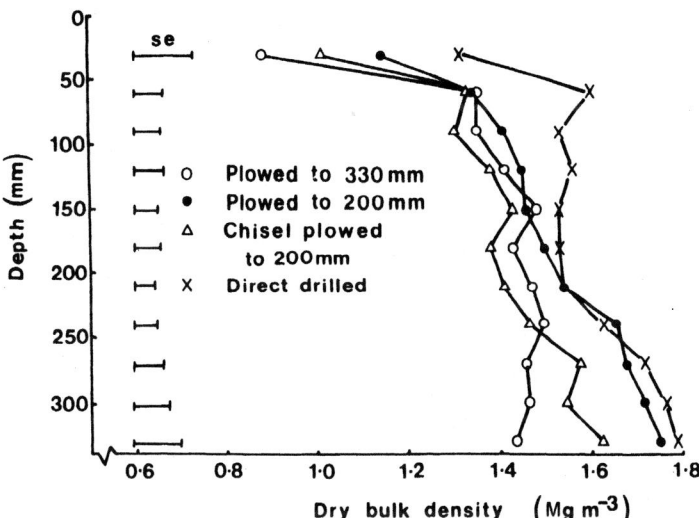

FIG. 11 Variation of soil bulk density with depth in a loam for four tillage treatments in the middle of a spring barley growing season. (Based on data from Ref. 59.)

In addition to measurements of the density of the bulk soil, it is sometimes appropriate to measure the bulk density of the aggregates or clods within the soil mass. For example, in studies of the movement of fluids through bulk soil, both inter- and intra-aggregate porosities may be of interest, since the large interaggregate pores dominate fluid movement [60]. Studies of root penetration may require the measurement of both aggregate and bulk soil densities since, although roots generally tend to grow preferentially in the interaggregate pore spaces, the extent to which roots do penetrate any aggregates present will be relevant in studies of, for example, crop use of water or nutrients if the latter are available within the aggregates. The bulk density of soil clods in potato ridges has been found to be relevant to problems in the harvesting of potatoes [61]. In measuring clod or aggregate bulk density, problems of variability associated with water status, bulk density gradients, and the range of clod sizes involved necessitates measurements on 50-100 clods per plot in replicated experiments. In such circumstances the older clod method [2], in which the clod is coated in wax and weighed in air and in water, is unacceptably slow, and even the more recent flotation method [22], which is 10 times quicker, is still tedious to use.

V. SUMMARY

Both direct and indirect measurements of soil bulk density are described. In the direct methods, the sample mass and volume are
determined. In the indirect methods, the effect of the sample on
gamma radiation is measured and related to bulk density by empirical calibration. The theory of the interaction of atoms of soil with
gamma photons is discussed in relation to photon energy and intensity together with soil chemical composition and bulk density. Basically, photons from a gamma source are absorbed or scattered during interaction with the electrons of the soil atoms such that the
number of photons incident on the detector in a given time is related to the sample bulk density.

Backscatter gauges detect only scattered photons, while transmission gauges are designed to detect unattenuated photons, provided the detector employs energy discrimination. Details of the
construction of gamma gauges are considered, as are calibration
procedures and the need for accurate water content measurements.

Both direct and indirect methods are detailed. Direct methods
discussed include the core sampling, rubber balloon, sand replacement, and clod methods. Indirect methods include both the backscatter and transmission gamma methods, which are described in
relation to problems associated with sample preparation, calibration,
operational safety, soil variability, and stones. Comparisons of
methods indicate that despite general agreement between the results
of direct and indirect methods, the latter tend to be more accurate,
especially the gamma-ray transmission method, which is particularly
suited to the layered soils usually found in agriculture, forestry,
and the natural environment. Examples are given of the use of the
gamma transmission method to detect changes in bulk density associated with soil compaction by wheels and soil loosening by tillage
implements.

REFERENCES

1. Freitag, D. R., Methods of measuring soil compaction, in *Com-
    paction of Agricultural Soils* (K. K. Barnes, W. M. Carleton,
    H. M. Taylor, R. I. Throckmorton, and G. E. Vanden Berg,
    Eds.), Am. Soc. Agric. Eng., St. Joseph, MI, 1971, pp. 47-
    103.
2. Anon., *Soil Mechanics for Road Engineers*, HMSO, London,
    1964.
3. Anon., *Determination of Depth and Volume of Backscatter Mea-
    surements*, Tech. Data Sheet 13, CPN Corp., Pacheco, CA.

4. Carlton, P. F., Application of nuclear soil meters to compaction control for airfield pavement construction, in ASTM Spec. Tech. Publ. No. 293, *Symp. Nuclear Methods of Measuring Soil Density and Moisture*, 1961, pp. 27-35.

5. Henshall, J. K., and D. J. Campbell, The calibration of a high resolution gamma-ray transmission system for measuring soil bulk density and an assessment of its field performance, *J. Soil Sci.*, 34: 453-463 (1983).

6. Coppola, M., and P. Reiniger, Influence of the chemical composition on the gamma-ray attentuation by soils, *Soil Sci.*, 117: 331-335 (1974).

7. Reginato, R. J., Gamma radiation measurements of bulk density changes in a soil pedon following irrigation, *Soil Sci. Soc. Am. Proc.*, 38: 24-29 (1974).

8. Soane, B. D., Gamma-ray transmission systems for the in situ measurement of soil packing state, in *Report for 1974-76*, Scottish Inst. Agric. Eng., 1976, pp. 59-86.

9. Gameda, S., G. S. V. Raghavan, E. McKyes, and R. Thériault, *Single and Dual Probes for Soil Density Measurement*, Paper No. 83-1550, Am. Soc. Agric. Eng., St. Joseph, MI, 1983.

10. Ferraz, E. S. B., and R. S. Mansell, *Determining Water Content and Bulk Density of Soil by Gamma-ray Attenuation Methods*, University of Florida Bull. No. 807, 1979.

11. Herkelrath, W. N., and E. E. Miller, High performance gamma system for soil columns, *Soil Sci. Soc. Am. J.*, 40: 331-332 (1976).

12. Lutz, J. F., Apparatus for collecting undistrubed soil samples, *Soil Sci.*, 64: 399-401 (1947).

13. Jamison, V. C., H. H. Weaver, and I. F. Reed, A hammer-driven soil core sampler, *Soil Sci.*, 69: 487-496 (1950).

14. Baver, L. D., W. H. Gardener, and W. R. Gardener, *Soil Physics*, Wiley, New York, 1972.

15. Buchele, W. F., A power sampler of undisturbed soils, *Trans. Am. Soc. Agric. Eng.*, 4: 185-187, 191 (1961).

16. Veihmeyer, F. J., An improved soil-sampling tube, *Soil Sci.*, 27: 147-152 (1929).

17. Erbach, D. C., *State of the art of soil density measurement*, Paper No. 82-1541, Am. Soc. Agric. Eng., St. Joseph, MI, 1982.

18. Blake, G. R., Bulk density, in *Methods of Soil Analysis*, Part I (C. A. Black, Ed.-in-Chief), Am. Soc. Agron., Madison, WI, 1965, pp. 374-390.

19. Cernica, J. N., Proposed new method for the determination of density of soil in place, *Geotech. Testing J.*, 3: 120-123 (1980).

20. Brasher, B. R., D. P. Franzmeir, V. Valassis, and S. E. Davidson, Use of Saran resin to coat natural soil clods for bulk density and moisture retention measurements, *Soil Sci.*, 101: 108 (1966).

21. Abrol, I. P., and J. P. Palta, Bulk density determination of soil clod using rubber solution as a coating material, *Soil Sci.*, 106: 465-468 (1968).

22. Campbell, D. J., A flotation method for the rapid measurement of the wet bulk density of soil clods, *J. Soil Sci.*, 24: 239-243 (1973).

23. Gill, W. R., Soil bulk density changes due to moisture changes in soil, *Trans. Am. Soc. Agric. Eng.*, 2: 104-105 (1959).

24. Greacen, E. L., D. A. Farrel, and J. A. Forrest, Measurement of density patterns in soil, *J. Agric. Eng. Res.*, 12: 311-313 (1967).

25. Chepil, W. S., Methods of estimating apparent density of discrete soil grains and aggregates, *Soil Sci.*, 70: 351-362 (1950).

26. Voorhees, W. B., R. R. Allmaras, and W. E. Larson, Porosity of surface soil aggregates at various moisture contents, *Soil Sci. Soc. Am. Proc.*, 30: 163-167 (1966).

27. Russell, E. W., and W. Balcerek, The determination of the volume and airspace of soil clods, *J. Agric. Sci. Camb.*, 34: 123-132 (1944).

28. Soane, B. D., D. J. Campbell, and S. M. Herkes, Hand-held gamma-ray transmission equipment for the measurement of bulk density of field soils, *J. Agric. Eng. Res.*, 16: 146-156 (1971).

29. Soane, B. D., A gamma-ray transmission method for the measurement of soil density in field tillage studies, *J. Agric. Eng. Res.*, 13: 340-349 (1968).

30. Van Bavel, C. H. M., R. J. Lascano, and J. M. Baker, Calibrating two-probe, gamma-gauge densitometers, *Soil Sci.*, 140: 393-395 (1985).

31. Burgess, T. M., and R. Webster, Optimal interpolation and isarithmic mapping of soil properties: I. The semi-variogram and punctual kridging, *J. Soil Sci.*, 31: 315-331 (1980).

32. McBratney, A. B., and R. Webster, How many observations are needed for regional estimation of soil properties? *Soil Sci.*, 135: 177-183 (1983).

33. O'Sullivan, M. F., J. W. Dickson, and D. J. Campbell, Interpretation and presentation of cone resistance data in tillage and traffic studies, *J. Soil Sci.*, 38: 137-148 (1987).

34. *The Ionising Radiations Regulations 1985*, HMSO, London, 1985.

35. Erbach, D. C., *Measurement of Soil Moisture and Bulk Density*, Paper No. 83-1553, Am. Soc. Agric. Eng., St. Joseph, MI, 1983.

36. Raper, R. L., and D. C. Erbach, *Accurate Bulk Density Measurements Using a Core Sampler*, Paper No. 85-1542, Am. Soc. Agric. Eng., St. Joseph, MI, 1985.

37. King, F. G., and A. W. Parsons, *Portable Radioactive Equipment for Measuring Soil Density*, Road Res. Lab., U.K., Res. Note RN/3628/FGK.AWP (1959).

38. Minaei, M., J. V. Perumpral, J. A. Burger, and P. D. Ayers, *Soil Bulk Density by Core and Densitometer Procedures*, Paper No. 84-1041, Am. Soc. Agric. Eng., St. Joseph, MI, 1984.

39. Schafer, G. J., P. R. Barker, and R. D. Northey, *Density of Undisturbed Soil Cores by Gamma-Ray Attenuation*, Rep. No. 67, New Zealand Soil Bureau, 1984.

40. Keisling, T. C., and D. A. Smittle, Soil bulk density corrections for providing a better relationship with root growth in gravelly soil, *Commun. Soil Sci. Plant Anal.*, 12:91-96 (1981).

41. Henshall, J. K., *The Calibration and Field Performance of a High Resolution Gamma-Ray Transmission System for Measuring Soil Bulk Density in Situ*, Unpubl. Dep. Note No. SIN/299, Scot. Inst. Agric. Eng., 1980.

42. Campbell, D. J., J. W. Dickson, B. C. Ball, and R. Hunter, Controlled seedbed traffic after ploughing or direct drilling under winter barley in Scotland, 1980-1984, *Soil Till. Res.*, 8: 3-28 (1986).

43. Campbell, D. J., and J. K. Henshall, Two new instruments to measure the strength and bulk density of soil in situ, in *Proc. 6th Int. Conf. Mechanisation of Field Experiments*, Dublin, 1984, pp. 338-344.

44. Campbell, D. J., and J. W. Dickson, Effect of four alternative front tyres on seedbed compaction by a tractor fitted with a rear wheel designed to minimize compaction, *J. Agric. Eng. Res.*, 29: 83-91 (1984).

45. Henshall, J. K., and D. L. O. Smith, An improved method for presenting comparisons of soil compaction effects below wheel ruts, *J. Agric. Eng. Res.*, 42: 1-13 (1989).

46. Dickson, J. W., and D. L. O. Smith, *Compaction of a Sandy Loam by a Single Wheel Supporting One of Two Masses Each at Two Ground Pressures*, Unpubl. Dep. Note No. SIN/479, Scot. Inst. Agric. Eng., 1986.

47. Soane, B. D., Techniques for measuring changes in the packing state and cone resistance of soil after passage of wheels and tracks, *J. Soil Sci.*, 24: 311-323 (1973).

48. Zimmerman, R. P., and L. T. Kardos, Effect of bulk density on root growth, *Soil Sci.*, 91: 280-288 (1961).

49. Edwards, W. M., J. B. Fehrenbacher, and J. P. Vavra, The effect of discrete ped density on corn root penetration in a planosol, *Soil Sci. Soc. Am. Proc.*, 28: 560-564 (1964).

50. Veihmeyer, F. J., and A. M. Hendrickson, Soil density and root penetration, *Soil Sci.*, 65: 487-493 (1948).

51. Wiersum, L. K., The relationship of the size and structural rigidity of pores to their penetration by roots, *Plant Soil*, 9: 75-85 (1957).

52. Douglas, J. T., M. G. Jarvis, K. R. Howse, and M. J. Goss, Structure of a silty soil in relation to management, *J. Soil Sci.*, 37: 137-151 (1986).

53. Ball, B. C., M. F. O'Sullivan, and R. W. Lang, Cultivation and nitrogen requirement for winter barley as assessed from a reduced-tillage experiment on a brown forest soil, *Soil Till. Res.*, 6: 95-109 (1985).

54. Pidgeon, J. D., and B. D. Soane, Effects of tillage and direct drilling on soil properties during the growing season in a long-term barley monoculture system, *J. Agric. Sci. Camb.*, 88: 431-442 (1977).

55. Soane, B. D., The effects of traffic and implements on soil compaction, *J. Proc. Inst. Agric. Eng.*, 25: 115-126 (1970).

56. Campbell, D. J., and M. F. O'Sullivan, The cone penetrometer in relation to trafficability, compaction and tillage, in *Soil Analysis: Physical Methods* (K. A. Smith and C. M. Mullins, Eds.). Marcel Dekker, Inc., New York, 1990, pp. 399-429.

57. Holmes, J. C., and D. A. S. Lockhart, Cultivations in relation to continuous barley growing: I. Crop growth and development, in *Proc. Int. Soil Tillage Conf.*, Silsoe, U.K., 1970, pp. 46-57.

58. Soane, B. D., D. J. Campbell, and S. M. Herkes, Cultivations in relation to continuous barley growing: II. Soil physical conditions, in *Proc. Int. Soil Tillage Conf.*, Silsoe, U.K., 1970, pp. 58-76.

59. O'Sullivan, M. F., *Soil Responses to Reduced Cultivations and Direct Drilling for Continuous Barley at South Road 1979-1982*, Unpubl. Dep. Note No. SIN/430, Scot. Inst. Agric. Eng., 1985.

60. Hillel, D., *Introduction to Soil Physics*, Academic Press, New York, 1982.

61. Campbell, D. J., The occurrence and prediction of clods in potato ridges in relation to soil physical properties, *J. Soil Sci.*, 27: 1-9 (1976).

8

Liquid and Plastic Limits

DONALD J. CAMPBELL *Scottish Centre of Agricultural Engineering, Penicuik, Midlothian, Scotland*

I. INTRODUCTION

Plasticity is the property that allows a soil to be deformed without cracking in response to an applied stress. It is common experience that a soil may exhibit plasticity, and hence be remolded, over a range of water contents. Above that range, the soil behaves as a liquid, while below that range it behaves as a brittle solid and will eventually fracture in response to increasing applied stress. The Swedish scientist Atterberg [1,2] was interested in quantifying the range of water contents over which a soil is plastic and, in so doing, he devised what are probably two of the oldest laboratory tests in soil mechanics. The gravimetric water content corresponding to the upper limit of plasticity, known as the liquid limit, was taken to be that at which a small slope, forming part of a groove in a sample of the soil, just collapsed under the action of a standardized shock force. The gravimetric water content corresponding to the lower limit of plasticity, known as the plastic limit, was that at which a sample of the soil, when rolled into a thread by the palm of the hand, split and crumbled when the thread diameter reached 3 mm. By convention, both water contents are expressed on a percentage basis. The numerical difference between the liquid and plastic limits was defined as the plasticity index. Remarkably, these simple empirical tests have been used, essentially unchanged, for more than three-quarters of a century by soil engineers and soil scientists [3].

Although Atterberg was concerned with the plasticity index of agricultural soils, his liquid and plastic limit tests found favor with those concerned with soil conditions for civil engineering. Such engineers found the limits, particularly the plastic limit, to be useful in the design and control testing of earthworks [4]. In addition, the tests became widely used for soil classification [4]. The widespread adoption of the liquid and plastic limit tests by civil engineers hinged on the development by Casagrande of apparatus to measure the limits [5]. Although his apparatus was based on that of Atterberg, Casagrande appreciated the need, where empirical tests are concerned, to specify closely every detail of the test procedure. In this way both the repeatability of the test by one operator and the reproducibility between operators was optimized [6]. Consequently, the Casagrande tests became widely adopted as the official standard by engineers in the United Kingdom [3], the United States [7], and elsewhere.

Soil scientists have tended to make less use of the Atterberg limits. The limits do not figure in soil survey or land capability classification systems but have been used mainly as indicators of the likely mechanical behavior of soil [8-10]. This has generally been done by establishing simple correlations between the plasticity limits or plasticity index and other properties considered important in determining soil behavior. An example of such a relation is shown in Table 1. It has been suggested, however, that the addition of liquid and plastic limit values would be a useful addition to soil particle size distribution in the classification of soils in the laboratory [11]. Since the Atterberg limits are related to the field texture, as determined in the hand, it is relevant that this method of classification is often preferred by soil scientists concerned with practical problems of soil workability in the field [12].

Two further index values may be derived from the Atterberg limits. The liquidity index, LI, is related to the percentage gravimetric soil water content, w%, the plastic limit, PL, and the plasticity index, PI, as follows:

$$LI = \frac{w\% - PL}{PI} \tag{1}$$

Thus, the liquidity index describes soil water content in relation to the range of water contents between the liquid and plastic limits. The activity A is the ratio of the plasticity index to C, the percentage by weight of soil particles smaller than 2 μm:

$$A = \frac{PI}{C} \tag{2}$$

The activity of a soil depends on the mineralogy of the clay fraction,

TABLE 1 Relation Between Potato Harvesting Difficulty, as Indicated by the Number and Strength of Clods in Potato Ridges, and Plasticity Index

Yield of 30-75 mm diameter clods, A (t/ha)	Crushing resistance of 30-45 mm diameter clods, B (N)	A × B	Plasticity index of soil
76.2	73.7	5615	12.8
95.0	17.6	1672	11.2
19.0	65.9	1252	10.3
60.5	40.4	2444	8.8
48.0	38.5	1848	8.1
29.2	26.8	782	6.2
26.8	19.4	519	5.1
1.4	52.2	73	3.6

Source: D. J. Campbell, unpublished data.

the nature of the exchangeable cations, and the concentration of the soil solution.

II. THEORIES OF PLASTICITY

In attempting to explain the mechanism behind the existence of the liquid and plastic limits, two basic approaches have been adopted. In the first, soil behavior is considered in terms of the cohesive and adhesive forces developed as a result of the presence of water between the soil particles. This is the traditional approach, and it has been summarized by Baver et al. [9]. The critical state theory of soil mechanics is used in the second approach. Critical state soil mechanics describes the stress-strain behavior of a soil in terms of the three-dimensional relationship of spherical pressure, deviatoric stress, and specific volume. Critical state theory is detailed by Schofield and Wroth [13], and it is mathematically complex. However, the basic concepts and their importance have been discussed by Kurtay and Reece [14].

A. Water Film Theory

Cohesion within a soil mass is due to a variety of interparticle forces [9]. Bonding forces include van der Waals forces, electrostatic forces between the negative charges on clay particle surfaces and the positive charges on the particle edges, particle bonding by cationic bridges, cementation effects of certain substances (e.g., iron oxides, aluminum, organic matter), and the forces associated with the soil water. Taken together, these forces will determine whether a soil will, when stressed, undergo brittle failure, plastic flow, or viscous flow.

At low water contents, most of the soil water forms annuli around the interparticle contacts [15-19]. Haines [15] showed, from consideration of an idealized soil comprising single-sized spheres, that the tensile force generated by the annulus of water between two spheres was related to the geometry of the annulus and the surface tension of the water. When the relation was extended to give the tensile force per unit cross-sectional area within the soil mass, it was seen that high interparticle forces are associated with small particle diameters. The validity of the relationship breaks down at higher water contents because the individual annuli of water start to coalesce. At a water content just above the plastic limit, the soil becomes saturated and, in a cohesive soil, the soil water tension and other bonding forces are in equilibrium with the repulsive forces due to the double-layer swelling pressure.

The theory developed by Haines [15] was extended by Nichols [20] to the laminar shapes of clay particles. Nichols showed that the interparticle force F was related to the particle radius r, the surface tension of the pore water T, the angle of contact between the liquid and particle α, and the distance between the particles d, as follows:

$$F = \frac{4k\pi rT \cos \alpha}{d} \tag{3}$$

where k is a constant. He also showed that for each of three soils, the product of the cohesive force and the water content was a constant at low water contents. At higher water contents, however, the cohesive force decreased rapidly with increasing water content.

It is important to note that although the existence of a relationship between water content and cohesion that exhibits a maximum has been demonstrated experimentally [21,22], the relation is only valid for dry soils that have been rewetted. When puddled soil is allowed to dry, cohesion increases with decreasing water content and reaches a maximum when the soil becomes dry. This effect probably arises because, in puddled soils, the number of interparticle contacts is maximized; hence cohesive forces other than those due to soil water are large.

Baver [23] has suggested that when a soil at the plastic limit is stressed, the laminar clay particles, each of which is surrounded by a water film and which were previously randomly orientated in the friable state, are rearranged such that they slide over one another. Thus, the cohesive forces associated with the tension effects in the water films are overcome and the soil deforms. When the stress is removed, the particles remain in their new position under the action of the cohesive forces and there is no elastic recovery. The soil has undergone plastic deformation or flow. Before the soil reaches the liquid limit, the water films have completely coalesced and the soil water tension has greatly decreased. Thus, cohesion decreases and the soil is capable of viscous flow. For a given applied stress, the difference in flow rate between viscous and plastic flow is due to the change in the magnitude of the cohesive forces.

A similar explanation for the difference in the nature of soil flow at the liquid and plastic limits has been offered by Sowers [24], who suggested that the water adsorbed on particle surfaces has a very high viscosity close to the particle surface. This effect has been attributed to increased hydrogen bonding [25]. With increasing distance from the particle surface, the viscosity decreases until ultimately it equals that of free water. Thus, in a soil that is just above the plastic limit, all the water is highly viscous and only plastic flow can occur. As the water content and particle separation further increase, the liquid limit is reached and the outermost layers have the viscosity of free water, allowing the soil to flow like a liquid.

Although the liquid limit is related to clay content and its surface area for most types of clay mineral, montmorillonite is an exception in that the liquid limit is controlled essentially by the thickness of the diffuse double layer. This was demonstrated by the linear relation between the liquid limit and the amount of exchangeable sodium ions present [26].

Although the interparticle forces associated with soil water may not provide a comprehensive explanation of the mechanism of plasticity, it is clear that soil particle sizes, their specific surface, and the nature of the clay minerals are all important. This is consistent with the common experience that, generally, both the liquid and the plastic limits depend on the type as well as the amount of clay in a soil [27].

B. Critical State Theory

Kurtay and Reece [14] have described how the critical state theory of soil mechanics gives a coherent picture of the stress-strain behavior of soil. Their description is free of any attempt to show the experimental justification of the theory and is therefore relatively simple in comparison with other descriptions, which include the experimental evidence in support of this complex theory [13].

If a relatively loose sample of soil is subjected to a progressively increasing uniaxial (deviatoric) stress while the confining stress
(spherical pressure) is kept constant, the soil volume will decrease
as the soil is compressed. This will occur for both unsaturated soil
and soil that is saturated but allowed to drain as it is compressed.
If the process continues, a point will be reached at which the soil
can be compressed no further. However, if the deviatoric stress is
maintained and the soil continues to distort without any change in
volume, the soil is said to be in the critical state. In terms of the
three-dimensional relationship of spherical stress, deviatoric stress,
and specific volume, the point describing this critical state is one of
the many possible critical state points, which together form the critical state line. The critical state line is an extremely important concept in that it allows, within the confines of a single theory, the
stress-strain behavior of a soil with any particle size distribution
to be explained, be it wet or dry, dense or loose, confined or
unconfined.

For the present purpose, however, the important quality of the
critical state line is that since it describes all conditions under which
a soil will undergo continuous remolding without a change in volume,
it follows that soil being prepared for either the liquid or the plastic
limit test must be described by a point on this line. Thus, the liquid
and plastic limit tests can give more than simple qualitative information about soil behavior.

During the liquid limit test, the soil water content, and hence the
specific volume, is adjusted by adding water and remolding the soil
until, in effect, the soil has a fixed undrained shear strength determined by the conditions of the test. Because the soil is continuously
remolded as water is added, it is in the critical state and under the
action of a negative pore water pressure.

When soil is prepared for the plastic limit test, it is continuously
remolded, and hence once again is in the critical state. However, since
the soil is much drier than in the liquid limit test, the pore water
pressure (matric potential) is even more negative. This negative
pore water pressure acts in the same way as if the soil were subject
to an additional externally applied stress and serves to increase the
shear strength of the soil. It is reasonable to speculate that the
plastic limit should, like the liquid limit, correspond to a state in
which the soil has a fixed undrained shear strength. Atkinson and
Bransby [28] reported that the undrained shear strength data obtained for four clay soils by Skempton and Northey [29] revealed
that all four soils had very similar undrained shear strengths at the
plastic limit. Perhaps more remarkably, the undrained shear strength
of each soil at the plastic limit was almost exactly 100 times the undrained shear strength at the liquid limit.

Knowing the ratio of the shear strengths at the liquid and plastic limits, it is possible to define the slope of the critical state line on a plot of the logarithm of the spherical pressure versus the specific volume in terms of the plasticity index [13,28]. Thus the plasticity index can be used as a direct indicator of soil compressibility.

The description of soil behavior at the liquid and plastic limits offered by critical state theory is, at first sight, quite different from that given by the water film theory and may give the impression that soil water content is irrelevant. However, the water content is important in critical state theory, but only insofar as it affects the pore water pressures.

III. DETERMINATION OF THE LIQUID AND PLASTIC LIMITS

The methods initiated by Atterberg [1,2] and developed by Casagrande [5], who introduced closely specified apparatus and test procedures, were adopted by the British Standards Institution, the American Society for Testing and Materials, and other organizations as the standard tests for those concerned with soil conditions for civil engineering. However, in 1975 in a revision of the relevant British Standard [3], a new test for the liquid limit, based on a procedure involving a drop-cone penetrometer, was introduced. The Casagrande tests were still included, but the cone penetrometer method was described as the preferred method for the determination of the liquid limit. Although various other methods of determining the liquid and plastic limits have been suggested, usually, but not always, based on correlation of the limits with other soil rheological properties, by far the most widely used methods are the Casagrande and, to a lesser extent, drop-cone tests.

A. Casagrande Tests

In the Casagrande liquid limit apparatus [3] (Fig. 1), the sample is contained in a cup that is free to pivot about a horizontal hinge and rests on a rubber base of specified hardness. A crank allows the operator to rotate a cam, which alternately raises the cup 10 mm above the base and allows it to drop freely on to the base. The test soil is mixed with distilled water to form a homogeneous paste, allowed to stand in an airtight container for 24 hours, and remixed; then a portion is placed in the cup. The sample is divided in two by drawing a standard grooving tool through the sample at right angles to the hinge. The crank is then turned at two revolutions per second until the two parts of the soil come into contact at the bottom of the groove over a length of 13 mm. The number of blows to the cup required to do this is recorded and the test repeated.

FIG. 1 The Casagrande grooving tool and liquid limit device showing a soil sample divided by the tool before testing.

If consistent results are obtained, a subsample of the soil is taken
from the region of the closed groove for the measurement of water
content. More distilled water is added to the test sample and the
procedure repeated. This is done several times at different water
contents to give a range of results lying between 50 and 10 blows.
The linear relation between the water content and the log of number
of blows is plotted, and the percentage water content corresponding
to 25 blows is recorded, to the nearest integer, as the liquid limit
of the soil.

A simplified test procedure for liquid limit determination using
the Casagrande apparatus is that known as the "one-point method."
Essentially the method involves making up a soil paste such that
the groove cut in the sample in the cup closes at a number of blows
as close as possible to 25, and certainly between 15 and 35. A correction factor, which varies with the actual number of blows, is applied to the water content of the soil to give the liquid limit [3].

The method has the advantage of speed, but this is at the expense of reliability [30].

For the Casagrande plastic limit test [3], the sample is mixed with distilled water until it is sufficiently plastic to be molded into a ball. A subsample of approximately 10 g is formed into a thread of about 6 mm diameter and the thread is rolled between the tips of the fingers of one hand and a flat glass plate until it is 3 mm in diameter. The thread is then remolded in the hand to dry the sample and again rolled into a thread. The operation is repeated until the thread crumbles as it reaches a diameter of 3 mm. A second subsample is similarly tested, and the mean of the two water contents (expressed as percentages) at which the threads crumble on reaching a diameter of 3 mm is recorded, to the nearest integer, as the plastic limit of the soil. Where the plastic limit cannot be obtained or where it is equal to the liquid limit, the soil is described as nonplastic.

Both these tests are undertaken on air-dry material passing a 425 μm sieve, although it has been suggested that when the bulk of the soil passes 425 μm, it may be more convenient to test whole soil [3]. However, it is generally agreed that the results for soils tested in the natural condition may differ from results of tests conducted on material that has been air-dried. This is particularly true of organic soils. Where an appreciable proportion of the soil is retained on the 425 μm sieve, removal of such material can influence the plasticity characteristics of the soil [31]. Because of these various aspects of the test procedures and because the tests are conducted on remolded soil, the results should be interpreted with caution in relation to the likely behavior of soil in the field.

B. Drop-Cone Tests

Casagrande was not satisfied with his apparatus for the determination of the liquid limit [32]. He considered that a simple direct shear test or an indirect shear test such as a penetration test would eliminate many of the deficiencies later summarized by Sherwood and Ryley [33]. Most of the shortcomings are related to the subjective aspects of the test and to the tendency for some soils to slide in the cup or liquefy from shock, rather than flowing plastically. After reviewing five alternative cone penetrometer tests, Sherwood and Ryley [33] concluded that a method developed by the Laboratoire Central des Ponts et Chaussées (58 Boulevard Lefebre, F-75732 Paris Cedex 15, France) [34] offered the possibility of a suitable method for liquid limit determination. The new method, which used apparatus already available in most materials testing laboratories, was shown to be easier to perform than the Casagrande method; moreover, it was applicable to a wider range of soils, less

dependent on the design of the apparatus, and less susceptible to
operator error. Largely as a result of the work of Sherwood and
Ryley [33], the drop-cone penetrometer test was adopted as the
preferred method for liquid limit determination by the British Standards Institution [3] in the United Kingdom. However, it has not
been adopted by the American Society for Testing and Materials.

The apparatus used in the drop-cone penetrometer test is shown
in Fig. 2. The mass of the cone plus shaft is 80 g and the cone
angle is 30°. The test soil, which is prepared to give a selection
of water contents in exactly the same way as in the Casagrande test,
is contained in a cup 50 mm deep and 55 mm in diameter. At each
water content, the soil is pushed into the cup with a spatula, to
ensure that air is not trapped, then leveled off flush with the top
of the cup. The cone is lowered until it just touches the soil surface, and the cone shaft is clamped. The shaft is then released
and the cone plus shaft allowed to fall freely for 5 seconds before

FIG. 2 The drop-cone penetrometer showing the cone position at
the start of a test.

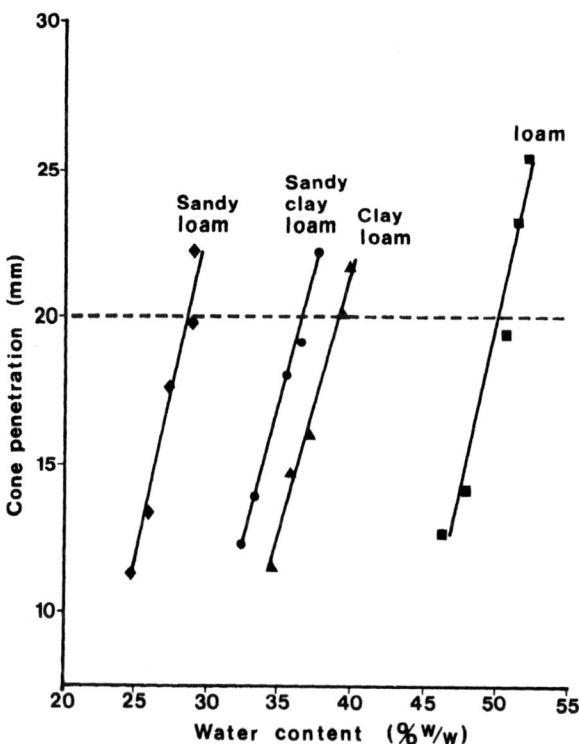

FIG. 3 The results of cone penetrometer liquid limit tests on four
arable topsoils of contrasting texture. The dashed horizontal line
indicates the cone penetrometer liquid limit. (Data from Ref. 42.)

the shaft is again clamped and the cone penetration noted from the
dial gauge. In many instruments the 5-second release is automatically controlled via an electromagnetic solenoid clamp as shown in
Fig. 2. A duplicate measurement is made and the procedure is then
repeated over a range of water contents. The linear relation between cone penetration and water content is plotted, and the percentage water content corresponding to a penetration of 20 mm is recorded, to the nearest integer, as the cone penetrometer liquid limit.
Typical test results for four soils are shown in Fig. 3.

Attempts have been made to develop a one-point cone penetrometer liquid limit test analogous to the one-point Casagrande test. As
with the latter, the method is a compromise between speed and accuracy but has been shown to be a satisfactory alternative [35]. The
one-point cone penetrometer test has been shown to be theoretically
sound and not based simply on statistical correlations [30].

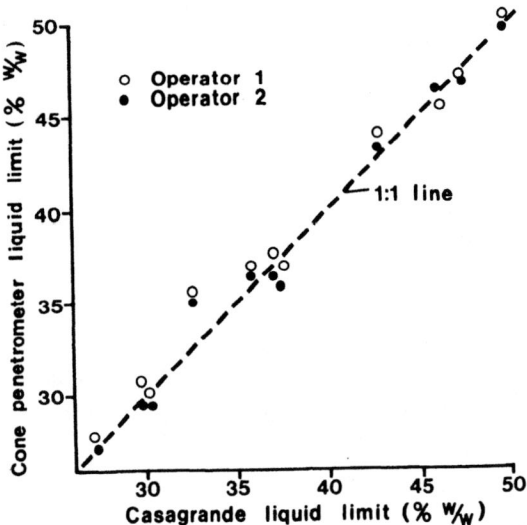

FIG. 4 The relation between the cone penetrometer liquid limit, as
determined by two operators, and the Casagrande liquid limit determined by operator 1 for some arable topsoils. (From Ref. 42.)

The drop-cone liquid limit method has been compared with the
Casagrande method for a range of soils used in civil engineering
[33,36-40] and agriculture [41-43]. Generally, the two tests were
found to give equivalent results. Occasionally, the drop-cone method gave rather greater values than the Casagrande method [44,45],
and occasionally the opposite was found [46,47]. Most of the differences in the results were attributed to effects of soil type and were
of the same magnitude as the operator errors associated with the two
methods (i.e., quite small). The two methods are compared in Fig.
4, which also shows the reproducibility of the drop-cone method.

With the widespread adoption of the drop-cone method for measuring the liquid limit, there were obvious advantages in using the
same apparatus to measure the plastic limit, if that were possible.
Scherrer [39] proposed a method of plastic limit determination that
involved extrapolation of the linear relation between water content
and cone penetration found in the region of the liquid limit but conceded that the necessary extrapolation implied possible sources of
inaccuracy in the method. In fact, Towner [48] showed that although
the water content-cone penetration relation is linear in the region of
the liquid limit, it becomes nonlinear at lower water contents, tend-
ing to show a minimum penetration. Campbell [49] made detailed
measurements of the water content-cone penetration relations for 18

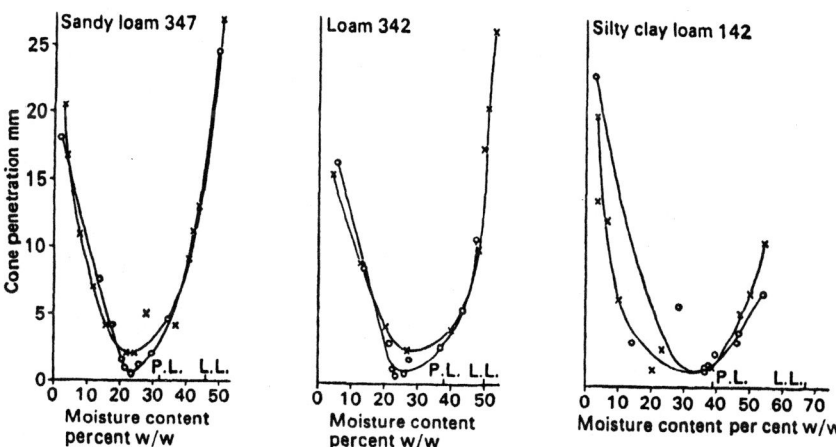

FIG. 5　Water content-cone penetration relations for three soils of contrasting texture in relation to the Casagrande liquid (LL) and plastic (PL) limits. Results obtained by two independent operators (x and o) are shown. (From Ref. 49.)

soils and found a pronounced minimum in the curve for each soil in the region of the Casagrande plastic limit. Results for three of the soils are shown in Fig. 5. The water content corresponding to the minimum of the curve was always numerically less than, but correlated closely with, the plastic limit.

It was suggested that the plastic limit be redefined as the water content corresponding to the minimum of the curve and that it be referred to as the cone penetrometer plastic limit. The possibility of the establishment of a fixed penetration value corresponding to the plastic limit was considered [48-50] but was dismissed because variation in penetration between soils was unacceptably high [49]. The cone penetrometer plastic limit was shown to offer reduced operator errors and to be a good indicator of soil behavior in an examination of the variation with water content of soil cohesion, soil-metal friction, susceptibility to compaction, implement draft, and the slope and intercept of the virgin compression line of critical state soil mechanics theory. For a given soil, all these relations were shown to exhibit turning points at a water content corresponding to the cone penetrometer plastic limit [22].

A distinct approach to the use of the cone penetrometer to measure the plasticity index was made by Wood and Wroth [51]. They suggested that the plastic limit be redefined to establish the undrained shear strength at the plastic limit as 100 times that at the liquid limit. The proposal was based on the assumption that all soils have the same strength at their liquid limits, and this was shown to be reasonable. Furthermore, it was shown that the proposal allowed a unique relation to be developed for remolded soil between strength and liquidity index and also between compression index and plasticity index [52].

C. Other Methods

Several workers have devised methods of measuring liquid and plastic limits that depend either on correlation with other soil physical or mechanical properties or on a revision of the definition of the limits that relates them more to changes in soil behavior. None of these methods has been widely adopted, but to a certain extent this is due to the difficulty of replacing long-established standard methods.

Faure [53] examined the water content-dry bulk density relation of several soils. In addition to the well-known peak corresponding to maximum dry bulk density, the density at very low water contents was found to be largely independent of water content until a threshold water content was reached. The latter was considered to correspond to the plastic limit and the water content at the maximum density to correspond to the liquid limit. These limits were thought to be constants for a given compactive effort and to depend only on the type of clay mineral in the soil. This is an interesting concept, which is quite consistent with the critical state theory of soil mechanics; this theory suggests that any soil, at any water content, may behave in a plastic fashion provided it can be subjected to a combination of stresses that correspond to a point on the critical state line [14].

Russell and Mickle [54] accepted that the Casagrande limits are well established, and they sought to devise a method of measuring them quickly and accurately. As a result of extensive experimental work on the water release characteristics of several soil textural groups, they found that for each group, the water tension corresponding to the liquid limit was a constant. However, the plastic limit was found to be at a fixed water tension only for clays, and the investigators recommended further work on plastic limit determination.

There have been attempts to relate the liquid and plastic limits to specific viscosities. Yasutomi and Sudo [55] used a forced oscillation viscometer to study soils at water contents in the region of the liquid limit, while Hajela and Bhatnagar [56] used a synchro-electric viscometer to measure the flow properties of clay pastes. Although the latter authors could predict the liquid limit with an error of only 1%, they conceded that the method was inappropriate for soils at water contents in the region of the plastic limit.

Vasilev [57] measured the water content remaining when a soil paste had been subjected to a standard stress. He found such a water content to be related to the plastic limit. He was also an early protagonist of the drop-cone apparatus for liquid limit determination, using a 76 g cone of 30° cone angle. His method gave results that were shown by Skopek and Ter-Stephanian [58] to correlate closely with those obtained by the Casagrande method.

Sherwood and Ryley [36] sought to eliminate the operator errors associated with the plastic limit test and considered as alternatives a compaction test, a cone penetration test, and two tests in which the soil was compressed under the action of a fixed stress to a residual water content. Although empirical correlations were found between the test results for each of the methods investigated and the plastic limit, it was considered that none of them represented a viable alternative to the Casagrande test where soil classification was concerned. However, these authors felt that the Casagrande test was unsuitable as a control test in earthwork construction and that an alternative test should be developed.

D. General Considerations

Since both the liquid and plastic limit tests are empirical, it is important that the test procedures be closely specified, if consistent results are to be obtained. The operator errors associated with any particular method can be a major factor in deciding whether the method should be adopted. It is also of interest to examine the effect of variations in the test procedure on the results obtained. If the test procedure can be simplified, in terms of either operator effort or time required for a test, then clearly such simplification is worthwhile.

Most test procedures specify that the soil should first be air-dried and then sieved through a 425 µm sieve [3], although wet sieving through a 425 µm sieve followed by air-drying has been proposed [59]. However, it has been suggested that in some circumstances either air-drying [50] or removal of any soil particle size fraction [3,31,46] can markedly affect the result obtained. The logical conclusion of such thinking is that ideally the tests should be conducted in situ if a way of doing this can be found [22]. The development of a practical in situ test is unlikely, however, because of the difficulty in obtaining an appropriate sequence of test water contents without the complication of hysteresis effects as the soil alternately wets and dries in a random way [60]. Such effects, probably together with cementation effects, have led to the need for samples prepared to a given water content to be thoroughly mixed [7] and allowed to cure for 24 hours before being tested [3], although the curing period is not universally agreed to be

TABLE 2 Cone Penetrometer Liquid Limit and Proposed Cone Penetrometer Plastic Limit [49] Determinations by Experienced and Totally Inexperienced Operators and the Corresponding Casagrande Limits for Some Arable Topsoils

Field texture[a]	Total organic matter (%)	Liquid limit (% w/w)			Plastic limit (% w/w)		
		Casagrande	Cone penetrometer		Casagrande	Cone penetrometer	
			Experienced operator	Inexperienced operator		Experienced operator	Inexperienced operator
SL	3.0	27	28	29	22	17	15
SL	3.9	30	31	31	26	17	18
SL	3.7	30	30	30	26	18	19
SCL	4.8	33	36	36	24	19	17
SCL	3.3	36	37	36	28	19	26
SCL	5.5	37	38	36	26	19	21
SL	7.4	37	37	38	31	25	22
SL	5.2	49	47	45	44	27	30

[a]U.S. Department of Agriculture criteria: SL, sandy loam; SCL, sandy clay loam.

Source: Based on data from Refs. 42 and 49.

necessary [47,61]. In addition, sample preparation may be complicated by the fact that some soils undergo irreversible changes on drying [50], while other soils may give index values that depend on the number of times the test sample is remolded and cured before the test, especially where the liquid limit is concerned [62,63]. The latter effect is thought to be due to particularly stable aggregates, which break down only with prolonged remolding [63,66].

Many other aspects of the test procedures have been investigated in detail. For example, although the standard test for the liquid limit using the drop-cone penetrometer includes a check on the sharpness of the cone used [3], an investigation by Houlsby [67] led him to conclude, in contrast to the work of Sherwood and Ryley [37], that the effect of variations in cone sharpness were very small in comparison with the effect of the roughness of the cone surface. As might be expected, the effects of both cone angle [68] and cone mass [60,68] influence the penetration obtained. The effect of temperature on the Casagrande liquid and plastic limits has been found to be appreciable where a wide range of ambient temperatures is not unusual [69], due to variation in the viscosity of the soil water with temperature.

The importance attached to the reproducibility of any empirical test result is reflected in the extent to which operator errors associated with liquid [31,42,43] and plastic [49,70,71] limit tests have been investigated. Although such investigations led to the development of the drop-cone test for the liquid limit, proposals involving either improvements to the Casagrande plastic limit test [71] or alternative test procedures [49] have not been widely adopted. This is no doubt partly because of the difficulties met in attempting to make changes to long-established methods. The reproducibility of the cone penetrometer liquid and plastic limit tests is shown for eight arable topsoils in Table 2.

When the Casagrande plastic limit either cannot be obtained or is greater than the liquid limit, the soil is described as nonplastic. However, it is common experience that such soils may indeed exhibit plastic behavior when subjected to the appropriate combination of stresses. In this respect, both the cone penetrometer plastic limit proposed by Campbell [49] and the plastic limit related to compactibility proposed by Faure [53] have the advantage that a plastic limit can be determined for all soils.

IV. APPLICATIONS OF TEST RESULTS

Probably the most widespread single application of the results of liquid and plastic limit tests is their use by engineers to classify soils [27]. This situation arose because engineers appreciated, at

an early stage in the development of soil mechanics, that the test
results were related to properties such as compressibility, permeability (i.e., saturated hydraulic conductivity), and strength [72].
Thus, the test results could indicate to the engineer the likely
mechanical behavior of the soil during the construction of earthworks. The use of remolded soils in the tests was entirely appropriate in this context.

However, to the scientist concerned with soil as a medium for
plant growth, remolding of the soil before testing has always been
considered to be a limitation to the value of the test result. Possibly because of this, soil classification has always placed more emphasis on soil particle size distribution, although it has been suggested that liquid and plastic limit values could usefully be added
to such classifications [11]. Although the precise prediction of soil
behavior in the field may be unlikely from such additional information, a more useful and objective grouping of soils could result.

Despite this limitation, extensive studies of the relation between
liquid and plastic limit values and other soil mechanical and physical
properties have been made. These studies have had two objectives:
(1) to understand soil behavior, and (2) to provide, via correlation,
a simple alternative to a difficult, expensive, or slow test procedure.
The following sections give some examples of the use of liquid and
plastic limits in soil classification and describe some of the relations
of the limits with other soil properties.

A. Soil Classification

Casagrande [72] developed a system of classifying soils based on
sieve analysis together with measurement of the liquid and plastic
limits on the fraction smaller than 425 μm. Developments of this
system now form the British Soil Classification System [4] and the
Unified Soil Classification System [73] (in the United States), both
of which have been widely accepted for classifying soils for engineering purposes. Casagrande plotted liquid limits against plasticity
indices to give what he called the plasticity chart (Fig. 6). An empirical boundary known as the A-line on the chart separated the
inorganic clays lying above the line from the silty and organic soils
lying below the A-line. Both above and below the A-line, the liquid
limit was used to divide soils into three classes of compressibility,
namely low, intermediate, and high, corresponding to liquid limits
of less than 35, between 35 and 50, and greater than 50, respectively. In the British Soil Classification System, the chart was extended to include soils with very high (70-90) and extremely high
(> 90) liquid limits, as shown in Fig. 6. Moreover, soils with liquid
limits below 20 were described as nonplastic, and it was recognized
that organic soils could occur both above and below the A-line.

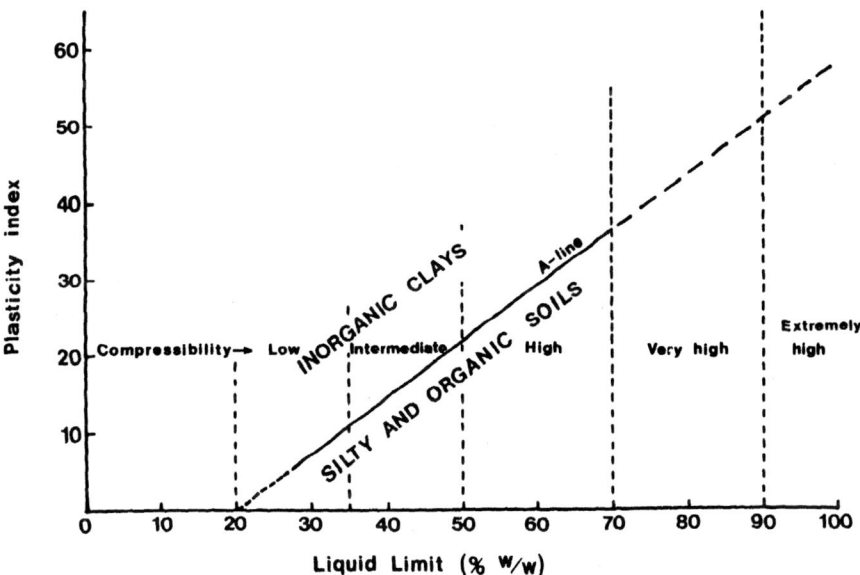

FIG. 6 The plasticity chart used in the British Soil Classification System. The original Casagrande system assigned all soils with liquid limits exceeding 50 to a single compressibility class.

Towner [48] has suggested, from the assumption that the liquid and plastic limits correspond to two fixed strengths, that a modified plasticity chart could be obtained for use in characterizing soil behavior.

Much can be deduced about the mechanical properties of a soil from its position on the plasticity chart. For a given liquid limit, the greater the plasticity index of a soil, the greater is its clay content, toughness, and dry strength, and the lower is its permeability. For a given plasticity index, soil compressibility increases with increasing liquid limit. The liquid and plastic limits are both dependent on the amount and type of clay in a soil. Kaolinitic clays generally lie below the A-line and behave as silts, while montmorillonitic clays lie just above the A-line. Most organic soils, whether silts or clays, lie below the A-line. Peats have very high liquid limits of several hundred percent, but a small plasticity index.

B. Relations with Other Soil Properties

1. Texture and Organic Matter

Bearing in mind the water film theory of plasticity and the importance of the specific surface of the soil particles, it is not surprising that plasticity characteristics are related to clay content.

Although this relation is referred to by many authors [8,74-78],
with, for example, several reporting a simple linear relation between
plasticity index and clay content [75,77,78], a closer relationship
was often found when other factors such as organic matter [74,78]
or silt content [75] were included.   O'Dell et al. [78] examined the
relation between the plasticity index and the silt and clay contents,
organic matter content, and clay mineralogy of Illinois soils.   They
found a very close correlation between plasticity index and a combination of clay percentage, clay percentage that is montmorillonite,
and percentage of organic carbon.   Where the relation between plasticity index and clay content was weak, the effect may have been
associated with particle sizes rather coarser than the clay fraction
[75], or it may have been attributable to the presence of strongly
aggregated clay-sized particles [63,79].   Baver [80] found that the
nature of the clay minerals also influenced plasticity and that montmorillonite in particular, because of its ability to undergo interlayer
swelling, exhibits high plasticity.   Moreover, the nature of the exchangeable cations was shown to be important, with sodium-satur-
ated soils having a much greater plasticity index than those saturated with potassium, calcium, or magnesium.   Both the number and
strength of soil clods found in potato ridges in a range of Scottish
soils have been found to be related to the plasticity indices of the
soils [10].   Variation in the plastic limit for a given site has also
been related to differences in aggregation, the plastic limit for soil
under permanent pasture being much higher than for soil under
arable cultivation [65].

In addition to the size distribution of particles and, particularly,
the percentage of clay, both particle shape and the percentage of
organic material in the soil have an effect on the plasticity characteristics.   Moreover, the effects of these factors usually interact.
Farrar and Coleman [81] found that the particle surface area, as
indicated by adsorption of water, was strongly related to the liquid
limit and rather less so to the plastic limit.   They found weaker correlations with the external surface area as measured by nitrogen
adsorption.   Hammel et al. [82] suggested that the liquid and plastic limits could be used as a less laborious method of measuring the
surface area of soils.   Although the liquid and plastic limits increase
with particle surface area, they may not do so in simple proportion,
since the water involved in filling soil pores may be involved in addition to that increasing the thickness of the water layer between particles [76].   Indeed, it has been suggested that soil-specific surface
determines the plasticity index and liquid limit only insofar as it determines the particle separation at the liquid and plastic limits [30].

Archer [8] reported the variations in liquid and plastic limits
with organic matter content for a wide range of soil textures.   He
found that both limits increased with organic matter content but felt

that the plasticity index might either increase or decrease with increasing organic matter, depending on the soil texture. The general trends in the data in Table 2 are consistent with his results. It has been suggested, however, that hydration of the organic matter in a soil must be fairly complete before water is available for film formation on the soil particles. Thus, although the plastic limit is increased, the quantity of water subsequently required to reach the liquid limit is unchanged and so the plasticity index remains the same [9]. Other workers have found that in general, organic matter influences the plasticity properties of a soil [74,78,83,84], but it may be that the role of organic matter in this context will vary with the nature of the organic material involved.

2. Workability in Relation to Tillage

The plastic limit has generally been taken to indicate the upper end of the range of water contents in which the soil is friable and most readily cultivated to produce a seedbed [85]. Although clod strength is low and breakage therefore relatively easy in the plastic range [8, 86], soils are also more susceptible to compaction and puddling and so clods are also easily formed [86,87]. Moreover, both soil adhesion to metal and tine draft are at their maxima within the plastic range [88], as is the angle of soil-metal friction [86]. Campbell et al. [22] have shown that both the angle of soil-metal friction (Fig. 7) and the draft force on a tine are at a maximum at the cone penetrometer plastic limit. Clearly there are disadvantages in attempting to cultivate the soil above the plastic limit; this applies also to subsoiling, which will be ineffective in loosening the subsoil unless it is drier than the plastic limit. Above the plastic limit, the soil will simply remold without shattering. In contrast, mole drainage channels can be satisfactorily established only when the soil at mole depth is above the plastic limit, although the soil immediately above the channel must remain friable enough to shatter and allow water access to the mole drain. Archer [8] has suggested that the plasticity index should be at least 22 if a soil is to be considered suitable for mole drainage.

3. Compressibility

When a soil wets up to water contents around the plastic limit, its resistance to compaction drops sharply [8]. Above the liquid limit, resistance to compaction can be very high, but relatively low compressive or shearing forces can easily destroy the pore structure of the soil, leaving it in a puddled state [89].

The optimum water content for compaction in the British Standard compaction test (2.5 kg rammer method) [3] has been shown to be correlated with the plastic limit [11,22,90]. However, it has been suggested that such a relationship is probably fortuitous,

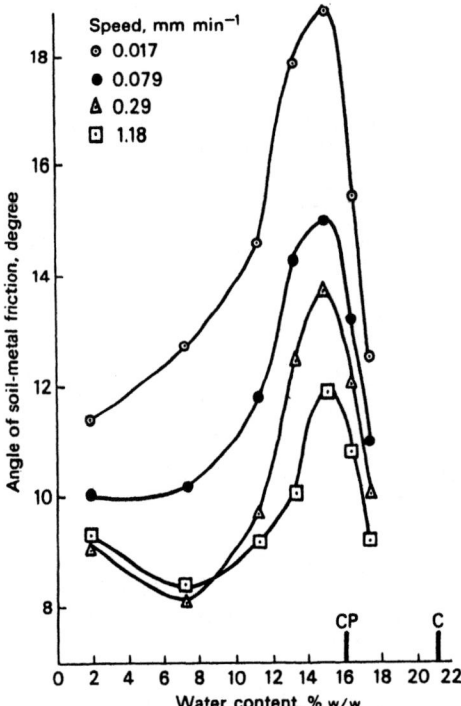

FIG. 7 The variation of soil-metal friction with water content at each of four sliding speeds for a sandy clay loam in relation to the cone penetrometer (CP) and Casagrande (C) plastic limits. (From Ref. 22.)

since the optimum water content for compaction decreases with increasing compactive effort [22]. Nevertheless, Bertilsson [91] found that the soil water content associated with the maximum slope of the virgin compression lines, for two of the four soils he studied, corresponded to the optimum water content for compaction. Similarly, Campbell et al. [22] found a maximum slope for the virgin compression lines of two soils at water contents lying between their Casagrande and cone penetrometer plastic limits. The water contents concerned were shown to correspond to the cone penetrometer "plastic limit" when this test was performed on intact aggregates of less than 10 mm diameter that had not been remolded. Since the maximum slope of the virgin compression line indicates the maximum susceptibility to compaction, it was suggested that a soil is much more likely to compact if subjected to tillage and traffic at water

contents close to the cone penetrometer "plastic limit," as determined on soil that has not been remolded but is in its natural state.

Since the plasticity index is related to the amount of clay in a
soil, it is not surprising that the compression characteristics have
been related to the plasticity index. Such relations can be empirical
[92], or developed, with the aid of critical state theory, from the
assumption that the strength at the plastic limit is 100 times that at
the liquid limit [52]. It has been suggested that although these relationships are not sufficiently precise to avoid the need to under-
take consolidation tests, they are adequate for preliminary design
purposes [92]. Furthermore, the ratio of the consolidation pressure
required to reduce soil water content to the plastic limit to that required to reduce it to the liquid limit has been shown to be a constant [30].

4. Water Regime

Uppal [93] determined the water release characteristics of nine remolded soils with plastic limits ranging from 17 to 34% w/w and found
that for all soils, the plastic limit corresponded to a matric potential
of -0.3 kPa on the wetting curve and -3 kPa on the drying curve.
His work was extended by Livneh et al. [94] to include a range of
bulk densities and water contents, and they found the plastic limit
to be in the range -6 to -60 kPa on the drying curve. However,
the latter investigators also concluded that the plasticity index was
more closely correlated with the logarithm of matric potential than
was the plastic limit. Rather higher values of -13 to -100 kPa were
found for the plastic limit on a drying curve by Stakman and Bishay
[95], who considered those of Uppal [93] to be unacceptably low in
relation to the thickness of the water layers surrounding soil particles at the plastic limit. Their results also confirmed those of Livneh
et al. [94], who found a linear relation between the logarithm of plasticity index and the logarithm of the corresponding matric potential
on the drying curve.

The value of field capacity relative to the plastic limit can affect
the behavior of a soil during cultivation. Where the plastic limit is
smaller than field capacity, the soil structure will be readily damaged
when worked at water contents between the plastic limit and field
capacity. A soil for which the plastic limit is greater than field capacity will have good workability. Similarly, susceptibility to slaking, which generally occurs above the liquid limit, depends on the
relative values of field capacity and the liquid limit [96]. Archer
[8] found that field capacity was close to and generally slightly
greater than the plastic limit for four contrasting soil textures (Fig.
8).

Saturated hydraulic conductivity (permeability) generally increases with liquid limit [27]. Sewell and Mote [97] made use of a

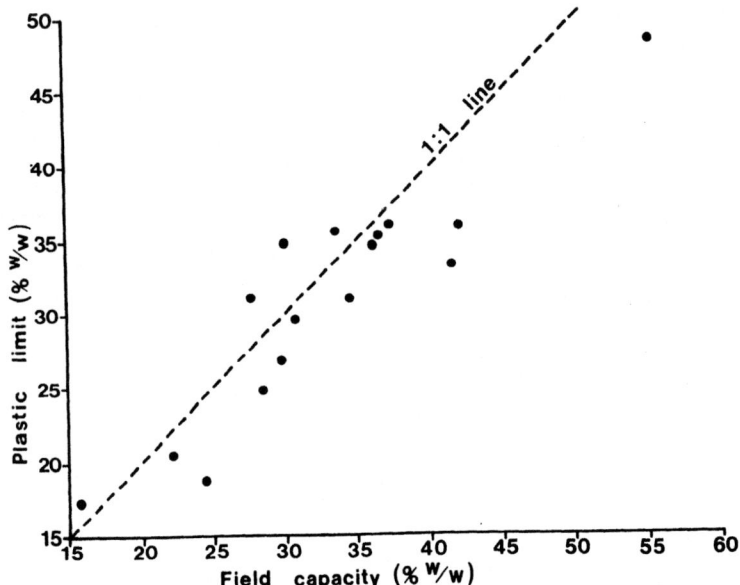

FIG. 8 The relation between plastic limit and field capacity for 16 soils. (Based on data from Ref. 8.)

relation between the logarithm of saturated hydraulic conductivity and the liquid limit to determine the effectiveness of various chemicals for sealing ponds without the necessity of making large numbers of conductivity measurements. Similarly, Carrier and Beckman [92] considered such simple correlations to be satisfactory for preliminary engineering design purposes.

5. Strength

Many researchers have reported empirical relationships between the plasticity index and either the shear strength [21,98], cohesion [99], or angle of internal friction of a soil [75,99,100]. Furthermore, the variation of cohesion with water content has been shown to have a maximum at the plastic limit [21], especially when the latter is determined with a cone penetrometer [22]. Wroth and Wood [52] suggested that the plastic limit be defined as that water content at which the soil has 100 times the strength it possesses at the liquid limit. On the assumption that all soils have the same strength at the liquid limit, they went on to use critical state soil mechanics theory to show that estimates of undrained shear strength depended only on the liquidity index of the soil.

V. SUMMARY

Plasticity is the property that allows a soil to be deformed without cracking in response to an applied stress. Such behavior can occur over a range of soil water contents, with the upper and lower limits of the range being referred to as the liquid and plastic limits, respectively.

The cohesive and adhesive forces associated with soil water and, especially, their variation with water content, determine whether a soil will, when stressed, undergo brittle failure, plastic flow, or viscous flow. At the plastic limit, the amount of water is just sufficient to surround each soil particle with a water layer that permits the laminar particles to slide over one another under stress and remain in their new positions when the stress is removed. At the liquid limit, the water layers between particles are thick enough to allow viscous flow to occur in response to an applied stress.

Dry soil to which water is added during continuous remolding to reach either the liquid or plastic limit is said to be in the critical state in terms of the critical state theory of soil mechanics. This theory describes the stress–strain behavior of any soil in relation to the three-dimensional relationship of spherical pressure, deviatoric stress, and specific volume. All points on the critical state line within this relationship correspond to states in which the soil can be continuously remolded without any change in volume.

The liquid limit has traditionally been determined with the Casagrande apparatus, but more recently a drop-cone test has become the preferred British Standard method.

The plastic limit is defined in the traditional method, which is still the British Standard method, as that water content at which a thread of soil, rolled between the fingertips of the operator and a flat glass plate, just crumbles when the thread reaches a diameter of 3 mm. More recently there have been attempts to redefine the plastic limit using tests based on the drop-cone apparatus. One proposal is that the minimum of the penetration-water content relation corresponds to the plastic limit. It has also been suggested that the plastic limit be defined such that the undrained shear strength of the soil at the plastic limit is 100 times that at the liquid limit.

Other methods of measuring liquid and plastic limits that have been proposed depend on either a correlation with other soil properties or a revision of the definitions of the limits so that they are more related to soil behavior.

The liquid and plastic limits have been widely used in soil engineering for soil classification because the limits are correlated with other important soil physical and mechanical properties. A possible objection to the tests as far as soils used in agriculture are concerned is that remolded soil is used. Nevertheless, the limits may provide a quicker, cheaper, or easier indication of other properties than their direct measurement where no great precision is required.

REFERENCES

1. Atterberg, A., Die plastizität der tone, *Int. Mitt. Bodenkd.*, 1: 10-43 (1911).
2. Atterberg, A., Die konsistenz und die bindigkeit der boden, *Int. Mitt. Bodenkd.*, 2: 149-189 (1912).
3. British Standards Institution, *Methods of Test for Soils for Civil Engineering Purposes*, BS 1377, British Standards Institution, London, 1975.
4. Dumbleton, M. J., *The Classification and Description of Soils for Engineering Purposes: A Suggested Revision of the British System*, Rep. No. LR182, Transport and Road Res. Lab., Crowthorne, U.K., 1968.
5. Casagrande, A., Research on the Atterberg limits of soils, *Public Roads*, 13: 121-130 (1932).
6. Sherwood, P. T., *The Reproducibility of the Results of Soil Classification and Compaction Tests*, Rep. No. LR339, Transport and Road Res. Lab., Crowthorne, U.K., 1970.
7. Sowers, G. F., A. Vesic, and M. Grandolfi, Penetration tests for liquid limit, ASTM Spec. Tech. Publ. No. 254: 216-226 (1968).
8. Archer, J. R., Soil consistency, in *Soil Physical Conditions and Crop Production*, Tech. Bull. No. 29, Ministry of Agriculture, Fisheries and Food, HMSO, London, 1975, pp. 289-297.
9. Baver, L. D., W. H. Gardner, and W. R. Gardner, *Soil Physics*, Wiley, New York, 1972.
10. Campbell, D. J., The occurrence and prediction of clods in potato ridges in relation to soil physical properties, *J. Soil Sci.*, 27: 1-9 (1976).
11. Soane, B. D., D. J. Campbell, and S. M. Herkes, The characterization of some Scottish arable topsoils by agricultural and engineering methods, *J. Soil Sci.*, 23: 93-104 (1972).
12. *Soil Textures*, Leaflet No. 895, Ministry of Agriculture, Fisheries and Food, London, 1984.
13. Schofield, A., and P. Wroth, *Critical State Soil Mechanics*, McGraw-Hill, London, 1968.
14. Kurtay, T., and A. R. Reece, Plasticity theory and critical state soil mechanics, *J. Terramech.*, 7: 23-56 (1970).
15. Haines, W. B., Studies in the physical properties of soils: II. A note on the cohesion developed by capillary forces in an ideal soil, *J. Agric. Sci., Camb.*, 15: 529-535 (1925).

16. Norton, F. H., Fundamental study of clay: VIII. A new theory for the plasticity of clay-water masses, *J. Am. Ceram. Soc.*, 31: 236-241 (1948).

17. Schwartz, B., Fundamental study of clay: XII. A note on the effect of surface tension of water on the plasticity of clay, *J. Am. Ceram. Soc.*, 35: 41-43 (1952).

18. Kingery, W. D., and J. Francl, Fundamental study of clay: XIII. Drying behavior and plastic properties, *J. Am. Ceram. Soc.*, 37: 596-602 (1954).

19. Vomocil, J. A., and L. J. Waldron, The effect of moisture content on tensile strength of unsaturated glass bead systems, *Soil Sci. Soc. Am. Proc.*, 26: 409-412 (1962).

20. Nichols, M. L., The dynamic properties of soil: I. An explanation of the dynamic properties of soils by means of colloidal films, *Agric. Eng.*, 12: 259-264 (1931).

21. Nichols, M. L., The dynamic properties of soils: III. Shear values of uncemented soils, *Agric. Eng.*, 13: 201-204 (1932).

22. Campbell, D. J., J. V. Stafford, and P. S. Blackwell, The plastic limit, as determined by the drop-cone test, in relation to the mechanical behaviour of soil, *J. Soil Sci.*, 31: 11-24 (1980).

23. Baver, L. D., The Atterberg consistency constants: Factors affecting their values and a new concept of their significance, *J. Am. Soc. Agron.*, 22: 935-948 (1930).

24. Sowers, G. F., Consistency, in *Methods of Soil Analysis*, Part 1 (C. A. Black, Ed.-in-Chief), Am. Soc. Agron., Madison, WI 1965, pp. 391-399.

25. Grim, R. E., Some fundamental factors influencing the properties of soil materials, in *Proc. 2nd Int. Conf. Soil Mechanics and Foundation Engineering*, Vol. 3, 1948, pp. 8-12.

26. Sridharan, A., S. M. Rao, and N. S. Murthy, Liquid limit of montmorillonite soils, *Geotech. Testing J.*, 9: 156-159 (1986).

27. *Soil Mechanics for Road Engineers*, HMSO, London, 1964.

28. Atkinson, J. H., and P. L. Bransby, *The Mechanics of Soils*, McGraw-Hill, Maidenhead, U.K., 1978.

29. Skempton, A. W., and R. D. Northey, The sensitivity of clays, *Géotechnique*, 3: 30-53 (1953).

30. Nagaraj, T. S., and M. S. Jayadeva, Re-examination of one-point methods of liquid limit determination, *Géotechnique*, 31: 413-425 (1981).

31. Dumbleton, M. J., and G. West, *The Influence of the Coarse Fraction on the Plastic Properties of Clay Soils*, Rep. No. LR36, Transport Road Res. Lab., Crowthorne, U.K., 1966.

32. Casagrande, A., Notes on the design of the liquid limit device, *Géotechnique*, 8: 84-91 (1958).

33. Sherwood, P. T., and M. D. Ryley, *An Examination of Cone-Penetrometer Methods for Determining the Liquid Limit of Soils*, Rep. No. LR233, Transport Road Res. Lab., Crowthorne, U. K., 1968.

34. *Determination rapide des limites d'Atterberg à l'aide d'un pénétromètre et d'un picnomètre d'air*, Dossier No. SGR/149, Laboratoire Central des Ponts et Chaussées, Paris, 1966.

35. Clayton, C. I., and A. W. Jukes, A one-point cone penetrometer liquid limit test? *Géotechnique*, 28: 469-472 (1978).

36. Sherwood, P. T., and M. D. Ryley, *An Investigation of Alternative Methods of Determining the Plastic Limit of Soils*, Tech. Note No. TN536, Transport Road Res. Lab., Crowthorne, U.K., 1970.

37. Sherwood, P. T., and M. D. Ryley, An investigation of a cone penetrometer method for the determination of the liquid limit, *Géotechnique*, 20: 203-208 (1970).

38. Stefanov, G., Discussion on liquid limit, in *Proc. 4th Int. Conf. Soil Mechanics and Foundation Engineering*, Vol. 1, 1958, p. 97.

39. Scherrer, H. U., Determination of liquid limit by the static cone penetration test, in *Proc. 5th Int. Conf. Soil Mechanics and Foundation Engineering*, Vol. 1, 1961, pp. 319-322.

40. Karlsson, R., Suggested improvements in the liquid limit test, with reference to flow properties of remoulded clays, in *Proc. 5th Int. Conf. Soil Mechanics and Foundation Engineering*, Vol. 1, 1961, pp. 171-184.

41. Towner, G. D., A note on the plasticity limits of agricultural soils, *J. Soil Sci.*, 25: 307-309 (1974).

42. Campbell, D. J., Liquid limit determination of arable topsoils using a drop-cone penetrometer, *J. Soil Sci.*, 26: 234-240 (1975).

43. Wires, K. C., The Casagrande method versus the drop-cone penetrometer method for the determination of liquid limit, *Can. J. Soil Sci.*, 64: 297-300 (1984).

44. Littleton, I., and M. Farmilo, Some observations on liquid limit values with reference to penetration and Casagrande tests, *Ground Eng.*, 10: 39-40 (1977).

45. Queiroz de Carvalho, J. B., The applicability of the cone penetrometer to determine the liquid limit of lateritic soils, *Géotechnique*, 36: 109-111 (1986).

46. Sivapullaiah, P. V., and A. Sridharan, Liquid limit of soil mixtures, *Geotech. Testing J.*, 8: 111-116 (1985).

47. Moon, C. F., and K. B. White, A comparison of liquid limit test results, *Géotechnique*, 35: 59-60 (1985).

48. Towner, G. D., An examination of the fall-cone method for the determination of the strength properties of remoulded agricultural soils, *J. Soil Sci.*, 24: 470-479 (1973).

49. Campbell, D. J., Plastic limit determination using a drop-cone penetrometer, *J. Soil Sci.*, 27: 295-300 (1976).
50. Allbrook, R. F., The drop-cone penetrometer method for determining Atterberg limits, *N.Z. J. Sci.*, 23: 93-97 (1980).
51. Wood, D. M., and C. P. Wroth, The use of the cone penetrometer to determine the plastic limit of soils, *Ground Eng.*, 11: 37 (1978).
52. Wroth, C. P., and D. M. Wood, The correlation of index properties with some basic engineering properties of soils, *Can. Geotech. J.*, 15: 137-145 (1978).
53. Faure, A., A new conception of the plastic and liquid limits of clay, *Soil Tillage Res.*, 1: 97-105 (1981).
54. Russell, E. R., and J. L. Mickle, Liquid limit values by soil moisture tension, *J. Soil Mech. Found. Eng. Am. Soc. Civ. Eng.*, 96: 967-989 (1970).
55. Yasutomi, R., and S. Sudo, A method of measuring some physical properties of soil with a forced oscillation viscometer, *Soil Sci.*, 104: 336-341 (1967).
56. Hajela, R. B., and J. M. Bhatnagar, Application of rheological measurements to determine liquid limit of soils, *Soil Sci.*, 114: 122-130 (1972).
57. Vasilev, Y. M., Rapid determination of the limit of rolling out, *Pochvovedenie*, 7: 105-106 (1964).
58. Skopek, J., and G. Ter-Stephanian, Comparison of liquid limit values determined according to Casagrande and Vasilev, *Géotechnique*, 25: 135-136 (1975).
59. Armstrong, J. C., and T. M. Petry, Significance of specimen preparation upon soil plasticity, *Geotech. Testing J.*, 9: 147-153 (1986).
60. Campbell, D. J., and R. Hunter, Drop-cone penetration in situ and on minimally disturbed soil cores, *J. Soil Sci.*, 37: 153-163 (1986).
61. Gradwell, M., and K. S. Birrell, Physical properties of certain volcanic clays, *N.Z. J. Sci. Technol.*, B36: 108-122 (1954).
62. Davidson, D. A., Problems in the determination of plastic and liquid limits of remoulded soils using a drop-cone penetrometer, *Earth Surface Processes Landforms*, 8: 171-175 (1983).
63. Coleman, J. D., D. M. Farrar, and A. D. Marsh, The moisture characteristics, composition and structural analysis of a red clay soil from Nyeri, Kenya, *Géotechnique*, 14: 262-276 (1964).
64. Sherwood, P. T., Classification tests on African red clays and Keuper Marl, *Quart. J. Eng. Geol.*, 1: 47-55 (1967).
65. Pringle, J., The assessment and significance of aggregate stability in soil, in Ministry of Agriculture, Fisheries and Food Tech. Bull. No. 29, *Soil Physical Conditions and Crop Production*, HMSO, London, 1975, pp. 249-260.

66. Blackmore, A. V., Subplasticity in Australian soils: IV. Plasticity and structure related to clay cementation, *Aust. J. Soil Res.*, 14: 261-272 (1976).

67. Houlsby, G. T., Theoretical analysis of the fall cone test, *Géotechnique*, 32: 111-118 (1982).

68. Budhu, M., The effect of clay content on liquid limit from a fall cone and the British cup device, *Geotech. Testing J.*, 8: 91-95 (1985).

69. Youssef, M. S., A. Sabry, and A. H. El Rami, Temperature changes and their effects on some physical properties of soils, in *Proc. 5th Int. Conf. Soil Mechanics and Foundation Engineering*, Vol. 1, 1961, pp. 419-421.

70. Ballard, G. E. H., and W. F. Weeks, The human factor in determining the plastic limit of cohesive soils, *Mater. Res. Stand.*, 3: 726-729 (1963).

71. Gay, C. W., and W. Kaiser, Mechanisation for remolding fine grained soils and for the plastic limit test, *J. Testing Eval.*, 1: 317-318 (1973).

72. Casagrande, A., Classification and identification of soils, *Proc. Am. Soc. Civ. Eng.*, 73: 783-810 (1947).

73. Tentative method for classification of soils for engineering purposes, in *Book of ASTM Standards*, Am. Soc. Testing Mater., Philadelphia, 1966, pp. 766-771.

74. De La Rosa, D., Relation of several pedological characteristics to engineering qualities of soil, *J. Soil Sci.*, 30: 793-799 (1979).

75. Humphreys, J. D., Some empirical relationships between drained friction angles, mechanical analyses and Atterberg limits of natural soils at Kainji Dam, Nigeria, *Géotechnique*, 25: 581-585 (1975).

76. Yong, R. N., and B. P. Warkentin, in *Developments in Geotechnical Engineering*, Vol. 5, *Soil Properties and Behaviour*, Elsevier, Amsterdam, 1975, pp. 62-68.

77. Mulqueen, J., Plasticity characteristics of some carboniferous clay soils in north central Ireland and their significance, *Ir. J. Agric. Res.*, 15: 129-135 (1976).

78. Odell, R. T., T. H. Thornburn, and L. J. McKenzie, Relationships of Atterberg limits to some other properties of Illinois soils, *Soil Sci. Soc. Am. Proc.*, 24: 297-300 (1960).

79. Sherwood, P. T., and B. G. Hollis, *Studies of Keuper Marl: Chemical Properties and Classification Tests*, Rep. No. LR41, Transport Road Res. Lab., Crowthorne, U.K., 1966.

80. Baver, L. D., The relation of exchangeable cations to the physical properties of soils, *J. Am. Soc. Agron.*, 20: 921-941 (1928).

81. Farrar, D. M., and J. D. Coleman, The correlation of surface area with other properties of nineteen British clay soils, *J. Soil Sci.*, 18: 118-124 (1967).

82. Hammel, J. E., M. E. Sumner, and J. Burema, Atterberg limits as indices of external surface areas of soils, *Soil Sci. Soc. Am. J.*, 47: 1054-1056 (1983).
83. Hendershot, W. H., and M. A. Carson, Changes in the plasticity of a sample of Champlain clay after selective chemical dissolution to remove amorphous material, *Can. Geotech. J.*, 15: 609-616 (1978).
84. McNabb, D. H., Correlation of soil plasticity with amorphous clay constituents, *Soil Sci. Soc. Am. J.*, 43: 613-616 (1979).
85. Russell, J. C., and F. M. Wehr, The Atterberg consistency constants, *J. Am. Soc. Agron.*, 20: 354-372 (1922).
86. Spoor, G., Fundamental aspects of cultivation, in Ministry of Agriculture Fisheries and Food Tech. Bull. No. 29, *Soil Physical Conditions and Crop Production*, HMSO, London, 1975, pp. 128-144.
87. Smith, N., Let's have traffic congestion in the potato field, *Farm Mech.*, 14: 137 (1962).
88. Nichols, M. L., Dynamic properties of soil affecting implement design, *Agric. Eng.*, 11: 201-204 (1930).
89. Koenigs, F. F. R., The puddling of clay soils, *Neth. J. Agric. Sci.*, 11: 145-156 (1963).
90. Weaver, H. A. and V. C. Jamison, Effects of moisture on tractor tire compaction of soil, *Soil Sci.*, 71: 15-23 (1951).
91. Bertilsson, G., Topsoil reaction to mechanical pressure, *Swed. J. Agric. Res.*, 1: 179-189 (1971).
92. Carrier, W. D., and J. F. Beckman, Correlations between index tests and the properties of remoulded clays, *Géotechnique*, 34: 211-228 (1984).
93. Uppal, H. L., A scientific explanation of the plastic limit of soils, *J. Mater.*, 1: 164-178 (1966).
94. Livneh, M., J. Kinsky, and D. Zaslavsky, Correlation of suction curves with the plasticity index of soils, *J. Mater.*, 5: 209-220 (1970).
95. Stakman, W. P., and B. G. Bishay, Moisture retention and plasticity of highly calcareous soils in Egypt, *Neth. J. Agric. Sci.*, 24: 43-57 (1976).
96. Boekel, P., The effect of organic matter on the structure of clay soils, *Neth. J. Agric. Sci.*, 11: 250-263 (1963).
97. Sewell, J. I., and C. R. Mote, Liquid-limit determination for indicating effectiveness of chemicals in pond sealing, *Trans. Am. Soc. Agric. Eng.*, 50: 611-613 (1969).
98. Voight, B., Correlation between Atterberg plasticity limits and residual shear strength of natural soils, *Géotechnique*, 23: 265-267 (1973).

99. Gibson, R. E., Experimental determination of the true cohesion and true angle of internal friction in clays, in *Proc. 3rd Int. Conf. Soil Mechanics and Foundation Engineering*, Vol. 1, 1953, pp. 126-130.

100. Kanji, M. A., The relationship between drained friction angles and Atterberg limits of natural soils, *Géotechnique*, 24: 671-674 (1974).

9

The Cone Penetrometer in Relation to Trafficability, Compaction, and Tillage

DONALD J. CAMPBELL and MICHAEL F. O'SULLIVAN *Scottish Centre of Agricultural Engineering, Penicuik, Midlothian, Scotland*

I. INTRODUCTION

Penetrometers include any device that can be forced into the soil and for which resistance to penetration can be measured. A wide variety of such instruments has been developed to measure either static penetration resistance (when the penetrometer is pushed into the soil at a constant rate) or dynamic penetration resistance (when it is driven by a series of blows). The penetration resistance so measured has been used as an index of a wide range of soil physical and mechanical properties in empirical soil studies in civil engineering, vehicle mobility, and agricultural soil mechanics. Penetration resistance will vary not only with the size and shape of the probe that penetrates the soil but also with a range of soil properties, including soil-metal friction, particle size distribution, water content, resistance to compression, and shear strength, which includes both internal friction and cohesion. Since the contribution of each of these properties to penetration resistance on a given occasion is usually unknown, care must be taken in the interpretation of penetration resistance measurements. Thus, while penetration resistance may be used as an empirical index of, for example, soil water content in an otherwise uniform soil (provided the device is calibrated in relation to water content for that soil), it cannot be used to compare water contents between soils.

Perhaps the most frequent use of penetrometers is to assess soil strength. Recognizing the need to standardize both the apparatus and test procedure where empirical tests are concerned, the

American Society of Agricultural Engineers have specified a penetrometer with a conical tip which has become widely used to give an
index of soil strength from a static penetration test [1]. The force
per unit base area of the cone required to push the cone into the
soil is considered to be an index of soil shear strength and is therefore referred to as the "cone index." The term "cone index" should
not be used to describe the result obtained with a nonstandard apparatus, but this is not always the case in practice. The terms
"cone resistance" and "penetration resistance," if considered as the
resistance to penetration of the soil by the cone, are, strictly,
forces. However, in practice the terms "cone resistance" and "penetration resistance" have been so widely used to refer to the force
per unit area of the cone base that this usage has been retained in
this chapter.

II. THEORY

A. Soil Penetration by Cones

Despite the wide use of cone penetrometers in soil studies, remarkably little work has been done on the theoretical basis of the operation. Farrell and Greacen [2], dissatisfied with the use of soil
mechanics theory of the bearing capacity of piles to explain cone
penetration, developed a new theory. Their treatment allowed not
only for soil cohesion and internal friction but also for soil compressibility, which is clearly important in agricultural soils but may be
irrelevant during pile driving in soft, saturated, cohesive soils.
Their theory is based on a model in which a spherical cavity is created in the soil at the penetrometer tip as penetration takes place.
A homogeneous, isotropic soil is assumed, with no allowance being
made for soil structure. The spherical cavity is created by two
types of soil deformation, each of which occurs in a distinct soil
zone. In the inner zone, the radius of which can be up to 6-10
times the radius of the probe, plastic failure occurs. Beyond this
zone is a second zone in which elastic compression occurs. The
total resistance to penetration is made up of a component associated
with formation of the cavity and a component due to soil-probe friction. The mean stress acting on the surface of the cone is assumed
to equal that required to form a cavity in the soil large enough to
accommodate the probe. The vertical component of that stress and
that due to soil-metal friction constitute the resistance to penetration of the soil by the probe.

   Assuming that any frictional resistance between the shaft of the
penetrometer and the soil can be neglected, the normal point resistance P_n (i.e., the soil resistance normal to the cone face) is given
by

$$P_n = \frac{P_r}{1 + \tan \delta \, \cot(\alpha/2)} \qquad (1)$$

where P_r is the total point resistance or cone resistance, δ is the angle of soil-metal friction, and α is the included angle of the cone. During penetration, the resultant stress on the soil is directed at an angle δ below the normal to the cone surface. The magnitude of this stress, the specific resistance P_s [3], is given by

$$P_s = \frac{P_n}{\cos \delta} \qquad (2)$$

Root extension rate has been found to correlate more closely with normal point resistance than with cone resistance [4]. Specific resistance was found to be better than cone resistance when predicting cultivation tine performance in a fine-grained soil, although cone resistance was the better predictor in a coarse-grained soil [3].

When a blunt probe is pushed into soil, a cone of soil builds up on the probe tip and subsequently acts as part of the probe. Koolen and Kuipers [5] have stated that this occurs once the cone angle exceeds $90° - \phi$, where ϕ is the angle of internal friction of the soil. This fact was used by Farrell and Greacen [2] to show the importance of the friction component of resistance to penetration. When a steel cone was rotated as it was pushed into the soil, there was a reduction of approximately 50% in resistance to penetration. No such reduction was found for a blunt probe, since the cone of soil did not rotate.

For three soils at three bulk densities and two water contents, experimentally determined resistances to penetration for a 3 mm diameter, 60° cone were found to be an average of 9% greater than the calculated values [2]. Much of the error was attributed to errors in measuring the compression index during rebound, which is one of the five soil parameters used in the calculations [2].

Since the theory assumes that the probe is surrounded by soil, prediction of resistance to penetration during the initial stages of penetration cannot be achieved. At that stage, the failure mode includes upheaval of the surface soil. In tests on sand, it has been shown that this effect ceases only when a penetration depth of about six times the cone height is reached [6].

In a further development of the theory [7], it was shown that for small cone angles, soil movement corresponds to the formation of a cylindrical rather than a spherical cavity. In tests with 60° and 10° cones, measurements of soil bulk density in the vicinity of the holes formed by the cones showed that for the 60° cone, density was higher at the edge of the hole and decreased more rapidly with

distance from the hole than it did for the 10° cone. From these results, together with those of tests on the effect on penetration resistance of the diameter of confined cores, it was concluded that
soil movement lay between the ideals of spherical and cylindrical
compression but, with small cone angles, was much closer to the
latter. Although both the theoretical and measured resistances to
penetration for the 10° cone were always higher than for the 60°
cone, the difference was usually small. In studies of the effect of
cone angle on cone index, Gill [3] found that cone index decreased
with cone angle until a minimum was reached, after which cone index increased with decreasing cone angle. The effect was attributed to the large frictional forces on the surface of sharp cones.

Although the relative proportions of shear, compressive, and
tensile strength that contribute to cone resistance vary with soil
water content [8], in saturated, fine-grained soils where the angle
of friction is zero, cone resistance increases with cohesion C [9].

$$P_r = aC \qquad\qquad\qquad (3)$$

The value of $\underline{a}$ may not be constant, but is likely to vary with soil
type between 10 and 20 [10]. Freitag [9] found a = 12.5, whereas
Reece and Peca [10] found a = 11. In air-dry sand, cone resistance
is empirically related to bulk density and angle of internal friction
[9], but the relationships differ between sands.

Steinhardt [11] deduced from soil mechanics theory that the increase in cone resistance with decreasing matric potential should be
approximately linear in wet soils. Such a relationship has been
found experimentally for a range of soils at matric potentials above
about -15 kPa [12-14]. The change in cone resistance with change
in matric potential, when both are measured in the same units, varies between about 20 and 50, depending on soil type and measurement depth [12,13]. It was greater in a cultivated silty clay loam
(53) than in the same soil under grass (47) and, in a clay loam,
was greater in unplowed land (40) than in plowed land (23) [12].

B. Relationships with Other Soil Properties

The soil strength parameters that determine cone resistance vary
with soil bulk density, water content, and structure, and with soil
type. The important soil type variables may include texture, particle shape, clay mineralogy, amorphous oxide content, organic matter content, stone content, and the chemical composition of the soil
solution [15-18]. Thus, relationships established using dried,
ground, sieved and rewetted samples may give some insights into
the mechanics of penetration but cannot be assumed to reflect relationships in the field.

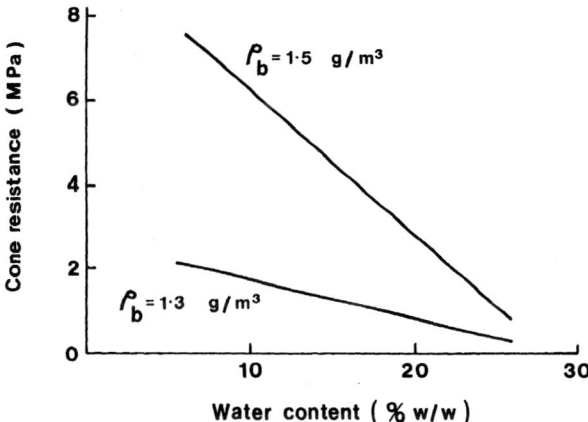

FIG. 1 Variation of cone resistance with water content at high and low bulk densities. (Based on data from Ref. 20.)

In field soils, cone resistance decreases with increasing soil water content or matric potential and increases with bulk density. Matric potential and volumetric water content may change during penetration because of soil compression and particle rearrangement, whereas gravimetric water content remains constant. Thus, gravimetric water content is probably a more useful choice as independent variable than matric potential or volumetric water content [5]. The rate of change of cone resistance with water content is less at low than at high bulk density, whereas the rate of change with bulk density is greater at low than at high water contents. Different functions have been proposed to account for these relationships [19]. For a given soil, the simplest suitable function is

$$P_r = A + Bm + C\rho_b + D\rho_b m \qquad (4)$$

where m is gravimetric water content, ρ_b is bulk density, and A, B, C, and D are empirical constants [20]. This relationship is illustrated in Fig. 1, using values of the constants for a loess soil, and appears to be widely applicable [21,22]. In some soils, however, the changes in cone resistance with bulk density and water content are not linear: cone resistance changes most rapidly at high bulk densities and low water contents [19]. The linear model (Eq. 4) may still be appropriate if the ranges of bulk density and water content are small or soil variability is high, but a more generally valid model may be

$$P_r = Am^B (\rho_b)^C \qquad (5)$$

where A, B, and C are empirical constants. The main problem in deriving and applying such empirical relationships is that strength changes with time, even if bulk density and water content remain constant [23]. Differences in structure associated with soil management may also limit the applicability of any empirical relationships [24].

In remolded soils, which have been dried, ground, rewetted to a range of water contents, and compacted to different bulk densities, cone resistance may increase with water content to a maximum and, thereafter, decrease [25,26]. The maximum cone resistance tends to occur at a lower water content than the maximum bulk density. This discrepancy is attributed to changes in structure with increasing compaction [9].

At constant water content and bulk density, cone resistance tends to increase with decreasing soil particle size [18,27]. At low or moderate organic matter contents, soil strength tends to increase with organic matter [18]. van Wijk and Beuving [28] compacted artificial mixtures of sand, clay, and organic matter and observed that cone resistance increased more rapidly with bulk density and decreasing matric potential at higher organic matter contents.

The shape of sand particles may influence cone resistance [17] through their influence on both soil internal friction and soil-metal friction. Effective stress is related to matric potential and the pore size distribution, which may, therefore, influence cone resistance [16,29,30]. Calcium carbonate encourages soil aggregation while an excess of sodium leads to slaking. Thus, cone resistance was found to increase with calcium carbonate content in fine-grained soils [18], while breaking strength of soil briquettes increased with sodium content because dispersion led to closer packing [15].

III. PENETROMETER DESIGN

Many users have designed and built their own penetrometers for specific purposes. Details of a selection of commercially available penetrometers are given in Table 1.

A. Cone and Shaft

Cone resistance tends to decrease with increasing cone angle to a minimum at about 20-40° and, thereafter, to increase [3,4,31]. The high cone resistance at small cone angles is associated with high soil-metal friction and at large cone angles with soil compaction and high strength ahead of the cone [3,8]. Figure 2, which was derived from measurements made in 67 agricultural fields [31], shows the relationship between cone resistance and cone angle for a fixed cone base area. Specific resistance (Eq. 2), however, increases with cone

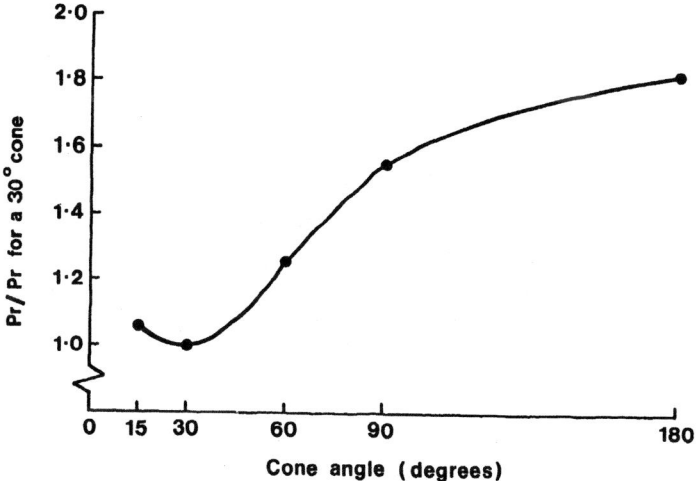

FIG. 2 Variation of cone resistance with cone angle for a fixed cone base area. (Reproduced with permission from the *Journal of Agricultural Engineering Research* [31].)

FIG. 3 The Farnell penetrometer, in which the spring used to measure insertion force is calibrated in terms of cone index and California bearing ratio, both of which are indicated on rotating drums.

TABLE 1 Details of Some Commercially Available Cone Penetrometers

Supplier	Model	Cone angle	Cone base diameter (mm)
Wykeham Farrance Engineering Ltd., Weston Road, Slough, Berkshire SL1 4HW, U.K.	Vicksberg	30°	15.9, 9.5
Findlay Irvine Ltd., Bog Road, Penicuik, Midlothian, U.K.	Bush	30°	12.8, 20.3
Soil Test Inc., 2205 Lee St., Evanston, IL 60202	Proving ring penetrometer	30°	28.4
	Corps of Engineers cone penetrometer	30°	20.3
Eijkelkamp BV, Nijverheidsstraat 14, 6987 EM Giesbeek, The Netherlands	Penetrometer Eijkelkamp	60°	11.3-35.7
	Penetrograph Stiboka	60°	11.3-25.2
	Pezograph	30° 60°	8.0-20.3 11.3, 16.0
Leonard Farnell & Co. Ltd., North Mymms, Hatfield, Hertfordshire AL9 7SR, U.K.	Cone penetrometer	30°	12.8, 20.3
Werner Gloor CH-3294, Buren, a.A., Switzerland	Penetrograph	30°	11.3, 13.8, 16.0
Rimik Pty Ltd, 4 Mabel St., Toowoomba, Queensland 4350, Australia	CP10	30°	12.8 20.3

Shaft diameter (mm)	Force measurement	Data recording	Approx. 1987 price ($)	Special features
8.7	Proving ring	Manual	580	Shaft scribed at 76 mm intervals
9.5	Electronic transducer	Digital recording at 15 preset depths	3910	Optional data storage unit for 500 penetrations; depth increment variable from 10 mm minimum
19.0	Proving ring	Manual	625	Shaft scribed at 152 mm intervals
15.9	Proving romg	Manual	450	
11.0-34.7	Hydraulic pressure gauge	Manual	470	Maximum reading indicator
11.0-24.6	Spring	Chart recorder	1170	
6, 10	Spring	Chart recorder	4820	Also measures shear strength
9.5	Spring	Manual	540	Calibrated in cone index and CBR units; shaft scribed at 75 mm intervals
	Spring	Chart recorder	3240	Hydraulically damped
9.5 15.9	Electronic transducer	Digital recording at 30 preset depths and storage for 300 penetrations	3180	Monitors penetration rates; additional storage available for a total of 750 penetrations; depth increment 15 mm

angle [3]. Soil tends to be displaced laterally at small cone angles, whereas the direction of displacement becomes more vertical with increasing cone angles [3,32]. Lateral soil displacement relates more closely to the mechanics of root growth than does the more vertical displacement produced by relatively blunt cones [7], and results obtained from sharp penetrometers correlate better with root extension rate than results from blunt penetrometers [4]. On the other hand, the load-bearing characteristics of the soil may be more closely related to soil resistance to blunt cones. Commercially available cone penetrometers (see, e.g., Fig. 3) tend to be fitted with 30° or 60° cones, but cones are easily changed by the user.

Although cone resistance is expressed as force per unit base area, it tends to increase with decreasing base area [9]. The cone penetrometer standard of the American Society of Agricultural Engineers [1] allows cone base areas of 320 mm^2 (0.5 in.^{2}) and 130 mm^2 (0.2 in.^{2}), both with a 30° angle. A 3% decrease in diameter is allowed for cone wear. In Europe, cones of 100 mm^2 base area are common, but cones with base areas up to 500 mm^2 have been used. Needle penetrometers, used in root growth studies, may have base areas of a few square millimeters only. Whiteley and Dexter [33] examined the effect on cone resistance of penetrometer diameters between 1 and 2 mm for a cone angle of 60°. They found that

$$P_r = b\left(1 + \frac{x}{d}\right)^2 \tag{6}$$

where d is diameter and x and b are constants. A large (d >> x) cone meets a resistance b, and x is a measure of the extra effective diameter of the cone. The constant x varied in the 0-0.7 mm range with soil type, water content, and structure. Equation 6 indicates that the effect of cone size decreases with increasing diameter, which is consistent with the results of other work using larger cones [9,34, 35]. Assuming a value of 0.6 mm for x, Eq. 6 indicates that cone resistances measured with 100, 130, 320, and 500 mm^2 cones are in the ratio 1.01:1.00:0.97:0.96, which is broadly in agreement with the data of Freitag [9]. These variations are within the 5% associated with acceptable wear on standard cones [1]. Thus, cone diameter may be ignored when comparing results from different field penetrometers with the same cone angle and with diameters in the 11-25 mm range. With needle penetrometers, however, diameter is important, and this must be taken into account when comparing results from different instruments.

In soils with well-developed structure, the mechanism of penetration may differ between cones of different sizes. A cone with a small diameter, relative to the size of structural units, may penetrate aggregates or planes of weakness between aggregates, whereas a large cone will tend to penetrate or deform aggregates [36].

Adhesion of soil to the penetrometer shaft and friction between shaft and soil both act to increase cone resistance. If the shaft diameter is small, compared with that of the cone, movement of soil into the cavity behind the cone may decrease cone resistance by relieving the pressure on the cone face [9]. Interaction between soil and shaft can have a significant effect on a standard ASAE penetrometer, especially in wet clay [8,9]. Freitag [9] found that the cone resistance of a clay was approximately constant with depth when a 320 mm^2 (20.3 mm diameter) cone was used with a small (9.5 mm) diameter shaft, but increased with depth below 130 mm to three times its original value at 300 mm depth when the standard (15.9 mm) diameter shaft was used (Fig. 4). An 8 mm diameter shaft was used with a 320 mm^2 cone by Reece and Peca [10] to eliminate the clay-shaft interaction: they suggested that the 8 mm shaft be adopted as standard. The difference in radius between the cone (6.4 mm) and the shaft (4.8 mm) on the smaller ASAE penetrometer is only 1.6 mm, decreasing to 1.5 mm at maximum acceptable cone wear. Thus, a reduction in standard shaft diameter may be desirable to avoid problems of soil-shaft interaction. However, a very small diameter shaft may not withstand the stresses imposed on it without bending.

Rough cones increase cone resistance [9], whereas lubrication of the cone decreases cone resistance by decreasing soil-cone friction, and reduces the downward component of soil movement ahead of the cone [3,32]. Specific resistance (Eq. 2), however, tends to be independent of cone finish [3]. The similarity in resulting soil

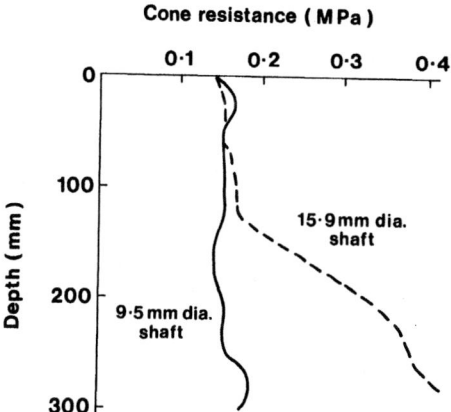

FIG. 4 Variation of cone resistance with depth for a 20.3 mm diameter cone, for two shaft diameters. (From Ref. 9.)

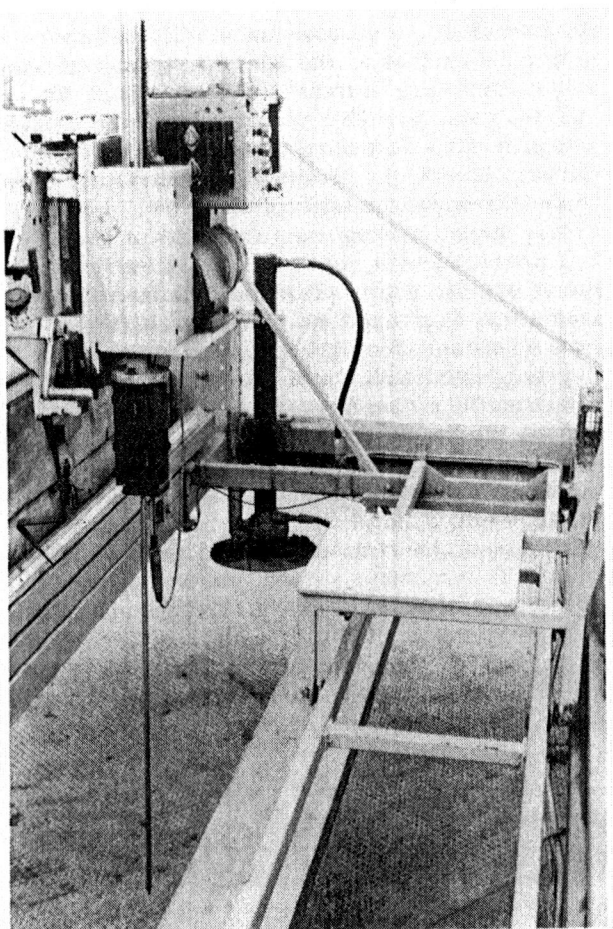

FIG. 5 A trailer-mounted, electrically-driven penetrometer. The
penetrometer assembly may be moved laterally by means of a rack
and pinion.

movement between that induced by a lubricated cone and by a growing root led Tollner and Verma [32] to advocate the use of a lubricated penetrometer in root growth studies. Such a penetrometer
may also have a use in studying soil-machine systems [37]. The
advantage of simplicity would be lost, however, and the mechanics
of penetration of a lubricated cone are as poorly understood as
those of one that is not lubricated. Cone surface roughness is not

an important factor in penetrometer design because abrasion by soil quickly removes irregularities.

B. Insertion and Measurement

Increasing penetration speed increases cone resistance in fine-grained soils [9] in which strength depends on strain rate [38]. Thus, a mechanically driven, constant-rate penetrometer would appear to have advantages over a manually inserted version. In most soils, however, cone resistance is fairly insensitive to penetration rate within the range expected from different operators aiming for the standard speed [1] of 30 mm s^{-1} [28,39,40]. Thus, the constant penetration rate possible with mechanically-driven penetrometers is not a significant advantage. Exceptions are saturated clay [34], where strain rate effects can be important [38], and soils with a strong layer overlying a weak layer. The large force required to penetrate the strong layer may cause an excessive penetration rate in the underlying layer.

A penetrometer may be mounted on a rack, which allows easy and accurate lateral movement (Fig. 5) [41,42]. This feature is advantageous when measurements are being made on a regular, closely spaced grid. On the other hand, hand-held penetrometers are more portable, and this can clearly be an advantage in some circumstances.

The device to measure force on a penetrometer must respond quickly to abrupt increases and decreases in force [40]. Thus, electronic transducers are more suitable than springs. Automatic recording of the force is also advantageous because accurate recording of force by the operator at a series of specific penetration depths during penetrometer insertion is difficult. A penetrometer that has an electronic transducer and automatic recording of the force is shown in Fig. 6. Analog recording allows rapid changes with depth to be detected and maxima and minima to be identified. The data must be digitized, however, for statistical analysis or input to mathematical models.

Digitization is tedious when data are recorded on some form of x-y plotter, and a fairly sophisticated computer is required to display and digitize electrically recorded analog signals [43,44]. Digital recording facilitates analysis of data, but maxima and minima will be missed on some or all penetrations, particularly when the depth increment between readings is large. This can be a disadvantage when cone resistance changes abruptly with depth or when the depth of a cultivation pan varies between penetrations. Averaging of data at predetermined depths may disguise the existence of such a variable cultivation pan, particularly if the depth increment is large relative to the thickness of the pan.

FIG. 6 The Bush electronic recording penetrometer and associated data storage unit.

IV. PROBLEMS IN USE

A. Variability

Cone resistance results are highly variable, even when measurements are made closely together [45]. This variability is influenced by cone size: small cones tend to produce more variable results than large cones [36]. Furthermore, the data may not be normally distributed. McIntyre and Tanner [46] and also Cassel and Nelson [47] found that a log transformation was necessary to normalize their

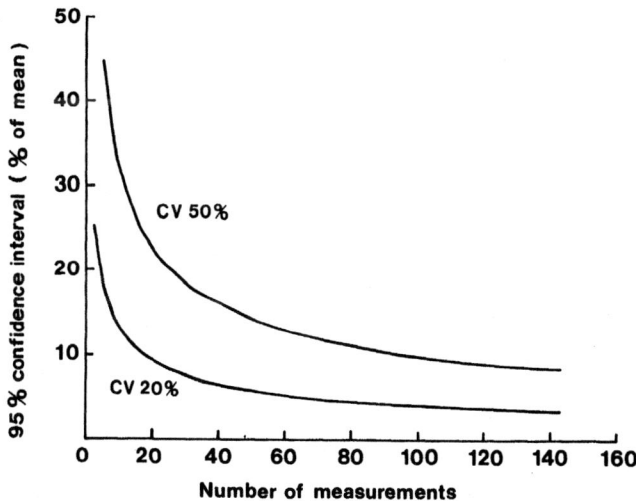

FIG. 7 Variation of the 95% confidence interval about the mean with the number of cone resistance observations, for two different coefficients of variation.

data, whereas Mitchell et al. [48] used a square-root transformation. Other workers have found that most of their data at individual depths were normally distributed [49,50]. Since the variance of cone resistance tends to increase with depth, a log transformation may be required if depth is included as a factor in analyzing results, even if the data at individual depths are normally distributed.

The number of measurements required depends on the required precision and on the coefficient of variation, which tends to be between 18 and 50% [49-52] but may be as large as 76% near the soil surface [49]. Assuming a normal distribution, the number of measurements required, N, may be estimated from

$$N = \left[\frac{2(CV)}{L}\right]^2 \tag{7}$$

where L is the 95% confidence interval, expressed as a percentage of the mean and CV is the coefficient of variation (%) [53]. This relationship is illustrated in Fig. 7 for representative values of CV for high and low variability. The size of the confidence interval L decreases (i.e., precision increases) with increasing numbers of measurements, but the rate of decrease declines at around 50 measurements when variability is low (20% CV) and at around 100 measurements when variability is high (50% CV). This result suggests that 50-100 measurements per treatment are required for maximum efficiency. If seven measurements were made, as recommended by ASAE [1], the 95% confidence interval would be between 15 and 38% of the mean. This contrasts with a maximum 5% error in individual penetrations associated with allowable cone wear [1]. The numbers of measurements required per treatment in field experiments may be less than indicated in Fig. 7 because results from all treatments and replications may be pooled to estimate error [53]. For example, Pidgeon and Soane [54] found that for adequate precision, 10 penetrations per plot were required in each of the eight replicates of four treatments in a tillage experiment.

The foregoing considerations of the number of penetrations required assume that all measurements are independent. O'Sullivan et al. [45] concluded that measurements made more than about a meter apart were independent, unless there were spatial trends in other soil properties, such as stone content, which influence cone resistance. If measurements are not independent (i.e., if they are spatially dependent), fewer may be needed [55].

B. Stones

The presence of stones increases the mean and the standard deviation of cone resistance data and may introduce unrepresentative high values and increase shaft friction. Although the influence of stones on bulk soil strength and compactibility should not be ignored, cone resistance is effectively a point measurement and, thus, cannot be used as an indicator of bulk strength. Therefore, the elimination from penetrometer data of outliers due to stones is desirable. Such outliers may bias treatment comparisons, though they are unlikely to affect treatment rankings [45]. In some soils, however, all penetrations are affected to some extent by stones. Thus, the effect of stones cannot be isolated from the effect of the soil between stones, and the combination of strength properties influencing cone resistance is likely to differ between stony and stone-free soils. Therefore, mathematical models based on cone resistance data from stone-free soils cannot be applied to stony soils.

Some penetrations fail to reach the required depth because of obstruction by stones. These penetrations may be abandoned and ignored with consequent bias or the successful part may be recorded. In the authors' laboratory, it has been shown that where penetrations fail to reach the desired depth, it is worthwhile to replace the first missing value in each penetration by the maximum measurable value.

C. Depth Effects

The cone resistance measured near the surface of a soil tends to be less than the value measured at depth, even when other properties are similar, because the mechanism of soil failure is different [3,6]. This surface effect may be unimportant when comparing treatments or soils at equal or equivalent depths, but it could be important when comparing cone resistances averaged over different depth increments or when choosing measurement depths for input to mathematical models.

Measurements made at different depths in one penetration are not independent. O'Sullivan et al. [45] showed that there was a high degree of correlation between measurements made at depths separated by up to 250 mm. They pointed out that one of the implications of this is that a significant treatment effect at one depth is likely to be accompanied by significant effects at adjoining depths.

The likelihood of shaft friction increases with depth, thus increasing measured cone resistance. Overburden effects will also increase with depth and influence cone resistance [56]. Shaft friction may be increased by bending of the shaft when high-strength layers or stones are encountered at depth. Thus, the relative magnitudes of the strength properties contributing to cone resistance may vary with depth, and so the interpretation of cone resistance values depends on the depth of measurement. Furthermore, a possible increase in the variance of cone resistance with depth means that simple averaging of cone resistance over a number of depths may be misleading, and the geometric mean may be more appropriate than the arithmetic mean.

Misleading conclusions may be drawn when comparing the cone resistances following compaction or tillage treatments, which cause large changes in surface levels. Detailed bulk density measurements beneath a wheel rut may be used to establish the original depth of each layer in the compacted soil [57]. This calculation cannot be made when only cone resistance is recorded, but a good approximation is to assume that each layer moves vertically by the same amount [57]. An example of this depth correction in a tillage experiment is shown in Fig. 8. The average bulk densities of the plowed and direct-drilled soils were 1.2 and 1.5 Mg m^{-3}, respectively, and the plowing depth was 250 mm, with respect to the plowed soil surface. Thus, the equivalent depth of direct-drilled soil was $250 \times 1.2/1.5 = 200$ mm, and the scale factor to convert the actual depth in plowed soil to the equivalent depth in direct-drilled soil was $200/250 = 0.8$. Figure 8 shows that an apparent cultivation effect below the depth of plowing was merely a consequence of the greater depth of topsoil in the plowed than in the direct-drilled land. Such depth corrections are essential when differences in

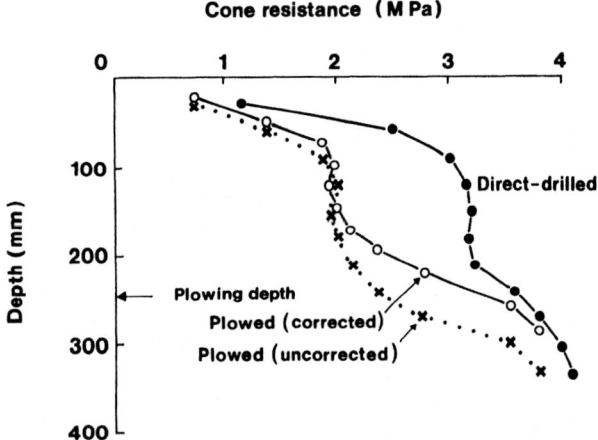

FIG. 8 Variation of soil cone resistance with depth for plowed and direct-drilled soils, before and after correction for the difference in surface level between treatments due to compaction.

surface level between treatments are large and the investigation is concerned with the mechanisms or processes that led to the measured values. A depth correction is unnecessary if only the results are of interest (e.g., when cone resistance is being related to root growth).

V.   APPLICATIONS

A.   Trafficability

In this section we are concerned with the ability of the soil to provide adequate traction for vehicles. Trafficability may also refer to the ability of the soil to withstand traffic without excess compaction or structural damage, which may influence subsequent crop growth or cultivations. The latter aspects are dealt with in Section V.B.

The cone penetrometer has been widely used for assessing soil trafficability [58-60] and for predicting the performance of tires [60,61] and cultivation implements [62]. The Waterways Experiment Station (WES) of the U.S. Army Engineers developed a trafficability assessment system for fine-grained soils in which "rating cone index" was the soil parameter [58]. Rating cone index is the average cone index of a 150 mm thick "critical layer" after empirical correction for the reduction in strength that occurs under the action of wheels in fine-grained soils. The critical layer was between 150 and 300 mm depths for most military vehicles. For very light or very heavy vehicles, the critical alyer was 75 mm shallower or deeper, respectively. The cone index required to allow 50 passes of a given vehicle, the vehicle cone index, was estimated from empirical formulas that took into account several factors, such as vehicle weight, tire-soil contact stress, engine power, and transmission type.

Paul and de Vries [63] plotted cone resistance against the subsequent wheelslip of a tractor pulling a manure spreader and used the cone resistance at 20% wheelslip as a criterion of trafficability. This value was combined with empirical relationships between cone resistance and water table depth [13] and a numerical simulation model of the drainage process [64] to investigate the effects of drain spacing on soil trafficability. Good agreement was found between model output and farmers' assessments of trafficability [64].

A dimensional analysis of tire-soil and cone-soil interaction by WES led to the development of dimensionless mobility numbers for dry, cohesionless sands and saturated, frictionless clays [59]. The clay and sand mobility numbers N_C and N_S are given by

$$N_C = P_r \frac{bd}{W} \left(\frac{\delta}{h}\right)^{1/2} \frac{1}{2 + b/2d} \tag{8}$$

and

$$N_S = G \left(\frac{bd}{W}\right)^{3/2} \frac{\delta}{h} \tag{9}$$

where b, d, and h are unloaded tire width, diameter, and section height, respectively, δ is the tire deflection under load, W is the vertical load on the tire, P_r is the cone index, and G is the gradient of cone index with depth. These mobility numbers were used as independent variables in empirical predictions of aspects of tire performance such as tire sinkage and torque, and hence drawbar pull [60]. Reece and Peca [10] concluded that the clay mobility number is a useful basis for performance prediction in saturated remolded clays but that predictions from the sand mobility number could be seriously in error because the value of G did not sufficiently reflect the variation in compactibility and strength between sands. The calculation of G has subsequently been refined [65] to take account of differences in compactibility and internal friction between sands.

A similar approach was adopted by Wismer and Luth [61], who were interested in the performance of agricultural tractors. They recognized that wheel behavior differed between the unsaturated, cohesive-frictional soils, which are usual in agriculture, and the saturated clays for which Eq. 8 was developed, but prediction errors were larger than predicted differences in performance [59]. Thus, they proposed empirical equations to predict the towing force on an undriven wheel, the pull generated by a driven wheel, and tractive efficiency [61] for agricultural soils from the "wheel numeric," C_n, of Eq. 8:

$$C_n = P_r \frac{bd}{W} \tag{10}$$

The average cone resistance of the top 150 mm was suggested [61] as input to Eq. 10 if the tire sinkage was less than 75 mm. Otherwise, the average cone resistance of the 150 mm layer, which included the maximum sinkage of the tire, should be used, although no guidance was given for estimating this sinkage in advance. Another difficulty with this procedure is the tendency of agricultural soils to compact, with a large but unpredictable change in strength, during the passage of a wheel. Traction is therefore more closely related to the properties of the compacted than the uncompacted soil. Consequently, predictions from the cone resistance after compaction were found to agree more closely with measurements in compactible soils than predictions from cone resistances before compaction [61]. Thus, the method is of limited predictive value in compactible soils.

The main theoretical objections to the prediction of tire performance from cone resistance are that cone resistance alone is insufficient to characterize the strength of soils [8] and that a penetrometer and a wheel induce markedly different soil strains [38]. Other problems are that the calibration data limit the accuracy of predictions and that the effects of soil compaction on cone resistance are not yet predictable. In common with all other empirical methods, results cannot be extrapolated to soils that have not been included in the calibration, and the method gives no insight into the processes involved. The advantages are that the penetrometer is simple and fast to use, and simple, practical relations may be developed between cone resistance and wheel performance. A theoretical treatment is complicated, and the effects of varying parameters can be studied only with difficulty [66].

Predictions of whether a soil is trafficable [58,63] may be adequate for the limited range of vehicles and soils used in deriving the prediction relationships. Predictions of the effects of varying soil and wheel parameters on properties such as tractive efficiency should be used only to rank treatments or make low-precision comparisons.

B. Compaction and Traffic

The use of the cone penetrometer for assessing soil compaction under wheels has been reviewed by Soane et al. [67], and some of the problems were discussed by O'Sullivan et al. [45]. Cone resistance effects due to compaction may be much larger than those due to bulk density [51]. However, water content effects may confound compaction effects. Thus, measurement at a standard water content is desirable even if not usually possible. If all measurements are made at a high water content, such as field capacity, bulk density effects on cone resistance are likely to be small relative to the effects at low water content (Fig. 1). Treatment effects may then be masked by soil variability. In addition, reproducible, low water contents are difficult to achieve in the field. Measurements at field water content should be made as soon as possible after the passage of wheels, because changes in matric potential and hydraulic conductivity associated with compaction will eventually lead to changes in water content below the wheel track. In soils to which Eq. 5 applies, differences in cone resistance between treatments may be small if the average bulk density is low. Depth effects, as discussed earlier, may also complicate treatment comparisons, even when a depth correction is made. Dickson and Smith [68] measured both cone resistance and bulk density below the ruts of a wheel supporting one of two loads at each of two ground pressures. After depth corrections were made, bulk density results confirmed the theoretical predictions that ground pressure is important to compaction at shallow depths while wheel load is more important at greater depths. In contrast, although cone resistance data were consistent with bulk density data at shallow depths, no treatment effects were detected at greater depths.

Despite these problems, the penetrometer is a useful, rapid method for assessing compaction, particularly near the soil surface and in poorly structured soils. Measurements made on transects across wheel ruts may be displayed as contour diagrams of cone resistance. Parameters may be derived from such diagrams to compare the relative depth, extent, and intensity of compaction under different wheels [45].

A recording penetrometer is particularly suited to the investigation of the distribution of compaction across a field [45] because of its good depth and lateral resolution and the high speed of data collection. This type of information may be displayed in the form of contour diagrams [45,69] such as Fig. 9, in which zones of high cone resistance lie below wheel track locations.

In addition to its use for empirical comparisons of compaction, cone resistance appears to be related to compactive effort [45]. Thus, the penetrometer has also been used to estimate stresses and their distribution under wheels and other loads [5,70,71]. Blackwell and Soane [70] categorized soil as hard, firm, or soft on the basis of penetrometer measurements, before applying the appropriate stress prediction equation. Bolling [71] calibrated cone resistance after the passage of a wheel with the stresses measured during compaction. For a standard ASAE penetrometer, it was found that

$$P_b = 3.1 - 0.6 \ P_r^{1/2} \tag{11}$$

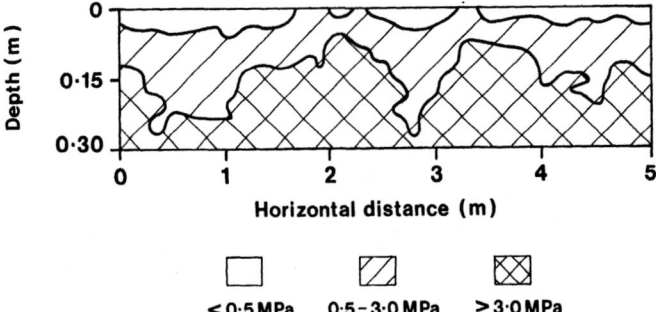

FIG. 9 Variation of cone resistance with depth across a field of conventionally grown winter barley. Zones of high cone resistance lie below wheel track locations.

where P_b is the measured stress in the soil. Koolen and Kuipers [5] calibrated a penetrometer by measuring the resistance to penetration of samples that had been subjected to different levels of compressive stress. This calibration was then used to estimate the lateral distribution of stresses that had existed under a loaded plate. Raghavan and McKyes [72] proposed an empirical equation, which related previous traffic intensity to cone resistance in a field of maize:

$$P_r = a + b \ln(nP) + c \ln(m) + dy + fD \tag{12}$$

where n is the number of wheel passes, P is the contact pressure, m is the water content, y is the depth, D is the number of days since seeding, and a, b, c, d, and f are empirical constants. These results suggest that prediction of cone resistance after compaction, which has been suggested as appropriate for the prediction of trafficability in compactible soils, may be possible from knowledge of the applied load, the cone resistance before compaction, and an estimate of the distribution of stress with depth. However, the relationahip between final cone resistance and applied stress may vary with soil type.

C. Tillage

The usefulness of the penetrometer in tillage studies has been summarized by O'Sullivan et al. [45]. They listed the following important applications: detection of compact layers; assessment, between treatments, of the relative depth, intensity, and persistence of loosening or compaction; and detection of changes in strength with time. Pentrometers have also been used to assess the suitability for root growth of the soil after different tillage treatments [20,22]. Such assessments, however, may be confounded by differences in struc-

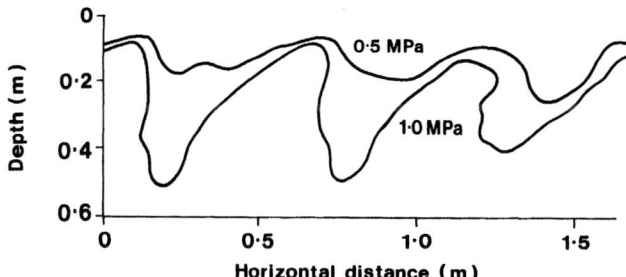

FIG. 10 Variation of cone resistance with depth across the direction of travel of a slant-leg subsoiler, showing the 0.5 and 1.0 MPa contours.

ture between treatments. For example, Ehlers et al. [20] found that the limiting cone resistance for root growth was greater in untilled soil (4.9 MPa) than in tilled soil (3.6 MPa). This difference was attributed to the ability of roots to exploit structural cracks and fissures that were destroyed by tillage.

The high degree of spatial resolution available with the penetrometer makes it suitable for investigating the spatial distribution of the effects of tillage implements [45,73-75]. An example is given in Fig. 10, which shows the results of penetration measurements made in a section across the direction of travel of a Paraplow, which is a slant-leg subsoiler. The pattern of loosening and the distribution of unloosened soil can be seen clearly.

Comparisons between tillage treatments may be complicated by water content differences, particularly if cone resistance is being used as an indirect measure of compaction. Measurements may be made at field capacity to eliminate water content effects, but bulk density effects may then be hidden, as mentioned earlier. If the penetrometer is being used to measure strength on a particular occasion, differences in water content, structure, and bulk density between treatments may cause difficulties in interpretation. The interpretation of cone resistance data is therefore often facilitated by simultaneous measurements of bulk density. Although the cone penetrometer is useful for making empirical comparisons between tillage treatments on the same soil, comparisons between soils are not possible because of the complicated effects of soil type on cone resistance.

The penetrometer has also been used to predict plow draft [62] and the performance of cultivator tines [3]. However, these predictions suffer from the same limitations as predictions of trafficability. Soil deformation around a cone will almost certainly differ from that around a tine and, thus, the cone is not a good analog of cultivator performance [76,77].

D. Other Applications

Although small-diameter needle penetrometers are usually used in
studies of root growth, the cone penetrometer ($\geq$ 100 mm^2 base area)
has also been used [20,22]. In heterogeneous soils, the correlation
between root growth rate and cone resistance tends to be low and to
differ for different depths and tillage treatments [20]. Therefore,
the cone penetrometer has little predictive value. A measure of bulk
soil strength, such as is given by a shear box, may be more relevant
to root growth than a point measurement such as is given by the
penetrometer [78]. Similarly, Scholefield and Hall [79] considered
cone resistance to be an inappropriate measurement in studies of
soil poaching by cattle, and instead devised a purpose-built penetrometer that both simulated the vertical and horizontal stresses
applied to the soil by a walking cow and measured the resulting soil
deformation.

The high correlations between cone resistance and many other
soil physical properties allow the penetrometer to be used for both
the investigation of soil spatial variability [69] and the estimation
of other soil physical properties [80]. For example, site uniformity
may be examined before locating the plots in a field experiment [81].
Furthermore, the California bearing ratio (CBR), which is an empirical bearing capacity measurement used in designing roads and run-
ways, may be estimated from cone resistance. The relationship for
Buckshot clay [80] is

$$CBR = 4.4P_r - 1 \tag{13}$$

with P_r expressed in megapascals. However, such empirical relationships, though sometimes useful, should be treated with caution: the
sample sizes and the stress and strain distributions differ between
the two tests. Soils with similar values of maximum penetration resistance may have different CBR values because the CBR test is
also sensitive to the rate of increase of penetration resistance with
depth [82]. Other civil engineering applications of the penetrometer
were reviewed by Perumpral [19].

VI. SUMMARY

Penetrometers may be used to make an empirical assessment of soil
strength. "Cone resistance" is the term used to describe the results obtained with a penetrometer. When a test procedure standardized by the American Society of Agricultural Engineers is
adopted, the test result can also be expressed as the soil cone index. Theoretically, a cone has been considered to penetrate soil
by creating a cavity, the mean stress normal to the cone surface

being that required to create such a cavity. Plastic failure is considered to occur in the soil in the region of the cone tip, while elastic failure occurs in a second region outside the first.

In field soils, cone resistance varies with soil properties including bulk density, water content, structure, organic matter, clay mineralogy, and stone content. Where only bulk density and water content vary, and only over limited ranges, a linear model of cone resistance may apply, but any more generally valid model is likely to be nonlinear. As the relative contributions of the various forces acting on a cone vary with soil water content, any relation between cone resistance and water content is usually applicable to one soil only.

Penetrometers in common use normally have cone angles of 60° or 30° with base areas of between 100 and 320 mm^2. Cone resistance is little affected by cone base diameter within this range, for a given cone angle, except possibly in well-structured soil. In most soils penetration rate has little effect on cone resistance. Penetrometers are usually hand-held but may be mounted on a horizontal rack for close-spaced measurements made in a regular grid. Force transducers are required, which can respond quickly to abrupt changes in penetration force. Automatic analog recording of force is necessary to avoid missing maxima and minima, but digitizing is necessary before statistical analysis or input to mathematical models can take place.

Cone resistance data are usually variable and possibly not normally distributed. Typically, 10 penetrations are required to characterize plots in replicated field experiments. Stones can cause penetrations to be incomplete, but ignoring such incomplete penetrations may introduce bias into comparisons.

Despite their limitations, cone resistance measurements are widely used because the measurements can be quickly and easily made. Simple, empirical relations with other soil properties such as trafficability may be derived and treatments compared. Bulk densities and water contents often have confounding effects when compaction is assessed in terms of cone resistance. Cone resistances under a wheel rut have been used to estimate the stresses in soil under a wheel. In tillage studies, the principal uses of cone resistance measurements are to detect compact layers and the extent of soil loosening and its persistence with time and subsequent traffic. Although the use of cone resistance in empirical predictions of soil behavior may be of limited value, detection of zones or layers of differential compaction, and soil heterogeneity generally, may usefully be made.

REFERENCES

1. Soil cone penetrometer, in Recommendation ASAE R313, *Agricultural Engineering Yearbook*, Am. Soc. Agric. Eng., St. Joseph, MI, 1969, pp. 296-297.
2. Farrell, D. A., and E. L. Greacen, Resistance to penetration of fine probes in compressible soil, *Aust. J. Soil Res.*, 4: 1-17 (1966).
3. Gill, W. R., Influence of compaction hardening of soil on penetration resistance, *Trans. Am. Soc. Agric. Eng.*, 11: 741-745 (1968).
4. Voorhees, W. B., D. A. Farrell, and W. E. Larson, Soil strength and aeration effects on root elongation, *Soil Sci. Soc. Am. Proc.*, 39: 949-953 (1975).
5. Koolen, A. J., and H. Kuipers, *Agricultural Soil Mechanics*, Springer-Verlag, Heidelberg, 1983.
6. Rohani, B., and G. Y. Baladi, Correlation of mobility cone index with fundamental engineering properties of soil, in *Proc. 7th Int. Conf. Int. Soc. Terrain-Vehicle Systems*, Vol. 3, Calgary, 1981, pp. 959-990.
7. Greacen, E. L., D. A. Farrell, and B. Cockroft, Soil resistance to metal probes and plant roots, in *Trans. 9th Int. Congr. Soil Science*, Vol. 1, Adelaide, 1968, pp. 769-779.
8. Mulqueen, J., J. V. Stafford, and D. W. Tanner, Evaluating penetrometers for measuring soil strength, *J. Terramech.*, 14: 137-151 (1977).
9. Freitag, D. R., Penetration tests for soil measurements, *Trans. Am. Soc. Agric. Eng.*, 11: 750-753 (1968).
10. Reece, A. R., and J. D. Peca, An assessment of the value of the cone penetrometer in mobility prediction, in *Proc. 7th Int. Conf. Int. Soc. Terrain-Vehicle Systems*, Vol. 3, Calgary, 1981, p. A1.
11. Steinhardt, R., Evaluating penetration resistance and wheel sinkage response to soil water suction changes in a draining clay soil, *Soil Sci. Soc. Am. Proc.*, 38: 518-522 (1974).
12. Steinhardt, R., and B. D. Trafford, Some effects of sub-surface drainage and ploughing on the structure and compactibility of a clay soil, *J. Soil Sci.*, 25: 138-152 (1974).
13. Paul, C. L., and J. De Vries, Prediction of soil strength from hydrologic and mechanical properties, *Can. J. Soil Sci.*, 59: 301-311 (1979).
14. van Wijk, A. L. M., Soil water conditions and playability of grass sportsfields: II. Influence of tile drainage and sandy drainage layers, *Z. Vegetationstechnik*, 3: 16-22 (1980).
15. Gerard, C. J., The influence of soil moisture, soil texture, drying conditions and exchangeable cations on soil strength, *Soil Sci. Soc. Am. Proc.*, 29: 641-645 (1965).

16. Byrd, C. W., and D. K. Cassel, The effect of sand content upon cone index and selected physical properties, *Soil Sci.*, 129: 197-204 (1980).

17. Stitt, R. E., D. K. Cassel, S. B. Weed, and L. A. Nelson, Mechanical impedance of tillage pans in Atlantic coastal plains soils and relationships with soil physical, chemical and mineralogical properties, *Soil Sci. Soc. Am. J.*, 46: 100-106 (1982).

18. Horn, R., Die Vorhersage des Eindringwiderstandes von Böden anhand von multiplen Regressionsanalysen (The prediction of the penetration resistance of soils by multiple regression analysis), *Z. Kulturtechn. Flurbereinig.*, 25: 377-380 (1984).

19. Perumpral, J. V., *Cone Penetrometer Application—A Review*, Paper No. 83-1549, Am. Soc. Agric. Eng., St. Joseph, MI, 1983.

20. Ehlers, W., U. Kopke, F. Hesse, and W. Bohn, Penetration resistance and root growth of oats in tilled and untilled loess soil, *Soil Tillage Res.*, 3: 261-275 (1983).

21. Voorhees, M. L., and P. N. Walker, Tractionability as a function of soil moisture, *Trans. Am. Soc. Agric. Eng.*, 20: 806-809 (1977).

22. Whiteley, G. M., and A. R. Dexter, Root development and growth of oilseed, wheat and pea crops on tilled and untilled soil, *Soil Tillage Res.*, 36: 379-393 (1982).

23. Davies, P., Influence of organic matter content, soil moisture status and time after reworking on soil shear strength, *J. Soil Sci.*, 36: 299-306 (1985).

24. Vomocil, J. A., and W. J. Flocker, Degradation of structure of Yolo loam by compaction, *Soil Sci. Soc. Am. Proc.*, 29: 7-12 (1965).

25. Mirreh, H. F., and J. W. Ketcheson, Influence of soil bulk density and matric pressure on soil resistance to penetration, *Can. J. Soil Sci.*, 52: 477-483 (1972).

26. Ayers, P. D., and J. V. Perumpral, Moisture and density effect on cone index, *Trans. Am. Soc. Agric. Eng.*, 25: 1169-1172 (1982).

27. Ball, B. C., and M. F. O'Sullivan, Soil strength and crop emergence in direct drilled and ploughed cereal seedbeds in seven field experiments, *J. Soil Sci.*, 33: 609-622 (1982).

28. van Wijk, A. L. M., and J. Beuving, Relation between soil strength, bulk density and soil water pressure head of sandy top-layers of grass sportsfields, *Z. Vegetationstechnik*, 1: 53-58 (1978).

29. Williams, J., and C. F. Shaykewich, The influence of soil water matric potential on the strength properties of unsaturated soil, *Soil Sci. Soc. Am. Proc.*, 34: 835-839 (1970).

30. Vepraskas, M. J., Cone index of loamy sands as influenced by pore size distribution and effective stress, *Soil Sci. Soc. Am. J.*, 48: 1220-1225 (1984).

31. Koolen, A. J., and P. Vaandrager, Relationships between soil mechanical properties, *J. Agric. Eng. Res.*, 29: 313-319 (1984).

32. Tollner, E. W., and B. P. Verma, Modified cone penetrometer for measuring soil mechanical impedance, *Trans. Am. Soc. Agric. Eng.*, 27: 331-336 (1984).

33. Whiteley, G. M., and A. R. Dexter, The dependence of soil penetrometer pressure on penetrometer size, *J. Agric. Eng. Res.*, 26: 467-476 (1981).

34. Turnage, G. W., Influence of viscous-type and inertial forces on the penetration of saturated, fine-grained soils, *J. Terramech.*, 10: 63-76 (1973).

35. Woo Jung La, F. Callebout, D. Gabriels, and M. de Boodt, Dependence of the cone index on the cone type and the soil structure, *Meded. Fac. Landbouw. Rijksuniv. Gent*, 50: 55-64 (1985).

36. Bradford, J. M., The penetration resistance in a soil with well-defined structural units, *Soil Sci. Soc. Am. J.*, 44: 601-606 (1980).

37. Tollner, E. W., and B. P. Verma, *Using Similitude Principles to Study Soil-Penetrometer Interactions*, Paper No. 84-1044, Am. Soc. Agric. Eng., St. Joseph, MI, 1984.

38. Yong, R. N., C. K. Chen, and R. Sylvestre-Williams, A study of the mechanisms of cone indentation and its relation to soil-wheel interaction, *J. Terramech.*, 9: 19-36 (1972).

39. Carter, L. M., Portable recording penetrometer measures soil strength profiles, *Agric. Eng.*, 48: 348-349 (1967).

40. Anderson, G., J. D. Pidgeon, H. B. Spencer, and R. Parks, A new hand-held recording penetrometer for soil studies, *J. Soil Sci.*, 31: 279-296 (1980).

41. Soane, B. D., Techniques for measuring changes in the packing state and cone resistance of soil after the passage of wheels and tracks, *J. Soil Sci.*, 24: 311-323 (1973).

42. Billot, J. F., Use of penetrometer for showing soil structure heterogeneity application to study of tillage implement impact and compaction effects, in *Proc. 9th Conf. Int. Soil Tillage Research Organization*, Osijek, Yugoslavia, 1982, pp. 177-182.

43. Wells, L. G., L. O. Lewis, and R. J. Distler, Remote electronic acquisition of soil cone index measurements, *J. Terramech.*, 18: 201-207 (1981).

44. Spencer, H. B., R. K. Hendrie, and G. Gilfillan, *A Hand Operated Recording Penetrometer*, Unpubl. Dep. Note No. SIN/221, Scot. Inst. Agric. Eng., 1977.

45. O'Sullivan, M. F., J. W. Dickson, and D. J. Campbell, Interpretation and presentation of cone resistance data in tillage and traffic studies, *J. Soil Sci.*, 38: 137-148 (1987).

46. McIntyre, D. S., and C. B. Tanner, Anormally distributed soil physical measurements and non-parametric statistics, *Soil Sci.*, 88: 133-137 (1959).

47. Cassel, D. K., and L. A. Nelson, Variability of mechanical impedance in a tilled one-hectare field of Norfolk sandy loam, *Soil Sci. Soc. Am. J.*, 43: 450-455 (1979).

48. Mitchell, C. W., R. Webster, P. H. T. Beckett, and B. Clifford, An analysis of terrain classification for long-range prediction of conditions in deserts, *Geog. J.*, 145: 72-85 (1979).

49. O'Sullivan, M. F., and B. C. Ball, A comparison of five instruments for measuring soil strength in cultivated and uncultivated cereal seedbeds, *J. Soil Sci.*, 33: 597-608 (1982).

50. Gerrard, A. J., The use of hand-operated soil penetrometers, *Area*, 14: 227-234 (1982).

51. Voorhees, W. B., C. G. Senst, and W. W. Nelson, Compaction and soil structure modification by wheel traffic in the northern corn belt, *Soil Sci. Soc. Am. J.*, 42: 344-349 (1978).

52. Kogure, K., Y. Ohira, and H. Yamaguchi, Basic study of probabilistic approach to prediction of soil trafficability—Statistical characteristics of cone index, *J. Terramech.*, 22: 147-156 (1985).

53. Snedecor, G. W., and W. G. Cochran, *Statistical Methods*, Iowa State University Press, Ames, 1967.

54. Pidgeon, J. D., and B. D. Soane, Effects of tillage and direct drilling on soil properties during the growing season in a long-term barley monoculture system, *J. Agric. Sci., Camb.*, 88: 431-442 (1977).

55. McBratney, A. B., and R. Webster, How many observations are needed for regional estimation of soil properties? *Soil Sci.*, 135: 177-183 (1983).

56. Bradford, J. M., D. A. Farrell, and W. E. Larson, Effect of soil overburden pressure on penetration of fine metal probes, *Soil Sci. Soc. Am. Proc.*, 35: 12-15 (1971).

57. Henshall, J. K., and D. L. O. Smith, An improved method for presenting comparisons of soil compaction effects below wheel ruts, *J. Agric. Eng. Res.*, 42: 1-13 (1989).

58. Knight, S. J., and D. R. Freitag, Measurement of soil trafficability characteristics, *Trans. Am. Soc. Agric. Eng.*, 5: 121-132 (1962).

59. Freitag, D. R., *A Dimensional Analysis of the Performance of Pneumatic Tires on Soft Soils*, U.S. Army Waterways Exp. Stn. Rep. No. 3-688, 1965.

60. Turnage, G. W., Tire selection and performance prediction for off-road wheeled vehicle operations, in *Proc. 4th Int. Conf.*

Int. Soc. Terrain-Vehicle Systems, Vol. 1, Stockholm, 1972, pp. 61-82.

61. Wismer, R. D., and H. J. Luth, Off-road traction prediction for wheeled vehicles, J. Terramech., 10: 49-61 (1973).

62. Gee-Clough, D., M. McAllister, G. Pearson, and D. W. Evernden, The empirical prediction of tractor-implement field performance, J. Terramech., 15: 81-94 (1978).

63. Paul, C. L., and J. de Vries, Effect of soil water status and strength on trafficability, Can. J. Soil Sci., 59: 313-324 (1979).

64. Paul, C. L., and J. de Vries, Soil trafficability in spring: 2. Prediction and the effect of subsurface drainage, Can. J. Soil Sci., 63: 27-35 (1983).

65. Turnage, G. W., Prediction of in-sand tire and wheel vehicle drawbar performance, in Proc. 8th Int. Conf. Int. Soc. Terrain-Vehicle Systems, Vol. 1, Cambridge, 1984, pp. 121-150.

66. Gee-Clough, D., Selection of tyre sizes for agricultural vehicles, J. Agric. Eng. Res., 25: 261-278 (1980).

67. Soane, B. D., P. S. Blackwell, J. W. Dickson, and D. J. Painter, Compaction by agricultural vehicles: A review: I. Soil and wheel characteristics, Soil Tillage Res., 1: 207-237 (1981).

68. Dickson, J. W., and D. L. O. Smith, Compaction of a Sandy Loam by a Single Wheel Supporting One of Two Masses Each at Two Ground Pressures, Unpubl. Dep. Note No. SIN/479, Scot. Inst. Agric. Eng., 1986.

69. Hartge, K. H., H. Bohne, H. P. Schrey, and H. Extra, Penetrometer measurements for screening soil physical variability, Soil Tillage Res., 5: 343-350 (1985).

70. Blackwell, P. S., and B. D. Soane, A method of predicting bulk density changes in field soils resulting from compaction by agricultural traffic, J. Soil Sci., 31: 51-65 (1981).

71. Bolling, I. H., How to predict the soil compaction of agricultural tires, in Proc. Int. Conf. Soil Dynamics, Vol. 5, Auburn, 1985, pp. 936-952.

72. Raghavan, G. S. V., and E. McKyes, Effect of vehicular traffic on soil moisture content in corn (maize) plots, J. Agric. Eng. Res., 23: 429-439 (1978).

73. Cassel, D. K., H. D. Bowen, and L. A. Nelson, An evaluation of mechanical impedance for three tillage treatments on Norfolk sandy loam, Soil Sci. Soc. Am. J., 42: 116-120 (1978).

74. Threadgill, E. D., Residual tillage effects as determined by cone index, Trans. Am. Soc. Agric. Eng., 25: 859-867 (1982).

75. Billot, J. F., Use of penetrometry in tillage studies, in Proc. Int. Conf. Soil Dynamics, Vol. 2, Auburn, 1985, pp. 213-218.

76. Freitag, D. R., R. L. Schafer, and R. D. Wismer, Similitude studies of soil-machine systems, *Trans. Am. Soc. Agric. Eng.*, 13: 201-213 (1970).

77. Johnston, C. E., R. L. Jensen, R. L. Schafer, and A. C. Bailey, Some soil-tool analogs, *Trans. Am. Soc. Agric. Eng.*, 23: 9-13 (1980).

78. Collis-George, N., and J. E. Lloyd, Description of seedbeds in terms of shear strength, in *Modification of Soil Structure* (W. W. Emerson, R. D. Bond, and A. R. Dexter, Eds.), Wiley, Chichester, U.K., 1978, pp. 111-117.

79. Scholefield, D., and D. M. Hall, A recording penetrometer to measure the strength of soil in relation to the stresses exerted by a walking cow, *J. Soil Sci.*, 37: 165-176 (1986).

80. Karafiath, L. L., and E. A. Nowatzki, *Soil Mechanics for Off-road Vehicle Engineering*, Trans Tech Publications, Clausthal, Germany, 1978.

81. Ball, B. C., and M. F. O'Sullivan, Cultivation and nitrogen requirements for drilled and broadcast winter barley on a surface water gley (gleysol), *Soil Tillage Res.*, 9: 103-122 (1987).

82. Freitag, D. R., Methods of measuring soil compaction, in *Compaction of Agricultural Soils*, Am. Soc. Agric. Eng., St. Joseph, MI, 1971, pp. 47-103.

10

The Penetrometer in Relation to Mechanical Resistance to Root Growth

A. GLYN BENGOUGH* *University of Aberdeen, Aberdeen, Scotland*

I. INTRODUCTION

Regions of high mechanical resistance in the soil sometimes arise as natural soil features, or can be caused by compaction by heavy farm machinery or by the formation of plow pans. Compact soils of high strength can reduce crop root growth rates and thus limit the availability of water and nutrients to the plant. This may result in lower crop yields, which can be of considerable economic importance. Subsoiling or deep loosening of compact soil layers can sometimes increase crop yields [1], although it may be difficult to assess what factor was originally limiting crop growth. Jarvis [2] found that loosening of compacted soil layers by deep ripping resulted in yield increases in 88% of sites on sandy soils in southwest Western Australia. Such deep cultivations will have many effects on soil conditions, but at least some of the yield increase may be due to the alleviation of mechanical impedance to root growth [3]. By careful soil management, it is often possible to avoid excessive soil compaction and structural damage.

Adverse soil physical conditions, therefore, often limit the growth and development of plant root systems and may reduce crop growth and yield. Compaction usually reduces the volume of large pores in the soil and may restrict root growth because of increased

---

*Current affiliation: Scottish Crop Research Institute, Dundee, Scotland

mechanical resistance and/or poor aeration. Roots can decrease
their diameter to grow through narrow holes in the soil, although
this capability is limited by the size of the root cap and stele [4-6].
If a root cannot make use of existing pores, it must exert sufficient
pressure to push its way through the soil. This results in a reduction in the root elongation rate and can change the morphology of
the root system [7].

The stunting of the root system caused by soil of high mechanical resistance may reduce crop yield if it limits the supply of
water and nutrients to the shoot. If sufficient water and nutrients
are available to a stunted root system, shoot growth may not be reduced [8].

There are many difficulties in studying the effects of mechanical
resistance on root growth through soil, due to the complicating factors of soil heterogeneity and the possible effects of restricted aeration or restricted nutrient or water supply. To simplify the situation, root growth has often been studied in artificial systems that
permit careful control of the root environment [9,10].

Probably the best indirect method of estimating soil resistance
to root growth involves measuring soil resistance to a metal probe
or penetrometer. Although "resistance" is generally used to mean
a force, in the context of roots and penetrometers the term commonly refers to the force exerted by the penetrometer or root divided
by its cross-sectional area; the latter usage is adhered to here.
Direct comparisons between root and penetrometer resistance suggest
that penetrometers experience between two and eight times greater
resistance in soil than do roots [11].

This chapter describes how soil resistance to root growth can
be estimated by measuring the resistance encountered when pushing
a small-diameter "needle" penetrometer through the soil. Theoretical
aspects of soil resistance to penetrometers and plant roots are dealt
with briefly, followed by practical aspects of penetrometer design.
Finally, the interpretation of penetrometer resistance measurements
is discussed with respect to the resistance experienced by plant
roots growing through the soil, and how that resistance may affect
root growth rates.

II. THEORY OF SOIL PENETRATION BY
METAL PROBES AND PLANT ROOTS

Attempts have been made to model penetration resistance to roots
and to pentrometers using the bulk mechanical properties of the soil.
These models are generally based on the theoretical pressure required to expand a cavity in the soil.

Greacen et al. [12,13] developed models to predict soil resistance
to 'blunt' (large cone angle) and to 'sharp' (small cone angle) conical-

tipped penetrometers. The following paragraphs summarize and discuss these models.

According to these authors, penetrometer resistance consists of two components: the pressure required to expand a cavity in the soil, and the frictional resistance on the surface of the penetrometer tip. Expansion of the soil cavity is assumed to occur through compression of the surrounding soil to accommodate the probe. Two main zones of compression are identified: a zone of compression with plastic failure surrounding the probe, and a further zone of elastic compression immediately outside it. The "cavity" pressure depends on the pattern of soil deformation creating the cavity, which is governed by the shape of the penetrometer tip. Sharp penetrometers (e.g., 5° cone semiangle) were suggested to cause cylindrical soil deformation, and blunt probes (30° cone semiangle) spherical soil deformation.

The calculation of the cavity pressure required measurement of a large number of soil mechanical properties. Cavity pressures were predicted for three soils at different bulk densities and matric potentials. The calculations of Greacen and coworkers suggest that considerably less resistance is required to expand a soil cavity by cylindrical soil deformation than by spherical deformation.

To calculate total point resistance Q, the coefficient of soil-metal friction U was determined separately and then substituted into the equation

$$Q = N(1 + U \cot a) \tag{1}$$

where 2a is the cone angle and N is the normal stress on the probe surface (equal to the cavity pressure).

The predictions of penetrometer resistance were tested experimentally by measuring soil resistance to blunt and sharp penetrometers. The difference between the calculated and measured resistances, expressed as a percentage of the measured resistance, was typically about 20%, but sometimes more than 35%. Predicted cavity pressure for cylindrical soil deformation was only 0.25-0.45 of the pressure required for spherical cavity expansion. However, the higher frictional resistance to the sharp probe caused the total resistance to a sharp probe to be higher than the total resistance to a blunt probe. Frictional resistance often constituted more than half the total penetrometer resistance, according to their calculations. Penetrometer resistance has been shown to decrease considerably upon rotation of the probe tip [12,14,15], thus confirming that much of the resistance to a penetrometer is frictional.

To explain the difference between root and probe penetration resistance, Greacen et al. [13] suggested that roots deform the soil cylindrically and experience much less friction than a penetrometer.

Experimental evidence for cylindrical soil deformation by roots was provided by using X-radiography to examine soil density patterns surrounding roots and sharp probes [16]. Patterns of soil particle displacement around a pea root resembled more closely those around a sharp probe than those around a blunt probe [17].

Abdalla et al. [18] presented a simpler model of root growth, considering the deformation of a cylinder of soil in front of the root tip. The model predicts that for any given strain, less stress is required to deform the soil radially than axially. Experiments were performed in which a penetrometer with sides capable of radial expansion (by inflation of a rubber membrane) was placed under a static load in a container of sand. Inflation of the membrane resulted in the probe penetrating to a greater depth. This suggested that some mechanism exists whereby radial expansion behind the penetrometer tip had reduced axial penetration resistance. Since a root grown under mechanical stress does expand radially behind the tip, this may cause a reduction in the axial resistance to root growth.

Hettiaratchi and Ferguson [19] considered the theoretical pressure required for cylindrical and spherical deformation modes in a frictionless cohesive material. The pressure required for cylindrical deformation was always less than for spherical deformation, the difference increasing with cohesion. Experiments were performed in cohesionless sand and in a poorly graded fine silt that was slightly cohesive when moist. Using a flat-ended punch capable of both axial displacement and independent radial expansion, the investigators found that radial enlargement required less pressure than axial displacement and that radial expansion reduced the pressure required for axial displacement.

Further evidence was provided by photoelastic analysis of the stresses around a slit cut into a block of epoxy resin. The bottom of the slit was loaded to simulate the "punch indentation mode," while radial deformation was simulated by loading the sides of the slit. The resulting stress patterns suggested that the radial deformation mode indirectly increases the peak shear stresses under the punch, bringing this zone closer to failure. No theoretical explanation of this phenomenon was attempted, but it was suggested that radial expansion behind the root tip might ease the pressure of roots through the soil.

Root growth has also been studied in artificial systems where it was intended that a constant, uniform, and known value of mechanical impedance could be imposed, while simultaneously maintaining a carefully controlled supply of aerated nutrient solution to the roots. Roots have been grown through ballotini contained in flexible-sided cells, which could be subjected to an external confining pressure [8,18,20]. However, resistance to root elongation in the ballotini considerably exceeds the external confining pressure [21,22], and thus root growth can be severely slowed by very low confining pressures [8]. This type of study is therefore more useful for observing changes in root morphology, differences between species or varieties, and interactions between mechanical impedance and aeration, than it is for studying the effect of absolute values of resistance on root elongation rate.

III. NEEDLE PENETROMETER DESIGN

Penetrometers may differ from roots in diameter, shape, and frictional properties, in rate of penetration, and in tip flexibility (see Table 1). Additionally, only the few millimeters of root nearest the apex is pushed through the soil [23], and the root extracts water from the surrounding soil as it grows. It is therefore important to consider how the choice of probe design may affect penetrometer resistance in relation to root growth.

A. Cone and Shaft

1. Diameter

Root tip diameters generally lie in the 0.1-2 mm range, while penetrometer tip diameters cover an even greater range, from large field penetrometers (> 10 mm) [24] down to small-needle penetrometers (< 0.2 mm) [25]. Even in homogeneous soil, penetration resistance can depend on probe diameter as soil particles of finite size must be displaced. Any diameter dependence should therefore be most noticeable for very small probes, which may have to displace particles of comparable size.

Studies of the effect of probe diameter on penetration resistance have produced a variety of results. Dexter and Tanner [26] found the average resistance to pushing spheres through soil increased with decreasing diameter in the 40-10 mm range. However, Barley et al. [27] found no difference in resistance between 30° semiangle conical probes of 1, 2, and 3 mm diameter penetrating remolded sandy loam. Similarly, Bradford [28] found no significant difference between the average resistance experienced by probes of 3.8 and 5.1 mm diameter, penetrating undisturbed field cores. Whiteley et al. [11] also found no significant effect of probe diameter on mean penetration resistance, in undisturbed clods or remolded soil, for diameters ranging from 1 to 2 mm.

In contrast, in a study of the effect of diameter on penetrometer resistance in remolded soil cores with textures ranging from heavy clay to sand, resistance to a 1 mm probe was typically 45-55% greater than to a 2 mm probe [29]. Gooderham (1973, cited in Ref. 11) also found that resistance to a 1 mm diameter probe was 35-74%

TABLE 1 Main Differences Between Plant Roots and Penetrometers

Characteristic	Roots	Penetrometers
Diameter	Generally 0.1-2 mm	Generally 0.1-20 mm
Shape	Approximately para-boloid, but may expand radially if mechanically impeded	Usually conical
Friction	Unknown; probably small due to mucilage secretion and cells sloughing off root cap	Considerable friction on probe tip and on shaft (if nonrelieved)
Penetration rate	$< \sim 1$ mm h^{-1}	Often > 1 mm min^{-1}
Flexibility	Can follow cracks or planes of weakness through the soil	Rigidly mounted; follow a linear path through the soil
Water uptake	Extract water from the soil as they grow	Do not extract water

greater than to a 2 mm probe. As a further complication, a recent theoretical model [21] of resistance to root growth in granula media implies that small roots may experience less resistance than larger ones.

Much of the evidence appears to be contradictory, and there is a need for a comprehensive study over a wide range of penetrometer diameters and soil textures. Furthermore, penetrometer resistance in a laterally confined soil core may be increased due to confinement, if the ratio of penetrometer diameter to core diameter exceeds about 1:20 [30]. Diameter can also influence probe resistance through friction on the shaft of the penetrometer. The surface area of a penetrometer shaft is directly proportional to its diameter, whereas the force on the penetrometer tip is proportion to the square of the tip diameter. Thus, shaft friction is relatively more important for smaller probes, and this has been confirmed by experiment [27]. To reduce soil-metal shaft friction, a relieved shaft (i.e., a shaft with a smaller diameter than the probe tip) is commonly used.

2. Shape and Friction

Penetrometer shapes vary from cylinders [25] to shapes resembling plant root tips [23], but the most widely used design has a conical tip. Shape of the probe or root tip determines both the mode of soil deformation and the amount of frictional resistance on the tip. Greacen et al. [13] suggested that plant roots and narrowly tapered ("sharp") probes cause cylindrical soil deformation, which theoretically requires less pressure than the spherical deformation caused by blunt probes (Section II). Radial expansion occurring behind the tips of impeded roots [18] may aid root penetration by bringing soil in front of the root tip closer to failure [19].

It has been predicted that in the absence of shaft friction, frictional resistance on the penetrometer tip can account for more than half of the total penetrometer resistance [13]. Movement of a root tip through soil is confined mainly to the 5 mm or so nearest the apex [23]. Sloughing off of root cap cells and the secretion of mucilage [31] may have a lubricating action, reducing root penetration resistance [17].

B. Insertion and Measurement

1. Resistance Measurement

The commonest and most easily interpreted penetrometer results are from measuring the resistance to a probe driven into soil at a constant rate. Other designs measure the magnitude or rate of probe penetration under different constant loads [14]. In this chapter, only the constant-rate penetrometers are considered.

To obtain a constant rate of penetration, it is necessary either to drive the probe downward into the soil with some sort of motor [27] or to raise the soil sample on a moving platform toward a stationary probe [23]. Using a force transducer to measure the resisting force to probe penetration, the movable crosshead of a strength-testing machine provides a convenient drive capable of a wide range of speeds [15,32]. Proving rings, strain gauges, and electronic balances have all been used to measure the resisting force to probe penetration [23,27,33]. The advantage of an electronic balance or force transducer is that it permits automatic digitization and transfer of data to a microcomputer at regular depth intervals. Proving rings may have an unacceptably high flexibility, causing the ring to unload when a void is encountered, with the result that the void is not clearly detected.

2. Rate of Penetration

Roots typically grow at a rate of 1 mm or less per hour, which is an inconveniently slow rate at which to conduct penetrometer tests. Most needle penetrometer measurements are performed at rates of penetration between 1 and 3 orders of magnitude greater than root growth rates [11]. Eavis [23] found no effect of rate of penetration on the penetrometer resistance of a silty clay loam at rates between 5 and 0.1 mm min^{-1}. At lower rates of penetration, however, the resistance decreased, but by only 13% at a penetration rate 20 times slower. A small decrease in the penetrometer resistance of sandy loam and clay was noted at rates below 0.02 mm min^{-1} [34]. In saturated clay, penetrometer resistance is more obviously linked to penetration rate, as a result of its interaction with pore water pressure [17].

From the current limited evidence, it appears that penetrometer resistance is only weakly dependent on penetration rate in the range of water contents and penetration rates normally used for measurements. At lower rates, penetration resistance decreases slightly, and at much higher rates it may increase considerably. The slower rate of root growth as compared with rate of probe penetration is likely to be of greatest importance in relatively wet impermeable soil and of much lesser importance in dry sandy soil.

IV. INTERPRETATION OF PENETROMETER RESULTS

A. Problems in Use

Most penetrometer designs that take small-diameter probes are not suitable for field use, although some small-probe field penetrometers have been constructed [22,28]. Large-diameter field penetrometers have been used with some success in root growth studies [24,35,36], but such devices are even less representative, in terms of their diameter and rate of penetration, of growing roots.

When collecting soil samples for laboratory needle penetrometer measurements, care must be taken that the soil is compressed as little as possible during coring. Where cores are collected too close together, or are of small diameter, the soil is likely to be compacted, resulting in higher bulk density and penetrometer resistance. Soil core diameter may affect penetrometer resistance if the core diameter is less than about 20 times that of the probe; this is because of edge effects due to the soil confinement. Where more than one penetration is performed into each core, penetrometer resistance may be decreased if tensile failure cracks form between the penetration holes [30].

Stones that are encountered during penetration may cause a rapid increase in resistance. This can damage a sensitive force transducer if an overload cutoff is not included in motor-driven penetrometers. Because roots can grow around stones and other localized regions of high resistance, such readings should be specially identified or removed from the data set [37]. Penetrometer readings taken after a stone has been pushed aside by the probe tip may also have to be discarded in case the stone rubs against the penetrometer shaft, creating higher frictional resistance.

Penetrometer readings obtained as the probe is entering the surface of the soil (at depths less than three times the probe diameter) should be discarded: the values of resistance will be anomalously low because the soil failure mechanism near the soil surface is different from that in the bulk of the soil [38].

B. Comparisons Between Penetrometer Resistance and Root Resistance

The ideal way to determine the relationship between penetrometer resistance and root resistance is to measure both independently in the same soil. Because of experimental difficulties in measuring root resistance, however, very few such studies have been made. These studies involve measurement of the force exerted by a root penetrating a soil sample. Force measurements must be taken after the root has extended at least three times its diameter into the soil, but before the tip becomes anchored by root hairs.

To convert the force-to-root resistance requires measurement of the root cross-sectional area. Root tips often swell in response to mechanical impedance and, as a continuous record of root force and diameter normally cannot be obtained, it is not obvious whether it is most relevant to measure the initial or final root diameter. Indeed, because root tips are parabolically tapered, the distance behind the root tip at which diameter is measured is of considerable importance. The diameter measurement should be at the position on the root that is level with the soil surface and should be recorded at the same time the force measurement is made. The root resistance then calculated should correspond to the normal stress on the surface of the tip if it is uniformly distributed over the root tip surface. However, if radial deformation requires less pressure than axial deformation, the normal stress may not be uniform over the root surface.

Stolzy and Barley [39] found that root resistance was only 80% of the resistance to a 3 mm diameter, 30° semiangle penetrometer in a remolded sandy loam. This result was based, however, on the penetration of only two pea radicles, and it included a generous allowance for root skin friction. Root skin friction was estimated by measuring the force required to pull the root out of the soil and was added to the measured force exerted by the root. Root diameter was measured 3-5 mm behind the tip.

Eavis and Payne [23,40] compared root resistances with the resistance to a metal probe of similar size and shape, in a remolded sandy loam. The penetrometer required a pressure between four and eight times greater than the root to penetrate the soil. Root tip diameter was measured about 5 mm from the root tip, behind the zone of elongation.

Whiteley et al. [11] compared resistance to root penetration with resistance to 1-2 mm diameter, 30° semiangle penetrometers, in remolded soil cores and in undisturbed field clods of a sandy loam.

Penetrometer resistance was between 2.6 and 5.3 times greater than root resistance, based on root diameter about 4 mm behind the root tip. Recently, Misra et al. [41] found resistance to a 1 mm diameter, 30° semiangle penetrometer was between 1.8 and 3.8 times higher than resistance to roots penetrating clay loam aggregates of various sizes.

Thus penetrometers experience a resistance between about two and eight times greater than roots. Further indirect evidence of this difference comes from comparing studies of root elongation rate and penetrometer resistance with measurements of the maximum pressures roots can exert. Critical values of penetrometer resistance at which root elongation ceases are in the 0.8-5.0 MPa range, depending on the soil and crop [30]. Maximum axial pressures a root can exert vary between about 0.24 and 1.45 MPa, depending on species, but are mostly in the 0.9-1.3 MPa range [42]. Such maximum pressure is ultimately limited by the cell turgor pressure near the root tip. Thus root elongation is halted in soil with a penetrometer resistance up to 6 or more times greater than the maximum pressure the root can exert. There are many possible reasons for this large difference between penetrometer and root resistance in soil (Section II), and there is considerable evidence that the friction on the penetrometer tip is a major factor [15,22], but the relative importance of other factors is unclear.

C. Correlation with Root Elongation Rate

Root elongation rate decreases approximately inversely with increasing soil penetrometer resistance (Fig. 1) [24,43]. A similar form of relationship between applied pressure and root growth has also been obtained in studies using pressurized cells of ballotini [8,18]. Voorhees et al. [34] found that root elongation rates correlated better with the resistance to a sharp probe after the frictional component of resistance (estimated by measurement of the coefficient of soil-metal friction) had been subtracted.

D. Small-scale Variations in Soil Strength

Unlike roots, penetrometers follow a linear path through the soil and are unable to follow biopores, cracks, or planes of weakness in the way that roots have been observed to do [7]. This limits the usefulness of penetrometers in structured soil, where the average resistance measured by large penetrometers will overestimate the resistance to root growth. Soil structure exists as a hierarchy [44], and so even macroscopically "unstructured" soils contain spatial variations in strength on a much smaller scale, which a root may be able to exploit. Ehlers et al. [24] found that roots grew through untilled soil that presented a high penetrometer resistance, whereas root growth

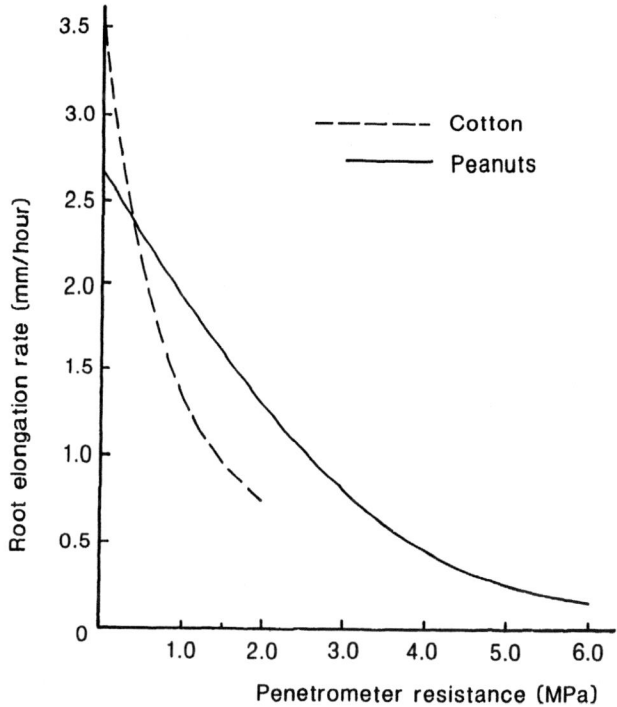

FIG. 1 Root elongation rate for peanuts and cotton versus soil
penetrometer resistance. [Reproduced from H. M. Taylor and L. F.
Ratcliff, Root elongation rates of cotton and peanuts as a function
of soil strength and water content, *Soil Science*, 108: 113-119 (1969).
© by Williams and Wilkins, Baltimore, MD.]

was halted in the tilled soil with the same penetrometer resistance.
They concluded that the untilled soil contained more cracks and
biopores, which were available for root growth but were not detected by the large (11 mm diameter) field penetrometer.

Although large penetrometers are of little use in coarsely structured soil, individual peds can be considered to be continuous even
though the soil itself is structured on a larger scale [30]. Dexter
[45] used this idea together with the probability of roots penetrating peds to model root growth through a bed of aggregates. Bradford [28] found that the variability of readings was greater for the
smaller of two penetrometers, even though the average resistance
was the same. Groenevelt et al. [25] took this idea further, using
a very small (0.15 mm diameter) needle penetrometer to examine what
fraction of the soil was penetrable by roots. The "percentage" linear

penetrability" decreased with increasing soil bulk density. Significant correlations have been obtained between the percentage linear penetrability and root length in a sandy soil [36]. Spectral analysis of penetrometer data has been attempted [46] but has not yet been applied directly to root growth.

V. SUMMARY

At present there is no reliable method of estimating soil resistance to root growth, short of direct measurement of root force. Penetrometers provide the best estimates of root resistance but experience between two and eight times greater resistance than a root in unsaturated soil. Much of the difference between root resistance and penetrometer resistance may be accounted for by a higher component of frictional resistance on the penetrometer tips. Factors such as radial expansion of mechanically impeded root tips may also be important.

Root elongation rate decreases as soil penetrometer resistance increases, but the relationship depends on the particular soil and crop. Needle penetrometers are of use in detecting small-scale strength variations within the soil. Care must be taken to choose the probe shape, diameter, and penetration rate most appropriate for any given situation, and then to standardize the procedure.

REFERENCES

1. Bennie, A. T. P., and F. J. P. Botha, Effect of deep tillage and controlled traffic on root growth, water use efficiency and yield of maize and wheat, *Soil Tillage Res.*, 7: 85-95 (1986).
2. Jarvis, R. J., in Plant Res. Div. Annual Report, Western Australia Dept. Agric., 1983, pp. 39-42 (cited in Ref. 3).
3. Hamblin, A. P., The influence of soil structure on water movement, crop root growth, and water uptake, *Adv. Agron.*, 38: 95-158 (1985).
4. Scholefield, D., and D. M. Hall, Constricted growth of grass roots through rigid pores, *Plant Soil*, 85: 153-162 (1985).
5. MacDonald, I. R., Seedbed and related investigations, in Macaulay Institute for Soil Research Annual Report, 1984, pp. 98-101.
6. Dexter, A. R., Model experiments on the behaviour of roots at the interface between a tilled seed-bed and a compacted subsoil, *Plant Soil*, 95: 135-147 (1986).
7. Russell, R. S., *Plant Root Systems: Their Function and Interaction with the Soil*, McGraw-Hill, London, 1977.

8. Goss, M. J., Effects of mechanical impedance on root growth in barley (*Hordeum vulgare* L.): I. Effects on elongation and branching of seminal roots, *J. Exp. Bot.*, 28: 96-111 (1977).

9. Barley, K. P., The effects of mechanical stress on the growth of roots, *J. Exp. Bot.*, 13: 95-110 (1962).

10. Gill, W. R., and R. D. Miller, A method for study of the influence of mechanical impedance and aeration on the growth of seedling roots, *Soil Sci. Soc. Am. Proc.*, 20: 154-157 (1956).

11. Whiteley, G. M., W. H. Utomo, and A. R. Dexter, A comparison of penetrometer pressures and the pressures exerted by roots, *Plant Soil*, 61: 351-364 (1981).

12. Farrell, D. A., and E. L. Greacen, Resistance to penetration of fine probes in compressible soil, *Aust. J. Soil Res.*, 4: 1-17 (1966).

13. Greacen, E. L., D. A. Farrell, and B. Cockroft, Soil resistance to metal probes and plant roots, in *Trans. 9th Int. Congr. Soil Science*, Vol. 1, 1968, pp. 769-779.

14. Waldron, L. J., and G. K. Constantin, Soil resistance to a slowly moving penetrometer, *Soil Sci.*, 109: 221-226 (1970).

15. Bengough, A. G., and C. E. Mullins, The use of a low-friction penetrometer to estimate soil resistance to root growth, in *Proc. 11th Int. Conf. Int. Soil Tillage Research Organization*, Vol. 1, 1988, p. 1-6.

16. Greacen, E. L., D. A. Farrell, and J. A. Forrest, Measurements of density patterns in soils, *J. Agric. Eng. Res.*, 12: 311-313 (1967).

17. Cockroft, B., K. P. Barley, and E. L. Greacen, The penetration of clays by fine probes and root tips, *Aust. J. Soil Res.*, 7: 333-348 (1969).

18. Abdalla, A. M., D. R. P. Hettiaratchi, and A. R. Reece, The mechanics of root growth in granular media, *J. Agric. Eng. Res.*, 14: 263-248 (1969).

19. Hettiaratchi, D. R. P., and C. A. Ferguson, Stress deformation behaviour of soil in root growth mechanics, *J. Agric. Eng. Res.*, 18: 309-320 (1973).

20. Barley, K. P., Influence of soil strength on growth of roots, *Soil Sci.*, 96: 175-180 (1963).

21. Richards, B. G., and E. L. Greacen, Mechanical stresses on an expanding cylindrical root analogue in granular media, *Aust. J. Soil Res.*, 24: 393-404 (1986).

22. Bengough, A. G., The use of penetrometers in estimating mechanical resistance to root growth, Ph.D. thesis, University of Aberdeen, Aberdeen, U.K., 1988.

23. Eavis, B. W. Mechanical impedance to root growth, Paper No. 4/F/39, in *Agricultural Engineering Symp.*, 1967, Silsoe, U.K., 1967, pp. 1-11.

24. Ehlers, W., U. Kopke, F. Hesse, and W. Bohm, Penetration resistance and root growth of oats in tilled and untilled loess soil, *Soil Tillage Res.*, 3: 261-275 (1983).

25. Groenevelt, P. H., B. D. Kay, and C. D. Grant, Physical assessment of a soil with respect to rooting potential, *Geoderma*, 34: 101-114 (1984).

26. Dexter, A. R., and D. W. Tanner, The force on spheres penetrating soil, *J. Terramech.*, 9: 31-39 (1973).

27. Barley, K. P., E. L. Greacen, and D.A. Farrell, The influence of soil strength on the penetration of a loam by plant roots, *Aust. J. Soil Res.*, 3: 69-79 (1965).

28. Bradford, J. M., The penetration resistance of a soil with well defined structural units, *Soil Sci. Soc. Am. J.*, 44: 601-606 (1980).

29. Whiteley, G. M., and A. R. Dexter, The dependence of soil penetrometer pressure on penetrometer size, *J. Agric. Eng. Res.*, 26: 467-476 (1981).

30. Greacen, E. L., K. P. Barley, and D. A. Farrell, The mechanics of root growth in soils with particular reference to the implications for root distribution, in *Root Growth* (W. J. Whittington, Ed.), Butterworths, London, 1969, pp. 256-268.

31. Oades, J. M., Mucilages at the root surface, *J. Soil Sci.*, 29: 1-16 (1978).

32. Callebaut, F., D. Gabriels, W. Minjauw, and M. De Boodt, Determination of soil surface strength with a needle-type penetrometer, *Soil Tillage Res.*, 5: 227-245 (1985).

33. Misra, R. K., A. R. Dexter, and A. M. Alston, Penetration of soil aggregates of finite size: I. Blunt penetrometer probes, *Plant Soil*, 94: 43-58 (1986).

34. Voorhees, W. B., D. A. Farrell, and W. E. Larson, Soil strength and aeration effects on root elongation, *Soil Sci. Soc. Am. Proc.*, 39: 948-953 (1975).

35. Barraclough, P. B., and A. H. Weir, Effects of a compacted subsoil layer on root and shoot growth, water use and nutrient uptake of winter wheat, *J. Agric. Sci.*, 110: 207-216 (1988).

36. Jamieson, J. E., R. J. Morris, and C. E. Mullins, Effect of subsoiling on physical properties and crop growth on a sandy soil with a naturally compact subsoil, in *Proc. 11th Int. Conf. Int. Soil Tillage Research Organization*, Vol. 2, 1988, pp. 499-503.

37. O'Sullivan, M. F., J. W. Dickson, and D. J. Campbell, Interpretation and presentation of cone resistance data in tillage and traffic studies, *J. Soil Sci.*, 38: 137-148 (1987).

38. Gill, W. R., Influence of compaction hardening of soil on penetration resistance, *Trans. Am. Soc. Agric. Eng.*, 11: 741-753 (1968).

39. Stolzy, L. H., and K. P. Barley, Mechanical resistance encountered by roots entering compact soils, *Soil Sci.*, 105: 297-301 (1968).

40. Eavis, B. W., and D. Payne, Soil physical conditions and root growth, in *Root Growth* (W. J. Whittington, Ed.), Butterworths, London, 1969, pp. 315-336.

41. Misra, R. K., A. R. Dexter, and A. M. Alston, Penetration of soil aggregates of finite size: II. Plant roots, *Plant Soil*, 94: 59-85 (1986).

42. Misra, R. K., A. R. Dexter, and A. M. Alston, Maximum axial and radial growth pressures of plant roots, *Plant Soil*, 95: 315-326 (1986).

43. Taylor, H. M., and L. F. Ratliff, Root elongation rates of cotton and peanuts as a function of soil strength and water content, *Soil Sci.*, 108: 113-119 (1969).

44. Dexter, A. R., Advances in characterization of soil structure, *Soil Tillage Res.*, 11: 199-238 (1988).

45. Dexter, A. R., A stochastic model for the growth of roots in tilled soils, *J. Soil Sci.*, 29: 102-116 (1978).

46. Grant, C. D., B. D. Kay, P. H. Groenevelt, G. E. Kidd, and G. W. Thurtell, Spectral analysis of micropenetrometer data to characterize soil structure, *Can. J. Soil Sci.*, 65: 789-804 (1985).

11

Root Growth: Methods of Measurement

DAVID ATKINSON   *The School of Agriculture, University of Aberdeen, Aberdeen, Scotland*

LORNA A. MACKIE-DAWSON   *The Macaulay Land Use Research Institute, Aberdeen, Scotland*

I. INTRODUCTION

In different genotypes and under varied growing conditions (e.g., with different soil physical conditions in the growing medium), plant roots and root systems show variation over a range of their characteristics. This chapter reviews methods that can be used in the field or laboratory to describe the root system of plants. In many cases such measurements may be used to assess whether a particular physical factor or group of factors has influenced the root system. The major soil physical factors that influence roots are temperature, aeration, water potential, and mechanical impedance. Böhm [1] gives a more complete description of methods that can be used in all circumstances. Before describing methods that might be used to study the effects on plant roots of soil physical conditions, it is essential to consider which root properties seem likely to be sensitive to soil physical conditions bearing in mind that other factors such as the growth of the aerial parts of the plant and soil biological and chemical factors will also influence root growth. All these topics have been reviewed in general terms by Russell [2].

This chapter aims to describe methods and to illustrate, using selected papers, the uses to which particular methods have been put and some of the limitations to their use. In both field and laboratory many methods can give information on a range of parameters. For example, when a root system is directly observed (in solution culture, adjacent to a mini-rhizotron tube, or through an observation panel) measurements can be made of length, diameter, longevity, and other

variables. Furthermore, a number of methods involve specialized
growth media or controlled growing conditions. It is therefore more
logical to describe studies of root systems by major method types
rather than by root system property. Consequently, this chapter
is laid out rather differently from the others in that the major methods and the significance of the measurements they facilitate are discussed together in the context of determining how plant function
can be influenced by soil physical conditions.

A. Root System Properties

All root systems are composed of a number of individual roots. Root
systems are branched structures with several orders of branching
(usually up to 4). Individual roots themselves are made of large
numbers of cells, and the size, shape, and form of these cells and
the numbers in a particular tissue (e.g., xylem or cortex) can be
altered by the growing environment, as can the function of these
individual cells. Major soil physical factors, such as soil water potential and soil mechanical resistance, can affect root processes
(e.g., cell wall extensibility and wall pressure) in a number of ways
[3]. Cell wall pressure is closely related to the rate of root growth,
while osmoregulation is quantitatively related to changes in soil water
potential but may be less completely related to changes in mechanical
resistance. As a consequence of these effects, the length of individual roots, their rate of extension, and increases in root diameter can
be changed by soil physical condition. Changes in these properties
affect the overall volume of soil exploited by roots, via effects on
horizontal spread and the depth of penetration which, in turn, influence the extent of the resources of water and nutrients available to
the plant. Other parameters that can vary include the angle at which
roots grow through soil (e.g., their susceptibility to geotropism).

The root system present in the soil at any given time is the sum
of all the roots that thus far have been produced and have remained
alive. The longevity of roots varies between species [4] and between
root types in a species. The rate of production together with longevity will result in a given total root length and an average root length
density—that is, the length of root (LA) under a given area of soil
surface or the length (LV) in a given volume of soil. These factors
are important to the ability of the root system to obtain nutrients for
plant growth. In addition to possible effects of soil factors on morphology, root function (e.g., nutrient uptake per unit root length,
surface area, or volume) may be altered as a consequence of effects
on the types and ages of root present. However, the exact effects
of physical conditions on all the parameters above are not yet fully
described or understood. Roots are normally considered in relation
to their ability to supply water and nutrients to the plant, but they

TABLE 1   Root System Characteristics that can be Affected by Soil Physical Condition

Major category	Characteristics affected
Anatomy	Cell size, cortex width, balance of xylem cell types, epidermal wall form, root diameter, root shape
Characteristics of individual roots	Diameter, growth rate, angle, length, mass, root longevity, root hair length and density, mycorrhization, pressure
Branching pattern	Amount, density, number of orders, position, distance between branches
Characteristics of whole root system	Horizontal distribution, vertical distribution, length, mass, absolute and relative distribution
Function	Absorption of nutrients and water, anchorage, production of biologically active molecules (e.g., enzymes, phenolics)

are also required to anchor the plant [5] and to produce hormones, which may regulate the growth and performance of both root and shoot [6]. The root system of most plants exists in nature in a symbiotic association with fungi (mycorrhizas), and so assessments of effects of soil conditions on roots should also consider effects on both endo- and sheathing mycorrhizas.

B. Potential Effects of Physical Properties

Root system parameters that are able to vary and so may be influenced by a change in soil physical conditions are listed in Table 1.

The following case studies of the effects of physical impedance indicate how some of these parameters can vary and illustrate that the root system is extremely plastic in its response.

1. Case Studies

In a study of the effect of zones of contrasting bulk density on root system development in oats, the effect of a given value of bulk density varied according to its relation to the density of other areas in the soil column [7]. Roots growing in surface soil with a bulk density of 1.24 Mg m^{-3} did not penetrate deeper soil with a density of 1.52, while roots in a surface soil with a density of 1.38 did. Increases in bulk density at depth decreased the depth of rooting. Compaction did not reduce branching, although it did influence root survival. The length of branch roots, which was normally highest

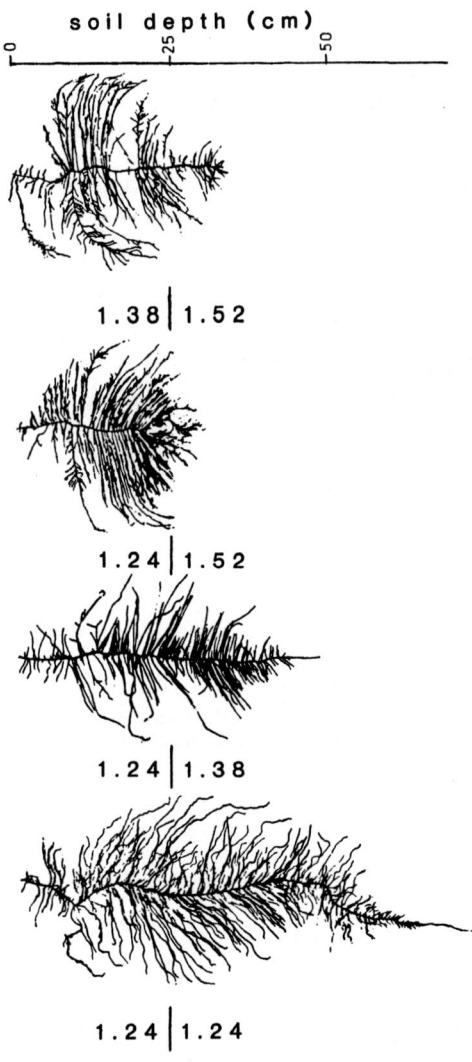

soil depth (cm)

1.38 | 1.52

1.24 | 1.52

1.24 | 1.38

1.24 | 1.24

FIG. 1 Effect of soil density in the topsoil and subsoil on development of oat root systems. Bulk densities above and below the boundary are given in megagrams per cubic meter. (From Ref. 7.)

in the surface, was affected (Fig. 1). Where the elongation of the main axis was reduced by a dense subsoil, its diameter increased and there was stimulated branching.

A quantitative assessment of the response of roots to mechanical stress was carried out in a system in which roots were grown within

TABLE 2   The Relative Effect on Barley Root Growth of Pore
Diameter and Externally Applied Pressure in Cells of Ballotini

	Pressure (kPa)		
	0	50	50
	Pore diameter (μm)		
Variable	160 or 70	160	70
Root type affected	None	Axis	All
Seminal root, mean length (cm)	8.6	2.0	2.0
Distance apex to laterals (mm)	30	4	4
Lateral roots (no. per axis)	19	10	9
Lateral roots, mean length (cm)	4.8	9.3	2.2

Source: Goss, M. J., Effects of mechanical impedance on root growth
in barley (*Hordeum volgare* L.): I.   Effects on the elongation and
branching of seminal root axis, *J. Exp. Bot.*, 28: 96-111 (1977).

flexible-sided cells filled with ballotini (glass spheres), to which
known pressures were applied while root nutrition was maintained [8].
This work showed that the extension of the seminal axes of barley
(410 μm diameter) was inhibited by pressures as low as 10 kPa
when the pore diameter of the medium was 160 μm or less.   The degree of reduction in the size of the root system varied between species, with barley being among the more sensitive species.   With a
pore size of 160 μm, lateral roots (150 μm diameter) were able to
grow freely even though the main axis was inhibited (Table 2).
Where the main axis was impeded close to the cell surface, most
lateral development occurred close to the surface and the length of
secondary laterals increased: 3 cm length per primary lateral versus 0.4 cm in the control.   In these experiments lateral development
was able to compensate for the reduced main axis, so that total
growth and nutrient uptake and therefore root functioning were unaffected (Table 3).   It should be noted, however, that as discussed
in Chap. 10, it has recently been demonstrated that the actual pressure a root must exert in growing through uniform ballotini is considerably greater (by up to 10 times) than the externally applied
confining pressure on the ballotini.

In soil, root elongation was reduced by 60% by a mechanical resistance of 1.8 MPa in ryegrass (root diameter 250 μm) and 2.6 MPa
in pea (1 mm diameter) [9].   Roots prevented from elongating in-

TABLE 3 The Effect on Root Dry Weight and Nutrient
Uptake of Impedance of the Main Axis of Barley Plants
in a Situation Where Lateral Root Development was
Allowed

Axis de- velopment	Plant weight (mg)		Nutrient uptake (μmol)	
	Root	Shoot	Calcium	Phosphorus
Unimpeded	43	131	19	15
Impeded	43	131	19	14

Source: (See Table 2.)

crease their diameter, mainly by increasing the cross-sectional area
of the cortex [10]. Root diameter may therefore be important both
in terms of ability to penetrate the soil and as an indication of a
response to soil impedance.

2. Types of Response

In addition to changes in total root length or mass, there can be
alterations in the partitioning of dry matter both within individual
roots (e.g., by increasing diameter) and within system (e.g., by
increasing branching and thus changing system spread). Goss [8]
showed that even when root system mass is unchanged, length can
be reduced by 65% by mechanical impedance, while Logsdon et al.
[11] demonstrated that an increase in root diameter can compensate
in part for a reduction in total length. Appropriate measurements
are clearly needed to establish such effects. Effects on root system
morphology may or may not be associated with changes in crop performance, and effects under field conditions may well be different
from those in the laboratory. For example, an increasing concentration of roots at the "soil" surface in the laboratory had no effect on
nutrient supply [8], but in soil, in the absence of irrigation, it
might be expected to have adverse consequences because the surface
would quickly dry out.

C. Purpose of Measurements

The optimum method for the assessment of the root system will depend both on the characteristics of the root system itself and on the
purpose for which the assessment is being made. Root systems may
be studied for one or more of the following reasons:

1. To assess the significance of a given change in soil physical
    conditions on plant function.
2. To help to interpret the reasons for a given plant response to
    a particular soil treatment through an understanding of effects

on water and nutrient supply.

3. To improve the use of inputs (e.g., irrigation water and fertilizers) or to study and optimize the effects of tillage and other soil management practices.
4. To allow the development of better plant root systems by conventional breeding or genetic engineering.

D. Significance of Root Features

Criteria for the selection of methods of assessing root performance are limited by our incomplete understanding of the importance of particular root characteristics and the consequences of changes in them. However, there is information on the importance of some root system characters, and we can identify some of the circumstances in which particular measurements will be useful, such as the relation between the root system and the availability of water to the plant.

1. Characters Influencing Water Supply

Soil water flux is the product of the hydraulic potential gradient between soil and the root and the unsaturated hydraulic conductivity. Typical maximum flux rates seem to be around 2.5 mm d^{-1} [2]. This would allow an equivalent rate of root water uptake of around 160 µl d^{-1} cm^{-1}. On this basis, for example, a root density of around 2 cm of root per square centimeter of soil surface area should be sufficient to supply transpirational needs under conditions normally encountered in the United Kingdom. Both the uniformity of root distribution and the mean distance between roots influence water supply [12]. Thus if this average root length density is clustered in one small region, the flux to the root surface will be inadequate. In addition, as the soil water content decreases, the consequent decrease in unsaturated conductivity necessitates a greater root density to supply the same flux.

The total volume of soil exploited by a plant's root system is also important because it represents the total amount of water and nutrients available to that plant. This volume is usually represented by the horizontal spread and either the maximum rooting depth, the depth of the deepest root, or the depth containing 95% of roots. Therefore, both average root length density and root distribution within the soil volume exploited are important. However, to define a root system just in these terms gives a static picture. For a plant growing in the field, transpiration rate changes during a single season as a consequence of changes in weather and in the plant's needs for water. Thus it is important to be able to assess the root system as it grows, and to assess factors such as root death, especially since the longevity of the lateral roots of some species in soil can be measured in days rather than weeks [4].

TABLE 4 The Functional Significance of Some Commonly Measured Root System Parameters

Root parameter	Significance
Mass	Total root system size, standing crop
Length	Total root system size, ability to absorb water and mineral nutrients
Number	Hormone production
Density	Water inflow rates
Distribution: vertical	Soil volume exploited, stability
horizontal	Interactions with other species, stability
Specific root length	Carbon allocation within system
Diameter	Water stress regulation, soil penetration, response to soil impedance
Longevity	Soil carbon supply, root length change
Periodicity	Available length at particular times in the season
Growth rate	Ability to exploit within the soil volume
Ratio of root to shoot	Relative carbon partitioning
Branching	Nutrient uptake potential within a given volume
Root hair density	Effective root surface area
Mycorrhizal infection	Carbon allocation, effective surface area for nutrient uptake

E. Commonly Measured Characteristics

Despite the long list of root characteristics that might be measured (Table 1), researchers have commonly determined a smaller set of characteristics. These characteristics (which, with their functional significance, are listed in Table 4), may be measured in the laboratory, where with plants grown in solution culture, sand, or ballotini, it is easier to retrieve the root system, or in the field, where a range of additional factors apply. The properties that can be measured by various field methods are summarized in Table 5.

TABLE 5 Major Root Parameters That can be Measured by Different Field Techniques

Parameter	Methods[a]				
	Observation (e.g., root laboratory, rhizotron, minirhizotron)	Monolith (e.g., soil cores, monoliths needle board)	Profile (e.g., soil trench, pit, wall)	Excavation (e.g., total or partial)	Activity (e.g., water use, nutrient uptake, radioisotope injection)
Mass		+		+	
Length	+*	+	+*		
Number	+*	+	+		
Density	+*	+	+*	+	+
Distribution		+	+	+	+
Diameter	+	+	+		
Turnover	+	+			
Longevity	+				
Periodicity	+				+
Growth rate	+				+

[a]Asterisk indicates that a number of assumptions about depth of soil samples must be made to allow values to be calculated.

II. METHODS OF STUDYING ROOT SYSTEMS IN THE LABORATORY

Laboratory studies are normally concerned with assessing the effect of either a single or a limited range of soil physical properties on plant and root performance. Single-factor studies usually relate to soil temperature, water potential, osmotic potential, or aeration [13], whereas studies of the effect of soil impedance often involve simultaneous changes in other factors. For example, when bulk density was changed from 1.24 to 1.52 Mg m^{-3}, there was also reduction in the volume of pores that are filled with air at field capacity and an increase in the volume of pores from which water would be unavailable to plants [7]. Because the results of such experiments are likely to be influenced by the types of containers and media used, these are also briefly reviewed.

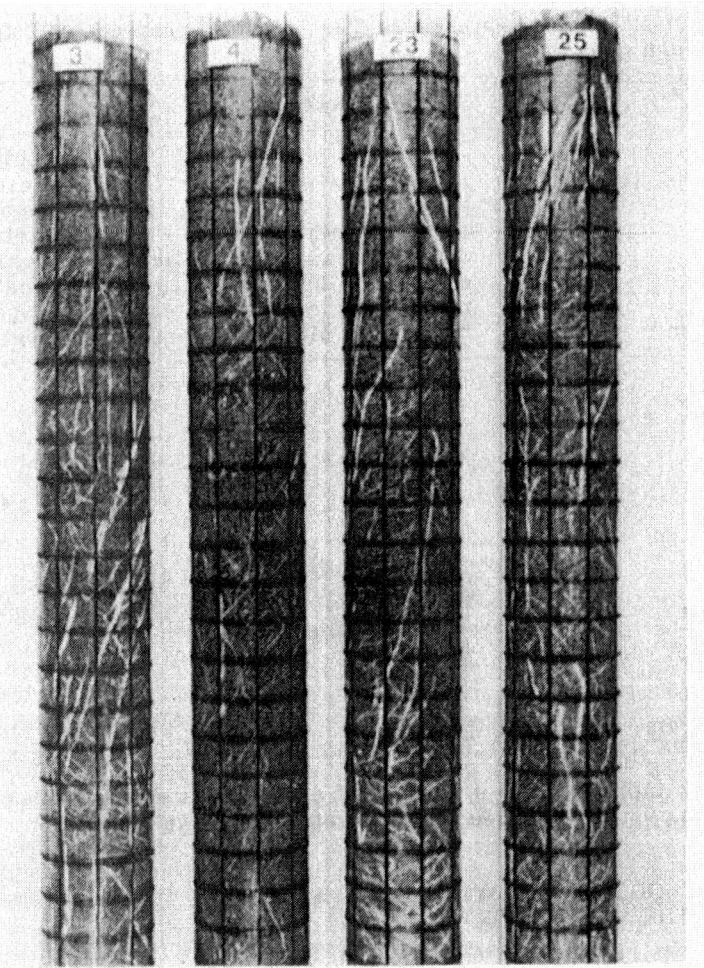

FIG. 2 Use of acrylic cylinders to observe differences in root form, density, and distribution in spring barley. To calculate information as total or average root lengths, the relationship between length at the observation surface and in the whole soil volume must be known. (From Ref. 19.)

A. Containers

1. General Factors

Container methods permit the isolation of individual environmental factors that will normally interact with other factors in influencing root growth in the field. Replication and "management" are easier than in the field, although container effects on root growth may be unnatural because of the absence of soil organisms (fungi, soil arthropods, etc.) and the restricted space. Container methods are best suited for studying plants with small root systems or for investigating the early stages of plant development.

2. Container Types

The *size* of a container determines the total volume of soil available to a plant. Conventional plastic or clay plant pots, Mitscherlich pots, glass pots, petri dishes, tubes made from glass or plastic, and cardboard cartons have all been successfully used as containers, but their limited volume frequently results in roots concentrating near the walls of the vessel and around its bottom. As a consequence of moisture and temperature differentials, the concentration of roots between the wall of a pot and the soil tends to be greater in a porous clay pot than in a plastic pot.

In container experiments designed principally to study root growth and distribution, the depth of the container needs to be large because of the root's tendency to grow downward when restricted by the walls of a container [1]. Boxes, 80 cm high, made from metal [14], wood [15], or plastic [16] have been used in this type of study, but cylindrical tubes are more common. Iron [17], clay [18], asbestos [7], plastic [15], acrylic [19], and glass [20, 21] have all been used. Where tubes are of a transparent material, roots can be directly observed and measured (Fig. 2). They may be buried in the soil or in insulated boxes to prevent the establishment of unrealistic temperature differentials. Boxes with glass observation panels set into their sides have also been used to allow continuous observations to be made [22,23].

Boxes of this type vary in size in relation to the type of plant being investigated. For studies of M.1 apple rootstocks, boxes 60 × 17.5 × 42.5 cm high were used [22], while studies of maize [23] used smaller boxes. Such boxes may be used simply as a means of observing the response of roots receiving particular treatments or as a means of assessing the uptake of radioisotopes incorporated into the soil adjacent to the soil-observation interface. The latter method has been used to observe the uptake of ^{86}Rb from the soil around individual roots [23].

Such boxes have been used to assess the effect of soil moisture and soil temperature on the root growth of glass and clover species [24]. Using the observation windows, it was possible to assess treatment effects on root system length, the elongation of individual roots, root diameter, and root number. All these parameters were affected by soil temperature.

Where plants are grown in containers that allow observation of the root system, the possible effects of light on root growth must be considered, although few studies have addressed this point. A comparison of the effect of a range of different light exposures [22], varying from total darkness to total light, on the growth of apple rootstocks showed that while continuous illumination severely checked growth, increased suberization, and reduced the development of lateral roots, the short periods of 3 × 20 minutes per week or 2 hours per 2 weeks needed for observation had little obvious effect. Effects were greatest during periods of maximum root growth, with length being more affected than root number. Given the paucity of data on the effects of light, it is prudent to reduce unnecessary exposure to a minimum.

3. Filling Containers

Care must be taken when filling containers, particularly when physical conditions (e.g., bulk density) are being controlled [7]. Soil should be sieved when nearly air-dry, the appropriate fertilizer added, and the soil then moistened to a friable condition and mixed. It should then be put in the containers and compressed, layer by layer, several centimeters at a time, with the top few millimeters of each layer loosened before adding the next layer, to avoid stratification. All containers should normally be watered and allowed to stand for a period from a week to a month before the experiment begins, to allow the soil to settle. Where short-lived radioisotopes are being used [23], however, this may not be possible. These procedures are needed both for tubes and boxes. Where measurements are to be made beside an observation surface it is essential not to smear the viewing surface.

B. Media

For experimental purposes, plants can be grown in a solid, liquid, or gaseous rooting medium, the type of medium used depending on the scope and aims of the experiment.

1. Solid

Soil. Soil is the most realistic growing medium for terrestrial plants and for long-term experiments. However, the extraction of whole root systems from soils other than very sandy ones is difficult if not impossible, as is the complete removal of soil particles from recovered roots [25,26]. Although exact nutrient compositions cannot be produced easily, soil temperature, water content, and compaction levels can all be manipulated. For example, containers of soil have been produced in which one layer is varied in bulk density, thickness, and depth from the surface [27]. Maintenance of a given matric potential is usually made using tensiometers or by weighing [28]. A variation of this technique in which roots are grown in soil within porous membrane envelopes has been used successfully [29,30]. Here the root system was confined within the "envelope," and water and nutrients were able to move across the membrane.

Undisturbed Soil Columns. Undisturbed soil columns have great potential for the study of the effect of soil physical conditions on root growth, because here structure, texture, and water availability may be unaltered. Columns allow the experimenter to control certain soil conditions (e.g., water content), and plant growing conditions (e.g., temperature). Disadvantages of this type of method are that properties such as pore size distribution, structure, and bulk density cannot be precisely determined before experimentation. Undisturbed columns can be very large [31] and can be collected by hand coring, power coring, or by hydraulic sampling machines. The columns can be preserved with a coating of paraffin wax, plastic, foil, or liquid plastic material.

Sand. Sand is often used in nutrient experiments because of its low buffering capacity and available nutrient content, and the ease with which it can be manipulated. It is also relatively easy to wash sand from the roots, although some may still firmly adhere to the roots [19]. The physical properties of the medium can be altered by varying sand particle size. Fine sand gives a higher water-holding capacity, while coarse sand is more freely draining and so better aerated. Root systems obtained from experimental media of this type are generally similar to those grown in solution culture [1].

Other Solid Media. Perlite, which is composed of expanded volcanic rock fragments, is uniform and inert and so is suitable for studies of germination or seedling development. However, it is less suitable for long-term growth experiments and for studies of physical effects on roots. Perlite seems to result in root system development, which is similar to that in solution culture [1]. It is a good root growth medium where roots are to be used for studies of ultrastructure. Particles of perlite embedded in roots cause less damage to knives used in sectioning than do sand grains. The differing penetrating abilities of roots have been studied using agar, paraffin, and wax materials of differing hardness [32,33]. Vermiculite has also been used as a growing medium and seems to give growth comparable to that of roots grown in soil [1].

2. Special Techniques

Solution Culture. The major advantage of solution culture is that the ionic composition of the root environment can be defined, measured, and manipulated with precision and that the entire medium can be held under standard conditions (e.g., of temperature

and aeration). Because of ion uptake/efflux by plant roots, however, nutrient solutions are liable to rapid changes in ionic composition, and so require more routine maintenance than is needed for
soil-based systems.

Solution culture is only of limited use in soil physical studies.
The uniform medium, lack of physical resistance, and the absence
of soil flora and fauna make it difficult to compare root growth in
solution culture with that in soil. However, this approach has applications in studies of impeded aeration, water stress, and temperature. Roots can be maintained at precise temperatures by flowing
the solutions through a refrigeration or heating unit before entry
into the plant growth containers [34], or by immersing plant-growth
containers in thermostatically-controlled water-baths [35].

Split-root Techniques. Spatial variability in nutrient supply
can be controlled using split-root containers, in which isolated parts
of the root system receive different nutrient supplies, either in solution culture [36] or in solid media [37]. Individual roots can also
be separated out to study specific effects [38].

Water Stress Control. Osmotic control of plant water stress can
be obtained in solution culture using sodium chloride [39], polyethylene glycol 4000 (PEG) [40], or a range of other chemicals [28]. This
method allows plant water stress to be accurately maintained and more
easily reproduced than is possible in a soil growing medium. However,
the stress brought about by the two methods has a different physio-
logical bases in each case. The water stress to which roots are exposed normally in soils is primarily due to a substantial negative
matric potential, while in a PEG-modified solution the stress results
from substantial negative osmotic potential. Although stresses due
to both matric and osmotic potential have been shown to produce similar effects on plant growth [41], it must be remembered that the
matric potential at the root surface can be considerably less than
that measured in the bulk of the soil.

Pressure Control. Many workers have studied effects of mechanical stress by growing roots through which aerated nutrient solution is circulated. The cell walls consist of flexible impervious polyester membrane. A known hydrostatic pressure is applied by suspending the cells in water-filled vessels, to which an external pressure is applied [42-46]. Details of construction, use, and the types
of measurements that can be made are given by Goss [8].

3. Gas

Pressure has also been applied to roots grown behind thin rubber
diaphragms forced against the root by gas under controlled pressure
[47]; however, the actual pressure experienced by the root is uncertain.

C. Measurement of Roots in Laboratory Media

1. Measurement of Roots in Soil

Impregnated Sections. Roots can be studied in undisturbed samples by impregnating the soil with resin, using samples collected from field plots or pot experiments. The method involves removal of soil water in exchange for a solvent in which the concentration of resin is gradually raised. After addition of an accelerator or hardener, the cured soil blocks are sectioned for examination [48]. Several combinations of fixative and impregnating resin have been used, including a mixture of acetic acid-formaldehyde and ethanol and a polyester resin for impregnation [49], acetone instead of ethanol in the preceding procedure [48], and glutaraldehyde-acetone with a resin for impregnation [50,51]. The best methods can preserve the form of delicate biological materials, such as root tissues and protozoa cells [48]. Staining roots in the blocks with methylene blue and basic fuchsin followed by sectioning can lead to good identification of the detailed structure of preserved materials [51]. Fluorochromes, such as acidine orange [52], can be used to increase the natural root fluorescence. Glutaraldehyde impregnation has been used to study the soil pore network available to protozoa and roots [50]. Although the method is expensive and labor-intensive, it allows detailed examination of the soil-root interface. Root-soil interactions can also be assessed using the scanning electron microscope (Fig. 3).

Nuclear Magnetic Resonance Imaging. Nuclear magnetic resonance (NMR) has been used as a noninvasive tool for studying roots in situ [53]. The technique was used to obtain images of root systems grown in a range of soil types, vermiculite, sand, perlite, fritted clay, potting soil, and "peatlite," but the clarity of image varied according to the magnetic properties of the medium examined. To observe relatively dry soil, hence to optimize the root image, measurements were made at the end of the watering cycles. Soil moisture levels more conducive to good growth reduced the clarity of the image. At present the images produced are not good enough to permit quantitative data to be obtained, but there is a possibility of obtaining three-dimensional images in the future.

Neutron Radiography. Neutron radiography has been used to produce two-dimensional images of plant root systems [54]. Plant roots grown in narrow (2.5-5.0 cm wide) boxes, with neutron-sensitive back plates, were irradiated with thermal neutrons, and photographic images were obtained from these plates. Roots were identified because of preferential neutron scattering by the roots. In this way the elongation rates of soybean and maize roots were obtained from sequential radiographs. Lateral roots (< 0.33 mm) were poorly visible. As with the NMR imaging method, there is a need

FIG. 3 A root of spring barley and the soil attached to it as seen
with the Stereoscan electron microscope.

for improved resolution before the method can be regarded as a
practical means of producing quantitative data. However, both
these methods have the advantages of being nondestructive and
usable with soils under relatively natural conditions.

Autoradiography. Radioisotopes have been used in a variety of
ways and in a large number of studies to observe roots or give a
measure of root activity in soil [23,55]. If two species grown in
mixed culture are injected, one with ^{32}P and one with ^{33}P, the roots
of the two species can be subsequently identified in a section of a
soil block containing the cut ends of the roots of both species. Mix-
ing radioisotopes into the soil and assessing depletion around roots
has been used by a number of workers [23] as a means of assessing
root activity in soil boxes (see also Section III.D.3). The technique
could also be used to assess the effects of soil impedance on uptake.

2. Measurements of Root Parameters

Number. The number of roots can be counted either in samples
of washed roots (from soil columns, pots, etc.) or in situ (glass-
faced columns, solution culture, etc.). The number of root tips per

unit volume of soil has been used as a criterion of root distribution
in soil [56]. It has been suggested that root number is closely related to leaf number [57].

Mass. For determination of root mass, roots are washed free
of soil, then oven-dried (usually at about 80°C) for 24-48 hours.
Mass can be measured in all methods that permit roots to be destructively sampled (i.e., pots, columns, solution culture). Mass is a
good measurement for characterizing the total amount of root material
in soil but is neither a good nor a unique indicator of the absorbing
potential of the roots in soil. It seems to be less sensitive to soil
factors than root length, so that treatments with major effects on
root length may have no effect on mass [58]. If the fresh weight
of roots is determined, it is also important to measure oven-dry mass,
since the water content of roots is variable.

Surface Area. Surface area is a useful measurement that can be
related to water or nutrient uptake. A crude estimate of root surface area can be obtained from root length and diameter if root hairs
and the extramatrical hyphae of mycorrhizas are ignored. A range
of direct methods have been used to estimate the root surface on
roots washed free of soil. These include photoelectric attenuation
[59,60], dye adsorption [61], and the retention of calcium on the
external surface of the root following a brief immersion in a concentrated solution of calcium nitrate and centrifugation [62].

Strength. The tensile strength of single washed roots has been
obtained from the force required to break 5-10 cm lengths of root of
known diameter [63].
A system for measuring the buckling stress of clamped, excised
roots has been devised. This technique was modified to characterize
root elasticity by hanging weights from one end of a 10 mm length of
root, which was clamped at its other end. The root was guided between microscope slides, and elasticity was related to the deflection
caused by a known weight [64].

Diameter. The diameter of newly washed root samples from soil
cores or solution culture, or from roots in impregnated soil blocks,
can be measured directly. This method has been used to obtain estimates of root surface area and in estimating length from volume
[65]. Large numbers of measurements are needed to characterize
diameter accurately. The effects of external pressure on root diameter have been studied in beds of ballotini [8]. By varying the size
of the glass beads, it becomes possible to examine interactions between pore size distribution and root diameter. Root diameter is
usually measured directly using a microscope with a micrometer eyepiece. For larger roots, calipers or a micrometer screw can be used.

For more detailed measurements, roots can be inserted in a pith block, then sectioned and examined under the microscope.

Length

(1) Basic calculations. Length can be measured directly using samples of wet roots in a water-filled dish using calipers or by placing roots on graph paper and counting squares. There is a tendency to use this time-consuming method on small numbers of roots only, but for large roots (> 5 mm diameter) it is usually the easiest method [66]. For samples with a greater root length, measurements can be made more rapidly using some type of sampling method, usually by counting the number of intersections between roots and a random or regular pattern of lines. Total root length R can be estimated using

$$R = \pi \frac{AN}{2H} \tag{1}$$

where R is the total length of roots, A is the area of the field of view, and N is the number of intersections between the roots and a set of randomly oriented straight lines whose total length is H [67].

Newman [67] applied this principle to a system where a number of fields of view were examined using a microscope with a hairline in the eyepiece, which was randomly reoriented before each new examination. Using this method, the time for root length measurement was reduced to less than half that required by direct measurement (e.g., 24 minutes to measure 3.4 m of root with a CV of 4.3% versus 67 minutes by direct measurement).

Since Eq. 1 requires only that the orientation between roots and a set of lines be random, this equation can be used equally well for regular arrangements of lines such as parallel straight lines or a grid, provided there is no preferred orientation of the roots in relation to these lines. Furthermore, where the line spacing or the spacing of a square grid is equal to the unit in which root length is measured, it is easy to show that A/H is 1 (for lines) and 1/2 (for a grid), or in general for a square grid

$$R = \frac{\pi Nk}{4} \tag{2}$$

where k is the grid spacing. (This theory has been used by Marsh [68] and Tennant [69], who used the approximation $\pi = 22/7$, and a similar result was obtained empirically by Head [70].)

Such a grid counting method has several advantages over Newman's original suggestion: there is no necessity for the roots to be uniformly spread out over the counting area, and time is saved because of the simpler measurement procedure [68,69]. Experimental details of a procedure that enable a single set of measurements to be obtained within 6 minutes and with a coefficient of variation of 5% or less have been given by Tennant [69], who showed that is is appropriate to match the grid size to the length of root to be measured, to keep N between about 100 and 500. Time can be saved using hand counting methods because organic debris does not have to be carefully separated from the roots.

(2) Automated methods. The theory already presented may also be used in systems in which the scanning of the root sample and counting of the number of times roots intersect some regular pattern are automated. In one system [71] (commercially available from Commonwealth Aircraft Corp. Ltd., 304 Lorimer St., Port Melbourne, Victoria 3207, Australia), roots are spread out on a transparent rotating turntable, which is traversed by a light beam and detector, and the number of times that roots interrupt the beam is converted to give a direct readout in meters. This machine works best with 20-40 m of total root lengths on the turntable. Greater root lengths tend to be underestimated because of the overlapping of roots, while lengths less than 20 m tend to be overestimated, although a suggested correction curve is supplied with the instrument. The measurement time is about 10 minutes, and the coefficients of variation are about 3%, or less for root lengths exceeding 10 m.

Length can also be measured using a high-resolution scanning camera [72] and an image-analyzing computer (such as the Quantimet) [73]. A computerized scanning system is commercially available from Delta-T-Devices Ltd. (128 Low Road, Burwell, Cambridge CB5 OEJ, U.K.). With those techniques roots can also be stained with methyl violet to allow ease of detection. Because of limitations of the size of field of view, however, only small samples can be measured, and time must be taken to carefully separate organic debris from the root sample. Roots can also be stained with a fluorescent dye to permit them to be distinguished from other debris when illuminated under ultraviolet light [74].

(3) Indirect methods. Root length can be calculated from counts of root numbers found in sectioned impregnated blocks of soil. Formulas are available to convert root number to root length for random [75] and strongly anisotropic [76,77] root distributions. For random root distributions, provided a reasonably large sample of randomly orientated sections are taken, assumptions inherent in the calculations are met. The equation used for random root distributions is

$$L_T = 2N \qquad (3)$$

where L_T is the total length of root per unit volume of soil (cm cm^{-3})

and N is the number of roots intersecting a plane of unit surface area (no. cm^{-2}).

Elongation Rate. Root elongation rates have been studied in relation to external pressure and pore size distribution, using pressurized ballotini-filled cells. Root elongation may be recorded as the difference in length between successive measurements made directly or on still film shots or using time-lapse cinematography [78]. Time-lapse photography can also be used on glass-fronted observation boxes, in observation tubes, or in root laboratories. Time-lapse photography allows the quantification of detailed reactions between roots and soil. The method has been used to study root nutation [79], variation in root-soil contact [80], and variation in root diameter [78,81].

Root Age. Root color and morphology are the main identifying criteria for root age. For apple, anatomical changes with age have been described in detail [82]; the roots were considered to be alive if they were turgid, and white to light brown [56]. The fluorescence of roots has been shown to decrease with root age, disappearing when suberization begins, while a positive correlation has been found between the intensity of fluorescence and rate of new root growth and ion absorption [83]. Root activity can also be assessed using stains such as tetrazolium blue, fluorescene diacetate (FDA—for mycorrhizal roots), acridine orange, or one of the many stains that change in color as a result of changes in pH (e.g., bromocresol purple). As well as using these techniques, aging by visual identification in situ can be performed when roots are visible through a glass face [70,84].

Species Identification. It is possible to identify anatomical differences between plant roots [85-87], and these differences should be visible on roots observed in sections cut from resin-impregnated blocks of soil (Section II.C.1). In observation units, roots of different plants can be distinguished on the basis of a range of characteristics including color (which can vary from translucent white to pale brown), diameter, branching pattern, and root hair development.

Distribution. Root distribution can be studied in situ using glass-walled containers [20,88] or in impregnated blocks of soil [50, 75]. Nuclear magnetic resonance techniques [53] can also be used for qualitative assessments of root and water distribution in situ in relation to soil physical and chemical factors. If care is taken with washing and dissection, root distribution can also be studied using container-grown plants [7].

Branching. In solution culture, or with an easily washed-off potting medium, where the entire root system of a plant can be extracted, the main and lateral roots can be measured and identified. Root branching can also be studied in situ using either glass-walled observation chambers or tubes [20] or a soil impregnation technique [50]. Root branching has been studied in relation to external pressure and pore size using the ballotini-filled cell methods [8].

Hackett and Bartlett [89] have given a detailed description of the density of branching and of changes in lateral length along axes, for plants grown in solution culture.

Describing the branching of root systems and characterizing the amount of branching, the pattern of branching, and the different orders of branching has proved to be difficult. Fitter [90] has demonstrated the use of a topological ("tree") system of analysis that may be of use in characterizing the nature of root system branching. This type of analysis provides a way of assessing the morphological effects of differences in soil physical properties. However, it is not clear how well this system will work on root systems extracted, probably somewhat incompletely, from soils.

Volume. Root system volume can be calculated from measurements of length and mean root diameter. Asamoah [91] measured root volume to an accuracy of ±0.025 ml using a meteorological hook gauge. This fast, nondestructive method could be used to make a time sequence of measurements on roots grown in solution culture.

III. FIELD METHODS

A. Introduction

There are no field methods that allow roots to be directly observed without their removal from the soil or the establishment in the soil of an in situ observation surface. Where soil sampling is used, because observations on the same volume of soil cannot be repeated, temporal and spatial variation may become confused in data obtained from sequential samplings. Where spatial variation is very high, as it normally is with potatoes and with trees and other woody perennials [4,93], it may prevent the detection of temporal variation. The techniques that can be used to assess root growth in the field divide into three groups.

1. *Root system removal.* Either the complete root system (excavation) or part of the root system (soil monolith, soil cores, needleboards) is either removed from the soil or measured or assessed in situ (profile wall) [1].

2. *Observation methods.* A viewing surface is inserted into the soil, either as a small unit such as a small observation window [91] or an observation tube or mini-rhizotron [93], or as a large walk-in facility [94-96].

FIG. 4 Root system of 26-year-old apple tree (Fortune/M9) excavated by the skeleton method. (From Ref. 99.)

3. *Indirect methods.* The presence of the root system is inferred from its activity (e.g., the removal of soil water [97] or the uptake of radioisotopes injected at a particular depth) [98]. These methods allow the activity of the system under a given set of conditions to be assessed. However, they do not always predict what would happen under other conditions; for example, a low level of root activity may result from few roots being present or from inactivity for one of a large number of possible reasons, of the roots that are present.

The discussion of methods in this section deals with each of these techniques in turn and is illustrated by reference to a small number of selected papers. Listing of all published variants of the basic methods has not been attempted. Most methods used in the field are simple in concept, and the emphasis has been placed on the interpretation of results that can be obtained, the types of situation in which the methods may be used, and the factors that may limit their use in studies of soil physical conditions.

B. Root System Removal

1. Excavation

For small plants, the same methods are used for total or partial excavation of the root system [91]. For large plants (e.g., mature trees), total excavation involves the removal of much more soil than by any sampling method [99]. Total excavations are useful in determining the mass and distribution of large roots and also the biomass. The large loss of fine roots during excavation [4] means that excavation is not very suitable for estimating root length. However, this method can be useful for studying the effects of soil type and of soil features, such as impeded drainage and depth of induration, on the development and adaptation of the whole root system.

Total Excavation Methods. A study of apple root systems involved the excavation of 26 fully excavated apple trees grown on either a light sand, a sandy loam, or a heavy clay loam. A total of about 1000 Mg of soil had to be moved by hand [100]. Root systems were exposed by either a skeleton or a block method. In both cases an entry trench was dug beyond the rooting volume of the tree under investigation. Beginning from this trench, soil was removed in 50 cm sections, moving systematically across the ground occupied by the root system under study. Soil was brushed away from the side of the trench with a small hand fork, leaving the root system exposed.

In the skeleton method, the root system is kept as entire as possible to allow later reconstruction (Fig. 4), while in the block method the soil is removed in blocks of known size (Fig. 5) and the roots

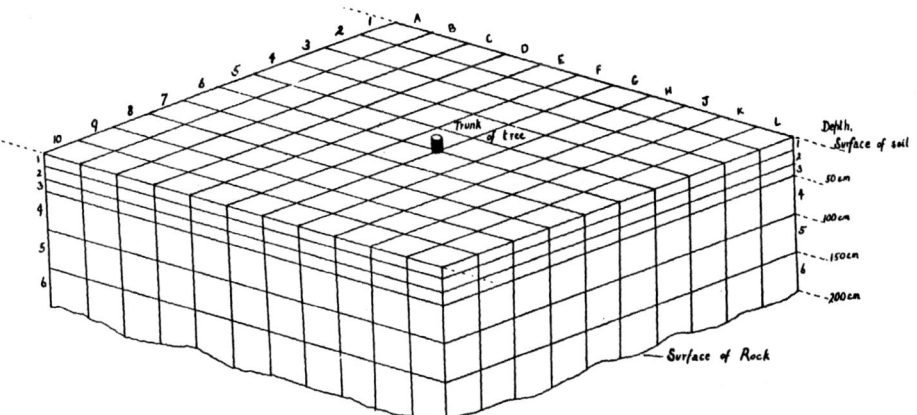

FIG. 5 Block method of tree excavation. *Source:* Rogers, W. S. and M. C. Vyvyan, Root studies: V. Rootstock and soil effect on apple root systems, *J. Pomol. Hortic. Sci.*, 12: 110-150 (1934).

TABLE 6 The Effect of Soil Type on the Root System of Apple
(Lanes Prince Albert/M1)

	Soil texture (series)		
Parameter	Sand (Wisley)	Sandy loam (Malling)	Clay loam (Wisborough Green)
Root system weight (kg)	4.2	16.3	8.5
Percentage of root weight in center 1 m of soil volume	48	34	49
Lowest percentage of root in 25% of soil volume		16	6
Percentage of root weight 50 cm depth	20	21	22
Ratio of stem to root weight	0.92	2.36	2.18
Root system spread (m)	4.4	6.0	5.1

removed, washed, and weighed [99,100]. Using one of these excavation methods, it is possible to compare horizontal and vertical root distribution, the total amount of root, and the uniformity of distribution on different soil types. Examples of the type of data that can be obtained are given in Table 6. In these studies it was found that a single quadrant (25% of the soil volume) could contain as little as 6% or as much as 51% of total root weight. This result has major implications for the choice of sampling methods and the accuracy of the data they will produce. Although time needed to sample a single individual tree, or even an individual cereal plant, will tend to limit the use of excavation, given the spatial variation inherent in tree root systems and the limited number of large roots on a tree that represent most root weight, excavation is probably the only method that will give a reliable estimate of the total biomass of a root system. The method works well for small trees and has been used as a means of studying temporal variation in root length and calculating nutrient inflow rates [91].

Partial Excavation Methods. Partial excavations have been used by a number of workers, again principally for trees [101,102] and may involve either the excavation of a section (usually one-quarter of the rooting volume [101] or of a combination of stump-pulling and root excavation [102]. It has been estimated that this latter procedure removed about 38% of root weight [66]. Partial excavation has been used to compare the effect of a number of soil types, some with marked soil physical limitations, on root weight and distribution [101].

2. Profile Wall Method

General Considerations. In this technique, a trench or pit is dug to expose a vertical soil profile from which records of partially exposed roots can be made. Horizontal areas, at different soil depths, can be prepared in the same way. It was first used by Weaver (1926) [103] and later developed by a number of other workers [104-106]. Unlike the pinboard and soil coring methods, this technique can be used on stony soils. The method is, however, labor-intensive and time-consuming, and it leads to extensive soil disturbance; in addition, it can be difficult to obtain a statistically meaningful number of replicates. Care also must be taken where root distribution is asymmetrical. Nevertheless, it has been suggested that this method gives among the most favorable ratios of information gained to labor expenses [107]. The trench is dug at a distance from the crop and can be cut back serially toward the base of the plants if information on horizontal distribution is needed. This is then similar to the "block excavation method [100]. For a row crop, the trench is usually dug across the rows. The trench can be dug by hand or using a mechanical digger, but it should be positioned so that a further layer of soil ($\sim$ 30-50 mm) can be later removed from the trench face to avoid changing roots. The exact size of the trench depends on a number of factors such as crop type and stage of development.

The Spiral Trench Method. Special considerations apply to tree crops, and for these the use of a logarithmic spiral trench has been suggested [104]. This gives an appropriate emphasis to soil at a distance from the tree trunk, where root density tends to be at a minimum but which nevertheless contributes a very large proportion of the total soil volume. In comparisons between the logarithmic spiral trench and a more conventional straight trench, the spiral trench encountered a higher average root density (0.017 roots cm^{-2} soil face area) than the straight trench (0.011 cm^{-2}) [105].

Face Preparation and Measurement. When the trench, regardless of type, has been dug, the working face of the profile is prepared using a profile knife (with a specially prepared long blade) to remove around 10-20 mm of soil, and in stone-free soils to smooth the face. In stony soils the preparation is best done with a spade, trowel, and knives. The roots exposed against the wall are then cut off with scissors and, starting at the soil surface, a further layer of soil is removed ($\sim$ 3-5 mm; 10 mm for trees), using a pick, trowel, fork, needle, screwdriver, or brush to expose the root systems [1]. Water pressure [108] and air pressure [109] have also been used. A frame containing a grid (the mesh size depending on root size and sampling strategy) is positioned over the prepared face, and the root system is recorded.

Measurements. Root numbers, length, diameter, and distribution may all be obtained from profile wall measurements and will be discussed in this order.

(1) Root number. The number of exposed roots visible in every square of the grid can be recorded, often onto a prepared sheet containing a matching array of boxes. This direct count is faster than the mapping of individual roots directly onto either graph paper or transparent sheets. However it does not show individual roots in their natural position relative to the profile wall. A direct record of this type can be used to derive the average number of roots per unit cross-sectional area, from which estimates of variation may be calculated.

A comparison of number of roots was obtained using the profile wall method by counting or draining onto foil [110]. In the counting procedure, after a soil profile had been sprayed with water, a 5 × 5 cm mesh was placed against the wall and exposed roots were counted [111]. To give the estimates a three-dimensional emphasis, visible parts of the roots (~5 mm long) were assumed to be equivalent to two roots. With the foil method, a transparent Plexiglas plate with a grid was hung over the profile. A transparent foil was fastened over the Plexiglas and each 5 mm long root marked. In general (Table 7), counts obtained with the foil method were higher than those obtained with direct counting [110]. Counts in densely rooted areas tended to be less than those obtained by the field method, although estimates of root length tended to be relatively similar (r = 0.94).

(2) Root length. Estimates of root length are based on the assumption that any root present in the soil will go back into the profile for at least the depth to which it has been exposed. In one variant of this method one root unit is set equal to a 5 mm root length for a profile that has been cut back by 5 mm, and roots 10 mm long have been counted as two root length units [111]. If root distribution is assumed to be uniform, root length per unit volume can be calculated. Estimates of root length obtained in this way are much lower than those obtained by washing roots from an undisturbed block of soil (Table 8). The cause of the underestimation is not clear, although the removal of the soil from the surface of the profile wall to expose the roots may result in the loss of some of the fine roots as well as roots growing parallel to the profile face. However, the method gives a good representation of root distribution. It seems to be most reliable for plants like trees, in which most roots are horizontally distributed. In grasses, where a large proportion of roots are vertical, many roots would be lost using the profile wall method, which would lead to an underestimation.

TABLE 7 Comparison of Two Methods of Recording
Root Data in a Profile Wall Study[a]

Soil depth (cm)	Method			
	Counting		Foil	
	no.	%	no.	%
0-15	599	61	791	64
15-30	258	27	330	27
30-45	75	8	79	6
45-60	39	4	33	3

[a]Both methods were used on the same profile wall
of 100 cm width, counting the numbers of 5 mm
root length units.

Source: Modified from Ref. 110.

(3) Root diameter. Using a small hand lens, a micrometer
screw, or calipers, the diameter of exposed roots can be measured
directly, in situ. The exposed roots may be distinguished by diameter on a root map (Fig. 6), or the number in each diameter class
may be recorded directly.

(4) Root distribution. The exact position of a root can be
mapped onto graph paper, or roots can be recorded directly onto a
transparent plastic sheet placed over the profile; in either case, root
position may be related to soil horizon, depth, boundaries, or special
features, such as large stones or earthworm channels [110,112].

TABLE 8 Comparisons of Estimates of Root Length
Density of Maize (*Zea mays*)[a]

Method	Root length density (cm cm^{-2})			
	0-20	20-40	40-60	60-100
Monolith	15.3	5.3	2.7	1.0
Profile wall	7.1	3.1	1.2	0.6

[a]Obtained from profile walls and soil monoliths
down to 100 cm depth.

Source: Böhm, W., In situ estimation of root length at natural soil
profiles, *J. Agric. Sci. Camb.*, 87: 365-368 (1976).

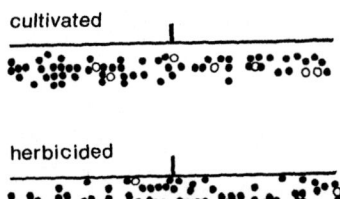

FIG. 6 Root distribution map obtained using a profile wall method and illustrating the difference in the root systems of mature apple trees, between 0 and 30 cm depth, when grown under cultivation or herbicide management: •, roots < 2 mm diameter; o, roots > 2 mm diameter. (From Ref. 106.)

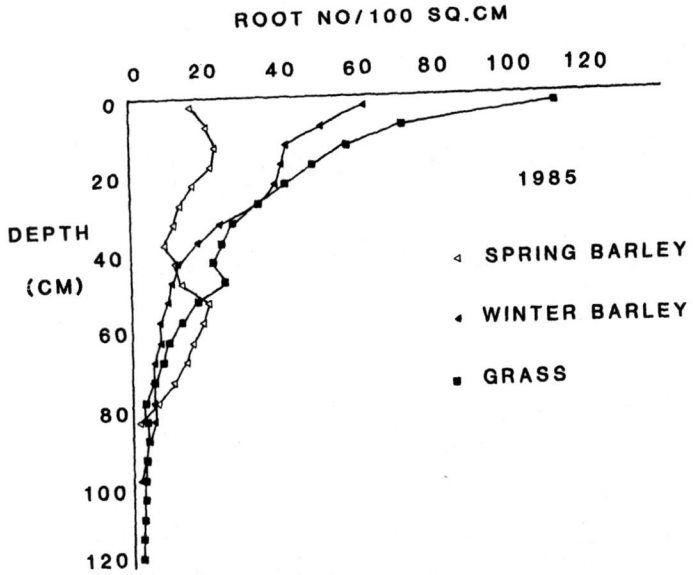

FIG. 7 Root distribution with depth, for three crops, determined by the profile wall method. (From Ref. 120.)

Pens of different colors can be used to distinguish between new and old roots.

Visual presentation of root counting can be used to demonstrate the effects of treatments that result in differences in soil physical condition (e.g., root distribution under herbicide treated or cultivated soil shown in Fig. 6). Counts can also be plotted versus depth (Fig. 7). This method has been used to study the effects of localized irrigation [113].

(5) Root volume. Root volume can be calculated from diameter and length as the volume of a cylinder [114].

3. Monolith Methods

A monolith is an undisturbed rectangular prism of soil isolated from the surrounding soil by excavation or by inserting a metal frame. Monolith methods involve the removal of a sample of soil to represent the whole or part of the rooting volume of a plant (depending on the size of the plant and of the soil volume removed). Following soil removal, samples may be washed to remove the roots from the whole or a part of the soil volume, or the roots may be held in something resembling their original position in the soil by a series of pins. This latter method is necessary where it is important to know how root system distribution is related to soil physical features.

Pinboard. The basic pinboard or needleboard method, which has been described in detail by Schuurman and Goedewaagen [15], has been modified to allow the easier extraction of roots from large soil sections and an improved photographic description. In the modified method, monoliths 152 × 41 × 91 cm deep were removed using a root extraction frame [115]. A monolith of this size is large enough to contain the root system of four corn (*Zea mays*) plants and weighs around 770 kg. After its removal, 6 mm diameter brass pins were driven through the monolith into a board on a 5 cm grid pattern, using a compressed-air gun. The monolith was then soaked, the side opposite from the pins removed, and the soil washed away, using around 7.6 kl of water per monolith at a pressure of 0.1 MPa. The root system was photographed under water, divided, dried, and weighed. This method has been used to assess the effects of treatments such as straw mulching on total root mass and on horizontal and vertical distribution [115]. A similar method was used to compare the root systems produced by *Zea mays* under a range of cultivation and compaction treatments using monoliths 70 × 25 × 60 cm deep [116].

Pinboards (35 cm long × 20 cm deep) have been used to assess effects of soil physical condition on the root system of winter wheat in a comparison between plowing and direct drilling [117]. A board sampling a volume 30 × 5 cm × 30 cm deep has been employed to study the cabbage root system [118]. Here the board with its 5 cm long pins placed at 5 cm intervals was hammered into a soil face and the board plus soil removed. Roots and soil were removed from the board as 36 samples of 5 cm^3 each, the roots were washed free of soil, and their length determined. Photographic records can be used to indicate the effects of treatments on branching, but these records can be difficult to quantify.

Modified Pinboard Methods. A method combining the relative ease of sampling given by soil coring and the preserved spatial geometry of the pinboard method has been developed [119]. Samples obtained by soil coring were encased in a perforated acrylic cylinder and the roots held in place using nylon fishing line sown through the holes in the cylinder with a needle. A 6 cm diameter tube pierced by a network of holes at 6 mm intervals was used. Soil was washed from the core and the remaining root system resuspended in 5% w/v gelatin. Like the more traditional pinboard system, this technique allows the root system's geometry to be related to soil physical characteristics. This method is more rapid and flexible than the more traditional pinboard method. However, both these methods may be superseded in time by more advanced techniques such as scanning NMR imaging [53] or neutron radiography [54] (Section II.C).

Soil Coring. Soil coring is the most frequently used method of root sampling. Coring is often used for sequential sampling of an experimental plot to give estimates of temporal changes in root length or weight, although any spatial variation may lead to misleading results. When samples are taken in relation to the planting geometry of the crops, information on the spatial distribution of their roots can be obtained. Soil core samples can be obtained from points immediately adjacent to the sites of soil physical measurements, or in some cases the same cores can be used for measurements of bulk density, pore size distribution, and root length. The published literature on this subject is very large (Ref. 1 contains an extensive listing), and the papers quoted have been selected only to illustrate some of the variations in technique that have been used and to indicate the types of information that can be obtained.

Welbank et al. [120] have described the use of a powered soil coring system to assess the growth and development of cereal root systems, and variations of this method have been used by a large number of subsequent workers. Samples were taken using 94 or 120 cm long coring tubes fitted with hardened cutting tips (Fig. 8).

To facilitate removal of the soil cores, the tubes were fitted with split liners. The coring tube was driven into the ground using a

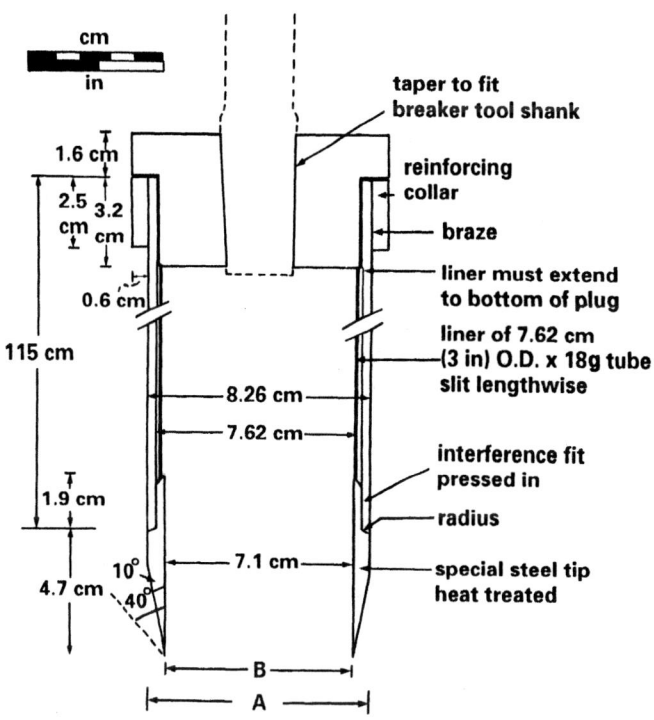

cm

in

taper to fit
breaker tool shank

1.6 cm

2.5 3.2
cm cm

reinforcing
collar

braze

0.6 cm

liner must extend
to bottom of plug

115 cm

liner of 7.62 cm
(3 in) O.D. x 18g tube
slit lengthwise

8.26 cm

7.62 cm

1.9 cm

interference fit
pressed in

radius

7.1 cm

special steel tip
heat treated

4.7 cm

10°

40°

B

A

FIG. 8 A design for a soil coring tube. (From Ref. 120.)

portable gasoline-powered motor hammer (Fig. 9a). A depth of 1 m
could be reached in moist, relatively stone-free sandy loam soil in
around 30 seconds. In dry soils and soils with stones, penetration
to this depth took much longer. Tubes were removed from the soil
using a tripod and chain hoist (Fig. 9b). After extraction, the soil
cores were divided into sections corresponding to different soil layers or soil depths. To avoid compaction of the soil core within the
liners, the internal diameter of the cutting tip was designed to be
slightly less than that of the liners.

The extent to which soil coring compresses the soil cores has
been assessed [120]. Bulk density of core samples usually showed
increases of 1-5% compared with undisturbed samples for a range of
depths, but in some soils, especially when wet, the compaction in a
1 m core could be as high as 25%. In this situation allowance must
be made for compaction before the cores are cut up. Injections of
paint to known depths have been used to assess the distribution of
soil compaction within the core and allow its correction [120].

FIG. 9 Apparatus for soil coring in the field: (above) corer insertion and (right) corer extraction.

After coring, soil is separated from the roots by washing on sieves [1].   A modification of the basic method has been described [121] as a hydropenumatic elutriation system.   This combined the kinetic energy of pressurized spray jets and the low energy of air flotation.   Air and water are used to isolate and deposit roots on a submerged sieve.   Washing times vary from 3 to 10 minutes per sample and are a function of soil type, plant species, the concentration of dispersing agent, and soaking time.   Using this equipment, nearly 100% root recovery was achieved in around 2 minutes for a sandy soil, 6 minutes for a loam, and 10 minutes for clay.   These units are available commercially from Gillison's (3033 Benzie Hwy, Benzonia, MI 49616) at a cost in 1989 of $3850 for an eight-chamber unit.

FIG. 9 (continued)

Welbank et al. [120] used root-washing cans (Fig. 10) coupled with the intersection method for measuring root length, to compare the effects of plant type and nutrition on root length, root weight, and specific root length (length per unit weight; see Table 9). Because specific root length can vary with crop variety, age, nutrition, depth, and soil physical conditions, use of some given value of specific root length to convert root mass to length is liable to systematic error (Table 9) and is inadvisable except where these values are obtained from representative subsamples taken from the actual samples being assessed. Using the extreme values given by Welbank et al., 1 g dry weight of roots could have a length as low as 33 m or as high as 199 m.

In all core-sampling studies, assessments of the proportion of recovered root that is estimated to be living can result in major

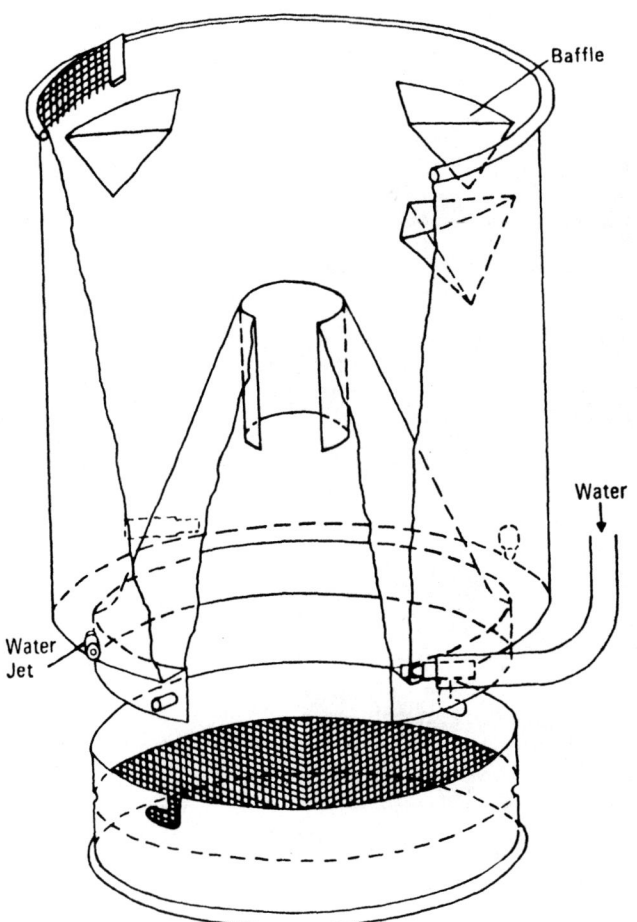

Baffle

Water

Water
Jet

FIG. 10   A can for separating roots from soil.   (From Ref. 120.)

errors in the assessment of the length (or weight) of functional
root.   A method to overcome these problems has been suggested by
Ward et al. [122].   These workers placed their soil cores in bags
made from 100 mesh (0.149 mm opening) cloth, which allowed clay,
silt, and fine sand to be washed out, but retained roots, organic
residues, and coarse sand when agitated in water.   Washed roots
and organic materials were separated from sand by flotation, stained
with 1% congo red, and saturated with 95% ethanol.   Under these
conditions, living but not dead roots stained dark pink to bright
red and were then measured by the grid intersect method (Section

TABLE 9   The Specific Root Length of Roots of Winter Wheat
(Mean of Cultivars Capelle-Desprez and Maris Ranger) in Two
Separate Years, at a Range of Depths

	Specific root length (m g^{-1} dry wt)			
	1969-1970		1970-1971	
Depth (cm)	9 Dec	14 Apr	15 Dec	28 Apr
0-15	155	198	226	217
15-25	101	187	128	200
25-35	40	105	62	82
35-45	33	49	84	84
45-55		52		89

Source: From Ref. 120.

II.C.2). These workers found that after staining by the method
above, the percentage length of a sample, measured using a grid
intersection method, increased by 18% compared with an assessment
made by laying the root system out on a dark background. The
increase was assumed to be due to an increase in the visibility of
small rootlets. A comparison of the staining of different species
indicated that monocotylendons stained more than dicotyledons.
Some dicot species (e.g., sugar beet) stained very poorly. Storage of root samples can affect the quality of staining. For example,
storage at 15°C for 35 days reduced the stainable root length of
wheat roots by 65%, whereas at 5°C the decrease was only 13% [122].

Special Factors in Relation to Trees. When sampling tree root
systems, a different range of problems is evident compared with
studies on either grassland or cereals. With graminaceous crops,
the horizontal component of the root system is relatively small, and
planting densities normally result in relatively complete soil exploita-
tion. The ratio of core volume to the soil volume exploited by an
individual plant is also relatively high. For tree crops, rooting
density varies greatly with depth, with distance from the tree, and
between samples taken at comparable positions. In one study, at a
distance of 150 cm from the trunk of apple trees and to 45 cm depth,
a high proportion, around 70%, of cores contained no roots. Even
close to the tree and near the soil surface the amount of root in a
core could be highly variable (e.g., 0-40 mg of roots <2 mm in diameter [92]. High spatial variability has also been described in the
root distribution of Douglas fir (see Table 10).

TABLE 10 Coefficients of Variation in Douglas Fir Root
Weights Obtained with a 4.5 cm Diameter Veihmeyer
Auger at Different Distances from the Tree and Root
Depths (cm).

Position in rela- tion to canopy	Coefficient of variation			
	0-15	15-46	46-77	77-107
Central 50%	95	51	40	18
110% of radius	37	84	21	31

Source: From Ref. 123.

        A count of the number of root tips recovered from 40 mm diame-
ter soil cores has been advocated as a means of assessing the distribution of fruit tree roots [56]. In this study as in the earlier
studies [92,123], spatial variation was high. Differences appeared
to be smaller in good soils compared with impeded soils. This type
of effect needs to be considered when comparing soils with different
physical characteristics.
        Overall, the large spatial variability in root distribution makes
it difficult to draw conclusions about the effect of treatments on tree
root density or distribution.

C. Observation Methods

1. Introduction

The methods described previously allow the root system to be quantified only at a single moment in time. The development of root systems in situ can be observed by creating a window into the soil.
This approach allows a sample of roots to be viewed but raises a
number of questions about the representativeness of the sample,
hence of the significance of deductions drawn from such observations. There are two main types of observation facility: large
permanent facilities and mini-rhizotrons.
        The earliest permanent root observation facilities were simple
pits with glass-lined walls [102]. These were ultimately developed
into large permanent root observation laboratories [96]. General
aspects of the methodology for this type of facility have been reviewed [124]. The alternative approach is to use an observation
tube or mini-rhizotron methods [93]. Both developments ultimately
depend on the observation of a sample of roots, and so common
criteria apply to evaluations of the significance of results.

2. Basic Criteria

The conditions that must be satisfied before data recorded from an observation surface can be converted into a root length per unit soil surface or root length per unit soil volume, are as follows [44].

1. The presence of the window must not result in an atypical root system.

2. Root density adjacent to the window either should be typical of that in a comparable volume away from the glass or should differ by a predictable amount.

3. The position of the observation panel in relation to the horizontal spread of the root system should give an acceptable representation (in respect to root density, periodicity of new growth, root turnover) of the whole soil volume exploited. This condition tends to be more important for perennial species (e.g., in trees, where there is extensive horizontal development of the root system).

4. The sample of the root system observed through the observation window is sufficiently large.

5. The periodicity of new growth at the window is similar to that elsewhere.

6. The spatial distribution of growth adjacent to the window is representative of growth elsewhere in the exploited soil volume.

The extent to which the roots being observed represent a valid sample of the growth, behavior, distribution, or density of the population as a whole will vary for different plant species and perhaps for different ages of plant material.

Calculation of Root Density. Calculations of root length density depend on the use of a conversion factor to give a generalized length from the "window length" [4,125]. This factor varies for roots of different diameter, is usually a function of species or variety, and will be influenced by the soil physical condition at the observation interface.

In a long-term study of apple, there was no difference in root density adjacent to the glass of a root laboratory and at a distance from it [4].

In a study of peas (*Pisum sativum* L.) grown in acrylic containers, rate of root elongation at the interface was lower than in the center of the container, but these differences decreased at higher soil bulk densities [126]. In another investigation with maize (*Zea mays*) and tomato (*Lycopersicon esculentum*), there was no clear difference between root weight per unit soil volume at the glass-soil interface and that in the bulk soil [127]. In the latter investigation, it was found that root length density varied widely over the observation surface, from 5 to 0.5 cm cm^{-1}. The frequency of high values increased with age. In 104-day-old plants, 60% of the sample area had a density between 0.5 and 1.9 cm cm^{-1}. In tomato, where total

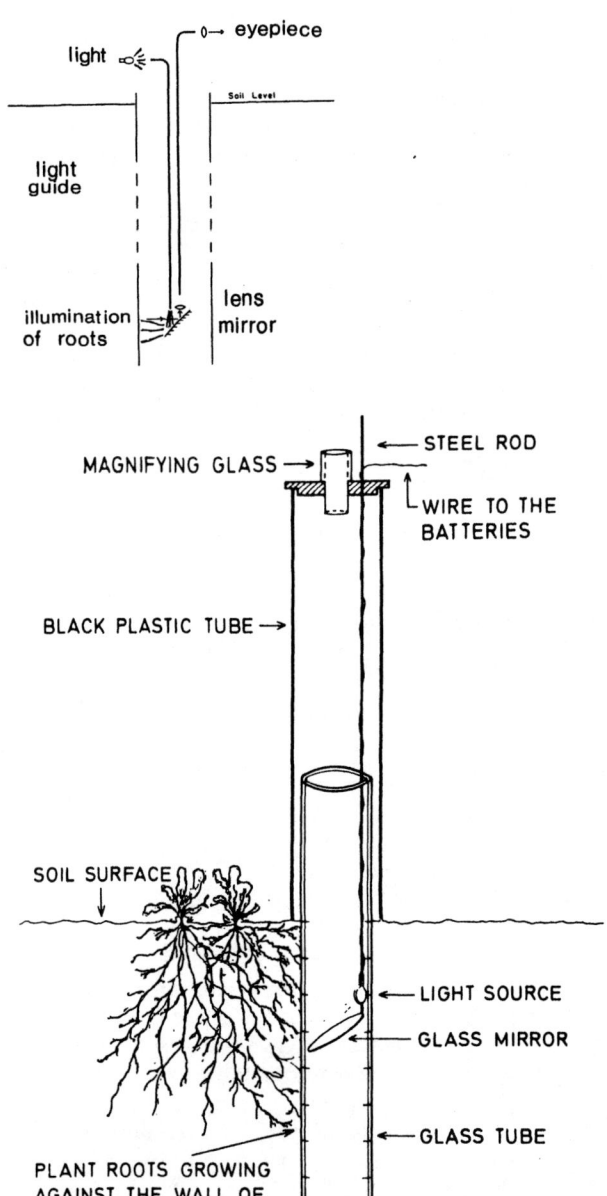

FIG. 11 Two mini-rhizotron methods using different equipment to measure root growth adjacent to the observation tube: (a) from Waddington [93] and (b) from Böhm [128].

root length was one-third of that in maize, the range of root length density was similar, but a high proportion of area had a density of less than 0.5.

3. Mini-rhizotrons

Waddington [93] used a square 5 × 5 cm glass or acrylic tube inserted into the soil at 45°, with a fiber-optics probe both to illuminate the soil and to observe the wall of the tube in a study of wheat roots (Fig. 11a). Böhm [128] used a similar system involving 64 mm diameter round glass tubes but with the soil adjacent to the tube illuminated by a bulb and a magnifying glass at the top of the observation tube (Fig. 11b). He found it necessary to pack air-dried soil around the tube to get good contact between glass and soil. Observations were made to improve the effects of cultivation on spring barley and oilseed rape.

Sanders and Brown [129] have used a medical duodenoscope to view root growth adjacent to 7.5 cm diameter tubes inserted at 45° to the soil surface and compared this method with soil coring for soybean. Root length was recorded by photographing the image of the root system, and length was estimated on the basis of an intersection method. Length at the tube-soil interface was converted into a length per unit volume, using the assumption that roots were seen up to 3 mm from the surface of the tube. A number of materials have been used for mini-rhizotron tubes. Glass gives good visibility but tends to fracture under cold conditions; acrylic tubes tend to be easier to insert and to last longer, but can scratch where angular sand grains are present.

Tube Insertion. It has been suggested that the angle of tube insertion has a significant effect on the interpretation of the results. In a comparison of vertical and 45° angle mini-rhizotrons with direct measurements from washed-out cores of oat roots, better agreement was found with the tubes at 45° [134]. Mini-rhizotron tubes have been used as access tubes for the neutron probe, so allowing comparisons of root density and soil water depletion [135]. In this study, comparison of mini-rhizotron measurements and soil core sampling showed poor correlations in the surface 20 cm.

Comparison with Other Methods. Table 11 shows the comparison between soybean root length and density estimates obtained by coring and by mini-rhizotron [129]. Estimates of root length were higher at all depths, other than 0-18 cm, using the mini-rhizotron method. Total estimated root length was similar for the two methods. The coefficient of variation for the core determinations was 53%, similar to the range quoted in the other publications reviewed [129], 61-95%. For the mini-rhizotron method, CVs were much lower. In contrast,

TABLE 11 Comparison of Root Length and Density in Soybeans
Determined by Soil Coring (SC) and the Mini-rhizotron (MR)
Methods

Depth (cm)	Root length (m)		Root distribution (%)		Root length density (cm cm^{-3})	
	SC	MR	SC	MR	SC	MR
0-18	38	25	68	46	10.6	6.8
18-36	9.5	13	17	24	2.6	3.6
36-54	2.8	4.9	5	11	0.6	1.4
54-72	5.5	10	10	19	1.5	2.8
Total	56	53				
CV%	53	14				

Source: From Ref. 129.

CVs of around 95% have been found for mini-rhizotron observations
of young apple trees [130].

In earlier work, Böhm et al. [131] compared mini-rhizotron results with those from soil water depletion, a pinboard, core sampling,
and the profile wall method for soybean on a loess soil (Tables 12
and 13). In this case the mini-rhizotron method gave a higher percentage of roots in the surface 15 cm, in contrast to the report of
Sanders and Brown [129]. In Böhm's study [131] the soil adjacent
to the tube was back-filled and the concentration of roots adjacent
to the tube was higher than in bulk soil. Gregory [132] used a

TABLE 12 The Distribution of Soybean Roots with Depth as
Determined by Four Different Methods

Soil depth (cm)	Distribution with depth (%)			
	Profile wall	Pinboard	Core samples	Mini-rhizotron
0-15	66	60	63	85
15-30	20	26	13	12
30-45	11	3	18	2
45-60	2	7	3	1
60-75	1	3	3	

Source: From Ref. 131.

TABLE 13   Root Length Density in Soybean at a Range of
Depths as Determined by Different Methods

Soil depth (cm)	Root length density (cm cm^{-3})			
	Profile wall	Pinboard	Core samples	Mini-rhizotron[a]
0-15	0.15	0.35	0.97	8.8
15-30	0.05	0.15	0.19	1.2
30-45	0.03	0.02	0.28	0.22
45-60	0.01	0.05	0.04	0.07
60-75	0.01	0.02	0.05	—

[a]Recalculated as in Ref. 129.

Source: From Ref. 131.

mini-rhizotron in a study of wheat and millet and estimated root
length by counting intersections with a 0.5 cm grid in the periscope
eye piece. Root length (R) was calculated as

$$R = \frac{\pi}{4} N \times 0.5 \times 1.18 \tag{9}$$

where N is the number of intersections, 0.5 the grid size, and 1.18
a scaling factor accounting for the difference between the field size
(40 mm) and the observed image (34 mm). As with most other studies, the mini-rhizotron method seemed to underestimate root length
at the surface compared with auger sampling. For wheat sampled at
0-60 cm depth between May and June, rhizotron results indicated a
constant rate of root growth, whereas coring showed a decrease.
Total length estimates for the two methods on a number of dates
were highly correlated ($r = 0.83$). However, the relationship between root length assessed using the two methods are different for
the two species under study [132].

For the mini-rhizotron method to be successful, a "settling in"
period following the installation of the tubes is needed. However,
measurements made using the mini-rhizotron method are more rapid,
taking around one-tenth of the time needed for other methods [129,
132,133].

TV Camera Systems.   Recently the mini-rhizotron method has
been combined with miniaturized television cameras; this combination
allows the soil adjacent to the tube to be seen on a television screen,
transferred to video tape, and later subjected to image analysis.

The method has been used by a number of workers [136-139]. The TV system increases the number of tubes that can be assessed, which is useful because results from a number of tubes are needed to give good correlations with other determinations of root length density [137]. A color video camera makes it possible to differentiate between old and new maize roots and to assess the effects of water stress [138] although again, root length near to the soil surface tends to be underestimated. Black and white equipment of this type can be obtained from Rees Instruments Ltd. (Old Woking, Surrey, GU22 9LF, U.K.). The cost for a complete system was around £5000 in 1989. Similar equipment is also available from Advance Visual Optics Ltd. (Hawk House, Halstead, Essex, 109 1HB, U.K.).

Borescopes. The alternative to the use of a TV system is to use a borescope, a rigid fibreoptic endoscopic system for remote visual inspection. This can be linked to a 35 mm camera or a video camera and the resulting film analyzed as above. A borescope allows direct observations to be made in color. Equipment of this type can be obtained from KeyMed (Southend-on-Sea, Essex, SS2 5QH, U.K.) and from ITI (Westfield, MA 01086).

Miniature Windows. The root laboratory method has been modified by the use of small observation windows set into the soil adjacent to trees receiving different trickle-irrigation treatments [91]. Rather than using an observation tube, a sheet of plate glass is held adjacent to the soil with a wooden frame. In this study soil was repacked beside the glass. Root growth can be estimated directly without the need for a periscope or other viewing device, which greatly reduces the time needed for measurements.

4.   Root Observation Laboratories

Introduction. Mini-rhizotron methods can be used in most field experiments with the observation system taken to the experiment. In contrast, root laboratories are permanent or semipermanent facilities around which experiments must be constructed. However, they allow a high degree of sophistication in measurement.

Simple Laboratories. Early observation laboratories were simply holes in the ground with glass walls and roof [94]. Simple designs of this type are still in use. The rhizotron [140] is 1.7 × 1.7 m square and 1.35 m deep, constructed largely of cement block with 12.5 mm thick glass. One of the interests of this laboratory is the study of the soil surface litter layer, and consequently the windows extend above ground level. This design necessitates a range of measures to prevent light piping from the surface down through the glass to deeper layers. The Brooms Barn observation pits [141] of the U.K. Rothamsted Experimental Station were 1.8 m × 1.2 m and 1.8 m deep with observation panels at the ends of the pits. A comparison of root growth in repacked and unrepacked soil showed that where soil was not repacked, severe soil slippage obscured root development. Comparisons were made of the root growth of sugar beet, potato, and barley, and of the effects of drought.

Large Permanent Laboratories

(1) The East Malling Laboratory. The first of the large root observation facilities was that built in 1961 in England at East Malling (now part of the Institute of Horticultural Research) [95]. A second facility to an improved design was completed in 1966 [96]. Both these laboratories provided 48 observation windows, each approximately 1 × 1.2 m in size. These windows were made from four panes of glass, approximately 50-60 cm, or from up to 24 smaller removable panes of glass. All windows used 6 mm unwired plate glass, which was engraved with 12.5 mm squares to allow root length to be assessed using an intersection method [70]. The soil was removed, before consolidation, in layers around 10 cm thick, It was air-dried, and later repacked at the same depths and approximate densities (Fig. 12). Repacking the windows in this way gave a good smear-free contact between glass and soil.

(2) Other facilities. Similar facilities, largely based on the East Malling design [96], have been constructed at sites in a number of countries [142-147], the most recent being at Pellston, Michigan [148]. Most of these have been used for studies of crop plants, but those in Michigan and in the United Kingdom (Cambridge), were built to study forest species [145,149].

(3) Window materials. In a comparison of the use of glass and acrylics in studies of soybean grown on a silt-loam soil, acrylic windows seemed to result in a greater concentration of roots at the observation surface, a poorer adhesion of soil to the surface, and shrinkage at high soil water potentials [150]. Glass therefore seems to be the best material for observation studies.

Advantages and Disadvantages of the Method. Because the capital cost of observation laboratories is higher than for any other method, it is important to characterize their main advantages and disadvantages. The principal disadvantages are as follows.
1. For good soil-glass contact, the soil adjacent to the glass must be replaced. In soils with a high sand content and in unstructured soils, this will have little effect, but the use of such materials in soil structural studies is limited.
2. Any experimental treatments that need to be investigated must be installed adjacent to the facility. The facility cannot be

FIG. 12 The East Malling root observation laboratory: (a) root
laboratory during excavation, (b) construction of the roof, (c) repacking of large windows, (d) small observation windows. (From
Ref. 95.)

used to study physical problems in the field (although the mini-
rhizotron method can be used in this way). With a limit on the
number of windows or compartments, as well as the need to guard
plants between treatments and for replication, the number of treatments that can be assessed at a given time is limited. This can be

circumvented by running root laboratory experiments and conventional well-replicated field trials in parallel. With this strategy, the observation laboratories are used to explain root effects.

3. The facilities are expensive to build and need continuous maintenance to keep the windows in good condition.

4. The ability to generalize from these measurements is incompletely understood.

5. To prevent the occurrence of light piping, the top of the observation window is usually sited below the soil surface. In addition, as with the mini-rhizotron method, there is some evidence that growth in the soil surface is underestimated [4].

6. While the laboratories can be used to assess the effects of differences in water supply and other variables, they are harder to use to assess the effect of cultivation treatments. The vibration occurring during cultivation may damage the windows.

The principal advantages of the method are as follows.

1. The ability to study the same volume of soil in a nondestructive manner, and so to document root longevity, turnover, and the times during the year when active growth occurs.

2. The opportunity to assess changes in soil and the roots of perennial crops over very long periods of time—for example, to assess the stability of channels in the soil (resulting from worm or root activity) over periods of a decade or more [151].

3. The ability, by detailed sampling behind windows, to relate root activity to soil conditions in the same place or to previously observed soil events.

Recording Methods. The commonest measurements made in observation facilities are of root length per unit area of window, the duration of root survival, and the relationship between individual roots and the soil. Root length is normally assessed by counting the numbers of intersections between roots and the grid system engraved on the windows and then converting to root length. For the 12.5 mm grid, which has often been used [76], the relationship is

$$1 \text{ insertion} = 1 \text{ cm length} \tag{5}$$

Taylor et al. [125] found the relationship

$$Y = 6.29 + 87.354x - 2.04x^2 \tag{6}$$

between the number of roots of maize (*Zea mays* L.) and tomato (*Lycopersicon esculentum* Mill.) crossing 100 cm horizontal transects at 15 cm depth intervals (y) and root length per square centimeter of viewing area (x). The viewing area was sampled at random in

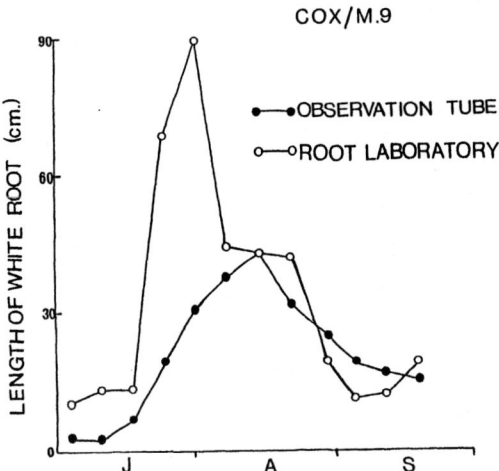

FIG. 13 Root growth of an apple tree during a 3-month period, observed using a mini-rhizotron observation tube and a root laboratory window. (From Ref. 106.)

only 7% of squares at the same date and depth (x). The correlation between the two methods of assessment was R = 0.97, suggesting that the transect method, which saves time, is adequate. A high correlation was also found between the number of intersections with selected horizontal lines and counts in the whole area for timothy (*Phleum nodosum*) [152].

In one 3-month study with apple, results by the laboratory window method diverged considerably from those obtained by observation tube for part of the period but then agreed closely over the final 7 weeks (Fig. 13).

Root survival, turnover, and longevity can be assessed by marking new roots at their initiation on strips of paper placed along particular horizontal or vertical grid lines or by tracings of whole or parts of windows. These records can then be compared with root presence on a series of successive weeks and the survival of particular roots assessed [4,96]. For perennial roots, changes in root diameter can be measured using the grid system to reidentify roots, using a microscope fitted with a micrometer eyepiece [4,84]. Relationships between individual root performance and soil features (e.g., root growth beside an indurated layer), can be assessed either by direct measurement or by the use of time-lapse cinematography, which allows detailed analysis of root development (Fig. 14) [78].

(a)

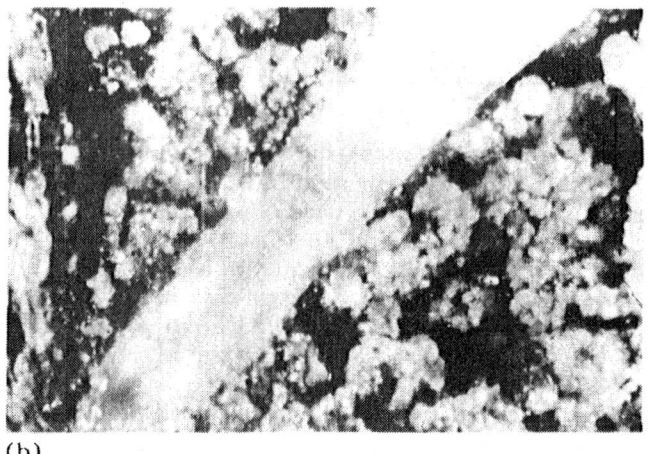

(b)

FIG. 14 Soil as seen in a root laboratory (a) before and (b) after
the growth through it of an apple root.

Use to Assess Effects of Soil Physical Properties. The number
of studies relevant to effects of soil physical properties are too numerous to document completely, although examples may indicate what
is possible. Root laboratories have been used to assess the effect
of varying rates of soil moisture depletion, induced by different
planting densities, on root growth and distribution [153]. Root

TABLE 14 The Effect of Soil Water Supply on Root Growth, Assessed Using Rhizotrons

Crop	Root measurement	Effect of drought/poor water supply	Ref.
Cotton	Depth	Increased	155
Corn (maize)	Rooting density	Inflow height at low root density	156
Cotton	Elongation rate	Stopped at water content, below 0.06 cm^3 cm^{-3}	157
Cotton	Rooting depth	Reduced new root growth	158
Cotton	Root distribution	Deepened	159
White oak	Elongation rate and number	Reduced below -0.3 MPa	160
Soybean	Depth	Increased water depletion depth	161
Soybean	Cumulative growth	Increased length, number, depth	162
Soybean	Daily growth	Increased growth	163

distribution was deeper where soil water deficits were higher. Similar results were obtained in studies where different soil moisture deficits arose from growing trees under bare or grassed soil [152]. In these studies [4,154], it was possible to relate root development at various depths in the soil to soil factors in the same layers. With sugar beet, potato, and barley, the root growth was usually 10-15 cm deeper than the maximum depth of soil moisture extraction [141]. A large number of studies have assessed the effects of soil water supply on root development [155-163] (Table 14).

Clearly, observation methods can be used to assess the effects of soil physical condition on root growth, although there may be problems in accurately assessing growth in the surface soil.

D. Indirect Methods of Assessing the Activity of Root Systems

1. Introduction

The major functions of roots in soil are to absorb water and nutrients. By assessing the depletion of water or the removal of nutrients from soil, the distribution of roots in soil or their activity can sometimes be inferred. Correlations may, however, be somewhat variable, and for many reasons root distribution may be only imperfectly related to activity.

2. Estimation from Soil Water Use

The rate of water depletion from the soil depends on both soil water potential, which influences the rate of water movement to the root surface, and root density (for the depth distribution of most root systems, axial resistance can be assumed to be small). When a root system begins to use water from moist soil, the pattern of water extraction and the rates of water use at various depths will reflect root density at those depths. However, as soon as roots begin to deplete soil water—and this will occur most rapidly in the zones with the highest root length density—the rate of depletion in these zones will begin to fall and the relationship between water depletion and root distribution will change. This relationship will be complicated by the redistribution of water in the soil profile, by rainfall, and by new root growth. Notwithstanding these considerations, a number of studies have attempted to relate roots and water. Water depletion, measured with a neutron probe, has been used as a means of estimating root distribution and density in a citrus orchard [164]. There was a good relationship between the cumulative moisture depletion with depth over the season and root distribution with depth. A higher percentage of root weight in the soil surface resulted in a higher depletion of soil moisture and it was concluded that "the rate of moisture depletion . . . indicated rather well differences in root density with depth among the soil profiles" [164]. In a soil that is not rewetted at depth during the growing season, the water content at any depth will remain constant after drainage has ceased until there is water extraction by roots. Thus, by monitoring profile water content at a range of depths with a neutron probe it has been possible to infer the progressive downward growth of the root system of annual crops from the dates on which there is a sudden drop in the water content at each depth [97]. It was also suggested that "the progress of the drying from down a soil profile allows the maximum depth from which quantities of water are extracted by crop roots," the "effective rooting depth" of the crop, to be specified.

With annuals, in the absence of rainfall or irrigation, there is progressive downward depletion of soil water as the root system extends. However, with perennials the main root system is already present, and water is initially lost more quickly in regions with greater rooting density.

Water depletion can act as a general guide to rooting in annual crops, but it is probably not suitable as a detailed indicator of root activity.

3. Estimation from the Uptake of Radioisotopes

Basic Principles. Estimate of root activity based on the uptake of a radioisotope, most usually ^{32}P, from a range of depths and horizontal positions, have been produced by a number of workers. The most usual basis of the method is that by taking a block of similar plants, a number of treatment plots can be established where the radioactive material is injected to a range of soil depths. To any one plant, all injections are made to a single depth (or a combination of relatively similar depths). At time intervals after the isotope injection, samples of foliage are taken and the amount of activity determined. After correction of values for dilution in the soil by unlabeled material, the relative root activity at a range of depths can be established.

The use of tracers to measure nutrient uptake depends on isotopic depletion [165], that is,

$$N_p = \frac{T_p \times N_s}{T_s} \qquad (7)$$

where T_p and N_p are the quantities of tracer and nutrient absorbed by the plant, T_s is the quantity of tracer added to soil, and N_s is the quantity of labile nutrient.

For N_p to be a valid estimate of nutrient absorption, the following conditions are important.

1. The application of tracer must not alter absorption from the zone to which tracer is added. The concentration of labile ions must not be appreciably altered. This is satisfied by using carrier-free tracer.
2. Labile nutrients in the soil must be uniformly labeled with added tracer. This is difficult to achieve in field experiments.

The requirements for satisfactory field experiments in both uniform and nonuniform soils have been detailed by Newbould and Taylor [165]. This method seems to have been fairly effective for annual crops, especially those with a high root length density, but rather less effective for perennial crops and plants with low root densities.

Radioisotope Placement in Soil. In a study of cotton root growth [166], a radioisotope was placed at 15 cm intervals over a distance of 122 cm along the planting row. Next carrier-free ^{32}P (4 ml) was placed in 7 ml gelatin capsules, which were dropped into each hole. Each capsule was dissolved in 100 ml of warm water poured down the access hole, which was then covered, and uptake was assessed by periodic sampling of young mainstem leaves. In this trial, plants were at a 102 cm spacing in the row, and the injections, 960 μCi (35 M Bq.) of ^{32}P per plot, were made at depths of 30, 61, 91, 122, 152, and 183 cm, 2 weeks after emergence. All placements were replicated three times. A similar method was used to assess the effect of cultivation and soil physical condition on root activity in barley [167]. Tomar et al. [168] used a similar method in a study of the effect of subsoil compaction on wheat with 500 µCi (18.5 M Bq.) of ^{32}P placed at depth but, in common with a number of other studies [169-171], variation between plots prevented definitive conclusions about the effects of treatments being drawn.

In studies of soil effects on fruit trees, similar problems relating to high variability were found [172]. Here coefficients of variation were often in excess of 50%. The distribution of activity within the tree canopy was also variable, with up to 43% of total activity being found in one-quarter of the canopy. Despite this variation, a good relationship between new root growth and the uptake of ^{32}P injected into the soil at a number of depths was found. The variation detected was considered to be a consequence of the limited movement of ^{32}P in soils and the low density of roots found on fruit trees. Where ^{15}N, which has a higher mobility in soil, was used as a tracer, sample-to-sample variability decreased greatly [152,169]. The use of ^{15}N as a tracer is complicated by the effects of the release of N by mineralization, giving variable dilution of added tracer. However, at times of the year when mineralization is likely to be small and for comparisons of similar soil depths, the method is effective. In a study where the CV for ^{32}P was 53%, it was only 27% for ^{15}N [169].

Clearly this method can be used to assess whether a plant obtains added tracer for a particular soil zone and so can be used to delimit the size of the soil volume being exploited. Where variation is not excessive, differences between soil physical treatments can be assessed.

Radioisotope Injection in Plants. In the method described above, the radioisotope is placed in soil and its uptake assessed by counting foliage. In the converse of this approach, a radionuclide is injected into a plant and its presence in the root system is detected by counting soil plus roots. Ellis and Barnes [173] have used this method to assess root activity. They injected 1-2 µl of ^{86}Rb (5-10 µCi (185-370 K Bq.) per plant) into the leaf sheath at the base of a ryegrass or barley plant. Above-ground material was removed 24 hours after injection, and 5 cm diameter cores of soil were taken, divided into 7.5 cm depth sections and bulked into 3 kg lots for counting. The soil was packed into 17 cm diameter polyethylene containers with re-entrant cavities in their bases, which fitted over a sodium iodide crystal activated with thallium. Samples were counted using the 1.08 MeV radiation of the ^{86}Rb. Repacking of samples normally affected results by no more than 3%. Calibration of this method against the root mass in core samples gave close agreement for both ryegrass and barley. Agreement was better 9 weeks after sowing than at maturity. Comparison of results obtained with the ^{86}Rb technique with soil

placement of ^{32}P showed a degree of similarity except that root
activity indicated using ^{32}P was relatively higher at 12.5-22.5 cm
depth.

IV. CONCLUSIONS

There is no single method of root measurement applicable for all
situations calling for assessments of the effects of soil physical condition. The principal factors influencing the choice of methods are
likely to be the availability of equipment and facilities, the crop
and/or soil to be investigated, and the type of root system effect
of interest. These considerations will essentially select the methods
to be used. Just as a range of measurements is made, in dealing
with the above-ground part of the plant, so with the below-ground
portion a range of measurements is needed to be able to characterize important factors such as the rate and type of growth, the
standing crop of roots, and their activity.

REFERENCES

1. Bohm, W., *Methods of Studying Root Systems*, Springer-Verlag, Berlin, 1979.
2. Russell, R. S., *Plant Root Systems—Their Function and Inter-action with the Soil*, McGraw-Hill, London, 1977.
3. Greacen, E. L., and J. S. Oh, Physics of root growth, *Nature (New Biol.)*, 235: 24-25 (1972).
4. Atkinson, D., Spatial and temporal aspects of root distribution as indicated by the use of a root observation laboratory, in *Ecological Interactions in Soil* (A. H. Fitter, D. Atkinson, D. Read, and M. B. Usher, Eds.), Blackwell, Oxford, 1985, pp. 43-65.
5. Coutts, M. P., Root architecture and tree stability, *Plant Soil*, 71: 171-188 (1983).
6. Blackman, P. G., and W. J. Davies, Root to shoot communica-tion in maize plants and the effects of soil drying, *J. Exp. Bot.*, 36: 39-48 (1985).
7. Schuurman, J. J., Influence of soil density on root development and growth of oats, *Plant Soil*, 22: 352-374 (1965).
8. Goss, M. J., Effects of mechanical impedance on root growth in barley (*Hordeum volgare* L.): I. Effects on the elongation and branching of seminal root axis, *J. Exp. Bot.*, 28: 96-111 (1977).
9. Gooderham, P., Some aspects of soil compaction, root growth and crop yield, *Agric. Prog.*, 52: 33-44 (1977).
10. Wiersum, L. K., The relationship of the size and the structural rigidity of pores to their penetration by roots, *Plant Soil*, 9: 75-85 (1957).

11. Logsdon, S. D., J. C. Parker, and R. B. Renau, Root growth as indicated by aggregate size, *Plant Soil*, 99: 267-275 (1987).

12. Tardieu, F., and H. Manichan, Characterization of the maize root system under field conditions as a water sink: I. Criteria for study, *Agronomie*, 6: 345-354 (1986).

13. Drew, M. C., Function of root tissues in nutrient and water transport, in *Root Development and Function* (P. J. Gregory, J. V. Lake, and D. A. Rose, Eds.), Cambridge University Press, Cambridge, 1987, pp. 71-102.

14. Partridge, N. L., A container for growing plants for root studies, *J. Am. Soc. Agron.*, 32: 907-908 (1940).

15. Schuurman, J. J., and M. A. J. Goedewaagen, *Methods for the Examination of Root Systems and Roots*, Pudoc, Wageningen, 1971.

16. Flocher, W. J., and H. Timm, Plant growth and root distribution in layered sand columns, *Agron. J.*, 61: 530-534 (1969).

17. Weaver, J. E., J. Kramer, and M. Reed, Development of root and shoot of winter wheat under field environment, *Ecology*, 5: 26-50 (1924).

18. Gliemeroth, G., Untersuchungen über Ausbildung und Herstung der Keim-und Kronenwurzehn bei Sommergetreide, *Z. Acker. Pflanzenbau*, 103: 1-21 (1957).

19. Atkinson, D., *Variation in Root Distribution in Spring Barley*, in Rep. Macaulay Inst. Soil Res. for 1986 (1987), pp. 174-181.

20. Nilsson, H. E., Method for the study of roots and root diseases under controlled culture conditions, *Swedish J. Agric. Res.*, 3: 79-88 (1973).

21. Wasterlund, I., and M. Johansson, Root observation technique including soil moisture measurement, in *Proc. Symp. Root Physiology and Symbiosis*, CNRF, Nancy, France, 1978, pp. 498-502.

22. Rogers, W. S., Root studies: IX. The effect of light on growing apple roots: A trial with root observation boxes, *J. Pomol. Hortic. Sci.*, 17: 131-140 (1939).

23. Walker, J. M., and S. A. Barber, Ion uptake by living plant roots, *Science*, 133: 881-882 (1961).

24. Garwood, E. A., Some effects of soil water conditions and soil temperature on the roots of grasses and clover: 2. Effects of variation in the soil water content and in soil temperature on root growth, *J. Br. Grassl. Soc.*, 23: 117-128 (1968).

25. McCully, M. E., Selected aspects of the structure and development of field-grown roots with special reference to maize, in *Root Development and Function* (P. J. Gregory, J. V. Lake, and D. A. Rose, Eds.), Cambridge University Press, 1987, pp. 53-70.

26. Atkinson, D., The nutrient requirements of fruit trees: Some current considerations, *Adv. Plant Nutr.*, 2: 93-128 (1986).

27. Baligar, V. C., V. E. Nash, F. D. Whisler, and D. L. Myhre, Sorghum and soybean growth as influenced by synthetic pans, *Commun. Soil Sci. Plant Anal.*, 12: 97-107 (1981).

28. Eavis, B. W., and H. M. Taylor, Transpiration of soybeans as related to leaf area, root length and soil water content, *Agron. J.*, 71: 441-445 (1979).

29. Brown, D. A., and A. ul-Haq, A porous membrane-root culture technique for growing plants under controlled soil conditions, *Soil Sci. Soc. Am. J.*, 48: 692-695 (1984).

30. McLaughlin, M. J., and A. M. Alston, Transformations and movement of P in the rhizosphere, *Plant Soil*, 97: 391-399 (1986).

31. Belford, R. K., Collection and evaluation of large soil monoliths for soil and crop studies, *J. Soil Sci.*, 30: 363-373 (1979).

32. Taylor, H. M., and H. R. Gardner, Relative penetrating ability of different plant roots, *Agron. J.*, 52: 479-581 (1960).

33. Taylor, H. M., and H. R. Gardner, Use of wax substrates in root penetration studies, *Soil Sci. Soc. Am. Proc.*, 24: 79-81 (1960).

34. Bhat, K. K. S., A low cost, easy to install flow culture system, suitable for use in a constant environment cabinet, *J. Exp. Bot.*, 31: 1435-1440 (1980).

35. Peterkin, J. H., Plant growth and nitrogen nutrition in relation to temperature, Ph.D. thesis, University of Sheffield, 1981.

36. Robinson, D., and I. Rorison, A quantitative analysis of the relationship between root distribution and nitrogen uptake from soil by two grass species, *J. Soil Sci.*, 36: 71-85 (1985).

37. Drew, M. C., Comparison of the effects of a localized supply of phosphate, nitrate, ammonium and potassium on the growth of the seminal root system and the shoot in barley, *New Phytol.*, 75: 479-490 (1975).

38. Brouwer, R., and E. A. Loen, Growth and uptake of individual roots of *Zea mays* L., *Mededeling 176 van Het IBS*, 19-25 (1962).

39. West, D. W., Water use and sodium chloride uptake by apple trees: 1. The effect of non-uniform distribution of sodium chloride in the root zone, *Plant Soil*, 50: 37-49 (1978).

40. Lawlor, D. W., Absorption of polyethylene glycols by plants and their effects on plant growth, *New Phytol.*, 69: 501-513 (1970).

41. Jarvis, P. G., and M. S. Jarvis, The water relations of tree seedlings: V. Growth of root respiration in relation to osmotic potential of the root medium, in *Water Stress in Plants* (B. Slavik, Ed.), Junk, The Hague, 1965, pp. 1167-1182.

42. Barley, K. P., The effect of mechanical stress on the growth of roots, *J. Exp. Bot.*, 13: 95-110 (1962).

43. Barley, K. P., Influence of soil strength on growth of roots, *Soil Sci.*, 96: 175-180 (1963).

44. Abdalla, A. M., D. R. P. Hettiaratchi, and A. R. Reece, The mechanics of root growth in granular media, *J. Agric. Eng. Res.*, 14: 236-248 (1969).

45. Goss, M. J., and M. C. Drew, *Effects of Mechanical Impedance on Growth of Seedlings*, Rep. ARC Letcombe Lab. for 1971, 1972, pp. 35-42.

46. Russell, R. S., and M. J. Goss, Physical aspects of soil fertility—The response of roots to mechanical impedance, *Neth. J. Agric. Sci.*, 22: 305-318 (1974).

47. Gill, W. R., and R. D. Miller, A method for study of the influence of mechanical impedance and aeration on the growth of seedling roots, *Soil Sci. Soc. Am. Proc.*, 20: 154-157 (1956).

48. Altemuller, H. J., Fluorescent light microscopy of soil/root interactions, in *Trans. 13th Int. Congr. Soil Science*, Hamburg, 1986, pp. 1546-1547.

49. Lund, Z. F., and H. O. Beals, A technique for making thin sections of soil with roots in place, *Soil Sci. Soc. Am. Proc.*, 29: 633-635 (1965).

50. Darbyshire, J. F., L. Robertson, and L. A. Mackie, A comparison of two methods of estimating the soil pore network available to protozoa, *Soil Biol. Biochem.*, 17: 619-624 (1985).

51. Tippkotter, R., K. Ritz, and J. F. Darbyshire, The preparation of soil thin sections for biological studies, *J. Soil Sci.*, 37: 681-690 (1986).

52. Altemuller, H. J., and T. Haag, Mikroskopinche untersuchungen on moiswurzeln im ungerstorten bodenver band, *Kali Briefe*, 16: 349-363 (1983).

53. Bottomley, P. A., H. H. Rogers, and T. H. Foster, NMR imaging shows water distribution and transport in plant root systems in situ, *Proc. Nat. Acad. Sci, U.S.A.*, 83: 87-89 (1986).

54. Willat, S. T., R. G. Struss, and H. M. Taylor, In situ root studies using neutron radiography, *Agron. J.*, 70: 581-586 (1978).

55. Baldwin, J. P., and P. B. Tinker, A method for measuring the lengths and uptake patterns of two interpenetrating root systems, *Plant Soil*, 37: 209-213 (1972).

56. Weller, F., A method for studying the distribution of absorbing roots of fruit trees, *Exp. Agric.*, 7: 351-361 (1971).

57. Richards, D., and R. N. Rowe, Root-shoot interactions in peach: The function of the root, *Ann. Bot.*, 41: 1211-1216 (1976).

58. Atkinson, D., and J. S. Chauhan, The effect of paclobutrazol on the water use of fruit plants at two temperatures, *J. Hortic. Sci.*, 62: 421-426 (1987).

59. Morrison, I. K., and K. A. Armson, The rhizometer—A device for measuring the roots of tree seedlings, *For. Chron.*, 44: 21-23 (1968).

60. Kemph, G. S., Measuring fibrous roots with a leaf area meter, *J. Range Manage.*, 29: 85-86 (1976).

61. Dunham, C. W., Use of methylene blue to evaluate rooting of cuttings, *Proc. Am. Soc. Hortic. Sci.*, 72: 450-453 (1958).

62. Carley, H. E., and R. D. Watson, A new gravimetric method for estimating root surface area, *Soil Sci.*, 102: 289-291 (1966).

63. Parlychenko, I. K., *Root Systems of Certain Forage Crops in Relation to the Management of Agricultural Soils*, Natl. Res. Comb. Can. Dominion, NRC 1088, Dept. Agric., Ottawa, 1942.

64. Goss, M. J., A. R. Dexter, and M. Evans, Mechanics of root elongation and the effects of 3,5-di-iodo-4-hydroxybenzoic acid (DIHB), *Plant Soil*, 99: 211-218 (1987).

65. Bhat, K. K. S., Nutrient inflows into apple roots, *Plant Soil*, 71: 371-380 (1983).

66. Atkinson, D., D. Naylor, and G. Coldrick, The effect of tree spacing on the apple root system, *Hort. Res.*, 16: 89-105 (1976).

67. Newman, E. I., A method of estimating the total length of root in a sample, *J. Appl. Ecol.*, 3: 139-145 (1966).

68. Marsh, B. a'B., Measurement of length in random arrangements of lines, *J. Appl. Ecol.*, 8: 265-267 (1971).

69. Tennant, D., A test of a modified line intersect method of estimating root length, *J. Ecol.*, 63: 995-1002 (1975).

70. Head, G. C., Estimating seasonal changes in the quantity of white unsuberized root on fruit trees, *J. Hortic. Sci.*, 41: 197-206 (1966).

71. Richards, D., F. M. Foubran, G. N. Garwali, and M. W. Daly, A machine for determining root length, *Plant Soil*, 52: 69-76 (1979).

72. Harris, G. A., *Root Length Measurements with a Modified Delta-T Area Meter*, Delta-T, Cambridge, U.K., 1986.

73. Baldwin, J. P., P. B. Tinker, and F. H. C. Marriott, The measurement of length and distribution of onion roots in the field and the laboratory, *J. Appl. Ecol.*, 8: 543-554 (1971).

74. McGowan, M., M. J. Armstrong, and J. A. Corrie, A rapid fluorescent dye technique for measuring root length, *Exp. Agric.*, 19: 209-216 (1983).

75. Melhuish, F. M., and A. R. G. Lang, Quantitative studies of roots in soil: I. Length and diameter of cotton roots in a clay-loam soil by analysis of surface ground blocks of resin-impregnated soil, *Soil Sci.*, 106: 16-22 (1968).

76. Melhuish, F. M., and A. R. G. Lang, Quantitative studies of roots in soil: II. Analysis of non-random populations, *Soil Sci.*, 112: 161-166 (1971).

77. Lang, A. R. G., and F. M. Melhuish, Lengths and diameters of plant roots in non-random populations by analysis of plane surfaces, *Biometrics*, 26: 421-431 (1970).

78. Atkinson, D., and J. K. Lewis, Time-lapse cinematographic studies of fruit tree root growth, *J. Photogr. Sci.*, 27: 255-257 (1979).

79. Head, G. C., Studies of diurnal changes in cherry root growth and nutational movements of apple root tips by time lapse cinematography, *Ann. Bot.*, 29: 219-224 (1965).

80. Atkinson, D., and S. A. Wilson, The root-soil interface and its significance for fruit tree roots of different ages, in *The Root-Soil Interface* (J. L. Harley and R. S. Russell, Eds.), Academic Press, London, 1979, pp. 259-271.

81. Huck, M. G., B. Klepper, and H. M. Taylor, Diurnal vatiation in root diameter, *Plant Physiol.*, 45: 529-530 (1970).

82. MacKenzie, K. A. D., The development of the endodermic and phi layer of apple roots, *Protoplasma*, 100: 22-32 (1979).

83. Dyer, D. J., and D. A. Brown, Relationship of fluorescent intensity to ion uptake and elongation rates of soybean roots, *Plant Soil*, 72: 127-134 (1983).

84. Head, G. C., Seasonal changes in the diameter of secondarily thickened roots of fruit trees in relation to the growth of other parts of the tree, *J. Hortic. Sci.*, 43: 275-282 (1968).

85. Schwaar, J., Lebende wurzeln in hoch und niedermooren, *Telma*, 2: 73-82 (1972).

86. Schwaar, J., Wurzeluntersuchungen auf moorboden, *Telma*, 3: 119-136 (1973).

87. Schwaar, J., Wurzeluntersuchungen ous neidermooren, *Ber. Dtsch. Bot. Ges.*, 84: 745-757 (1971).

88. Schumacher, R., F. Fankhouser, and E. Schlapfer, Development of apple roots; influence of the growth retardant Alar (2,2-dimethylhydrazide of succinic acid) on root development, *Schweiz. Z. Obst. Wein*, 107: 409-452 (1969).

89. Hackett, C., and B. O. Bartlett, A study of the root system of barley: III. Branching pattern, *New Phytol.*, 70: 409-413 (1971).

90. Fitter, A. H., Functional significance of root morphology and root system architecture, in *Ecological Interactions in Soil* (A. H. Fitter, D. Atkinson, D. J. Read, and M. B. Usher, Eds.), Blackwell, Oxford, 1985, pp. 87-106.

91. Asamoah, T. E. O., Fruit tree root system: Effects of nursery and orchard management and some consequences for growth, nitrient and water uptake, Ph.D. thesis, University of London, 1984.

92. Atkinson, D., and S. A. Wilson, The growth and distribution of fruit tree roots: Some consequences for nutrient uptake, in *Mineral Nutrition of Fruit Trees* (D. Atkinson, J. E. Jackson, R. O. Sharples, and W. M. Waller, Eds.), Butterworths, London, 1980, pp. 137-150.

93. Waddington, J., Observation of plant roots in situ, *Can. J. Bot.*, 49: 1850-1852 (1971).

94. Rogers, W. S., Root studies: VIII. Apple root growth in relation to rootstock, soil, seasonal and climatic factors, *J. Pomol. Hortic. Sci.*, 17: 99-130 (1939).

95. Rogers, W. S., and G. C. Head, *A New Root-Observation Laboratory*, Rep. E. Malling Res. Stn. for 1962, 1963, pp. 55-57.

96. Rogers, W. S., The East Malling root observation laboratories, in *Root Growth* (W. J. Whittington, Ed.), Butterworths, London, 1969, pp. 361-376.

97. McGowan, M., Depths of water extraction by roots, in *Isotopes and Radiation Techniques in Soil Physics and Irrigation Studies*, IAEA, Vienna, 1974, pp. 435-445.

98. Newbould, P., R. Taylor, and K. R. Howse, The absorption of phosphate and calcium from different depths in soil by swards of perennial ryegrass, *J. Br. Grassl. Soc.*, 26: 201-208 (1971).

99. Atkinson, D., *The Root System of Fortune/M9*, Rep. E. Malling Res. Stn. for 1971, 1972, pp. 72-78.

100. Rogers, W. S., and M. C. Vyvyan, Root studies: V. Rootstock and soil effect on apple root systems, *J. Pomol. Hortic. Sci.*, 12: 110-150 (1934).

101. Coker, E. G., Root studies: XII. Root systems of apple on Malling rootstocks on five soil series, *J. Hortic. Sci.*, 33: 71-79 (1958).

102. Dudney, P. J., *On the Estimation of Root Biomass in a Growth Pattern Experiment on Apples*, Rep. E. Malling Res. Stn. for 1971, 1972, pp. 66-67.

103. Weaver, J. E., *Root Development of Field Crops*, McGraw-Hill, New York, 1926.

104. Oskamp, J., and L. P. Batjer, Soils in relation to fruit growing in New York: II. Size, production and rooting habit of apple trees on different soil types in the Hitton and Morton areas, Monroe County, *Cornell Univ. Agric. Exp. Stn. Bull.*, 550: 1-45 (1932).

105. Huguet, J. G., A new method of studying the rooting of perennial plants by means of a spiral trench, *Ann. Agron.*, 24: 707-731 (1973).

106. Gurung, H. P., The influence of soil management on root growth and activity in apple trees, M. Phil. thesis, University of London, 1979.

107. Kopke, U., A comparison of methods for measuring root growth of field crops, *Z. Acker. Pflanzenbau*, 150: 39-49 (1981).

108. Kummerow, J., D. Krause, and W. Jon, Root systems of Chaparral shrubs, *Oecologia (Berlin)*, 29: 163-177 (1977).

109. Weir, L. C., The use of compressed air to escavate the roots of forest trees, *Can. Dept. Rural Dev. Res. Notes*, 22: 1-2 (1966).

110. Böhm, W., and U. Kopke, Comparative root investigations with two profile wall methods, *Z. Acker. Pflanzenbau*, 144: 297-303 (1977).

111. Böhm, W., In situ estimation of root length at natural soil profiles, *J. Agric. Sci. Camb.*, 87: 365-368 (1976).

112. Reijmerink, A., Microstructure, soil strength and root development of asparagus on loamy sands in the Netherlands, *Neth. J. Agric. Sci.*, 21: 24-43 (1973).

113. Levin, I., R. Assaf, and B. Bravdo, Soil moisture and root distribution in an apple orchard irrigated by tricklers, *Plant Soil*, 52: 31-40 (1979).

114. Bhaskaran, A. R., and D. C. Chakrabarty, A preliminary study on the variations in the soil binding capacity of some grass roots, *Indian J. Agron.*, 10: 326-330 (1965).

115. Nelson, W. W., and R. R. Allmaras, An improved monolith method for excavating and describing roots, *Agron. J.*, 61: 751-754 (1969).

116. Chaudhary, M. R., and S. S. Prihar, Root development and growth response of corn following mulching, cultivation and inter-row compaction, *Agron. J.*, 66: 350-355 (1974).

117. Finney, J. R., and B. A. G. Knight, The effect of soil physical conditions produced by various cultivation systems on the root development of winter wheat, *J. Agric. Sci. Camb.*, 80: 435-442 (1973).

118. Goodman, D., and D. J. Greenwood, Distribution of roots, water and nutrients beneath cabbage grown in the field, *J. Sci. Food Agric.*, 27: 28-36 (1976).

119. Gooderham, P. T., A simple method for the extraction and preservation of an undisturbed root system from a soil, *Plant Soil*, 31: 201-204 (1969).

120. Welbank, P. J., M. J. Gibb, P. J. Taylor, and E. D. Williams, *Root Growth of Cereal Crops*, Rep. Rothamsted Exp. Stn. for 1973, Vol. 2, 1974, pp. 26-66.

121. Smucker, A. J. M., S. L. McBurney, and A. K. Srivastava, Quantitative separation of roots from compacted soil profiles by the hydropneumatic elutriation system, *Agron. J.*, 74: 500-503 (1982).

122. Ward, K. J., B. Klepper, R. W. Rickman, and R. R. Allmaras, Quantitative estimation of living wheat root lengths in soil cores, Agron. J., 70: 675-677 (1978).

123. Reynolds, E. R. C., Root distribution and the cause of its spatial variability in Pseudotsuga taxifolia (Poir) Britt, Plant Soil, 32: 501-517 (1970).

124. Huck, M. G., and H. M. Taylor, The rhizotron as a tool for root research, Adv. Agron., 35: 1-35 (1982).

125. Taylor, H. M., M. G. Huck, B. Klepper, and Z. R. Lund, Measurement of soil grown roots in a rhizotron, Agron. J., 62: 807-810 (1970).

126. Voorhees, W. B., Root elongation along a soil-plastic container interface, Agron. J., 68: 143 (1976).

127. Cahoon, G. A., and E. S. Morton, An apparatus for the quantitative separation of plant roots from soil, Proc. Am. Soc. Hortic. Sci., 78: 593-596 (1961).

128. Böhm, W., Mini-rhizotrons for root observations under field conditions, Z. Acker. Pflanzerbau, 140: 282-287 (1974).

129. Sanders, J. L., and D. A. Brown, A new fibre optic technique for measuring root growth of soybeans under field conditions, Agron. J., 70: 1073-1076 (1978).

130. Atkinson, D., C. M. Crisp, and H. P. Gurung, The effect of mecoprop on shoot and root growth and mineral nutrition of young apple trees, in Proc. British Crop Protection Conf.— Weeds, 1982, pp. 281-284.

131. Böhm, W., H. Maduakor, and H. M. Taylor, Comparison of five methods for characterizing soybean rooting density and development, Agron. J., 69: 415-419 (1977).

132. Gregory, P. J., A periscope method for observing root growth and distribution in field soil, J. Exp. Bot., 30: 205-214 (1979).

133. Monteith, J. L., P. J. Gregory, B. Marshall, C. K. Ong, R. A. Saffell, and G. R. Squire, Physical measurements in crop physiology: I. Growth and gas exchange, Exp. Agric., 17: 113-126 (1981).

134. Bragg, P. L., G. Gobi, and R. Q. Cannell, A comparison of methods, including angled and vertical mini-rhizotrons for studying root growth and distribution in a spring oat crop, Plant Soil, 73: 435-440 (1983).

135. Upchurch, D. R., and J. T. Ritchie, Root observations using a video recording system in mini-rhizotrons, in Agronomy Abstracts, Am. Soc. Agron., Madison, WI, 1981.

136. Belford, R. K., and F. K. G. Henderson, Measurement of the growth of white roots using a TV camera system in the field, in Wheat Growth and Modelling (W. Day and R. K. Aiken, Eds.), Plenum Press, New York, 1984.

137. Upchurch, D. R., and J. T. Ritchie, Root observations using a video recorder system in mini-rhizotrons, *Agron. J.*, 75: 1009-1015 (1983).

138. Upchurch, D. R., and J. T. Ritchie, Battery operated colour video camera for root observations in mini-rhizotrons, *Agron. J.*, 76: 1015-1017 (1984).

139. Taylor, H. M., *Minirhizotron Observation Tubes: Methods and Applications for Measuring Rhizosphere Dynamics*, Spec. Publ. No. 50, Am. Soc. Agron., Madison, WI, 1987.

140. Carpenter, A., J. M. Cherrett, J. B. Ford, M. Thomas, and E. Evans, An inexpensive rhizotron for research on soil and litter-living organisms, in *Ecological Interactions in Soil* (A. H. Fitter, D. Atkinson, D. J. Read, and M. B. Usher, Eds.), Blackwell, Oxford, 1985, pp. 67-71.

141. Durant, M. J., B. J. G. Love, A. B. Missem, and A. P. Draycott, Growth of crop roots in relation to soil moisture extraction, *Ann. Appl. Biol.*, 74: 387-394 (1973).

142. Taylor, H. M., *The Rhizotron at Auburn, Alabama—A Plant Root Observation Laboratory*, Circular No. 171, Auburn University, 1969.

143. Hilton, R. J., D. S. Bhar, and G. F. Mason, A rhizotron for in situ root growth studies, *Can. J. Plant Sci.*, 49: 101-104 (1969).

144. Glover, J., The simultaneous growth of sugar cane roots and tops in relation to soil and climate, *Proc. S. Afr. Sugar Technol. Assoc.*, 1-16 (1967).

145. Freeman, B. M., and R. E. Smart, A root observation laboratory for studies with grapevines, *Am. J. Enol. Vitic.*, 27: 36-39 (1976).

146. Ovington, J. D., and G. Murray, Seasonal periodicity of root growth of birch trees, in *Methods of Productivity Studies in Root Systems and Rhizosphere Organisms*, Nauka, Leningrad, 1968, pp. 146-153.

147. Soileau, J. M., D. A. Mays, F. E. Khasauneh, and V. J. Kilmer, The rhizotron-lysimeter research facility at TVA Muscle Shoals, Alabama, *Agron. J.*, 66: 828-832 (1974).

148. Karnok, K. J., and R. T. Kucharski, Design and construction of a rhizotron-lysimeter facility at the Ohio State University, *Agron. J.*, 74: 152-156 (1982).

149. Fogel, R., The Michigan soil biotron, personal communication, 1987.

150. Taylor, H. M., and W. Böhm, Use of acrylic plastic as rhizotron windows, *Agron. J.*, 68: 693-694 (1976).

151. Atkinson, D., and G. C. White, The effects of weeds and weed control on temperate fruit orchards and their environment, in *Pests, Pathogens and Vegetation* (J. M. Thresh, Ed.), Pitman, London, 1981.

152. Atkinson, D., Some observations on the root growth of young apple trees and their uptake of nutrients when grown in herbicided strips in glassed orchards, *Plant Soil*, 49: 459-471 (1977).

153. Atkinson, D., The use of soil resources in high density planting systems, *Acta Hortic.*, 65: 79-89 (1978).

154. Atkinson, D., The growth, activity and distribution of the fruit tree root system, *Plant Soil*, 71: 23-36 (1983).

155. Klepper, B., H. M. Taylor, M. G. Huck, and E. L. Fiscus, Water relations and growth of cotton in drying soil, *Agron. J.*, 65: 307-310 (1973).

156. Taylor, H. M., and B. Klepper, Rooting density and water extraction patterns for corn (*Zea mays* L.), *Agron. J.*, 65: 965-968 (1973).

157. Taylor, H. M., and B. Klepper, Water relations of cotton: I. Root growth and water use as related to top growth and soil water content, *Agron. J.*, 66: 584-588 (1974).

158. Browning, V. D., H. M. Taylor, M. G. Huck, and B. Klepper, *Water Relations of Cotton: A Rhizotron Study*, Bull. Auburn Univ. Agric. Exp. Stn., No. 467 (1974.

159. Davis, J. M., and M. G. Huck, Identifying turgor responses of water-stressed cotton to rapid changes in net radiation using spectral analysis, *Crop Sci.*, 18: 605-612 (1978).

160. Teskey, R. O., and T. M. Hinkley, Influence of temperature and water potential on root growth of white oak, *Physiol. Plant*, 52: 363-369 (1981).

161. Kaspar, T. C., H. M. Taylor, and R. M. Shibles, Top-root elongation rates of soybean cultivars in the glasshouse and their relations to field rooting depth, *Crop Sci.*, 24: 916-920 (1984).

162. Huck, M. G., C. M. Peterson, G. Hoogenboom, and C. D. Burch, Distribution of dry matter between shoots and roots of irrigated and non-irrigated determinate soybeans, *Agron. J.*, 78: 807-813 (1986).

163. Hoogenboom, G., M. G. Huck, and C. M. Peterson, Root growth rate of soybean as affected by drought stress, *Agron. J.*, 79: 607-614 (1987).

164. Cahoon, G. A., and L. H. Stolzy, Estimating root density and contributions in citrus orchards by the neutron moderation method, *Proc. Am. Soc. Hortic. Sci.*, 74: 322-327 (1959).

165. Newbould, P., and R. Taylor, Uptake of nutrients from different depths in soil by plants, in *Trans. 8th Int. Congr. Soil Science*, Vol. 4, 1964, pp. 731-743.

166. Bassett, D. M., J. R. Stockton, and W. L. Dickens, Root growth of cotton as measured by ^{32}P uptake, *Agron. J.*, 62: 200-203 (1970).

167. Ellis, F. B., J. G. Elliot, B. T. Barnes, and K. R. Howse, Comparison of direct drilling, reduced cultivation and ploughing on the growth of cereals: 2. Spring barley on a sandy loam soil: Soil physical condition and root growth, *J. Agric. Sci. Camb.*, 89: 631-642 (1977).

168. Tomar, S. S., M. B. Russell, and A. S. Tomar, Effect of subsurface compaction on root distribution and growth of wheat, *Z. Acker. Pflanzenbau*, 150: 62-70 (1981).

169. Broeshart, H., and D. A. Nethsinghe, Studies on the pattern of root activity of tree crops using isotope techniques, in *Isotopes and Radiation in Soil-Plant Relationships, Including Forestry*, IAEA, Vienna, 1972, pp. 453-463.

170. Patel, R. Z., and A. M. Kabaara, Isotope studies on the efficient use of P fertilizers by *Coffea arabica* in Kenya: 1. Uptake and distribution of ^{32}P from labelled KH_2PO_4, *Exp. Agric.*, 11: 1-11 (1975).

171. Atkinson, D., The distribution and effectiveness of the roots of tree crops, *Hortic. Rev.*, 2: 424-490 (1981).

172. Atkinson, D., Some observations on the distribution of root activity in apple trees, *Plant Soil*, 40: 333-342 (1974).

173. Ellis, F. B., and B. T. Barnes, Estimation of the distribution of living roots of plants under field conditions, *Plant Soil*, 39: 81-91 (1973).

12

Gas Movement

BRUCE C. BALL *Scottish Centre of Agricultural Engineering,*
Penicuik, Midlothian, Scotland

KEITH A. SMITH *The Edinburgh School of Agriculture,*
Edinburgh, Scotland

I. INTRODUCTION

Soil aeration is maintained by movement of gases through the pore
space of the soil. Carbon dioxide produced by roots and respiring
organisms is exchanged for oxygen from the atmosphere. The mechanisms of exchange are diffusion, resulting in a net movement of
gas from a zone of higher concentration to one of lower concentration, and mass flow, where the whole gas mixture moves in response
to a pressure gradient. Buckingham [1] showed that most gas exchange is by diffusion. Mass flow is important only when pressure
differences develop because of changes in barometric pressure, temperature, or soil water content. Most gas movement occurs in the
air-filled pores, because diffusion in the gas phase is about 4 orders
of magnitude greater than through water. As air-filled porosity
varies with soil water content and compaction, these factors have a
major effect on the rate of gas exchange in soils.

Early aeration studies often involved the measurement of diffusion and its relationship with air-filled porosity. Hannen [2] was
probably the first to measure diffusion in soils. He measured CO_2
concentrations and found that diffusion rate was proportional to air-
filled porosity. By contrast, Buckingham [1] found that diffusion
rate was proportional to the square of air-filled porosity. Early
attempts at measuring diffusion rate in moist, undisturbed samples
by Smith and Brown [3], using carbon dioxide, failed because CO_2
was produced in the soil during measurement. Later workers succeeded by correcting for gas storage and loss or by tracing diffusion

with inert, radioactive gases or inert gases capable of detection by chromatography. Probably the first attempt to measure gas movement by diffusion in the field was made by Hutchins [4], who measured the rate at which oxygen was supplied to an oxygen-deficient absorber buried in the field. Buckingham [1] suggested in 1904 that a measurement of mass flow (viz., air permeability) could be used as a criterion of soil aeration; however, since aeration is governed more by diffusion than by mass flow, measurements of air permeability have become more relevant to the characterization of soil structure [5] or of intrinsic soil permeability [6]. Permeability, unlike diffusion, is sensitive to the size of individual pores, and Green and Ampt [6] showed how to calculate the permeability of "ideal" soils from the dimensions of the individual particles or pores.

Here we consider techniques to measure diffusion and flow of gases and air-filled porosity in the laboratory and in the field, and the relationships of diffusion and flow to air-filled porosity. We also consider the applications of these techniques to the description of soil aeration and soil structure in relation to tillage, compaction, drainage, irrigation, and fumigation.

II. GAS DIFFUSION

A. Principles

Gaseous molecules exhibit random movement as a result of their thermal energy. Where a gradient of partial pressure or concentration of a gas occurs, this random movement results in a net transfer, or *flux* of gas, along this gradient. This is the process of *gas diffusion*. In soils, gas diffusion is "counter-current;" that is, a flux of one gas is matched by a flux of another gas in the opposite direction. Oxygen is required by root cells for the metabolism involved in root growth and nutrient and water uptake, and also by microorganisms. The resulting consumption of oxygen causes a fall in concentration and the consequent creation of a concentration gradient between the soil and the atmosphere above. This is responsible for a net diffusive flux of oxygen into the soil. Carbon dioxide respired by the roots and microorganisms increases the soil concentration above that of the atmosphere, and, correspondingly, an outward flux occurs. Diffusion is also involved in the transfer of water vapor and soil gases (e.g., nitrous oxide, ethylene) produced under anaerobic conditions.

The diffusion coefficient of a particular gas is usually determined in the presence of another gas, commonly air. Pritchard and Currie [7] described a method to measure the countercurrent diffusion coefficients (D_0) of soil gases in air and gave values for carbon dioxide, nitrous oxide, ethylene, and ethane. Kirkham and Powers [8] presented the method of Loschmidt for measuring the countercurrent diffusion coefficients of oxygen and nitrogen. The coefficient D_0 depends on absolute temperature and pressure and can be calculated at the required values using the Boltzmann equation (see, e.g., Ref. 7).

In soil, gas diffusion coefficients are considerably less than in free air because of obstruction by soil particles and water. Water effectively blocks gas diffusion, since the diffusion coefficients of the two gases of main interest in soils, oxygen and carbon dioxide, are nearly 10,000 times greater in air than in water [9]. One-dimensional steady-state diffusion in soil is generally described by use of Fick's first law:

$$q_x = -D_s \frac{\delta C}{\delta x} \tag{1}$$

where q_x is the mass transfer rate of gas per unit area ($ML^{-2} T^{-1}$), D_s is the effective diffusion coefficient ($L^2 T^{-1}$), C is the gas concentration (ML^{-3}), and x is distance along the line of transfer (L). D_s is related to the diffusion coefficient in free air.

Currie [10] proposed the relationship

$$D_s = \alpha \varepsilon D_0 \tag{2}$$

where ε is the air-filled cross-sectional area of soil (equal to the air-filled porosity) and α is a factor to account for the reduction in the effectiveness of ε for diffusion because of deviations in pore direction from the overall direction of gas movement (tortuosity) and roughness of the pore surfaces. This aspect is considered in greater detail in Section V. Field soil is generally aggregated and contains roots; as pointed out by Currie [11], it cannot be regarded as homogeneous with randomly distributed pores. Currie suggested that soil contains two pore phases, the large pores between structural units (the intercrumb pores) and the small pores within the units (intracrumb pores). The diffusion coefficient within crumbs is considerably smaller than that between crumbs because of the greater complexity of the pore space within crumbs. Thus, diffusion in the soil profile as a whole consists of contributions from diffusion in crumbs, between crumbs, through the water films surrounding roots, and through the plant roots themselves [12].

To measure diffusion coefficients in soil, we need to identify the boundary conditions required (i.e., depth, soil compactness, and water content) and take into account factors that might cause errors, such as temperature, matric potential gradients, soil respiration, and changes in absolute pressure at the soil surface. Establishing boundary conditions in the field is difficult. Soil structure, bulk density, water content, and temperature can vary over only a few centimeters,

and in particular may differ markedly between soil horizons. This
makes the choice of appropriate diffusion solutions to Fick's law
uncertain. However, field methods can give useful indications of
gaseous exchange.

Laboratory methods of measurement of gas diffusion offer the
advantages that boundary conditions can be chosen, controlled, and
specified and that sample size and volume can be chosen to be representative of the soil layer(s) of interest. The main disadvantages
are soil disturbance during sampling and the problems associated
with relating measurements to field conditions.

Both laboratory and field methods rely on solving Fick's first
law. However, the application of this law to gases is empirical [13],
and only under special circumstances is the diffusion coefficient contained in Fick's law a constant, independent of the mole fraction and
the diffusive fluxes of other gases. In an atmosphere composed of
O_2, CO_2, and N_2, where N_2 is stagnant (a system similar to the soil
atmosphere), variations of about 10% from the tracer value of diffusion coefficient of O_2 and CO_2 are possible with variations in the
mole fraction [13]. Nevertheless, Fick's law is almost universally
used, and several solutions of it for different conditions are presented by Kirkham and Powers [8].

Techniques for measurement of diffusion in the gaseous state
are discussed in Section II.B. Methods of measurement of oxygen
diffusion rate (ODR) that use platinum electrodes [14] and relate
to the rate of supply of oxygen through water films, such as those
that occur at a root surface, are also dealt with in Section II.B.

B. Measurements

1. Laboratory Methods

Methods involve either steady-state diffusion (where a constant concentration gradient yields a constant flux of molecules) or non-steady-
state diffusion (where the concentration gradient and the flux of
molecules change with time). The method of Penman [15] is a relatively simple steady-state approach. Carbon disulfide or acetone
vapor evaporates from a dish sealed beneath the sample and diffuses
through it, the loss of liquid being measured by weighing. This
method suffers from error caused by mass flow, because the diffusion coefficient of the diffusing species is so different from that of
the counterdiffusing air [16] and because of the long time required
for establishment of steady-state diffusion (40-100 minutes).

Methods in current use involve non-steady-state diffusion. The
method recommended by Evans [17] and widely used in modified
forms was developed by Taylor [18]. In Taylor's simple apparatus,
oxygen from the atmosphere counterdiffuses with nitrogen from a
chamber sealed under the soil sample. Oxygen concentration is

measured regularly (5-minute intervals are typical) by temporarily isolating the sample and then allowing the gas in the chamber to enter a Beckman oxygen analyzer. Bakker and Hidding [19] modified the method by using a chamber initially filled with oxygen below the sample and by installing a membrane-covered electrode in the gas chamber to allow continuous recording of oxygen concentration without interrupting diffusion. They applied the method to minimally disturbed soil cores and included an oxygen storage correction factor. The most recent variant is that of Schjønning [20], where the chamber is initially filled with nitrogen and contains an electrode to monitor oxygen concentration. Data can be acquired automatically by microcomputer from up to 12 samples running simultaneously. The method is prone to error in moist, respiring topsoils, where up to 9% of the diffusing oxygen may be consumed by soil microorganisms [21].

A method that overcomes this problem employs hydrogen as the gas counterdiffusing with air [22]. The hydrogen is contained in a chamber sealed over the sample and the composition of the hydrogen-air mixture is measured using a katharometer (a thermal conductivity detector). This method is quick (the diffusion coefficient of hydrogen is four times greater than that of oxygen), and the gas composition can be measured frequently without interfering with the system. The method has been most recently applied to the assessment of gaseous diffusion within soil crumbs.

The problem identified by Rust et al. [23] is that these systems fail to take into account mass flow when the diffusivities of the two counterdiffusing gases differ significantly. In addition, such systems have one face of the sample open to the atmosphere, so that uniform boundary conditions of concentration, temperature, and pressure are difficult to maintain. Most recent methods overcome these problems by enclosing the sample between two gas-filled chambers and by using as the diffusing species gases at trace concentrations, to overcome the problem of mass flow. Such systems allow precise control of experimental conditions and can give accurate measurements of diffusion coefficients in soils of very low air-filled porosities (e.g., those that are nearly saturated and in which soil aeration is likely to limit plant growth). Three such methods used tracers of [85]Kr [24], sulfur hexafluoride [25], or Freon [26], respectively. These three tracers have very low solubility in water and are neither adsorbed on nor consumed in soil. In two of these methods [24,25], pressure differences are monitored by a micromanometer capable of detecting pressure differences between the end faces of the sample, and which could cause mass flow, of as low as 0.01 Pa. In the method of Reible and Shair [25], syringe samples of SF_6-air mixture are taken from each chamber at regular intervals and analyzed for their SF_6 concentration, using an electron capture

gas chromatograph. Samples of relatively small (2.54 cm) diameter
are tested. In the method of Sallam et al. [26], the size of the
chambers enclosing the sample varies according to the expected sample porosity. The air-Freon mixture is sampled at the beginning
and at the end of the diffusion measurement.

The method of Ball et al. [24] was designed for the use of minimally disturbed field samples held in their sampling cylinders. These
samples, 76 mm diameter and 50 mm long, can be inserted directly
into the apparatus in the field-moist condition or after equilibration
to a given matric potential. This method, with its self-contained
apparatus, which is relatively quick and easy to use, has been in
use in Britain and France for several years. It is briefly described
below; fuller details are given in Ref. 24.

In the apparatus (Fig. 1), two cylindrical gas chambers with
photomultipliers attached are sealed on to the ends of the stainless
steel sample holder. A mixture of ^{85}Kr and air, of activity of the
order of 400 GBq m^{-3} ($\sim$ 1 Ci m^{-3}) is injected into one gas cell and
diffuses through the soil until an equilibrium concentration is achieved
throughout the apparatus. The concentration of tracer is measured
by regular counting of β radiation at each photomultiplier. In the
latest version of this equipment, the counting data are recorded on
a small microcomputer (Epson HX20) and then transferred directly
into a larger microcomputer (Hewlett-Packard 9835A) for data analysis.

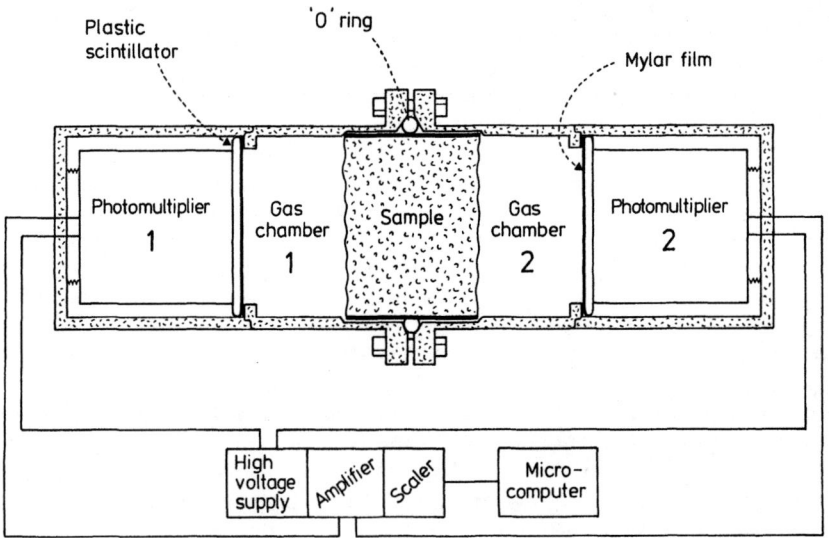

FIG. 1 Apparatus for measurement of gas diffusion and permeability. (From Ref. 24, with slight adaptation.)

It is assumed that after a short initial period (< 5 minutes), the relationship between the count rates in the two gas chambers is given by

$$C_I - C_R = 2C_e e^{-kt} \tag{3}$$

where C_I and C_R are the concentrations of gas in the injection and receiving gas chambers, respectively, t is time (s), C_e is the concentration in each gas chamber at equilibrium, and

$$k = \frac{2D_s A_s}{VL_s} \tag{4}$$

whence D_s may be found; A_s and L_s are the area and length of the sample, and V is the volume of the gas cell.

In practice, to speed up the measurements, diffusion is usually monitored only half-way to equilibrium. Values of $C_I - C_R$ and t, excluding those detected in the first 5 minutes, are fitted to Eq. 3, and k and $2C_e$ estimated by exponential regression. This modification allows the making of diffusion measurement on samples at or below field capacity, typically, in under an hour. Samples wetter than field capacity, particularly if they are compact or fine-textured, may require up to 15 hours for significant diffusion to occur and are best measured overnight. An alternative, more sophisticated, treatment of the count rates and times was recently proposed by Bruckler et al. [27]. In this treatment, an iterative finite-element method is used to solve Fick's second law of diffusion for both diffusion coefficient and effective porosity (ε_e). The latter is the portion of air-filled porosity effective for diffusion that is less than or equal to the total air-filled porosity [27]. Fick's second law is expressed as follows:

$$\varepsilon_s \frac{\partial c}{\partial t} = \frac{\partial}{\partial x} \left[D_s \frac{\partial c}{\partial x} \right] \tag{5}$$

Another method of measurement of diffusion using [85]Kr tracer was proposed by de Jong et al. [28]. In this method, different sizes of minimally disturbed field cores were mounted over a diffusion chamber containing a sodium iodide crystal detector. The rate of diffusion through the sample of radionuclide injected into this chamber was estimated by assuming an exponential decrease in tracer concentration with time. This technique allowed for use of a single detector; however, because the upper face of the sample was exposed to the laboratory atmosphere, some of the measurements gave enhanced diffusion coefficients; this was because of mass flow upward through the sample in response to pressure fluctuations in the

TABLE 1 Four Hypothetical Classes of Representative Elementary Volumes of Samples Relative to Soil Texture and Structure

Class	Texture	Structure	Hypothetical REV (cm^3)
1	Sandy	No peds	10^2
2	Loamy, silty	Small peds	10^3
3	Clayey	Medium peds, continuous macropores	10^4
4	Clayey	Large peds, continuous macropores	10^5

Source: From Ref. 30.

fume cupboard in which the measurements were made [28]. This effect was greatest at air-filled porosities exceeding 0.15 m^3 m^{-3}, particularly where cracks occurred within the sample or between the sample and its holder.

The choice of sample size, the degree of replication, and the extent of pretreatment depend on the objective of the experiment. Usually, disturbed samples can be small (say 20–30 mm diameter and 20–30 mm long), since the component particles or aggregates are small. This allows rapid measurement of diffusion. For such relatively inert samples, a simple technique such as Taylor's [18] would suffice. When minimally disturbed samples are required which are representative of the field condition, the choice of sample size and dimensions is difficult. Ideally, a sample should be of volume greater than or equal to the representative elementary volume (i.e., the smallest volume that contains a representative packing of particles that is repeated throughout the porous region) [29]. Bouma [30] recommended that a representative sample contain at least 20 peds and that the representative elementary volume (REV) be increased as the texture becomes finer and the structure becomes coarser. The REV classes suggested by Bouma for the sphere of influence relevant to individual plants are given in Table 1.

The sample size for most reported diffusion measurements (100–300 cm^3) corresponds to or is smaller than class 2 (Table 1). Thus the use of a technique like that of de Jong et al. [28], which can accommodate large samples, is desirable. Changing or measuring the temperature or water content of such large samples is difficult. In such cases, where a relatively large volume is to be considered, a field method of diffusion measurement (discussed below) may be more suitable. Many recently reported measurements of diffusion in minimally disturbed samples (see Section V) relate to the description of tillage and compaction treatments on specific soil layers, for example, around a germinating seed. Where these layers are narrow and well-defined, samples can be relatively small. Guidance for the construction of sampling equipment and for collection and preparation of minimally disturbed samples is given by McIntyre [31]. Such samples are commonly 35-75 mm deep [19,21,24,28]. Solutions for the diffusion coefficient generally need sample volume. Thus sample height and volume should be kept constant to minimize error.

Sampling distributions of relative diffusivities may be skewed and not normal (see, e.g., Ref. 32). In such cases, conventional parametric statistics do not strictly apply. Coefficients of variation of replicated relative diffusivities can be up to twice as great as those of the air-filled porosities measured on the same sample using conventional water release calculations [32]. Thus a greater number of samples may be required than for, say, assessment of soil water release.

Conventionally, samples are taken vertically so that gas diffusion corresponds to vertical movement in the profile. However, sampling in the horizontal direction may also require consideration. De Jong et al. [28] have shown that on drying the greater cracking in the vertical than in the horizontal direction of zero-tilled shrinking soils may cause significant anisotropy in diffusion coefficients.

Laboratory treatment of samples may also influence choice of size. If the matric potentials of samples have to be adjusted (e.g., if the intracrumb pores have to be blocked by water by wetting to field capacity to assess diffusion in the intercrumb pores), the time for attainment of equilibrium throughout the sample increases with sample height. The senior author commonly uses 50 mm deep samples for such experiments [32].

If samples are progressively dried and diffusion measurements made at each step with the core in its holder (see, e.g., Ref. 32), then shrinkage may occur from the walls of the holder. In such cases the gap can be filled with paraffin wax and the sample diameter remeasured; or, as suggested in Ref. 28, samples can be cut from larger blocks and the nondiffusing surfaces coated with wax.

2. Field Methods

Field methods of diffusion measurement overcome some of the problems of sampling, but have their own complexities. Raney [33] developed the first widely used method for measuring oxygen diffusion in the field. A nitrogen-filled probe was inserted to 0.3 m depth. Soil oxygen diffused radially into the probe and was measured by a Beckman oxygen analyzer; the diffusion coefficient was calculated from the rate of change of partial pressure of oxygen in the probe. Van Bavel [34] used a similar method but with an open-ended tube. He presented a simplified solution to radial diffusion by treating it as a linear problem, using shape factors related to the size of the tube inserted in the soil.

McIntyre and Philip [35] criticized the methods above for allowing no rigorous analysis because the geometry of the diffusion path was irregular and the boundary conditions not known. They developed a technique that measured soil surface gas exchange. A thin-walled brass cylinder was driven into the soil, the soil in the cylinder was flushed with air to give a known concentration initially, and then oxygen from a chamber placed on the cylinder was allowed to diffuse through the soil. The oxygen concentration in the chamber was measured with a membrane-covered oxygen cathode. This method takes into account errors due to temperature, relative humidity, and changes in soil porosity. Such a technique would be most useful where assessing diffusion through a crust or compact layer near the surface. However, as with the other techniques, it suffers from problems of oxygen storage and consumption. Such problems are discussed in the preceding section on laboratory methods.

These problems are perhaps of less significance in field techniques, which require realistic assessment of gas exchange, which includes respiration, though changing soil oxygen concentration as in these methods may change the rate of soil respiration. A rather laborious field technique overcame this problem by releasing the radioactive tracer ^{85}Kr from a buried container and measuring the activity received at various locations around the release point [36]. More recently, improvements in gas analysis techniques by chromatography techniques (extensively reviewed by Smith [37]) and developments in gas analysis by mass spectrometry [38] have enabled accurate assessment of oxygen concentrations in small (0.25-1 ml) soil gas samples.

A technique exploiting gas chromatography was proposed by Lai et al. [39]. Two principal advantages are claimed for their method: there is no removal or alteration of the soil from its natural state, and minimal instrumentation is required at the site of measurement. The method is based on the theory of radial diffusion of a finite quantity of a gas into a semi-infinite medium. Oxygen is injected through a needle inserted into the soil, small aliquots of soil air are withdrawn at regular intervals, and oxygen and nitrogen concentrations are measured using a portable gas chromatograph. Carbon dioxide fluxes to the soil surface predicted from the diffusion coefficients measured by this technique agreed well with independent field measurements [39].

A year later, Rolston and Brown [40] proposed a more complicated method for assessment of gaseous diffusion coefficients over a relatively large soil area (10^2-10^4 cm^2). A one-dimensional, transient-state diffusion solution was used. This method had the attraction that the block of soil in which the measurements were made,

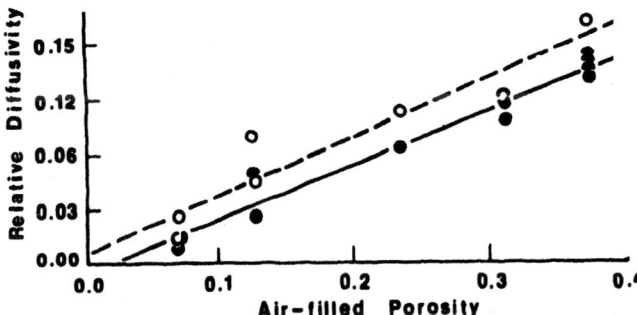

FIG. 2 Relationship between D_s/D_0 and volumetric air content for lawn soils: •, diffusion measurements with the injection method using the numerical model; ○, measurements with the core method. (From Ref. 41.)

though somewhat disturbed, was encased in plastic, and moisture tension was monitored at the same depths from which samples of soil atmosphere were withdrawn. The nitrogen concentration in these samples was measured and used in the claculation of diffusion coefficient.

In a recent modification of the method of Lai et al. [39], Jellick and Schnabel [41] used a numerical finite-difference model to allow the initial concentration profile within the sphere of injected gas to vary, based on experimental data. The diffusion coefficients compared favorably with those determined by Taylor's [18] laboratory method on minimally disturbed core samples (Fig. 2).

Dissolved soil oxygen can be determined in water samples using a polarographic oxygen electrode controlled by a purpose-built electronic analyzer [42]. Ray et al. [43] recently showed that the same method can be used successfully to determine the oxygen content of gaseous samples and that the analysis was more rapid and required less expensive equipment than is needed for gas chromatography.

Soil oxygen flux and redox potential can be measured in waterlogged soil using polarographic techniques. These methods are based on the reduction of oxygen at a platinum wire cathode buried in the soil. This is linked to a calomel or silver-silver chloride anode placed in electrical contact with the soil (Fig. 3). Redox potential is a measure of the intensity of reduction in soils containing no molecular oxygen. Oxygen flux (also termed oxygen diffusion rate, ODR) to a cathode is a measure of the rate of supply of oxygen from the air through the surrounding soil through a film of soil water. This flux is comparable to the maximum required by roots respiring in moist soil [42]. In measurement of ODR, Armstrong

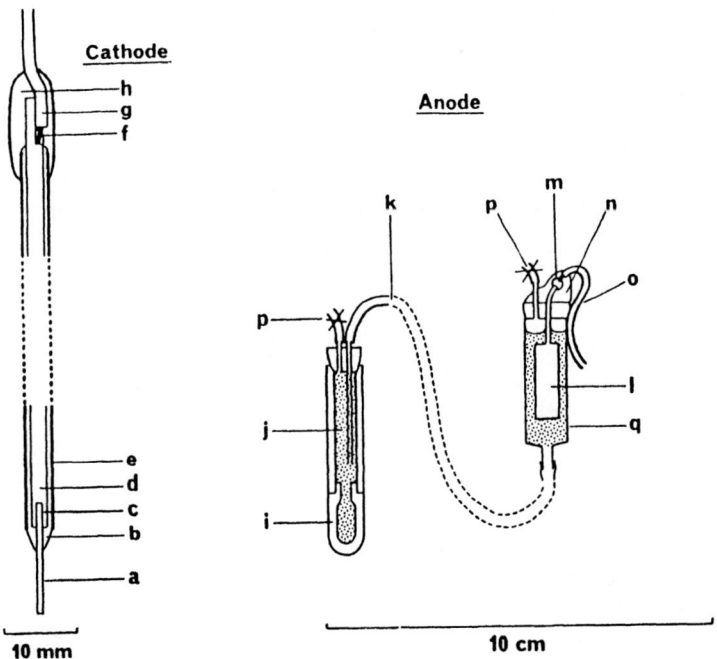

FIG. 3 Structure of the anodes and cathodes used for measurements
with bare platinum electrodes: a, platinum wire; b, epoxy resin; c,
crimped and soldered joint (Pt wire in hole at end of conductor); d,
mild steel conductor; e, heat-shrinkable insulating sleeve; f, soldered
connection; g, wire connected to cathode plug; h, self-amalgamating
insulation tape; i, porous pot (air-entry pressure ~ 100 kPa); j, saturated KCl solution; k, flexible connecting tube filled with solution;
l, prepolarized silver sheet (surface area ~ 25 cm^2); m, crimped and
soldered joint; n, silicone rubber cement; o, wire connected to anode
plug; p, tap for bleeding air from solution; q, syringe body. (From
Ref. 42.)

and Wright [44] recommended that the relationship between current
and voltage be established (polarogram). Where a plateau is reached
on the polarogram, current is related to the flux of oxygen to the
electrode.

      Both redox potential and ODR can be measured from the same
pair of electrodes. Blackwell [42] showed that platinum cathodes
can be left in the soil and remain functional for several months without removal for cleaning. These techniques work best in wet soil.
In unsaturated soil, variations in pH and aeration status alter the

shape of polarograms, and it may be necessary to measure soil electrical resistance before ODR can be calculated [45]. Reviews of the principles and the conditions under which this equipment can be used are given in Refs. 42, 44, and 45. Detailed descriptions of electrodes and electronic instrumentation required for multiple assessments of both measurements for a lysimeter installation and for field use are given in Refs. 42 and 44, respectively. These techniques, although presenting many operational limitations, were considered to be a relatively cheap means of identifying the influence of anaerobic microsites on soil aeration [45].

A novel field technique described as similar in principle to a tensiometer was proposed by Flühler and Läser [46]. This uses a hydrophobic membrane probe, which connects the soil atmosphere at a certain depth with a pressure transducer or oxygen electrode placed on the soil surface and was used to detect changes in total air pressure or oxygen partial pressure in the soil atmosphere during water infiltration and drainage. This method is limited by the use of sensors, which are not readily portable, though the use of a rotary fluid switch allows measurement of several probes sequentially by one sensor.

III. MASS FLOW

A. Principles

Mass flow is the movement of molecules in response to a pressure gradient. Thus mass flow can cause gas exchange between the atmosphere and the soil air only if there are changes in temperature, barometric pressure, wind, or water content [47]. Mass flow is a less important mechanism than diffusion [17] and accounts for only a few percent of the normal gas exchange [47].

Since mass flow is a smaller contributor to the maintenance of soil aeration than diffusion, fewer methods of measurement and data have been published. Interest revived recently in the use of air permeability measurements as an alternative to hydraulic conductivity measurements to describe soil structure, since air causes negligible sample disturbance [48]. Bowen [49] recently claimed that the technique allowed detection of limiting aeration conditions for cotton and of blockage of the pore system by water during rain or irrigation.

The flow of gases through soil is comparable to that of water, with certain restrictions. Darcy's law applies if flow is laminar or viscous, as it is when the flow rates are relatively small [48]:

$$q = -\left(\frac{K}{\eta}\right)\left(\frac{dp}{dx}\right) \tag{6}$$

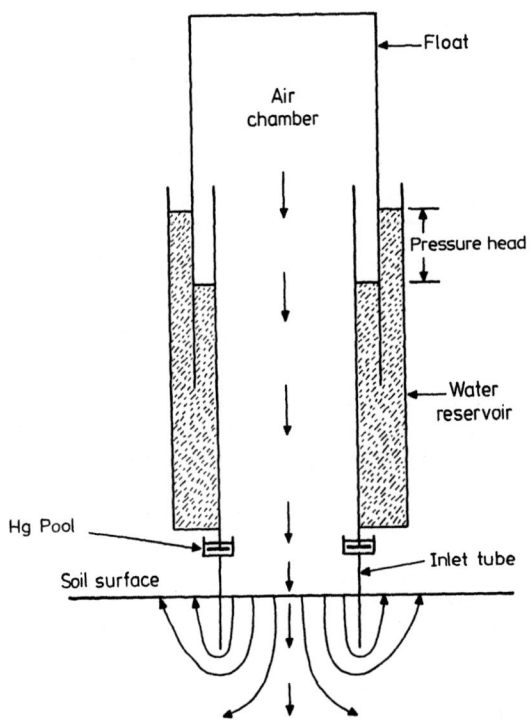

FIG. 4 Schematic diagram of air permeameter with mercury seal to allow quick attachment and good seal between permeameter and soil insert. See Ref. 50 for construction details.

where q is flow rate, p is pressure, x is distance, K is gas permeability, and η is viscosity. In a tube of radius r and length L, flow rate can be calculated from Poiseuille's law:

$$q = \frac{\pi r^4 \Delta p}{8 \eta L} \tag{7}$$

where Δp is the difference in pressure between the ends of the tube. It follows that flow rate, hence permeability in soils, depends on pore radius raised to the fourth power, whereas diffusion depends on pore radius raised only to the second power. Flow is less subject to errors due to small temperature differences than is diffusion, though ambient temperature, pressure, and humidity affect flow by their influence on gas viscosity [50]. Deviations from laminar or viscous flow occur when large pressure differences are applied to

samples containing pores of large enough radius to give flow velocities sufficiently high for a Reynolds number of about 2000 or greater;
under these conditions, flow becomes turbulent. (For an explanation
of the Reynolds number concept, see a fluid mechanics or physics
text, e.g., Ref. 51). Alternatively, gas slippage (i.e., gas moving
along pore surfaces) may occur in very small pores. As for gas
diffusion, air flow is blocked by water-filled pores, so that air permeability decreases as soil water content increases.

Field and laboratory techniques are available involving either
steady-state or non-steady-state flow. Steady-state measurements
of gas permeability are more generally applied than the non-steady-
state variety; this is the reverse of the situation relating to measurements of diffusion.

B. Measurements

Most methods are applicable both in the field and in the laboratory.
In the non-steady-state method of Kirkham [52], an air tank is pressurized, and the pressure released through a soil sample or through
a thin-walled cylinder pushed into field soil. Air permeability is
calculated from the rate of decrease in tank pressure as read from
a manometer attached to the side of the tank. In this technique,
the initial pressure applied to the sample may be high, up to about
3 kPa. In highly permeable samples, this may lead to the onset of
turbulent flow or change the spatial arrangement of the water in the
soil pores [48]. Most subsequent methods have been based on the
steady-state method proposed by Grover [50] and recommended by
Evans [17]. Grover devised a permeameter (Fig. 4). A float, a
thin-walled cylinder that can be suspended to keep it centered [48]
and is open only at the bottom forms an air chamber, which fits
over an annular water reservoir. The air pressure can be increased
by adding weights to the reservoir. The air is displaced directly
into field soil [50] or through a core sample sealed on the bottom
[48]. Bowen [49] proposed the latest version of this apparatus,
which incorporates a sensitive flowmeter and manometer, the latter
reading to a maximum of 0.5 kPa. The direct reading of flow and
pressure considerably speeds up measurements, since it is otherwise
necessary to time the fall of the float for a given distance to be
able to calculate permeability. The main advantage of this technique
is that constant, low pressures (0.03-1 kPa) can be applied. Kirkham [52] discussed in some detail the errors and the assumptions
involved in air permeability measurement, particularly that of neglecting gas compressibility. He integrated Eq. 6 into a form applicable to most air permeability measurement techniques:

$$q_V = \frac{K \Delta P A_s}{\eta L_s} \qquad\qquad (8)$$

where q_V is the volumetric flow rate $[L^3 \ T^{-1}]$, K is air permeability $[L^2]$, ΔP is the pressure difference across the sample $[M \ L^{-1} \ T^{-1}]$, A_s and L_s are the cross-sectional area and length of the sample, $[L^2]$ and $[L]$, respectively, and η is the dynamic gas viscosity $[ML^{-1} \ T^{-1}]$ corrected for temperature.

Evans and Kirkham [53] proposed an improved means of introducing air to the soil surface without disturbing the soil by soaking the soil surface with paraffin wax around a tube whose end contacts the soil surface. The outer perimeter of wax was contained in a ring pushed into the soil surface. Bowen [49] also suggested improved means of introducing the air to the soil below its surface.

Another practical field method that requires no float chamber was published by Green and Fordham [54]. In this apparatus, air compressed in a portable bottle is regulated via needle valves and passes through a flowmeter at a constant pressure indicated on a water manometer into cores of soil held in cylindrical tubes of 51 mm diameter. A wide range of soil permeabilities (typically 0-500 μm^2) can be covered by the three separate flowmeters and the several needle valves used. The tests can be carried out in less than a minute. Permeability is measured both with cores in situ and after digging them out.

Field techniques are often rather inconclusive because the variability of soil structure in the upper layers is large and is non-normally distributed. Permeability in such layers is also likely to be anisotropic. Janse and Bolt [48] measured air permeability on undisturbed cores using a Grover-type permeameter and found that "vertical" samples were about twice as permeable as "horizontal" samples, part of the effect being attributed to greater compression during horizontal sampling than during vertical sampling.

The final method discussed is that of Ball et al. [24] for minimally disturbed samples. It requires the same two-chamber apparatus as that used for the measurement of diffusion (Fig. 1). A differential micromanometer sensitive to pressure differences of at least 0.01 Pa is connected across the two gas chambers, and compressed air from a bottle is fed via a regulator and a flow controller to one gas chamber. Exhaust air is piped from the other gas chamber into a soap-film bubble meter or suspended-ball flowmeter. To preserve laminar flow, the pressure differences applied (0.15-300 Pa) and the resultant rates of flow (0.15-6 $cm^3 \ s^{-1}$) are kept small. For each sample, flow is measured at two or more pressure differences. This permits investigators to check their proportionality, giving two or more permeabilities, from Eq. 8. An advantage of this technique

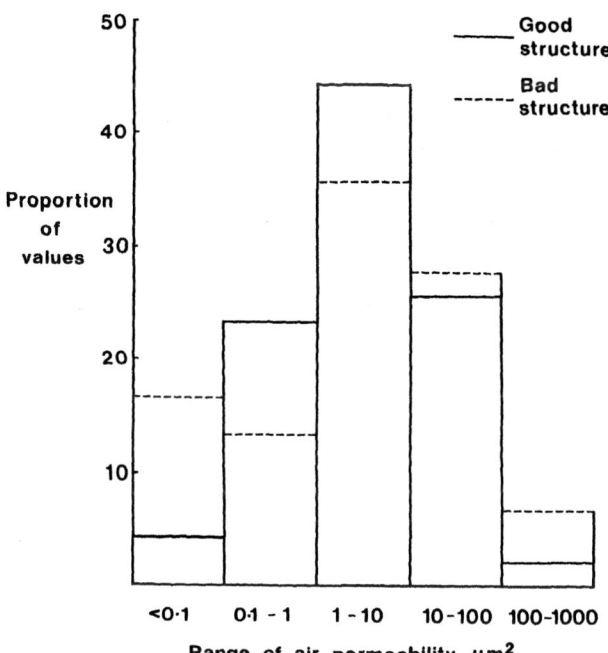

FIG. 5 Distribution of values (n = 90) of air permeability in a Belgian silt loam in areas of good and bad soil structure. (From Ref. 56.)

(which the senior author has used extensively) is that air permeability can be measured immediately after a diffusion measurement in 2-3 minutes without disturbing the sample. In addition, flow and pressure differences are measured with high accuracy. During the second measurement the ^{85}Kr tracer gas used for the diffusion measurement is flushed from the sample.

The comments made in Section II.B on choice of sampling size, intensity, technique, and preparation for minimally disturbed cores apply equally to air permeability measurements. The great sensitivity of air permeability to pore diameter means that sample disturbance such as cracking or shrinking from the sides of the holder has a greater effect than on measurement of diffusion. This sensitivity to pore and crack size also demands a greater requirement for samples to be as large as the representative elementary volume [30] than for diffusion. The use of smaller samples can be justified if the largest channels, such as those produced by earthworms and cracks, are avoided [32,55], provided these are not required in the assessment.

The statistical distribution of air permeabilities from a given depth and treatment within replicated field experiments is generally non-normal. The distributions are skewed and usually logarithmic normal [32,55,56]. Examples taken from data of Kirkham et al. [56] are shown in Fig. 5. The range of values is wider and the distribution less typically logarithmic normal for the soil of bad structure that that of good structure. The soil of bad structure had been compacted during an earlier sugar beet harvest. In such cases either nonparametric tests should be applied to reduce the data and get an indicator of statistical degree of spread, such as statistical rank analysis [56] or the Mann-Whitney U test [55] or parametric tests should be applied to log-transformed data [32]. Whichever technique is used, the variation among samples is large, with standard errors of replicated data often greater than the means [32,56]. Ball [32] attributed this variability to the great variation among replicates of the radius, length, and continuity of the largest air-filled pores. Thus a relatively large number of samples, usually 15-30 per treatment, is required for adequate assessment of air permeability [32,48,56].

IV. AIR-FILLED POROSITY

A. Principles

Since gases move almost exclusively in the air-filled pores, measurement of porosity is vital to the understanding of gas movement in soil. Air-filled porosity is often used as an indicator of the likely aeration status of the soil and its ability to conduct and store gases.

Air-filled porosity (ε_A) is that fraction of the total porosity (ε_T) which is occupied by air. If ε_T is the percentage of soil volume not occupied by solids, ε_A and ε_T are equal only in dry soils. ε_A is less than ε_T in moist soils because a fraction of the total porosity is occupied by soil water, this fraction being called the volumetric water content (θ). Thus air-filled porosity is:

$$\varepsilon_A = \varepsilon_T - \theta \tag{9}$$

ε_T may be calculated from the dry bulk density of the sample ρ_b, and from the particle density of the soil ρ_p, as follows [57]:

$$\varepsilon_T = 1 - \frac{\rho_b}{\rho_p} \tag{10}$$

ε_T in undisturbed soil cores may also be estimated as the volumetric water content at saturation θ_S. Air-filled porosity is most useful when determined at a given water potential. This can be readily achieved by equilibrating on tension tables as used for the determination of pore size distribution (see Chap. 3). Alternatively, samples may be taken at field moisture content. This is best taken as field capacity when ε_A is the "air capacity" and corresponds to the soil drainable porosity or macroporosity [57].

Air-filled porosity may also be determined by use of an air pycnometer. This apparatus uses the principle of Boyle's law. The volume of air in a sample is measured by observing the resulting pressure when a gas at a measured volume and pressure expands into a larger volume, which includes the sample [58]. This method excludes pores whose entrances are blocked by water films unless they are compressed by the change in pressure, when part of the volume is measured.

The air-filled porosity effective for diffusion may be calculated, using Fick's solution to the diffusion equation and measurements of concentration otherwise used for the calculation of diffusion coefficient [27]. This porosity is generally less than the air-filled porosity measured by other techniques because pores blocked by water films or soil are not measured.

B. Measurements

Sampling techniques relevant to the laboratory determination of ε_A using minimally disturbed cores are discussed in Chap. 3 (see also Refs. 31, 57, and 59). Sample size and sampling intensity are less than for assessment of gas movement because porosity and bulk density vary less than flow and diffusion properties within a soil horizon, since porosity does not depend on pore continuity. The Soil Survey of England and Wales recommended triplicate sampling of individual horizons [59]. If air-filled porosity at a specific water potential is required, the samples require equilibration on tension tables, as discussed in Chap. 3 (see also Ref. 57). To minimize equilibration times, sample heights no greater than 50 mm were recommended [57]. Methods of measurement of ρ_p and ρ_b, necessary for assessment of ε_T and thence ε_A, are given in Refs. 31 and 58. In cores of known volume, ρ_b is easily calculated from the weight of the soil core. In soils containing significant quantities of organic matter, the estimation of ρ_b by liquid pycnometry may overestimate the soil particle density because organic matter is destroyed in this technique. In such cases, a better estimate of ε_T may be the volumetric water content at saturation, θ_s. This may be determined after saturation either by capillary wetting and immersion or by vacuum saturation [31]. The first method may leave air trapped in the sample, thereby underestimating θ_s, and the second method may give structural breakdown and slaking in soils that are structurally unstable as trapped air is rapidly released from aggregates.

In his laboratory work, Ball (1986, unpublished) found that the best agreement between θ_S and ε_T resulted from saturation by capillary wetting and subsequent estimation of θ_S by weighing the sample immersed in water.

Field assessment of air-filled porosity is best achieved by using the gamma probe to measure bulk density and then making one or more assessments of water content by neutron moisture meter or auger sampler (see Chaps. 1, 7). Separate measurement of particle density is required. The neutron moisture meter requires calibration but is valuable in that, once an equilibrium bulk density has been reached and measured, repeated measurements with the neutron moisture meter give air-filled porosities that relate well to such changes in field aeration as occur, for example, during irrigation [60].

The air pycnometer can be used in the laboratory [58] or in the field [61]. Pidgeon [61] reported good agreement between ε_A of soil from contrasting tillage treatments determined by the air pycnometer and by oven drying. He found that the main limitation of this rapid technique was the requirement for accurately trimmed soil samples, which prevented its use on loose or stony soils. More recent experience with the air pycnometer indicated that its usefulness was also limited by adsorption on the soil colloidal fraction, even if the air is replaced by an inert gas, and by heating of the gas after compression, causing a long recovery time.

V. APPLICATIONS

A. Soil Structure and Aeration

The relationships between relative diffusivity D_S/D_0 and air-filled porosity ε_A have been widely used to estimate (1) the minimum porosity at which aeration limits plant growth, (2) the continuity or tortuosity of the air-filled pore system, (3) the content of air-filled pores blocked to entry of diffusing gas, and (4) the extent of soil aggregation.

These relationships are empirical only; the majority can be grouped under three generalized forms. The first is:

$$\frac{D_s}{D_0} = a(\varepsilon_A - b) \tag{11}$$

This is a straight line corresponding to a slope of unity, $a = 1$ (Fig. 6) for straight tubes aligned with the concentration gradient. If the tubes twist along the direction of the concentration gradient (i.e., are tortuous), then $a < 1$. A value of 0.66 for a was suggested for soils by Penman [15] (Fig. 6). Positive values of b are

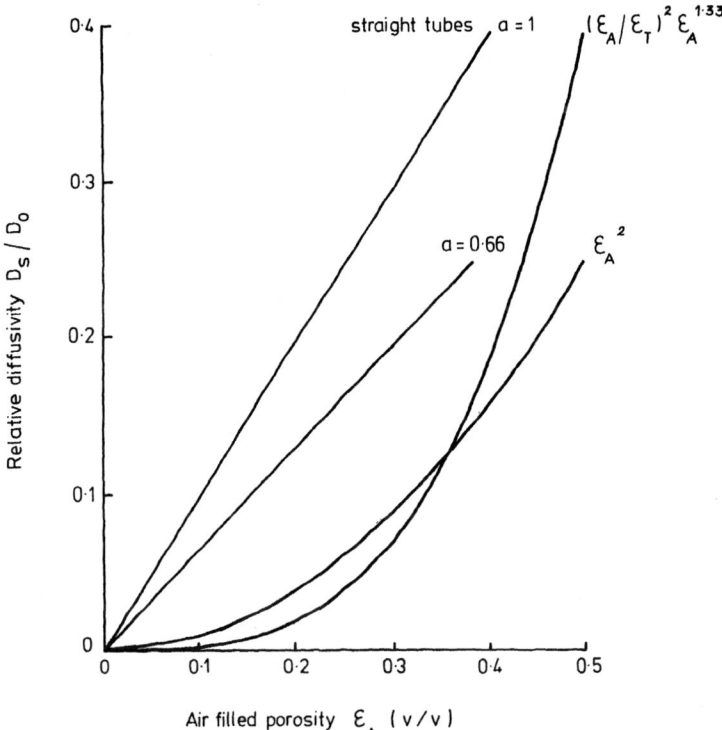

FIG. 6 Relationship between relative diffusivity and for tubes, $a = 1$, for soil where $a = 0.66$ [15], for soil pores interconnecting at random, ε_A^2 [66], and greater than random $(\varepsilon_A/\varepsilon_T)^2 \varepsilon_A^{1.33}$ [65].

measures of air-filled porosity blocked against diffusing gas; a ranges between approximately 0.25 and 0.5, with most values near 0.3, and b ranges up to about 0.1 v/v for minimally disturbed samples [21,62,63]. Authors reporting the above relationship may have data covering a restricted range; see Troeh et al. [64].

The second relationship, to which many published results on minimally disturbed cores have been fitted [10], is a curvilinear one:

$$\frac{D_s}{D_0} = \gamma \varepsilon_A^{\mu} \tag{12}$$

where $\gamma \leq 1$ and $\mu \geq 1$ are taken as measures of pore shape in dry materials [10]. In wet materials, γ and μ are interpreted as measures of pore continuity. μ has also been considered to equal the effective area for flow [65]. This relationship is similar to that

predicted from a model of pore space comprising a series-parallel
arrangement of pores arising from random apposition of planes within
the porous material. Marshall [66] found that if pore entrances and
exits are restricted to joint planes only (one entrance and exit per
pore), then

$$\frac{D_s}{D_0} = \varepsilon_A{}^2 \tag{13}$$

This relationship (Fig. 6) corresponds to the change in pore continuity in soil samples where pores interconnect at random. Millington and Quirk [65] assumed greater pore interconnection (smaller μ
value) and proposed

$$\frac{D_s}{D_0} = \left(\frac{\varepsilon_A}{\varepsilon_T}\right)^2 \varepsilon_A{}^{1.33} \tag{14}$$

where ε_T is the total porosity. Both relationships indicate that, as
expected, pore continuity or tortuosity improves as the soil dries
and ε_A increases.

Many authors express their data using Eq. 12; Troeh et al. [64]
reported that γ ranged from 1 to 5 and μ from 1.6 to 3.4. They
realized that the limitation of Eq. 12 is that no allowance is made for
the presence of blocked pores, since $D_s/D_0 = \varepsilon_A$ at the origin only.
Hence they proposed a combination of Eqs. 11 and 12 that allows for
blocked porosity and a power relationship between D_s/D_0 and ε_A:

$$\frac{D_s}{D_0} = \frac{(\varepsilon_A - c)^d}{1 - c} \tag{15}$$

where c has a similar physical significance to b of Eq. 11 and is a
measure of the blocked air-filled porosity, and the exponent d controls the degree of curvature of the relationship.

Ball et al. [63] applied Eq. 15 to results from undisturbed cores
from contrasting tillage and compaction treatments and found that c
was rarely significantly different from zero. They suggested that
the simple model

$$\frac{D_s}{D_0} = \varepsilon_A{}^f \tag{16}$$

would suffice for such applications, where f, like d, is related to
the increase in pore continuity with ε_A. A summary of the relationship between D_s/D_0 and ε_A for minimally disturbed samples is given

TABLE 2 Relationships Between Relative Diffusivity and Air-Filled Porosity in Minimally Disturbed Soil Cores

Soil texture	Soil depth and management	Soil moisture content	Range of ϵ_A	Relationship between D_s/D_0 and ϵ_A	Ref.
Silt loam	Compacted layer 4 cm deep at the surface of grass field	Field or equilibrated to 50 mbar tension	0–0.4	$0.27\ \epsilon_A$	27
Range	Range of distinct structures	Varied by changing soil water tension	0.05–0.35	$\sim 0.5\ \epsilon_A$	62
Sandy loams and silt loams	Arable topsoils nonpuddled puddled	Field	0.04–0.3	$0.85\ \epsilon_A^2$ $2.00\ \epsilon_A^2$	19
Range	Range: 5–140 cm depth	Varied by changing soil water tension	0–0.55	$\approx \epsilon_A^{1.5-3}$	67
Clay loam	Arable cultivation experiment, top 18 cm	Field	0.02–0.1	$\approx \epsilon_A^{2.5}$ (scattered)	68
Sand, silt, clay loams	Arable cultivation experiment, top 20 cm	Field	0–0.3	$0.17\ \epsilon_A$ or $\approx \epsilon_A^2$	69
Silt loam, clay loam	Arable cultivation experiment, top 20 cm	Varied by changing soil water tension	0–0.5	$\epsilon_A^{2-2.5}$ (ZL)[a] $\epsilon_A^{2.1-4.5}$ (CL)[a]	32
Loam	Compaction experiment	Varied by changing soil water tension	0–0.28 0.28–0.5	$3\ \epsilon_A^{2.75}$ $0.34\ \epsilon_A$	12
Sandy clay loam	Compaction and tillage experiments	At field capacity	0.05–0.3	$0.3-2.1\ \epsilon_A^{1.7-2.6}$	63

[a]ZL, silt loam; CL, clay loam.

in Table 2. Most of the data were collected to describe tillage or compaction treatments. The relationships were mostly curvilinear and were similar whether ε_A changed with field variation or with soil water potential on individual samples. More of these relationships correspond to Eq. 15 than to Eq. 16, where greater pore interconnection is assumed. Curved relationships were also reported in Refs. 28, 70, and 71. Approximate relationships derived from diffusion measurements in the field were $D_s/D_0 = \varepsilon_A^{1.7}$ [33] or $D_s/D_0 = 0.37 \, \varepsilon_A$ [41].

The third category of D_s/D_0 - ε_A relationships is that involving discontinuities. Relationships consisting of separate curved and straight line sections were reported in Ref. 12 (Table 2). Initial drainage of the large interaggregate pores (linear relationship) corresponded to maximum pore effectiveness for diffusion. Further drainage of the smaller intra-aggregate pores (curved relationship) corresponded to decreasing pore effectiveness [17].

The pore effectiveness for diffusion τ was defined as:

$$\tau = \left(\frac{D_s}{D_0}\right) / \varepsilon_A \qquad (17)$$

This corresponds to the index of pore continuity and tortuosity of Ball [32]. Currie [72] had previously discovered two-section or bimodal D_s/D_0 - ε_A relationships for packings of soil crumbs. He later used the relative positions and shapes of these sections for describing the influence of compaction on inter- and intra-aggregate porosity [73].

De Jong et al. [28] explored D_s/D_0 - ε_A relationships in swelling soils as they dried. They thereby distinguished structural shrinkage, which occurred first during drying from saturation, from normal shrinkage, which occurred upon further drying. During structural shrinkage $D_s/D_0 = \varepsilon_A^2$, but during normal shrinkage, $D_s/D_0 = \varepsilon_A$. The greater rate of change of D_s/D_0 with ε_A during normal shrinkage resulted from the appearance of soil cracks.

The content of blocked air-filled pores is important in describing the aeration status of a soil layer. Grable and Siemer [74] and the authors have detected air-filled porosities of up to 10% near saturation when gas diffusion through samples was zero. In contrast, recent data indicate significant D_s/D_0 in many soils at air-filled porosities of less than 10%, particularly in well-structured soils [19,32, 63,68,70]. Lower limits of aeration status might be better specified as relative diffusivities (Stepniewski [75] suggested D_s/D_0 of 0.005- 0.01), or as oxygen diffusion rates [45]. However, such limits depend on soil structure, water distribution, soil respiration rate, and crop. Glinski and Stepniewski [12] present a detailed discussion of

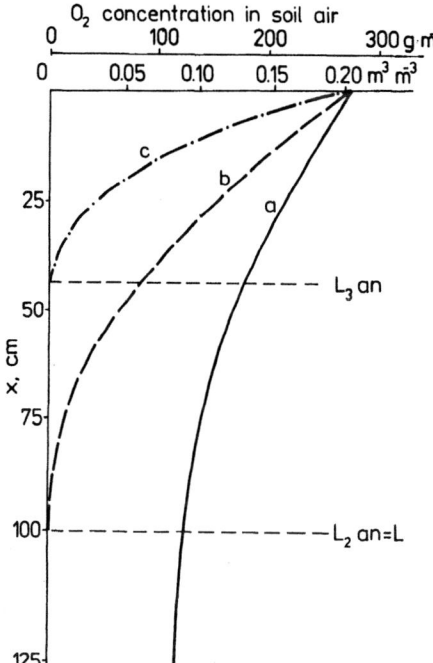

FIG. 7 Oxygen distribution in a homogeneous soil profile at different values of D_s, curves a, b, c, and constant rate of respiration at the soil surface. (From Ref. 12.)

the influence of these factors on soil aeration. These authors also discuss how the variation in air-filled porosity and D_s/D_0 with depth down the soil profile can be used to calculate the oxygen concentration at each depth by treating the soil as a stack of homogeneous layers. Similar methods to calculate the concentrations and fluxes of gases in the soil profile have been presented elsewhere [9,76].

Typical distributions of oxygen in a Polish soil profile are shown in Fig. 7. The curves show the large influence of oxygen diffusion coefficient D_s on soil oxygen concentration. Curve b gives D_s at the critical value (10^{-6} m^2 s^{-1}) for maintaining biological activity in the active zone, the top 100 cm of soil. Curves a and c represent D_s values of twice and half this value of 10^{-6} m^2 s^{-1}, respectively. Campbell [76] includes a suitable computer program written in BASIC for calculating profile oxygen concentrations.

No clear relationship exists between air permeability and air-filled porosity because permeability depends on the pore size distribution from which it may be calculated [77]. The lack of any

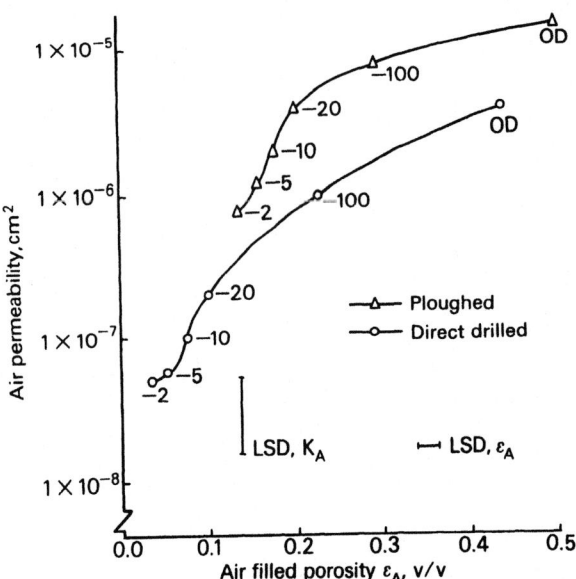

FIG. 8 Relationship between air permeability and air-filled porosity in silt loam sampled vertically at 30-80 mm depth. Each point is the mean of 25 measurements; the numbers beside the points are the potentials at which the samples were equilibrated. (From Ref. 32.)

consistent relationship between K_A and ε_A among replicate samples of minimally disturbed soil [54,55,63] is usually attributed to the presence of wormholes or cracks, which have a large effect on permeability owing to the dependence of K_A on the fourth power of the pore radius. Relationships between K_A and ε_A for a given soil that is compacted to several bulk densities, or for minimally disturbed cores, show more consistent relationships as ε_A is increased by drainage through successive increases in soil water suction. These curves are frequently S-shaped. Two such curves are given in Fig. 8 for minimally disturbed cores taken from a tillage experiment. Similarly shaped curves were reported by Schjønning [70]. Recently, this author has also shown that the air permeability of the macropores (> 30 μm equivalent diameter) is exponentially related to the macroporosity [78].

In addition to its dependence on ε_A and pore size distribution, K_A is influenced by the continuity and tortuosity of the air-filled pores. Groenevelt et al. [55] proposed two indices of macropore continuity and tortuosity, K_A/ε_A and K_A/ε_A^2, where ε_A is that corresponding to -5 kPa water potential and pores exceeding 60 μm effective pore diameter are open. They suggested that soils with similar values of K_A/ε_A have similar pore size distributions and pore continuities, whereas soils with similar values of $K_A/\varepsilon_A{}^2$ have similar pore size distributions only.

Finally, measurements of air permeability, gas relative diffusivity, and air-filled porosity made on a given sample can be combined by use of tube models [70,71,79] to give information on the average radius and tortuosity of the continuous pores in that sample. This is possible because of the different dependence on pore radius of diffusive and viscous flow. At low matric potentials between −1 and −5 kPa, when only the macropores are drained, these model results relate to the suitability of the soil pore system for plant roots.

B. Tillage, Compaction, and Organic Matter Amendments

Tillage generally increases [80] and compaction decreases [81] air-filled porosity and the rate of gas diffusion in soil. However, it may stimulate respiration [82], thus decreasing oxygen concentration. Additions of organic matter to the surface as mulches or incorporated in the topsoil improve soil structure and aeration [47] and increase respiration [33] because of the addition of microbial substrate.

The value of measurements of gas movement in assessing tillage and compaction effects lies in their sensitivity to the *arrangement* of air-filled pores. These effects may not be detected from measurement of air-filled porosity or bulk density. Relationships between D_s/D_0 and ε_A are extensively used.

1. Tillage

Early efforts to distinguish between seedbed tillage treatments used diffusion measurements both in the laboratory [83] and in the field [33,80]. Deep plowing and moldboard plowing to normal depth generally yielded the highest diffusion rates. Treatment differences were only broadly associated with differences in soil bulk density.

Laboratory measurements of air permeability have also been used to distinguish between tillage treatments, large differences in air permeability accompanying the small differences in air-filled porosity [48].

The importance of maintaining a nonpuddled and unsmeared soil surface has been shown from measurements of diffusion [19,84], which reveal that wet, puddled, or crusted surfaces, where ε_A is less than 0.2, can have a D_s/D_0 of only one-tenth that in nonpuddled soils [19], though dry crusts may actually improve gas exchange [84]. Air permeability measurements also are sensitive indicators of surface sealing [50], crusting [53], or capping [54], though Evans and Kirkham [53] stressed that care must be taken to avoid cracks

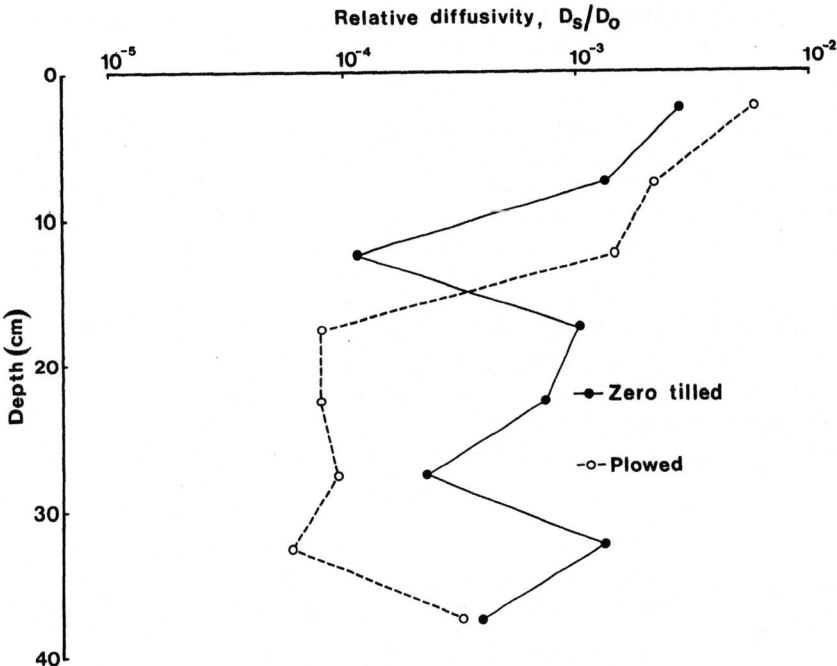

FIG. 9 Relative diffusivity at -1 kPa water potential in an English clay loam, after 6 years of contrasting cultivation treatment (Redrawn, with slight modification, from Ref. 86.)

or wormholes, since these can give very high permeabilities. Air permeabilities can also reveal the presence of crusts in dry conditions better than measurements of diffusion [53].

Zero tillage can result in the buildup of compaction by wheels and weathering. Nevertheless, natural agents, whether biological or environmental, may generate structural attributes that may moderate the effects of compaction or make the soil resistant to further compaction [85]. At similar air-filled porosity, the diffusion coefficient is often greater in zero-tilled than in plowed soil [68,69].

Depth of sampling is important. Douglas and Goss [86] found that at -1 kPa matric potential, D_S/D_0 and K_A were greater in plowed than in zero-tilled topsoil, but both were greater at the boundaries between topsoil and subsoil and in the upper subsoil after zero tillage (20-35 cm depth, Fig. 9). The latter effect was attributed to compaction below plow depth and to the disruption of the continuity of channel-type macropores.

Air permeabilities have been found to be greater in general in plowed than in zero-tilled topsoil but similar or less after plowing than in zero-tillage soil below plowing depth [32,48,86].

2. Organic Matter Amendments

The incorporation of organic matter as a green manure may give an improvement in soil aeration, assessed from D_S values, which is not reflected in measurements of oxygen concentration as partial pressures [33]. Richter and Jacobs [87] reported little effect on CO_2 concentration after incorporation of straw by plowing in sandy soils. The present authors also recently observed (unpublished data) that in both autumn and spring, the influence of omission of tillage was considered greater than the presence or absence of straw. Measurements of air permeability by Schjønning [78] and by the senior author in layers containing straw were approximately equal to those in layers containing no straw at corresponding depths.

3. Compaction

The influence of compaction on the content and the movement of soil gases was reviewed by Grable [81]. He identified the important influence of compact layers in the soil profile on the composition of the soil air. He showed that compaction has little influence on soil aeration until ε_A falls below, typically, 0.1 to 0.2 v/v and that compaction may be relatively severe without altering the composition of the soil air. Bertrand and Kohnke [88] found that oxygen diffused more slowly in compact subsoils than in noncompact subsoils and that diffusion rates corresponding to $D_S/D_0 < 0.16$ limited root growth. Grable and Siemer [74] indicated that plant growth was controlled by soil aeration only over a narrow range of relative diffusivities (< 0.005–0.009). These diffusivities corresponded to matric potentials less than -2 kPa (the air-entry value). At matric potentials of less than approximately -100 kPa, soil strength was the greater limitation to root growth than soil aeration. They suggested that the gas diffusivity limits corresponded to ε_A limits of less than 0.10–0.12 v/v. Similarly, Boone et al. [71] defined lower and upper critical aeration limits (LCAL, UCAL) for root growth calculated from relative diffusivities, respiration rates, and water release characteristics. LCAL ranged between -2 and -8 kPa, corresponding to ε_A's of 0.05–0.08 v/v. UCAL ranged between -6 and -20 kPa, corresponding to ε_A's of 0.15–0.20 v/v. D_S/D_0 and ε_A were related differently according to soil type.

The progressive increase in weight and frequency of use of tractors and wheeled implements has brought a new demand for techniques to reveal the influence of compaction on soil conditions for cropping. Gas movement measurements are suitable to define not only the influence of traffic on seedbed aeration but also the extent of recovery of soil from compaction [85].

C. Drainage

The influence of waterlogging on plant growth at various stages of development and the influence of drainage on aeration have been reviewed by several authors [10,89,90]. Soil oxygen stress has been shown to influence directly the growth of a number of crops (e.g., wheat [91], soybeans [92], and potatoes [93], though toxic or growth-modifying substances produced in the soil as a result of anaerobic activity may also impair root growth. For example, ethylene may influence potato tuber initiation [94] or the response of bean roots to physical impedance [95]. Soil nitrogen may also be lost as nitrous oxide and molecular nitrogen because of denitrification as a result of anaerobic activity [96]. The rate of supply of oxygen in waterlogged soil or aggregates is best measured polarographically [42,93]: see Section II.B. These techniques allow detection of microsites of poorly aerated soil in soil that is otherwise generally well aerated [45]. Callebaut et al. [45] also reported relationships between oxygen diffusion rate and oxygen concentration in the soil atmosphere and empirical relationships for calculating the aeration profile in soils from water suction measurements with tensiometers. The rate of loss of nitrogen as denitrification is best estimated by measurement of the nitrous oxide flux using gas chromatography, having inhibited the reduction of N_2O to N_2 by use of acetylene [97]. Gas chromatography may also be used to detect ethylene [37].

Experimentation with contrasting drainage treatments is difficult because of the problem of hydrological separation of drained and undrained plots [98]. Evidence of the improvement of soil oxygen status and reduction in denitrification losses is more readily found using large scale lysimeters [42,91,99], where waterlogging treatments can be applied with much greater control than is possible in the field.

Drainage in Scottish upland forests has been shown to improve soil aeration [100]. Different species of trees gave different depths of summer drying and of aerated layers in a peaty gley soil. The authors found that measurements of oxygen were more useful for estimating the depth of well-aerated soil than were measurements of carbon dioxide.

D. Fumigants

Fumigants are frequently used to control nematodes and fungi before horticultural crops are planted. They are applied as liquids or gases via injection chisels drawn through the soil. Transport of the gas away from the injection point is primarily by diffusion. Description of this transport and subsequent prediction of the distribution of fumigant in the soil needs to take into account adsorption and dissolution (reversible processes) and decomposition and chemical bonding [101,102]. Early estimates of the sorption coefficient of soils [103] showed correlations with soil surface area, organic matter, moisture, and clay contents. Low soil moisture contents result in rapid evaporative loss [104].

Gas chromatography may be used to analyze samples of the soil atmosphere taken to assess the field distribution of fumigants [37, 105]. As an alternative to such experimental investigation, simulation modeling may be employed [102,106].

VI. SUMMARY AND CONCLUSIONS

Many methods for measurement of diffusion, both laboratory and field, have been developed in the past 20 years. Diffusion rates are expressed commonly in the dimensionless form of relative diffusivity.

Laboratory techniques increasingly use minimally disturbed samples and inert tracer gases, which may be radioactive. Such gases are not consumed or stored in the sample, as is possible when CO_2 or O_2 is used to trace diffusion. Methods featuring the enclosure of the sample end faces by two chambers offer the advantage of precise control of boundary conditions, particularly pressure and temperature, and the possibility of additional measurement of effective porosity. Methods for measurement of mass flow of air, expressed commonly as air permeability, have been less extensively developed than methods for measurement of diffusion. The same techniques are generally applicable in the field and in the laboratory. Steady-state laboratory techniques offer the advantage of preserving laminar flow of gases by application of constant, low flow rates in controlled conditions.

All techniques have benefited from technological developments such as improved gas chromatographs, photomultipliers, electronic differential manometers, and digital flowmeters. Data may be collected and processed by microprocessors, which can scan detectors rapidly. Finite-element analysis is increasingly used to process data. Gas chromatography is valuable for field measurements of diffusion, by allowing analysis of field samples in laboratory chromatographs or in portable field detectors.

Air permeability measurements are used mainly to determine soil structure, whereas gas diffusion measurements may be used to determine both soil structure and aeration. Field diffusion measurements reveal aeration status more directly than laboratory measurements, particularly where soil layers are uniform. However, boundary conditions are hard to specify. Core samples can be taken from thin layers and can be measured under controlled conditions in the laboratory. Such data are useful for structure measurements, in addition to aeration assessment. The measurement of soil structure requires a knowledge of the relationship of relative diffusivity and air permeability to air-filled porosity. This relationship can be established readily with relative diffusivity and can be interpreted to yield information on pore continuity, tortuosity, soil aggregation, content of blocked porosity, and swelling and shrinkage. The relationship, once established, can allow soil aeration status to be monitored using field assessments of air-filled porosity, possibly using a neutron probe. Air permeability and air-filled porosity are not related well because air permeability is very sensitive to large pores and cracks. Air permeability is particularly useful for determination of structural anisotropy, surface sealing and capping, and intrinsic permeability. For laboratory assessments, the number and size of samples required to give a representative sample mean of a given soil layer increases in the following order: air-filled porosity, relative diffusivity, air permeability. The latter two may be distributed non-normally.

The application of gas movement measurements to tillage and compaction studies allows assessment both of short-term effects, such as on seedbed aeration, and long-term effects, such as buildup of soil structures under zero tillage or during recovery from compaction. Diffusion theory can be used to specify the method and conditions of soil fumigation. Gas movement measurements also help to quantify the improvement of aeration by drainage, which results directly from lowering of the water table and indirectly from amelioration of soil structure. Finally, a knowledge of the aeration required for a specific crop may allow prediction of the drainage and tillage requirement of that crop.

REFERENCES

1. Buckingham, E., *Contribution to Our Knowledge of the Aeration of Soils*, Bull. No. 25, U.S. Dept. Agric. Bur. Soils, 1904.
2. Hannen, F., Untersuchungen über den Einfluss der physikalischen Beschaffenheit des Bodens auf die Diffusion wer Kohlensaure, *Forsch. Geb. Agric. Phys.*, 15: 6-25 (1892).
3. Smith, F. B., and P. E. Brown, The diffusion of carbon dioxide through soils, *Soil Sci.*, 35: 413-423 (1933).
4. Hutchins, L. M., Oxygen supplying power of the soil, *Plant Physiol.*, 1: 95-150 (1926).
5. Buehrer, T. F., *The Movement of Gases Through the Soil as a Criterion of Soil Structure*, Tech. Bull. No. 39, Arizona Agric. Exp. Stn., 1932.

6. Green, H., and G. A. Ampt, Studies on soil physics: II. The permeability of an ideal soil to air and water, *J. Agric. Sci.*, *Camb.*, 5: 1-26 (1912).

7. Pritchard, D. T., and J. A. Currie, Diffusion coefficients of carbon dioxide, nitrous oxide, ethylene and ethane in air and their measurement, *J. Soil Sci.*, 33: 175-184 (1982).

8. Kirkham, D., and W. L. Powers, *Advanced Soil Physics*, Wiley-Interscience, New York, 1972.

9. Grable, A. R., Soil aeration and plant growth, *Adv. Agron.*, 18: 58-106 (1966).

10. Currie, J. A., Gaseous diffusion in porous media: 2. Dry granular materials, *Br. J. Appl. Phys.*, 11: 318-324 (1960).

11. Currie, J. A., Gaseous diffusion in the aeration of aggregated soils, *Soil Sci.*, 92: 40-45 (1961).

12. Glinski, J., and W. Stepniewski, *Soil Aeration and Its Role for Plants*, CRC Press, Boca Raton, FL, 1985.

13. Jaynes, D. B., and A. S. Rogowski, Applicability of Fick's law to gas diffusion, *Soil Sci. Soc. Am. J.*, 47: 425-430 (1983).

14. Stolzy, L. H., and J. Letey, Characterizing soil oxygen conditions with a platinum microelectrode, *Adv. Agron.*, 16: 249-276 (1964).

15. Penman, H. L., Gas and vapour movements in the soil: I. The diffusion of vapour through porous solids, *J. Agric. Sci.*, *Camb.*, 30: 437-462 (1940).

16. Evans, D. D., Effect of combined pressure and concentration gradients on gaseous flow through soils, *Iowa State Coll. J. Sci.*, 27: 165-166 (1953).

17. Evans, D. D., Gas movement, in *Methods of Soil Analysis*, Part 1 (C. A. Black et al., Eds.), Am. Soc. Agron., Madison, WI, 1965, pp. 319-330.

18. Taylor, S. A., Oxygen diffusion in porous media as a measure of soil aeration, *Soil Sci. Soc. Am. Proc.*, 14: 55-61 (1949).

19. Bakker, J. W., and A. P. Hidding, The influence of soil structure and air content on gas diffusion in soils, *Neth. J. Agric. Sci.*, 18: 37-48 (1970).

20. Schjønning, P., *A Laboratory Method for Determination of Gas Diffusion in Soil*, Tidss. Planteavls Specialserie, Beret. No. S1773, Statens Planteavlsforsog, Denmark, 1985.

21. Gradwell, M. W., A laboratory study of the diffusion of oxygen through pasture topsoils, *N.Z. J. Sci.*, 4: 250-270 (1961).

22. Currie, J. A., Gaseous diffusion in porous media: I. A non-steady state method, *Br. J. Appl. Phys.*, 11: 314-317 (1960).

23. Rust, R. H., A. Klute, and J. E. Gieseking, Diffusion-porosity measurements using a non-steady-state system, *Soil Sci.*, 84: 453-463 (1956).

24. Ball, B. C., W. Harris, and J. R. Burford, A laboratory method to measure gas diffusion in soil and other porous materials, *J. Soil Sci.*, 32: 323-333 (1981).

25. Reible, D. D., and F. H. Shair, A technique for the measurement of gaseous diffusion in porous media, *J. Soil Sci.*, 33: 165-174 (1982).

26. Sallam, A., W. A. Jury, and J. Letey, Measurement of gas diffusion coefficient under relatively low air-filled porosity, *Soil Sci. Soc. Am. J.*, 48: 3-6 (1984).

27. Bruckler, L., B. C. Ball, and P. Renault, Gaseous diffusion coefficient and porosity effective for diffusion using krypton-85 tracer and a finite element calculation method, *Soil Sci.*, 147: 1-10 (1989).

28. De Jong, E., J. T. Douglas, and M. J. Goss, Gaseous diffusion in shrinking soils, *Soil Sci.*, 136: 10-18 (1983).

29. Youngs, E. G., Soil physical theory and heterogeneity, *Agric. Water Manage.*, 6: 145-159 (1983).

30. Bouma, J., Use of soil survey data to select measurement techniques for hydraulic conductivity, *Agric. Water Manage.*, 6: 177-190 (1983).

31. McIntyre, D. S., in *Methods for Analysis of Irrigated Soils* (J. Loveday, Ed.), Tech. Commun. No. 54, Commonw. Agric. Bur., Australia, 1974, pp. 12, 21.

32. Ball, B. C., Pore characteristics of soils from two cultivation experiments as shown by gas diffusivities and permeabilities and air-filled porosities, *J. Soil Sci.*, 32: 483-498 (1982).

33. Raney, W. A., Field measurement of oxygen diffusion through soil, *Soil Sci. Soc. Am. Proc.*, 14: 61-65 (1949).

34. Van Bavel, C. H. M., Simple diffusion well for measuring soil specific diffusion impedance and soil air composition, *Soil Sci. Soc. Am. Proc.*, 18: 229-234 (1954).

35. McIntyre, D. S., and J. R. Philip, A field method for measurement of gaseous diffusion into soils, *Aust. J. Soil Res.*, 2: 133-145 (1964).

36. Robertson, J. R., *Diffusion from a Gaseous Source in a Porous Medium—A Field and Theoretical Comparison*, U.S. Geol. Survey Prof. Paper No. 650-D, 1969, pp. D265-D273.

37. Smith, K. A., Gas chromatographic analysis of the soil atmosphere, in *Soil Analysis: Instrumental Techniques and Related Procedures* (K. A. Smith, Ed.), Dekker, New York, 1983, pp. 407-454.

38. Robertson, G. W., and J. M. Bracewell, Rapid representative sampling of soil gases and their determination by mass spectrometry, *J. Soil Sci.*, 30: 681-689 (1979).

39. Lai, S. H., J. M. Tiedje, and A. E. Erickson, In situ measurement of gas diffusion coefficient in soils, *Soil Sci. Soc. Am. Proc.*, 40: 3-6 (1976).

40. Rolston, D. E., and B. D. Brown, Measurement of soil gaseous diffusion coefficients by a transient-state method with a time-dependent surface condition, *Soil Sci. Soc. Am. J.*, 41: 499-505 (1977).

41. Jellick, G. J., and R. R. Schnabel, Evaluation of a field method for determining the gas diffusion coefficient of soils, *Soil Sci. Soc. Am. J.*, 50: 18-23 (1986).

42. Blackwell, P. S., Measurements of aeration in waterlogged soils: Some improvements of techniques and their application to experiments using lysimeters, *J. Soil Sci.*, 34: 271-285 (1983).

43. Ray, D., D. G. Pyatt, and I. M. S. White, The effect of the frequency of sampling on the observed concentration of oxygen in an afforested peat soil, *J. Soil Sci.*, 38: 115-122 (1987).

44. Armstrong, W., and E. J. Wright, A polarographic assembly for multiple sampling of soil oxygen flux in the field, *J. Appl. Ecol.*, 13: 849-856 (1976).

45. Callebaut, F., D. Gabriels, W. Minjauw, and M. De Boodt, Redox potential, oxygen diffusion rate, and soil gas composition in relation to water table level in two soils, *Soil Sci.*, 134: 149-156 (1982).

46. Flühler, H., and H. P. Läser, A hydrophobic membrane probe for total and partial pressure measurement in the soil atmosphere, *Soil Sci.*, 120: 85-91 (1975).

47. Henderson, R. E., and W. H. Patrick, Soil aeration and plant productivity, in *CRC Handbook of Agricultural Productivity*, Vol. 1, CRC Press, Boca Raton, FL, 1982, pp. 51-69.

48. Janse, A. R. P., and G. H. Bolt, The determination of the air permeability of soils, *Neth. J. Agric. Sci.*, 8: 124-131 (1960).

49. Bowen, H. D., Air permeability measurement, in *Proc. Int. Conf. Soil Dynamics, Auburn, AL*, Vol. 3, 1985, pp. 481-489.

50. Grover, B. L., Simplified air permeameters for soil in place, *Soil Sci. Soc. Am. Proc.*, 19: 414-418 (1955).

51. Sears, F. W., M. W. Zemansky, and H. D. Young, *University Physics*, Addison-Wesley, Reading, MA, 1982.

52. Kirkham, D., Field method for determination of air permeability of soil in its undisturbed state, *Soil Sci. Soc. Am. Proc.*, 11: 93-99 (1946).

53. Evans, D. D., and D. Kirkham, Measurement of the air permeability of soil in situ, *Soil Sci. Soc. Am. Proc.*, 14: 65-73 (1949).

54. Green, R. D., and S. J. Fordham, A field method for determining air permeability in soil, in *Soil Physical Conditions and Crop Production*, Tech. Bull. No. 29, HMSO, London, 1975, pp. 273-288.

55. Groenevelt, P. H., B. D. Kay, and C. D. Grant, Physical assessment of soil with respect to rooting potential, *Geoderma*, 34: 101-114 (1984).

56. Kirkham, D., M. De Boodt, and L. De Leenheer, Air permeability at the field capacity as related to soil structure and yields, in *Proc. Int. Symp. Soil Structure*, Ghent, Belgium, 1958, pp. 377-391.

57. Hall, D. G. M., M. J. Reeve, A. J. Thomasson, and V. F. Wright, *Water Retention, Porosity and Density of Field Soils*, Tech. Monogr. No. 9, Soil Survey, Harpenden, U.K., 1975.

58. Vomocil, J. A., Porosity, in *Methods of Soil Analysis*, Part 1 (C. A. Black et al., Eds.), Am. Soc. Agron., Madison, WI, 1965, pp. 299-314.

59. Hodgson, J. H. (Ed.), *Soil Survey Field Handbook*, Tech. Monogr. No. 5, Soil Survey, Harpenden, U.K., 1976.

60. Jayawardane, N. S., and W. S. Meyer, Measuring air-filled porosity changes in an irrigated swelling clay soil, *Aust. J. Soil Res.*, 23: 15-22 (1985).

61. Pidgeon, J. D., *A Portable Air Pycnometer for Rapid Field Measurement of Air-filled Porosity in Soils*, unpublished Dept. Note No. SSN/169, Scot. Inst. Agric. Eng., 1974.

62. Ayres, K. W., R. G. Button, and E. De Jong, Soil morphology and soil physical properties: I. Soil aeration, *Can. J. Soil Sci.*, 52: 311-321 (1972).

63. Ball, B. C., M. F. O'Sullivan, and R. Hunter, Gas diffusion, fluid flow and derived pore continuity indices in relation to vehicle traffic and tillage, *J. Soil Sci.*, 39: 327-339 (1988).

64. Troeh, F. R., J. D. Jabro, and D. Kirkham, Gaseous diffusion equations for porous materials, *Geoderma*, 27: 239-253 (1982).

65. Millington, R. J., and J. P. Quirk, Transport in porous media, *Trans. 7th Int. Congr. Soil Sci.*, Madison, Wis., 1: 97-106 (1960).

66. Marshall, T. J., The diffusion of gases through porous media, *J. Soil Sci.*, 10: 79-82 (1959).

67. Flühler, J., *Oxygen Diffusion in Soils*, Beer & Co., Zurich, 1972.

68. Boone, F. R., S. Slager, R. Miedema, and R. Eleveld, Some influences of zero-tillage on the structure and stability of a fine-textured river levee soil, *Neth. J. Agric. Sci.*, 24: 105-119 (1976).

69. Richter, J., and A. Grossgebauer, Investigation of the soil gas regime in a tillage experiment: II. Apparent diffusion coefficients as a measure of soil structure, *Z. Pflanzenernaehr. Bodenkd.*, 141: 181-202 (1978).

70. Schjønning, P., Soil pore characteristics: I. Models and soil type differences, *Tidsskr. Planteavl*, 89: 411-423 (1985).

71. Boone, F. R., H. M. G. van der Werf, B. Kroesbergen, B. A. ten Hag, and A. Boers, The effect of compaction of the arable layer in sandy soils on the growth of maize for silage: I. Critical matric water potentials in relation to soil aeration and mechanical impedance, *Neth. J. Agric. Sci.*, 34: 155-171 (1986).

72. Currie, J. A., Gaseous diffusion in porous media: Part 3. Wet granular materials, *Br. J. Appl. Phys.*, 12: 275-281 (1961).

73. Currie, J. A., Gas diffusion through soil crumbs: The effects of compaction and wetting, *J. Soil Sci.*, 35: 1-10 (1984).

74. Grable, A. R., and E. G. Siemer, Effects of bulk density, aggregate size and soil water suction on oxygen diffusion, redox potentials and elongation of corn roots, *Soil Sci. Soc. Am. Proc.*, 32: 180-186 (1968).

75. Stepniewski, W., Oxygen diffusion and strength as related to soil compaction: II. Oxygen diffusion coefficient, *Pol. J. Soil Sci.*, 14: 3-13 (1981).

76. Campbell, G. S., *Soil Physics with Basic*, Elsevier, Amsterdam, 1985.

77. Marshall, T. J., A relation between permeability and size distribution of soil pores, *J. Soil Sci.*, 9: 1-8 (1958).

78. Schjønning, P., Soil permeability by air and water as influenced by soil type and incorporation of straw, *Tidsskr. Planteavl*, 90: 227-240 (1986).

79. Ball, B. C., Modelling of soil pores as tubes using gas permeabilities, gas diffusivities and water release, *J. Soil Sci.*, 32: 465-481 (1981).

80. Bruce, R. R., and L. R. Webber, The use of a diffusion chamber as a measure of the rate of oxygen supplied by a soil, *Can. J. Agric.*, 33: 430-436 (1953).

81. Grable, A. R., Effects of compaction on content and transmission of air in soils, in *Compaction of Agricultural Soils*, Am. Soc. Agric. Eng., St. Joseph, MI, 1971, pp. 154-164.

82. Grabert, D., Measurements of soil respiration in model experiment on deepening the mould, *Albrecht-Thaer-Arch.*, Müncheberg, 12: 681-689 (1968).

83. Blake, G. R., and J. B. Page, Direct measurement of gaseous diffusion in soils, *Soil Sci. Soc. Am. Proc.*, 13: 37-42 (1948).

84. Domby, C. W., and H. Kohnke, The influence of soil crusts on gaseous diffusion, *Soil Sci. Soc. Am. Proc.*, 20: 1-5 (1956).

85. Campbell, D. J., J. W. Dickson, B. C. Ball, and R. Hunter, Controlled seedbed traffic after ploughing or direct drilling under winter barley in Scotland, *Soil Tillage Res.*, 8: 3-28 (1986).

86. Douglas, J. T., and M. J. Goss, Modification of pore space by tillage in two stagnogley soils with contrasting management histories, *Soil Tillage Res.*, 10: 303-317 (1987).

87. Richter, J., and E. Jacobs, Field measurements on the CO_2-regime of a sandy soil, *Landbauforsch. Volkenrode*, 22: 61-68 (1972).

88. Bertrand, A. R., and H. Kohnke, Subsoil conditions and their effects on oxygen supply and the growth of corn roots, *Soil Sci. Soc. Am. Proc.*, 21: 135-140 (1957).

89. Cannell, R. Q., and M. J. Jackson, Alleviating aeration stresses, in *Modifying the Root Environment to Reduce Crop Stress* (G. F. Arkin and H. M. Taylor, Eds.), Monogr. No. 4, Am. Soc. Agric. Eng., St. Joseph, MI, 1981, pp. 141-192.

90. Luthin, J. N., *Drainage of Agricultural Lands*, Am. Soc. Agron., Madison, WI, 1957.

91. Cannell, R. Q., R. K. Belford, K. Gales, C. W. Dennis, and R. D. Prew, Effects of waterlogging at different stages of development on the growth and yield of winter wheat, *J. Sci. Food Agric.*, 31: 117-132 (1980).

92. Sojka, R. E., Soil oxygen effects on two determinate soybean isolines, *Soil Sci.*, 140: 333-343 (1985).

93. Cary, J. W., Potato tubers and soil aeration, *Agron. J.*, 77: 379-383 (1985).

94. Catchpole, A. H., and J. Hillman, Effect of ethylene on tuber initiation in *Solanum tuberosum* L., *Nature (London)*, 223: 1387 (1969).

95. Kays, S. J., C. W. Nicklow, and D. H. Simons, Ethylene in relation to the response of roots to physical impedance, *Plant Soil*, 40: 565-571 (1974).

96. Sahrawat, K. L., and D. R. Keeney, Nitrous oxide emission from soils, *Adv. Soil Sci.*, 4: 103-148 (1986).

97. Ryden, J. C., L. J. Lund, and D. D. Focht, Direct measurement of denitrification loss from soils: I. Laboratory evaluation of acetylene inhibition of field methods, *Soil Sci. Soc. Am. J.*, 43: 104-110 (1979).

98. Cannell, R. Q., M. J. Goss, G. L. Harris, M. G. Jarvis, J. T. Douglas, K. R. Howse, and S. LeGrice, A study of mole drainage with simplified cultivation for autumn-sown crops on a clay soil: I. Background, experiment and site details, drainage systems, measurement of drainflow and summary of results, *J. Agric. Sci., Camb.*, 102: 539-559 (1984).

99. Cannell, R. Q., R. K. Belford, P. S. Blackwell, G. Govi, and R. J. Thomson, Effects of waterlogging on soil aeration and on root and shoot growth and yield of winter oats (*Avena sativa* L.), *Plant Soil*, 85: 361-373 (1985).

100. King, J. A., K. A. Smith, and D. G. Pyatt, Water and oxygen regimes under conifer plantations and native vegetation on upland peaty gley soil and deep peat soils, *J. Soil Sci.*, 37: 485-497 (1986).

101. Hemwall, J. B., Theoretical considerations of several factors influencing the effectivity of soil fumigants under field conditions, *Soil Sci.*, 90: 157–168 (1960).

102. Rolston, D. E., R. D. Glauz, and B. D. Brown, Comparisons of simulated and measured transport and transformation of methyl bromide gas in soils, *Pestic. Sci.*, 13: 653–664 (1982).

103. Call, F., Soil fumigation: IV. Sorption of ethylene dibromide on soils at field capacity, *J. Sci. Food Agric.*, 8: 137–142 (1957).

104. Smelt, J. H., M. Leistra, M. C. Sprong, and H. M. Nollen, Soil fumigation with dichloropropene and metham-sodium: Effect of soil cultivations on dose pattern, *Pestic. Sci.*, 5: 419–428 (1974).

105. Abdalla, N., D. J. Raski, B. Lear, and R. V. Schmitt, Distribution of methyl bromide in soils treated for nematode control in replant vineyards, *Pestic. Sci.*, 5: 259–269 (1974).

106. Siebering, H., and M. Leistra, Computer simulation of fumigant behaviour in soil, in *Soil Disinfestation* (D. Mulder, Ed.), Elsevier, Amsterdam, 1979, pp. 135–161.

13

Soil Temperature Regime

GRAEME D. BUCHAN   *Lincoln University, Canterbury, New Zealand*

I. INTRODUCTION

Temperature has a fundamental control on almost all biological, chemical, and physical processes in the environment. In cool temperate and colder climates, it demarcates growing and "nongrowing" seasons. Heat flow in soil controls the temperature of both the soil and, via earth's surface energy balance, the lower atmosphere, thus affecting the whole terrestrial biosphere. Yet until recently soil temperature and its effects were poorly patronized areas of research. Soil and plant scientists were more preoccupied with water, mainly because, with adequate temperature established within the growing season, water frequently becomes the major and often erratic determinant of growth, while being more controllable than temperature, either in deficit (via irrigation) or in excess (via drainage). More recently, a wider need has arisen to either measure or model the "soil temperature regime," which is defined here to include the depth and time variations of temperature and heat flux. Thus the literature shows increased attention to effects of soil temperature on root and shoot growth and development; root biochemistry; microbiological activity; nutrient, fertilizer, and pesticide transformations; and physical processes within soil. In addition, the increasing sophistication of crop growth and evapotranspiration models requires improved submodels or measurements of soil temperature regime. Similarly, in climatic modeling and remote sensing, there is a requirement for more accurate data, both on heat flow in soil as a component of the earth's surface energy balance and on soil (especially surface) temperature.

Over the past two decades, significant advances in theory have
been made in the analysis of coupled flows of heat and water, including the application of irreversible thermodynamics, and in the
analysis of flow and phase-change processes in freezing soils. Significant progress in applications has also been made in (1) more
realistic modeling of heat flow, or simultaneous heat and water flows,
by inclusion of the surface energy balance as the governing boundary condition, and (2) measurement and recording techniques for
temperature, heat flux, and thermal properties.
    The basic mechanisms of coupled heat and water flows in soil
were first described in 1957 in the pioneering work of Philip and
de Vries [1]. Despite this, the large impact this coupling can have
on heat flow in the field is not yet fully appreciated. Thus, on the
one hand, recent models of simultaneous water and heat flows in
field soils have correctly incorporated the coupled flow equations.
On the other hand, in the design of experimental techniques and
interpretation of field measurements, the assumption is frequently
made that the heat flow equation can be viewed as "uncoupled"
from the moisture flow equation (i.e., that heat flow in soils is
"conductive" and equal to a thermal conductivity λ times a temperature gradient, where λ implicitly contains the thermal vapor flux
induced by the temperature gradient). While this assumption is
valid in a uniformly moist soil, it can fail badly in the presence of
a strong moisture (i.e., water potential) gradient, which drives an
isothermal vapor flux. This both contributes to the total soil heat
flux and implies latent heat demand at the sites of vaporization.
In the field, this occurs in drying soils, where much of the total
soil evaporation into the atmosphere can derive from "subsurface
evaporation," which exerts a strong influence on heat flux and the
temperature profile. Neglecting such effects can lead to large errors in measurements of heat flux and thermal properties [2].
    This chapter therefore has a dual role. First, it reviews under-
lying theory, then experimental methods. Second, because many of
these methods have been devised or used on the misguided assumption that heat flow can always accurately be described by an "uncoupled" conduction equation, it presents a new clarification of the
potentially large effects coupled flows can have on field measurements. The vital concept is the correct interpretation of the soil
heat flux, including its surface value appearing in the energy balance equation.
    A review of solutions of the "uncoupled" conduction equation
includes periodic solutions and Fourier methods; an overview of the
basic characteristics of the diurnal and annual waves, including noncyclic effects imposed on the former by the latter; "transient" solutions from Laplace transform and other methods; and numerical methods. The calculation of thermal properties from physical composition

is described. A brief section reviews both simplified and more recent, mechanistic theories of freezing soil. The measurement section is a state-of-the-art review of techniques of measuring soil temperature, heat flux, and thermal properties. Sampling criteria and data smoothing are also considered.

There is a remarkable dearth of collected works on soil temperature regime, with a few exceptions [3,4], and notably in the Soviet literature [5,6], though several texts devote sections to basic theory and applications (see, e.g., Refs. 7 and 8). It is hoped that this chapter will help both to remedy this deficiency and to correct some prevalent misconceptions.

Because theory and measurement of soil temperature regimes are so intimately related, a large part (Section II) of this chapter concerns the theory underlying experimental measurements and the extension to modeling of soil temperature regime. The reader concerned solely with field measurements may therefore wish to go straight to Section III. However, to understand the principles and potential pitfalls of measurement methods for soil heat flux and thermal properties, as well as the use of measurements in modeling, the theory of Section II is necessary. To help the reader, a list of symbols is given at the end of the chapter.

II. THEORY

A. Surface Energy Balance

The most powerful and physically revealing models of soil heat flow incorporate its fundamental driving mechanism, the energy balance at the soil surface. The net radiation R_n received per unit area of the soil surface is

$$R_n = (1 - \alpha)R_s + \varepsilon L_d - L_u \qquad (1a)$$

where R_s and L_d are incident solar and longwave radiation, and α and ε are the short-wave reflection coefficient and long-wave emissivity of the soil surface. The term $L_u = \varepsilon \sigma T_0^4$, where σ is the Stefan-Boltzmann constant, represents long-wave emission, detected during infrared thermometry of the surface temperature T_0 [9]. R_n is partitioned at the soil surface according to the energy balance equation

$$R_n = H + L_v E + G_0 \qquad (1b)$$

where H is the sensible heat flux from soil to air, $L_v E$ is the latent (evaporative) heat flux (L_v being the latent heat of vaporization) and G_0 is the heat flux into the soil. For vegetated soil, L_d "seen"

by the surface will include plant as well as sky emissions, H will
include a small stem heat conduction term as well as the usual convection to air, and E, the *soil* evaporation, will be only a portion
of total evapotranspiration. Note that the term "sensible" is applicable to any heat flow causing a local change of temperature. Thus
most of G_0 produces sensible heat (i.e., temperature) change, but
in a drying soil some supplies the latent heat required for evaporation taking place within the bulk of the soil.

The dominant solar radiation term, R_s in Eq. 1a, whose main
cyclical periods are diurnal and annual, drives similar cyclical variations in surface temperature T_0 and air temperature T_a, while $L_v E$,
H, and L_d are all closely controlled by atmospheric temperature and
vapor pressure. Thus Eq. 1b provides a mechanistic basis for relating soil temperature to more commonly measured meteorological
variables, and many previous empirical relationships—for example,
between soil and air temperature (see, e.g., Refs. 10 and 11)—
could be more satisfactorily explained using Eq. 1b, even though
under tall vegetation complex modeling of intracanopy exchanges
would be required. Equation 1b also provides a mechanistic basis
to understand the practical alteration of temperature regime, for
example, by mulching.

1. Components of the Total Soil Heat Flux G_{tot}

In practice the "surface" for the energy exchanges represented by
Eqs. 1 will be a thin layer, with thickness dependent on the surface
microprofile, but typically some few millimeters for a crumb-struc-
tured surface. However, this layer is not necessarily the site of
total soil evaporation E_{tot}. In drying soils the evaporation sites
retreat, at least partially, into subsurface layers [2]. This has
important consequences for interpretation of both Eq. 1b and of the
soil heat flux $G(z,t)$, a function of soil depth z, whose surface value is G_0. As shown in Fig. 1a, E_{tot} is partitioned here as follows:

$$E_{tot} = E_0 + E_{s0} \tag{2}$$

$$E_{s0} = \int_0^\infty E_s(z) \, dz \tag{3}$$

Here E_0 is the evaporation sourced at the "surface" (which is replaced by liquid flow from below), and E_{s0} derives from "subsurface
evaporation." $E_s(z)$ (kg m^{-3} s^{-1}) is the vapor source strength per
unit volume at depth z, associated with vapor flow toward the surface driven by the *moisture* gradient. (Vapor distillation induced
by the temperature gradient is separately accounted for in the effective thermal conductivity, see Section II.C.) E_{s0} will be the dominant component in a soil with a dry surface. Equation 1b may then

(a)

E_o = rate of evaporation from sites <u>at</u> surface, replaced by liquid flow from below

E_{so} = rate of vapor flow <u>through</u> surface, derived from subsurface evaporation

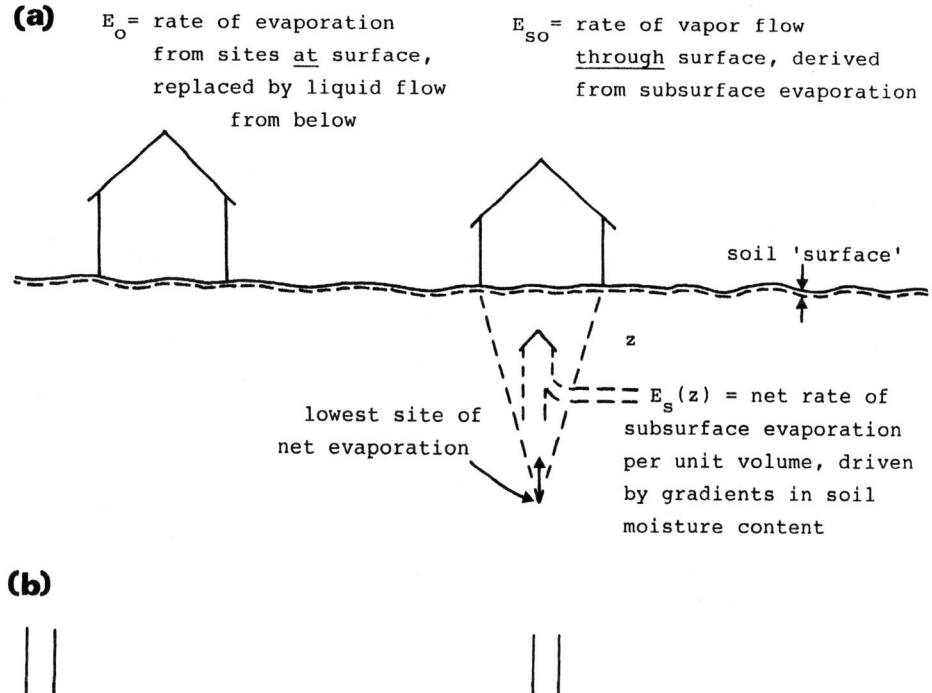

soil 'surface'

z

lowest site of net evaporation

$E_s(z)$ = net rate of subsurface evaporation per unit volume, driven by gradients in soil moisture content

(b)

R_n H $L_v E_o$

$G_o = G_T$, conductive or thermally driven heat flux

R_n H $L_v E_{tot} = L_v (E_o + E_{so})$

$G_o = G_{tot}$
$= G_T$ + (negative) G_{vp}

G_T G_{vp}

Here G_o is the conductive heat flux plus the (negative) isothermal latent heat flux

FIG. 1 (a) Partitioning of total evaporation $E_{tot} = E_0 + E_{s0}$ at the surface of a drying soil. (b) The two possible interpretations of the terms in Eq. 1b, shown under typical daytime conditions. At night the direction of G_T will usually reverse.

be interpreted in two ways, as shown in Fig. 1b.   First, if $E = E_0$, then $G_0 = G_T(0,t)$ [i.e., the surface value of the conductive or thermally driven heat flux, $G_T(z,t)$, Section II.C].   Divergence in $G_T(z,t)$ (i.e., variation of G_T with depth) within the soil will then result from both changes in temperature and the subsurface phase change $E_s(z)$ (corresponding to evaporation or condensation at depth z).   Second, if as is normally assumed, $E = E_{tot}$, then G_0 must be reduced by an amount $L_v E_{s0}$, corresponding to the subsurface evaporative energy demand.   Then G_0 becomes the surface value of the *total* soil heat flux G_{tot} (see Section II.C) given by

$$G_{tot} = G_T + G_{vp} \tag{4}$$

The term "isothermal latent heat flux" is introduced here for G_{vp} $(= -L_v E_{s0})$, which represents the latent heat carried from evaporating subsurface layers by the isothermal vapor flux (i.e., by vapor diffusion under a moisture gradient).   For example, during daytime heating of a drying soil, G_T at the surface will be positive (into the soil), but $G_{tot} = G_T + G_{vp}$ will be reduced by the negative G_{vp}. Then subsurface divergence in G_{tot} (i.e., its variation with depth) is required to fuel only changes in soil temperature.   Thus, in the customary use of Eq. 1 to calculate *total* soil evaporation E_{tot}, it is vital in soil undergoing subsurface evaporation to identify G_0 with G_{tot}.   However earlier practice frequently entailed the erroneous identification of G_0 with the "thermal soil heat flux" G_T, which (Section III.C) is the heat flux obtained by methods measuring the temperature gradient (e.g., the heat flux plate).

B. Heat Conduction: Uncoupled Equations

Conduction of heat down a temperature gradient dT/dz is governed by the Fourier equation

$$G_T = -\lambda \frac{dT}{dz} \tag{5}$$

where in soil the thermal conductivity λ (W m^{-1} K^{-1}) includes a vapor distillation term (Section II.D).   Divergence in the heat flux G_T (W m^{-2}) causes heat changes, both sensible and latent, and is thus subject to the energy conservation equation

$$C \frac{\partial T}{\partial t} = -\frac{\partial G_T}{\partial z} + S(z,t) \tag{6}$$

where C (J m^{-3} K^{-1}) is the volumetric heat capacity of the soil.   S (W m^{-3}) represents local heat sinks or sources, important in field soils only in the presence of phase changes of water (Sections II.C and II.F). Neglecting both spatial variations in λ and the phase-change term S (the latter to be considered in the following section), combination of Eqs. 5 and 6 gives the simple "uncoupled" heat diffusion equation

$$\kappa^{-1} \frac{\partial T}{\partial t} = \frac{\partial^2 T}{\partial z^2} \tag{7a}$$

$$= \frac{\partial^2 T}{\partial r^2} + r^{-1} \frac{\partial T}{\partial r} \tag{7b}$$

where $\kappa = \lambda/C$ (m^2 s^{-1}) is the thermal diffusivity. The second equation, in cylindrical coordinates, is relevant to the use of cylindrical conductivity probes (Section II.E). Equation 7 is "uncoupled" in the sense that, with the thermal vapor flux implicit in λ, it can be solved independently of the moisture diffusion equation. Its use implies a "no-coupling assumption," invalid in soil undergoing aqueous phase changes, in particular subsurface evaporation.

The thermal properties λ, C, and κ are functions of physical composition and also are relatively weak functions of T itself at typical field temperatures. The former implies complex variations with position and time, so that analytic solutions require simplifying assumptions (Section II.E). The latter dependence implies that Equations 7 are, strictly, weakly nonlinear. These relations are readily extended to three-dimensional form, with $\partial^2 T/\partial z^2$ replaced by $\nabla^2 T$.

C. Heat Flow: Moisture Coupling

Heat and water flows can interact strongly in soil. This interaction is small in the extreme cases of soil close to absolute dryness or saturation, but important at intermediate states of wetness. The main coupling of flows is by the following mechanisms: (1) the influence of gradients of temperature on the flow of water, both in the liquid phase by its effect on surface tension, and more importantly in the vapor phase, by its much stronger effect on vapor pressure (i.e., thermally driven water flow), and conversely (2) the influence of gradients of water potential, driving liquid and vapor water flow, on the flow of heat (i.e., water-potential-driven heat flow). The interactions of heat and *liquid* water flow may often be considered to be negligible [12], with a few important exceptions. Examples corresponding to mechanisms 1 and 2 are, respectively: the often rapid migration of liquid water under temperature gradients toward a freezing front, leading under some circumstances to frost heave or formation of "ice lenses;" and heat convection by intense infiltration of water.

By contrast, heat and water *vapor* flows may be strongly coupled, causing conductive heat flux in soils to be accompanied by an often large latent heat flux. The source of this coupling can be seen by considering the one-dimensional (vertical) vapor flux J_v in soil [13], which is the sum of the thermal (J_{vT}) and isothermal (J_{vp}) vapor fluxes.

$$J_v = -D_v \frac{de}{dz} = J_{vT} + J_{vp} \tag{8}$$

$$J_{vT} = -\eta D_v hs \left[\frac{dT}{dz}\right] \tag{9}$$

$$J_{vp} = -D_v e_s(T) \left[\frac{dh}{dz}\right] \tag{10}$$

Here, e is the actual vapor pressure in the air phase, $e_s(T)$ is the saturation vapor pressure (svp), $s = de_s/dT$ is the slope of the svp curve, and $h = e/e_s$ is relative humidity. $D_v = \alpha\theta_a\nu D_{va}$ is the apparent vapor diffusivity (kg m^{-1} s^{-1} Pa^{-1}) in soil air, where D_{va} is the diffusivity in bulk, still air, θ_a is air-filled porosity, and α is a pore space tortuosity factor. The "mass flow factor" $\nu = p/(p - e)$ ≈ 1 (where p is the total air pressure in soil) accounts for a small mass flow contribution to vapor transfer [1]. In Eq. 9, the added "enhancement factor" η is required to give the effective thermal vapor diffusivity ηD_v [1,13,14].

Thus the vapor flux (Eq. 8) is partitioned into two components. The *thermal vapor flux* J_{vT} [15] represents thermally driven vapor transfer. This carries latent heat from hotter (higher vapor pressure) to cooler (lower vapor pressure) regions, contributing to the effective thermal conductivity λ. Conversely, the *isothermal vapor flux* J_{vp} represents a water-potential-driven latent heat transfer, $L_v J_{vp}$. Thus, neglecting osmotic effects, a moisture gradient controls humidity h in Eq. 10 according to

$$h = \exp\left(\frac{\psi_m M_w}{RT}\right) \tag{11}$$

where ψ_m (J kg^{-1}) is the matric potential and $M_w = 18.016 \times 10^{-3}$ kg mol^{-1} is the molecular weight of water. Equation 11 implies h > 0.99 for $\psi_m > -13$ bar. Thus as an approximate rule of thumb, J_{vp} will be relatively small in soils wetter than the wilting point. Then only the thermal vapor flux (already inherent in λ) need be considered. However J_{vp} is significant in the presence of strong moisture gradients—for example, in the upper layers of drying field soils.

Following Eq. 8, we may define a *total soil heat flux* G_{tot}

$$G_{tot} = G_c + G_{vT} + G_{vp} = G_c + L_v J_{vT} + L_v J_{vp} \tag{12}$$

containing a "pure" conduction component G_c, a "thermal latent heat flux" G_{vT}, and an "isothermal latent heat flux" G_{vp}. In reality, pure conduction and the thermally induced distillation (G_{vT}) are intertwined as complex series-parallel processes, hence are not strictly additive. However this difficulty can be circumvented, since both processes are proportional to -dT/dz and may be combined into a single "thermal soil heat flux"

$$G_T = G_c + G_{vT}$$

$$= -\lambda \frac{dT}{dz}$$

(13)

where λ is the apparent thermal conductivity (i.e., as calculated by the Philip-de Vries model discussed below).

The uncoupled heat diffusion equation (Eq. 6) then becomes the coupled heat diffusion equation [1]

$$C \frac{\partial T}{\partial t} = - \frac{\partial G_{tot}}{\partial z}$$

$$= \frac{\partial(\lambda \partial T/\partial z)}{\partial z} - L_v \frac{\partial J_{vp}}{\partial z}$$

(14)

where the last term accounts for phase change induced by a moisture gradient. Thus divergence in J_{vp} represents a heat sink (a site of net evaporation) or source (a site of net condensation). In field soils undergoing subsurface evaporation, the heat sink effect will tend to increase divergence in G_T, hence the curvature of the temperature profile. We return to this effect and its practical impact on heat flux measurement in Section III.C.

The concept of an effective thermal conductivity, enhanced by the thermal vapor distillation effect, can be treated theoretically in two distinct ways. The first method is to solve simultaneously the coupled equations for transfer of water and of heat (see, e.g., Refs. 13 and 16). Thus Eq. 14 is the heat transfer equation (but strictly requires for completeness an additional, usually negligible, liquid water flow term [12]). However this method, while more comprehensive and accurate, requires complex, numerical modeling.

The second method is that of Philip and de Vries [1], and it essentially accounts for the thermal vapor flux equation (Eq. 9) by building it into the de Vries thermal conductivity model [17]. This model, outlined in the next section, calculates λ from the conductivities of individual soil components. As vapor transfer occurs in the air-filled pores, with net distillation from warm to cold ends, the air phase conductivity becomes:

$$\lambda_{av} = \lambda_a + h\lambda_{vs} \tag{15}$$

Here λ_a is the conductivity of still air and

$$\lambda_{vs} = L_v \nu D_{va} \frac{de_s}{dT} \tag{16}$$

is the vapor distillation term for saturated air, ν is the "mass flow factor" discussed in connection with Eq. 8. Note that Eq. 16 represents essentially the same thermal vapor flux effect as Eq. 9 but contains the simple bulk air diffusion coefficient rather than an effective one for a complex pore space. The latent heat term $h\lambda_{vs}$ can be "very effective in increasing the thermal conductivity of soils since it multiplies the conductivity of the air-filled pores by a factor ranging from 2 at 0°C to 20 near 60°C" [12].

The advantage of this second albeit more approximate method is that it incorporates thermal vapor transfer into a single macroscopic conductivity λ, effectively decoupling the equation of heat flow from that of moisture flow. It does not, of course, account for heat transfer induced by a moisture gradient.

The theory of coupled flows in porous media can be approached from a more abstract, thermodynamic viewpoint, based on the thermodynamics of irreversible processes [12,18,19]. Essentially this provides only an overlying formalism for the first, coupled-transport, approach described above. Phenomenological transport coefficients are introduced, but these still need to be derived using the mechanistic ideas of that approach.

The coupling of flows in soil is demonstrated dramatically under prolonged steady-state heat flow. This can lead to marked thermally induced redistribution of moisture (e.g., in laboratory determination of λ (Section III.D.2) or around underground cables or pipes).

D. Calculation of Thermal Properties

The thermal conductivity and heat capacity of a soil depend on its physical composition, especially moisture content, so that single measurements are of limited use. A theoretical basis for estimating the variation of these properties with moisture content is thus required.

1. Volumetric Heat Capacity C

The heat capacity C of a unit volume of soil is calculated simply and exactly as the sum of the heat capacities of its phases; it can be simplified to [12]

$$\begin{aligned} C &= x_m C_m + x_o C_o + x_w C_w \\ &= 4.18 \times 10^6 (0.46 x_m + 0.60 x_o + x_w) \quad \text{J m}^{-3} \text{ K}^{-1} \end{aligned} \tag{17}$$

TABLE 1 Thermal Properties of the Principal Soil Phases (Solids at 10°C, Ice at 0°C)

Material	Volumetric heat capacity, C ($MJ\ m^{-3}\ K^{-1}$)	Thermal conductivity, ($W\ m^{-1}\ K^{-1}$)
Quartz	2.0	8.8
Clay minerals	2.0	2.9
Organic matter	2.5	0.25
Water	4.2	$0.552 + 2.34 \times 10^{-3} T$ $- 1.10 \times 10^{-5} T^2$
Ice	1.9	2.2
Air	1.25×10^{-3}	$0.0237 + 0.000064 T$[a]

[a]T in degrees Celsius.

Source: Refs. 12 and 20.

where x denotes the volume fraction and C the volumetric heat capacity of a phase, with subscripts m, o, and w designating mineral solids, organic matter, and liquid water, respectively. The small contribution of the (moist) air phase has been neglected. Thermal properties of the principal phases are given in Table 1.

3. Thermal Conductivity λ

The macroscopic conductivity λ of Eq. 5 summarizes a heat flow that is spatially averaged over microscopically complex paths via irregular particle and pore phases, and thus cannot be calculated exactly. An approximate "dielectric analog" model has been developed by de Vries [17], based on the application to a granular medium of "potential theory," which treats systems in which an induced response (here, a flow of heat) at any point is proportional to the local gradient of a "potential" (here temperature). Typical variations of thermal conductivity with water content are shown in Fig. 2 for sand, loam, and peat soils.

The model views soil as a continuous medium, denoted by subscript c (either liquid water in sufficiently moist soil or air in drier soil), with volume fraction x_c and conductivity λ_c, in which are dispersed regularly shaped "granules" of the other four soil components (either air or water, respectively; plus quartz, clay, and organic matter). The overall conductivity is then a weighted mean of the conductivities of the components (Table 1)

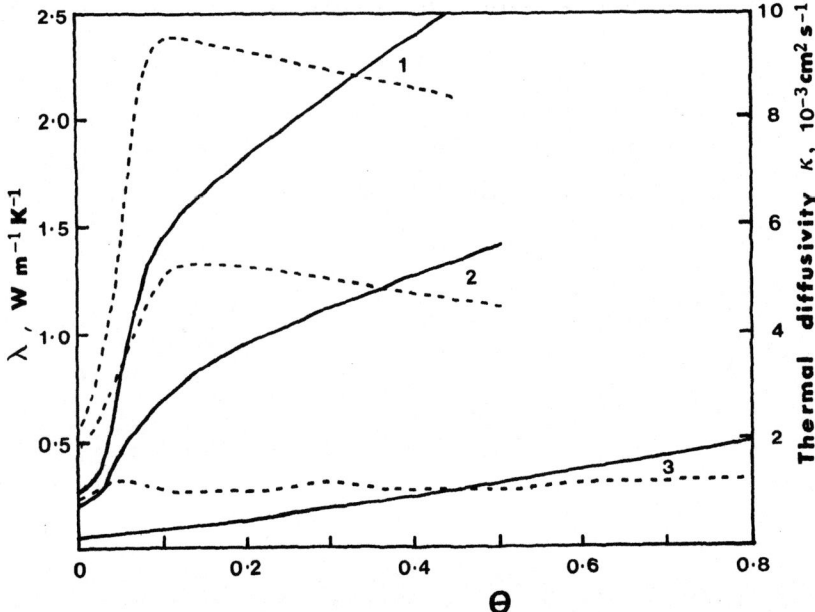

FIG. 2 Variation of soil thermal conductivity (solid curves) and diffusivity (dashed curves) with volumetric water content θ for: 1, quartz sand (x_m = 0.55); 2, loam (x_m + x_o = 0.50; 3, peat (x_o = 0.20). (From Ref. 12, courtesy of Hemisphere Publishing Corp.)

$$\lambda = \frac{x_c \lambda_c + \Sigma \; k_j x_j \lambda_j}{x_c + \Sigma \; k_j x_j} \tag{18}$$

Physically, each weighting factor k_j is the ratio of the average temperature gradient in a "granule" of phase j to that in the background phase. Assuming spheroidal granules, potential theory gives, to a good approximation

$$k_j = \frac{2}{3}\left[1 + \left(\frac{\lambda_j}{\lambda_c} - 1\right)g_1\right] + \frac{1}{3}\left[1 + \left(\frac{\lambda_j}{\lambda_c} - 1\right)(1 - 2g_1)\right] \tag{19}$$

where g_1 is a shape factor for phase j. The assumption that all granules of phase j, though varying in scale, are geometrically similar spheroids, with principal axes in the ratio $a_1 = a_2 = na_3$, allows use of a *single* shape factor g_1. A single k_j factor (along with the factors of 1/3 in Eq. 19) emerges from averaging over random orientations of the granules.

TABLE 2   Weighting Factors k_j for Thermal Conductivity: Eq. 18

Continuous medium	k_j			
	Quartz	Clay	Organic matter	Air
Water (moist soil)	0.267	0.523	1.30	See text
Air (dry soil, $x_w < x_{crit}$)	0.0161	0.047	0.36	1.0

For both sand and clay soils, de Vries [17] deduced representative average values $n = 5$ and $g_1 = 0.125$ for the soil particles. The model, summarized as follows, subdivides the entire moisture range into four regions [20].

Dry Soil. Here air is the continuous medium, and the large ratios λ_j/λ_c (Table 1) require λ evaluated from Eq. 18 to be increased by an empirical multiplication factor of 1.25. Table 2 shows weighting factors calculated from Eq. 19 with $g_1 = 0.125$ and data of Table 1.

Moist Soil Between Saturation and the Permanent Wilting Point: $x_{PWP} < x_w < x_{sat}$. Water is now the continuous medium, so that $x_c = x_w$, and above the permanent wilting point (PWP) h can be set approximately equal to 1 in Eq. 15. With progressive water desorption, the representative air spheroids become increasingly elongated, and de Vries suggested a linear interpolation for the air shape factor, $g_a = 0.035 + (x_w/x_{sat})(0.333-0.035)$, between the values 0.333 for spherical bubbles close to saturation and 0.035 for dry soil. This formula, along with temperature-dependent λ_{av} in Eq. 15, gives k_j for air in Eq. 19. The k_j for the other, solid phases, again using $g_1 = 0.125$ and data of Table 1, are shown in Table 2.

Moist Soil Below the Permanent Wilting Point: $x_{crit} < x_w < x_{PWP}$. With progressive drying below wilting point, both the air shape factor g_a and humidity h will decrease, the latter from close to one, to zero at absolute dryness. Consequently, de Vries suggested a linear interpolation for g_a between a value 0.013 at $x_w = 0$ and the value at PWP derived above, and a linear approximation $\lambda_v = (x_w/x_{PWP})\lambda_{vs}$ to the vapor term $h\lambda_{vs}$ in Eq. 15.

Soil Below a "Critical" Water Content: $x_w < x_{crit}$. The transition from water to air as the continuous medium was suggested by de Vries to occur at a critical water content x_{crit} of about 0.03 for coarse-textured soils, and 0.05-0.10 for fine-textured soils. Below this he recommended a simple linear interpolation of λ versus x_w,

between its dry value as calculated above in the first subsection
and the value at x_{crit} (from the third subsection).

The model predicts λ values "with an accuracy of usually better
than 5%, except in the interpolation range, where the error becomes
of the order of 10%" [12].

The air shape factor is determined in the model in a "somewhat
ad hoc manner" [2]. However the resultant errors should be small
for the following reasons. First, there is a partial cancellation of
error in calculating k_a from g_a via Eq. 19, and in turn λ from k_a
via Eq. 18. In essence, the relative conductivity of a phase matters much more to the overall conductivity than small variations in
the shape of its granules, particularly when their orientations are
randomized. Second, the air phase contribution to λ is in any case
proportionately small, except in two cases: (1) in very dry soil,
when results are more reliant on the calculation of the first subsection
with air as the continuous phase, for which no g_a is required, and
(2) at higher temperatures (T > ~30°C), when λ_{av} is large. (In
fact $\lambda_{av} = \lambda_w$ at T = 59°C [17].) However the reduced contrast
between λ_{av} and λ_w must then act to reduce the sensitivity to
shape factor.

Hence fastidious computation of g_a is unwarranted. The model's
greatest limitations are its use of (1) the assumption that intergran-
ule spacing is sufficient to avoid disturbance of intragranule temperatures in potential theory and (2) idealized spheroidal granules for
pore-occupying phases.

In summary, the model accounts well for the strong moisture
dependence of conductivity, and also for its density dependence.
It has also been applied successfully to swelling soils, with soil
solids as the continuous medium [21]. Temperature dependence,
due almost entirely to vapor distillation, may be considered to be
weak over restricted ranges of temperature, particularly below 30°C.

A curve found empirically to represent the moisture dependence
of conductivity is [22,23]

$$\lambda(x_w) = A + B\left(\frac{x_w}{x_{sat}}\right) - (A - D) \exp\left[-C\left(\frac{x_w}{x_{sat}}\right)^E\right] \tag{20}$$

where A, B, C, D, E are parameters to be determined by curve
fitting, either to experimental values or to the results of the de
Vries model (D is the dry soil conductivity). Alternatively, for
use in numerical models of nonuniform soils, approximate relationships of these parameters to soil composition and density have been
developed [22]. Earlier empirical formulas for estimation of thermal
conductivity from density and water content were developed by
Kersten [24].

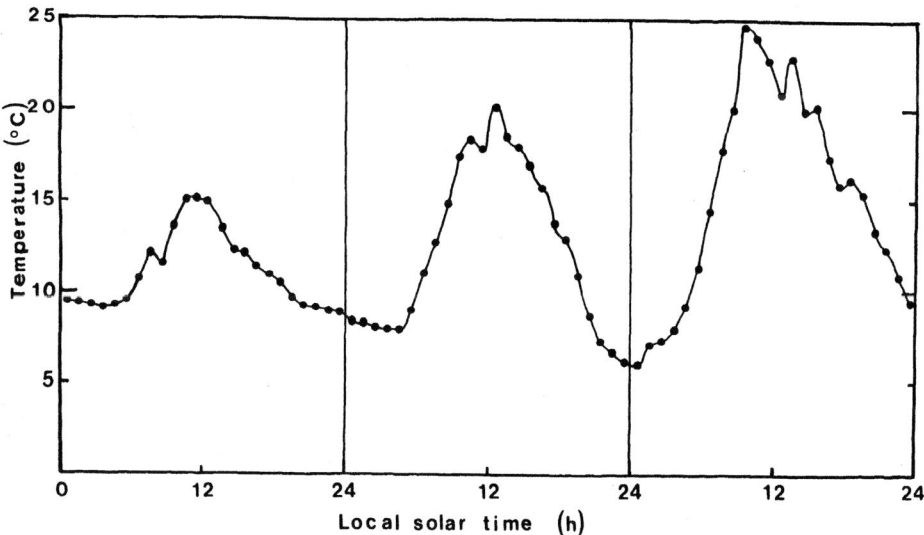

FIG. 3 "Soil surface weather": hourly measured bare soil surface temperature T_0 over 3-day period (8-10 June 1979) at Aberdeen, Scotland. Note noncyclic changes.

While curves of λ and κ against moisture content differ markedly between soils (Fig. 2), they become very similar when plotted against matric potential [25].

E. Solutions of the Conduction Equation

This section deals with solutions of the uncoupled heat conduction equations of Section II.B, primarily Eqs. 7. As discussed in later sections, these solutions have considerable practical application, both in the experimental determination of thermal properties from measured temperatures and heat flows, and in the modeling of soil temperature regimes from restricted sets of field measurements (see, e.g., Refs. 26-28). In the field, complex variations of soil thermal properties and surface weather, hence $T_0(t)$, require numerical simulation methods for greatest accuracy. Figure 3, showing $T_0(t)$ measured hourly over a 3-day period, illustrates "weather" complexity. However, simplifying assumptions are often justified, enabling analytic solutions. These include neglect of the weak T variation of thermal properties, uniformity or analytic variations of thermal properties with depth, and analytic boundary and initial conditions.

1. Analytic Methods

Analytic theory deals with two main types of time variation: periodic variations, and simple nonperiodic variations (i.e., transient or short-term heat flow).   Correspondingly, the two main methods are Fourier transform (FT) and Laplace transform (LT), respectively. Their common advantage is that via an integral transform, they remove the time dependence in $T(r,t)$, so that the partial differential equation (Eq. 7) becomes an ordinary differential equation in the space (r) coordinates only.   We consider only one-dimensional solutions, mainly for vertical (z) variations, since these are dominant in the field, but also the radial (r) solution for the cylindrical probe (Section III.D).

Periodic Variations.   The Fourier method has the appeal of analyzing the temperature variation into a set of harmonics of the dominant diurnal or annual waves.   According to FT theory, an irregular, *continuous* signal of limited duration can be broken down into an infinite sum of harmonics [29].   However, in practice, temperature data are usually available as a *discrete* sequence of N (evenly spaced) points in time, called a "time series" (e.g., with N = 24 for hourly data over one day).   Then the infinite sum becomes a finite sum of M = N/2 harmonics (assuming N even), the so-called discrete Fourier transform (DFT).   For example, a periodic N-point surface variation can be transformed to

$$T_0(t) = \bar{T}_0 + \sum_{n=1}^{M} A_n \sin(n\omega_1 t + \phi_n) \qquad (21)$$

where $\omega_1 = 2\pi/\tau$ is the fundamental angular frequency, with period τ = 24 hours or 12 months for the diurnal or annual wave, respectively.   The N parameters (i.e., $\bar{T}_0$ plus amplitudes A_n and phases ϕ_n) can be determined from the N measured data [26,29].   Assuming that Eq. 7 is linear, the depth penetration of $T_0(t)$ is simply the sum of the penetrations of each harmonic [30,31].

$$T(z,t) = \bar{T}_0 + \Sigma A_n \exp\left(-\frac{z\sqrt{n}}{D_1}\right) \sin\left(n\omega_1 t + \phi_n - \frac{z\sqrt{n}}{D_1}\right) \qquad (22)$$

For Eq. 22 to apply in the field, three implicit assumptions should be satisfied, at least approximately:

1.   The uniform soil assumption, that thermal properties are constant with depth.

2.   An initial condition assumption, that the actual initial T profile equals $T(z,0)$ given by Eq. 22; this implies an "isothermal assumption," that temperatures at all depths vary around the same average value $\bar{T}_0$.

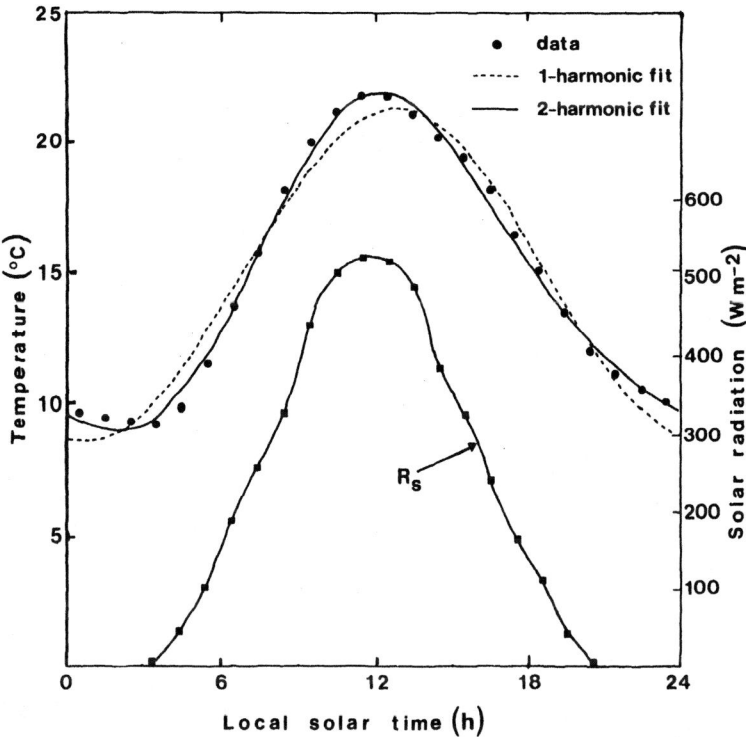

FIG. 4 "Soil surface climate": 15-day average diurnal variations
of bare soil surface temperature T_0, showing measured data and
one and two harmonic fits to data, and solar radiations R_s. Note:
Period (6-20 June 1979) includes days of Fig. 3.

3. The assumption that $T(z,t)$ is approximately periodic, or
equivalently that the "noncyclic change," defined as the difference
between successive midnights (or between a given month in successive years for the annual wave) is close to zero.

Conditions 2 and 3 can be satisfied using a "superposition trick"
(i.e., by exploiting the linearity of Eq. 7 to subtract out, and solve
separately for, the difference between the measured T variation and
that required by the condition). For example, periodicity in a noncyclic diurnal variation (e.g., Fig. 3) can be achieved by subtract-
ing a linear "ramp" variation from single-day data [28]. Also, by
averaging the diurnal variation over several days, a smoother, periodic variation is achieved (Fig. 4) [26,27].

Equation 22 represents a damped, phase-delayed progression of
each harmonic, into the soil, illustrated in Fig. 5. $D_1 = \sqrt{2\kappa/\omega_1}$ is

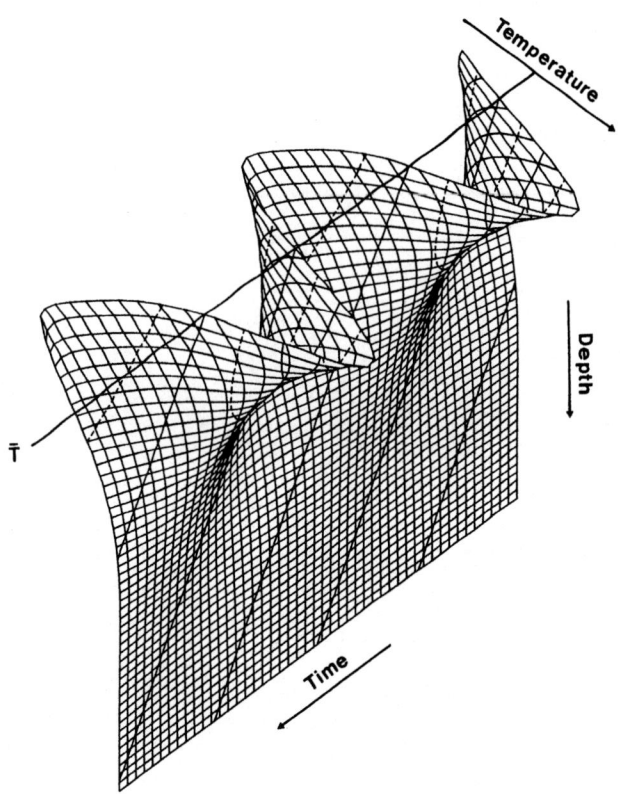

FIG. 5 Three-dimensional plot of soil temperature, showing decay of amplitude and increasing phase lag with depth. Plot shows a two-harmonic springtime wave at Aberdeen, Scotland, with A_1 = 4.8 K, A_2 = 1.1 K, ϕ_1 = -17°, ϕ_2 = -89° in Eq. 22. Note wave asymmetry due to second harmonic. (After G. S. Campbell, *An Introduction to Environmental Biophysics*, Springer-Verlag, Berlin, 1977.)

the "damping depth" of the fundamental (n = 1), with values ranging between about 8 and 16 cm for the diurnal wave ($\omega_1 = 2\pi/86{,}400$ s^{-1}) in mineral soils [12]. Higher harmonics are more rapidly damped, with damping depth decreasing according to $D_n = D_1/\sqrt{n}$. The amplitude is attenuated to 5% of A_n at depth $3D_n$ and 0.7% at $5D_n$, representing an approximate limit of penetration. For the annual wave, the $\sqrt{n}$ rule implies a damping depth $\sqrt{365}$ = 19 times the diurnal value. Thus a typical diurnal damping depth $D_d = D_1 = 0.12$ m gives an annual value $D_a = 2.29$ m.

From Eq. 22, the conductive soil heat flux $G_T = -\lambda \partial T/\partial z$ is:

$$G_T(z,t) = \sum_{n=1}^{M} A_n \sqrt{(\lambda C n \omega_1)} \, \exp\left(-\frac{z}{D_n}\right) \sin\left(n\omega_1 t + \phi_n - \frac{z}{D_n} + \frac{\pi}{4}\right) \tag{23}$$

At the surface, we have:

$$G_T(0,t) = \sum_{n=1}^{M} A_n \sqrt{(\lambda C n \omega_1)} \, \sin\left(n\omega_1 t + \phi_n + \frac{\pi}{4}\right) \tag{24}$$

Thus for each harmonic, the temperature variation lags behind the heat flux by phase $\pi/4$ (i.e., a time lag of $\pi/4n\omega_1 = \tau/8n$). For the fundamental, this is 3 hours for the diurnal and 1.5 months for the annual variation. However this does not represent the lag of extrema in T_0 behind extrema in the solar irradiation, because higher harmonics contribute to $T_0(t)$ and because extrema in G_0 are determined by the *total* surface energy balance (see Figs. 3, 4, and 6). For a typical diurnal wave on moist bare soil, maximum T_0 occurs at about 1300 hours local solar time [27,31], and minimum T_0 around sunrise. There are additional lags under vegetation, typically about a half-hour for short grass, and an hour for cereal crops.

A simple model of the diurnal or annual wave (subscripts a and d, respectively) assumes a single harmonic for each. Their combination is

$$T_0(t) = T_0 + A_a \sin(\omega_a t + \phi_a) + A_d \sin(\omega_d t + \phi_d) \tag{25}$$

where ω_a, ω_d are the fundamental frequencies. Thus $\omega_a = \omega_d/365$. Hence, on average, noncyclic change is an integral feature of the diurnal wave, with a net 24-hour heat gain (or loss) by the soil in the warming (cooling) half of the year. Averaged over each such semiannual period, the noncyclic change in heat storage, drawn from the annual wave every day, is [2]:

$$\Delta_a S = \frac{2\lambda A_a}{\omega_d D_a} \qquad (\text{J m}^{-2} \text{ d}^{-1}) \tag{26}$$

For the diurnal cycle, the net flow into (out of) the soil during the warming (cooling) semidiurnal period is [2]:

$$\Delta_d S = \frac{2\lambda A_d}{\omega_d D_d} \qquad (\text{J m}^{-2} \text{ d}^{-1}) \tag{27}$$

For a cool-temperate bare soil, the annually averaged diurnal amplitude A_d is typically about 5 K, and the annual amplitude A_s about 9 K (author's data), implying $\Delta_a S/\Delta_d S = 0.19$.

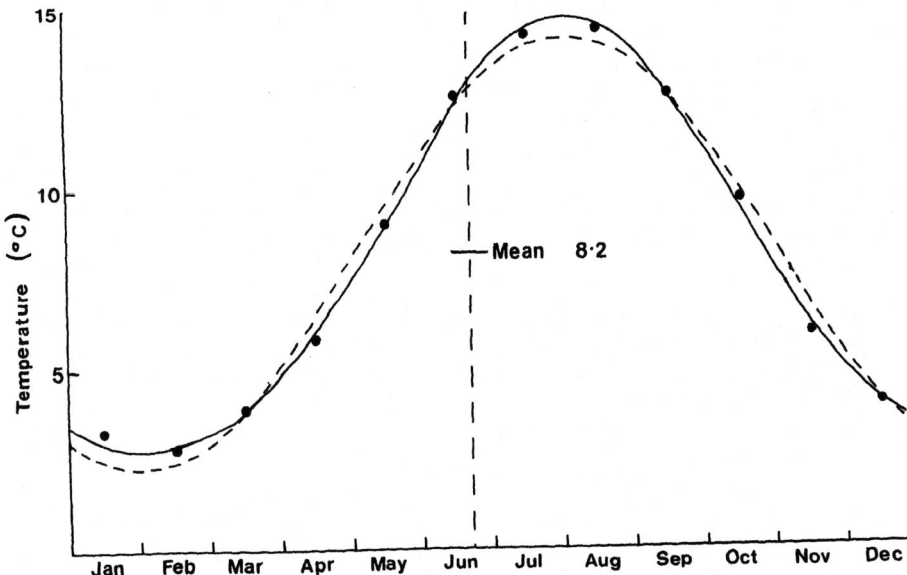

FIG. 6 Annual wave of soil temperature at 30 cm depth, Aberdeen, Scotland (1966-1975, 10-year mean): •, observed data; dashed and solid curves: one and two-harmonic fits to data, respectively. Center vertical line marks midsummer day.

Turning attention to noncyclic change in surface temperature $T_0(t)$, it can be shown that its semiannual average is $\Delta_a T = \pm 2\omega_a A_a/\pi = \pm 4A_a/365$ (K d^{-1}). Assuming $A_a = 9$ K gives an average noncyclic change of only 0.1 K per day. Thus while vagaries of weather may produce large (e.g., 5 K or more) single-day noncyclic changes, its average over many days may usually be neglected [26]. Note that $\Delta_a T/A_d = 2\%$ per day for the above data, so that relative to intradiurnal variations, the noncyclic change in heat storage (19% above) is much greater than that in surface temperature. This reflects the greater effective depth of the annual wave heat reservoir.

However a single 24-hour harmonic is inadequate to represent the diurnal wave. For irregular single-day variations, at least six harmonics are required [28,15]. For multiday average variations, two harmonics are often adequate [11,27]; the amplitude ratio A_2/A_1 is typically around one-quarter or less in the summer months [27,32] but may reach values up to 0.8 in winter [32]. Figure 4 shows a 15-day average diurnal variation of $T_0(t)$ for bare soil. The asymmetry is typical and contrasts with the nearly symmetrical solar radiation curve $R_s(t)$. Three characteristic stages can be identified: (1) steep morning rise, (2) slower afternoon decline, (3) even slower nocturnal cooling. The asymmetry of stages 1 and 2 is due to heat storage in soil and atmosphere causing decline of heat-removing fluxes in the afternoon. The slow cooling of stage 3 is due to the dominant control of nighttime microclimate by, first, net longwave exchange (the difference between surface and effective sky radiation temperatures being reduced compared to daytime), and, second, by the upwelling soil heat flux. The pronounced second harmonic is due to the compound effect of (1) a strong second harmonic in the driving solar radiation $R_s(t)$ [27], imposed mainly by abrupt nighttime zeroing of the R_s curve (Fig. 4), and (2) the soil and atmosphere storage effects noted above. While these soil and atmospheric storage effects produce asymmetry, they in fact lessen the strength of the second harmonic in $T_0(t)$ compared to $R_s(t)$. Thus in Fig. 4 the amplitude ratio A_2/A_1 is 0.14 for T_0 but 0.24 for R_s.

For the smoother annual wave (Fig. 6) a two-harmonic fit is adequate for both soil [31,33] and air [34] temperature. In soil, A_2/A_1, typically 0.12-0.15 [31,33], is less than for the diurnal wave. This reflects the smoother annual progression of R_s, with no analog of abrupt nighttime darkening, except at very high latitudes. Also, the asymmetry in the annual wave is less than in the diurnal wave [35].

The rate at which heat is absorbed into the soil under given surface conditions will clearly increase with both λ and C, and this rate is measured by the term $\sqrt{(\lambda C)}$ in Eq. 24. This term is given various names, the most meaningful—from the analogy with electrical theory—being "thermal admittance" [36,37]; it controls daytime heat absorption and nighttime heat release by the soil. The strong control of the latter over nighttime microclimate explains why soils with lower λ and C (e.g., peats) can exacerbate frosts [12]. The insulation provided by plant cover has similar effects. Admittance, a measure of the rate of surface heat absorption, may be contrasted with the thermal diffusivity (λ/C), which is a measure of the rate at which soil attempts to equalize its temperature by internal diffusion of heat.

Nonperiodic Variations. The Laplace time transform of $T(z,t)$ is given by [38]

$$L<T(z,t)> = \int_0^\infty T(z,t)e^{-st}\, dt \qquad (28)$$

and is a function of z and s only, where s is the dimensionless Laplace parameter. Thus while the FT method decomposes $T(z,t)$

into a set of harmonics and their parameters, the LT employs only one parameter and so is more useful for analyzing simple transient (e.g., rising or decaying) variations. It can be shown [38] that the LT of the heat diffusion equation (Eq. 7) is the following ordinary differential equation:

$$\kappa \frac{d^2 L<T(z,t)>}{dz^2} - sL<T(z,t)> + T(z,0) = 0 \qquad (29)$$

There are two distinct uses of the LT in the analysis of soil heat flow:

1. The conventional or "analytic" use, which involves solution of Eq. 29 for $L(z,s)$, followed by the inverse L^{-1} of the transform, to obtain an explicit solution for $T(z,t)$. Here s plays a purely algebraic role: no numerical value is assigned. The LT is rarely used in this way. One example, however, is solution of the cylindrical heat flow equation (Eq. 7), with r replacing z in Eq. 28, for the case of a heated, hollow, cylindrical probe used for conductivity measurement [39].

2. The predominant "numerical" use, employed when the propagation of a transient perturbation in soil heat flow is analyzed only as a means of deriving thermal properties (λ or κ), with no requirement for detailed solutions for $T(z,t)$. This requires only the forward, numerical transform of measured data: in essence, $L<T>$ is used in lieu of T itself [38]. The precise numerical value for s is now important, as exp(-st) "weights" the temperature record in Eq. 28. The choice $s \geq 5.0/t_{max}$, where t_{max} is the duration at the record, has been recommended, ensuring exp(-st) < 0.007 beyond t_{max} [40].

Assuming initially isothermal soil, we may write $T'(z,0) = 0$ where $T'(z,t)$ is the difference between $T(z,t)$ and the initial isothermal value. Then a solution to Eq. 29 for a semi-infinite soil subject to some surface boundary condition is

$$L<T'(z,t)> = \text{const} \exp[-z\sqrt{s/\kappa}] \qquad (30)$$

where the "constant" is actually a function of s, depending on the exact boundary condition applied [38]. However, given data for two depths z_1, z_2, this drops out in the ratio:

$$L_1/L_2 = \exp[-(z_1 - z_2)\sqrt{(s/\kappa)}] \qquad (31)$$

Thus κ can be determined from temperature records for two or more depths (see Section III.D).

This method can be applied even without initially uniform temperature, by using a "superposition trick." Then $T(z,t) = T_b(z,t) + T'(z,t)$ is viewed as the superposition of the transient T' on the "background" course T_b that T would have taken in the absence of the transient [38]. In practice this requires interpolation on longer records to estimate T_b.

The "cylindrical probe" for measuring thermal conductivity is essentially a line heat source inserted in the soil. The radial solution for this problem is simpler than for the finite-radius probe mentioned above [39]. For a probe in initially isothermal soil, with constant heating rate per unit length Q ($W\ m^{-1}$) switched on at $t = 0$, solution of Eq. 7b gives for probe temperature rise [41]

$$T_2 - T_1 = \frac{Q}{4\pi\lambda}\ \ln\left(\frac{t_2}{t_1}\right) \tag{32}$$

Details of additional analytic techniques developed for homogeneous, inhomogeneous, and layered soils may be found elsewhere in the literature [4,42,43].

2. Numerical Methods

The advantages of numerical methods include their ability to deal with nonuniform soils, with irregular boundary and initial conditions, with multidimensional flows, and with strong nonlinearities (e.g., in the moisture flow equation, if this is solved simultaneously). The soil volume is divided into a set of discrete volume *elements*, separated by boundary interfaces or *nodes*. The case of horizontal layering is depicted in Fig. 7. Local average temperature and conductivity values and heat storage (equivalently a heat capacity value C_i) are attributed to either the elements or the nodes, indexed here by i. The differential heat flow equation is then transformed into a set of algebraic equations, one for each i, including the upper (soil surface) and lower boundaries. Computer solution is by matrix algebra. The key to numerical methods is replacement of analytic time integration by time-stepping from t_j to $t_{j+1} = t_j + \Delta t$. Temperatures are updated using

$$T_i^{j+1} = T_i^j + \frac{\Delta Q_i}{C_i}$$

where ΔQ_i is the net heat flow toward i from nodes (or elements) $i - 1$ and $i + 1$ over time step Δt. To obtain improved approximations to the true *average* ΔQ_i, various interpolation schemes for either the temperature or "heat content" of i can be used, bridging both backward and forward in time. For temperature, a simple linear weighting can be used ($0 \leq \eta \leq 1$):

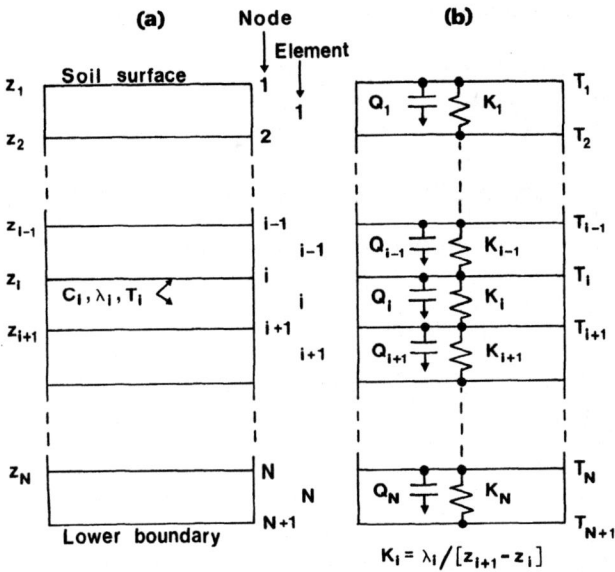

FIG. 7 Schematic layering of soil for numerical simulation of heat flow. (a) Finite-difference and finite-element methods. Values of T_i, λ_i, and centers of heat storage (with heat capacities C_i) are variously attributed to either nodes or elements, according to method used. (b) Network analysis method, showing equivalent resistors and capacitors [22].

$$\bar{T} = \eta T_i^{j+1} + (1 - \eta) T_i^j \tag{33}$$

Thus $\eta = 0$ computes the net heat flow at the "new" time t_{j+1} from temperatures and their gradients at the previous time t_j, the so-called forward-difference scheme, which gives a direct or "explicit" expression for T_i^{j+1} in terms of the known T_i^j at $i - 1$, i, and $i + 1$. With $\eta > 0$, this simplicity is lost. Then T_i^{j+1} depends in an "implicit" way on spatially adjacent temperatures at t_{j+1} [22]. An assumed exponential decay or rise of $T_i(t)$ over the time step corresponds to $\eta = 0.57$ [44]. More sophisticated interpolations exist [45,46].

There are three main numerical methods, differing in the ways they divide up the space-time "grid," attribute variables (to either nodes or elements), and refine the time integration. They are:

1. Finite difference, which assumes that node and time spacings are so small that parameters within them can be considered constant, and differentials may be replaced by their finite-difference forms [30,47].

2. Finite element, which uses elements of *finite* size and pre-
scribes the variation of key parameters across the element (e.g.,
a constant heat flux, or a linear variation of temperature) [44,19].
This reduces the number of nodes, hence computation time.

3. "Network analysis" [22,13], a recent method, developed for
general flow processes in soil. Network analysis also uses finite-
sized elements, but with a physically based analysis of flow and
storage analogous to that used for resistance-capacitance networks
in electrical circuit theory. Each element is attributed a conductivity K_i (the analog of a resistance), while a heat capacity and a
temperature are ascribed to each node (the capacitance analog)
(see Fig. 7). The method is recommended for its comparative simplicity, accuracy, and retention of physical insight [22].

A fourth alternative is the use of ready-made computer simulation packages (e.g., CSMP [45,48] or ACSL) obviating the need to
write detailed numerical algorithms.

For computational economy, grid spacings can be expanded in
approximate inverse proportion to local rates of change of temperature. For example, node spacing can be progressively increased
away from the soil surface [49]; or the algorithm can automatically
increase Δt as simulation of a transient progresses. Algorithms are
usually "calibrated" by comparing their output with exact, analytic
results for simpler problems. Element and time-step sizes are subject to two constraints: absolute values must be less than certain
"coarsest" values, determined by trial variation, above which there
is loss of accuracy [50]; and their relative values may be constrained
to ensure numerical stability—for example, $\Delta t < (\Delta z)^2/2\kappa$ for a layered model (Fig. 7) with $\eta < 0.5$ in Eq. 33 [22].

F. Freezing and Frozen Soil

Soil water freezes either as polycrystalline ice within the soil matrix
or as separate "ice lens" inclusions, which accrete when water mi-
grates toward a slowly moving freezing front. Freezing brings a
large reduction in hydraulic conductivity and a large increase in soil
strength. Frost heave, which can lift soil, roots, and overlying
structures, occurs only at or close to saturation, and usually only
in "frost-susceptible" soils [i.e., those with texture dominated by
silt or noncolloidal (> 0.2 μm) clay fractions] [51]. Upon melting,
holdup of surface water makes the thawed layers greatly susceptible
to mechanical damage or erosion. The prediction of freezing temperature and frost and thaw penetration in soil is important for coping
with frost heave, and for minimizing direct freezing damage to roots,
underground pipes, cables, etc.

This section summarizes the theory of freezing-point depression
(ΔT), heat flow, and thermal properties. An approximate distinction

can be made between *freezing* (or thawing) and *frozen* soil. In the
former, phase change is an ongoing process, accompanied by freezing-induced redistribution of moisture and by large effects on
"apparent" thermal properties [52]. In frozen soil, ice formation
has effectively ceased and thermal properties have stabilized.

The depression of freezing point, a shift in the ice-water equilibrium is due primarily to the lowering of the free energy (i.e.,
water potential) of soil water. It is given by [51]

$$\frac{L_f \, \Delta T}{273.15} = \frac{\psi_m + \pi}{\rho_1} - \frac{P_i}{\rho_i} \tag{34}$$

where $L_f = 3.33 \times 10^5$ J kg^{-1} is the latent heat of fusion of ice,
ψ_m and π are the matric and osmotic components of the liquid water
potential, ρ_1, ρ_i are the densities of liquid water and ice, and P_i
is the ice pressure. For soil with low heave pressure, or unsatur-
ated soil [52], $P_i = 0$, and then $\Delta T = 8.2 \times 10^{-7} \, (\psi_m + \pi)$. Thus
with $\pi = 0$ and $\psi_m = -15$ bar (PWP), onset of freezing will occur at
$T = -1.23°C$. As T is lowered beyond the onset of freezing, the
ice phase grows progressively, initially in larger pores, possibly as
water-drawing lenses, and later into surface-adsorbed layers. The
persistence of liquid is explained mainly by the lower energy (hence
ψ_m) of adsorbed water under the influence of particle surfaces, and
partly by the tendency of water to freeze as pure ice, concentrating the solutes and lowering π in the remaining liquid. The former
effect will clearly increase with clay content. Thus while most water
freezes between 0 and -2°C in soils low in clay [52], the unfrozen
water content in clay soils can be large at very low temperatures
[e.g., as much as 10% (by weight) at -20°C] [53-55].

The theory of heat flow in freezing soil exists at two levels.
Earlier work, aimed at practical prediction of frost (or thaw) penetration, was dominated by the "moving boundary" approach, in which
the freezing (thawing) zone is simplified to a sharp, moving front at
depth $z_f(t)$, and the rate of latent heat production $L_f \, dz_f/dt$ is balanced by net conduction away from the front [53,54,56,57]. Recent
work has advanced toward more mechanistic models, based on simultaneous solution of the transport equations for water and heat, including the phase transformation [51,52,58]. Striking features of
the models include large thermally induced water flux and dramatic
increases of thermal properties due to the phase change. Two major
problematic quantities of the theory requiring more accurate description are the "ice formation characteristic" dx_i/dT and the thermally
driven water flux responsible for freezing-induced redistribution.

Thermal properties of freezing soil exceed those of frozen soil
by up to several orders of magnitude, due to phase-change effects.

In freezing soil, continuing ice formation requires introduction of an apparent heat capacity [51,52]:

$$C_{app} = C - \rho_i L_f \frac{dx_i}{dT} \tag{35}$$

where C is the volumetric heat capacity of Eq. 17 with an added ice fraction term, $x_i C_i$. The second, latent heat term causes C_{app} to "increase abruptly by several orders of magnitude as soon as ice is formed" [52], and, though diminishing as T decreases, it dominates C_{app} down to a texture-dependent lower temperature at which ice formation slows to a negligible level ($\sim$ -2°C for the silt loam of Ref. 52). The temperature range between onset of freezing and this lower limit defines a freeze-thaw zone of finite thickness, in contrast to the sharp front assumed in the simpler moving-boundary models. The "apparent" thermal conductivity λ_{app} of freezing soil is similarly increased, by the contribution of thermally driven water flow. This transports latent heat of fusion in a manner analogous to transfer of latent heat of vaporization by thermally driven vapor flow in ice-free soil [52]. For *frozen* soil, C may be calculated using Eq. 17 with an ice term $x_i C_i$, and conductivity can be obtained from the theory of de Vries (Section II.D) with about the same accuracy as for unfrozen soil [55,59].

III. MEASUREMENT TECHNIQUES

A. Temperature

1. Sensor Characteristics

An understanding of the general characteristics of temperature sensors is essential for proper initial choice and subsequent use of a suitable probe type. These characteristics are as follows. First, the type of output Q (e.g., displacement, voltage, current): electrical output is essential for automatic electronic recording. *Range* must embrace the temperature span to be measured. Likely near-surface extremes are -30 to +50°C, though bare surface temperatures can exceed 60°C in hotter regions [35]. There are two sources of measurement error. *Precision* is a measure of a sensor's ability to reproduce a given value; it can be defined as the standard deviation of a set of repeated measurements of a fixed temperature. *Accuracy* represents the deviation of the measured mean of the set from the true temperature on an established standard scale. It depends on care of calibration, including choice of interpolation formulas relating measured output to true temperature. Thus "accuracy" cannot be less than "precision" but can be made close to it by careful calibration. *Stability* refers to drift in accuracy with time.

TABLE 3   Typical Error Requirements and Suitable Sensors

Objective	Allowable error (°C)	Most suitable sensors
Plant response and function [60]	±0.5	Any
Validation of soil T prediction models	±0.2	Any electrical sensor
Temperature gradients, spatial variability, physical management contrasts	≤ ±0.2	Thermocouple, thermistor, resistance

The *uniformity* of a sensor group or manufacturing method is the maximum expected difference in accuracy between sensors; it determines their interchangeability. (The term *tolerance* is also used, denoting typical or maximum deviation from a theoretical Q-T relationship.) Table 3 summarizes typical maximum-error requirements for various measurement objectives. The *resolution* of a device is the smallest difference in temperature it can detect. Thus precision cannot be less than resolution, though often the two are identical. "Resolution" is most commonly used to describe the readability of a total thermometer system (e.g., electrical sensor plus meter or recorder). It is typically a fraction (e.g, one-half) of a scale graduation, or one digit of a digital display.

Other priority characteristics include robustness, especially to exposure in soil, and, for electrical sensors, degree of immunity to error signals (e.g., spurious connection emf's for thermocouple wires, or interference pickup).

Additional features are summarized in Table 4, which compares the principal sensor types. The *temperature coefficient* $Q^{-1}dQ/dT$ (or dQ/dT) is a measure of output sensitivity, important for choice of range and precision of a meter or recorder for connection to the sensor. *Nonlinearity* is the maximum deviation from linear response over a chosen range (e.g., Fig. 8). Nonlinearity can be handled using linearizing bridge circuitry—for example, for the thermocouple [60] or the strongly nonlinear thermistor [61]. This was a favored practice when direct-readout instruments prevailed, but, with modern, programmable logging and data-processing methods, numerical conversion of unconditioned signal data is now preferable, and more accurate. The *time constant* (or response time) τ measures the delay in response to a step change in ambient T. It is the time taken for sensor T to reach $(1 - e^{-1}) = 63\%$ of the step change [60,61], and it determines sensor frequency response (Section III.B). Self-heating occurs in current-carrying sensors: the *dissipation constant*

TABLE 4 Comparison of Popular Sensor Characteristics

Characteristic	Thermocouple (type K)	Resistance (Pt)	Thermistor	Semiconductor junction[b]	Mercury in glass
T-sensitive property	emf	Resistance	Resistance	I–V characteristic	Volume
Range (°C)	-50 to 400	-260 to 800	-50 to 150	-55 to 150	-10 to 55
T coefficient (°C^{-1})	~ 40 µV	0.39%	Variable, ~ 4%	1 µA	0.016%
Nonlinearity (°C)[a]	0.5	0.4	Large	0.8 max, 0.1 with trim	
Typical accuracy (°C)	±0.1	±0.1	±0.2	±0.2	≥ ±0.2
Stability	Good	< 0.05°C per year	< 0.015°C per year	Long-term drift, < 0.1°C	Good
Uniformity	Excellent for single spool of wire	±0.3	±0.2	±0.2	
Time constant(s)	0.2	1	0.8	> 1	30
Dissipation constant (mW °C^{-1})		50	20		

[a]Nonlinearity over approximate range -10 to 50°C.

[b]Characteristics quoted refer to Analog Devices AD590 (Section III.A.2).

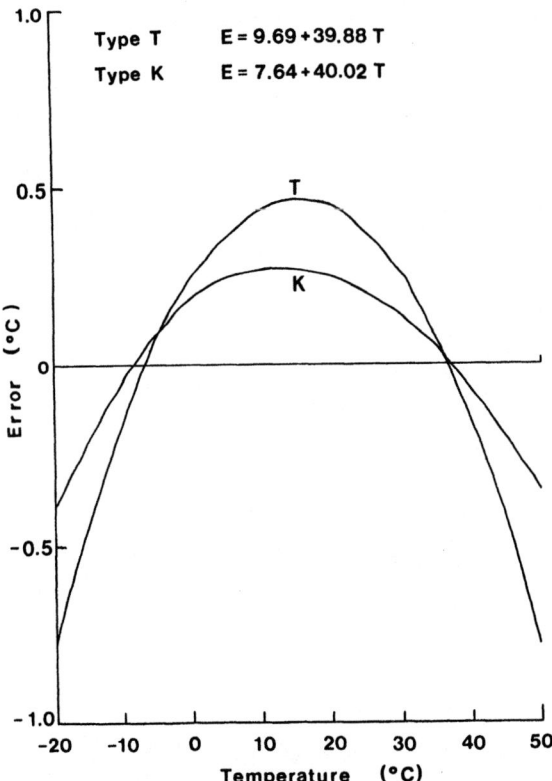

FIG. 8 Nonlinearity errors for popular type T and K thermocouples arising from the assumption of linear relationships between emf and temperature. See Table 4.

k is the power (mW) required to raise the sensor 1°C above ambient. Both τ and k depend on the thermal properties of the sensor's environment. Values quoted in Table 4 are typical for a sensor in soil. In air, k may be up to 50 times smaller, and τ will be larger.

2. Sensor Types

Only the main features of sensors are discussed here. Further information can be found in several excellent reviews of temperature measurement [60-65] and of basic measurement circuitry [60,66]. Table 4 summarizes the most popular types. Some of the characteristics quoted (e.g., range) are for a specific, good commercial sensor and may clearly vary for other available probes.

Liquid-in-Glass. Liquid-in-glass thermometers remain the standard soil probes in the meteorological services of most countries. Spirit-in-glass types placed over the surface (e.g., grass, bare soil) are used to measure minima. In-soil probes are invariably *mercury-in-glass*, with two main types [62]. The World Meteorological Organisation standard depths are 5, 10, 20, 50, and 100 cm [67]; in the United Kingdom, 30 cm (~ 1 ft) is included. The first, right-angled type is designed for depths to 20 cm, though it is also available for 30 cm. The stem has a right-angled bend and, with the bulb at the required depth, readings are taken from the graduated horizontal portion that lies along the ground surface. The second, "sheathed pattern" type, for measurements at 30 cm and deeper, has a straight stem fitted inside a sheath of stout glass tubing and is suspended by a chain inside a hollow steel tube in the soil. The bulb is embedded in wax, filling the bottom 40 mm of the glass sheath. This increases the time constant, enabling the thermometer to be lifted out of the steel tube for reading. A variant of the right-angled type, used in some countries for measurements from the surface to 1 m depth, has a stem bend of 30° just below the scaled portion. With the lower portion inserted vertically, the scale, sealed in an "insulating" glass sheath, projects into the air, supported by an angled (60°) metal stand.

Electrical and Electronic. These sensors are essential for intensive, automated measurements. Only the thermocouple is self-energized; others (i.e., resistance, thermistor, and semiconductor junction types) require an external power supply.

(1) Thermocouples. These are the most popular devices, low in cost and easily constructed. They are differential instruments, based on the Seebeck effect: a temperature-dependent contact emf is developed at the junction of two dissimilar metals. If two junctions at different temperatures are connected in a circuit (Fig. 9), the emf imbalance E increases as $T - T_r$ increases. Thermocouples can be used for accurate measurement of temperature differences, but more usually for absolute measurement of T (°C), either by separate measurement of T_r or by arranging $T_r = 0°C$ for the reference or "cold" junction. This can be achieved with an ice-water mix. Much more convenient is an ice-point electronic reference junction (ERJ), which is basically a resistance bridge (with battery or ac power supply) containing a resistance that varies with ERJ temperature T_j. As T_j varies with ambient temperature, the bridge generates a compensation voltage equal and opposite to the thermoelectric emf of an ordinary reference junction at T_j, thus forcing the ERJ to simulate the ice-point [60,61]. To ensure correct compensation, the ERJ and its connections should form an isothermal mass. Even then, imperfect compensation introduces an ERJ error, typically

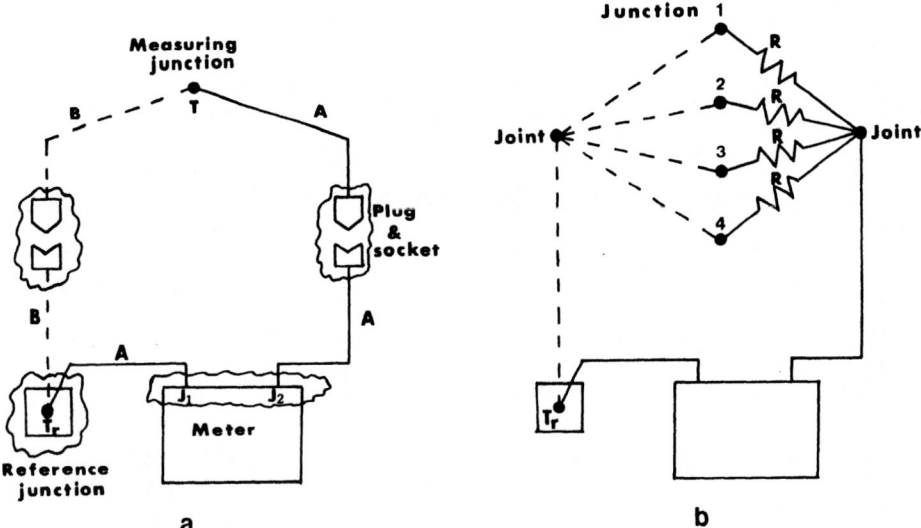

FIG. 9 Schematic thermocouple circuits. (a) Junction pair formed
from metals A and B. Areas enclosed in wavy lines must form isothermal masses to avoid "parasitic" emf's. (b) Parallel arrangement
for temperature averaging, with matched (e.g., 200 Ω ± 1%) swamping resistors R.

about ±0.1°C [63]. With T_r = 0°C, thermocouple output E(T) is
slightly nonlinear (Fig. 8), though the temperature or Seebeck coefficient is approximately constant over narrow ranges (e.g., about
40 μV °C^{-1} for the most popular T or K thermocouple types over
0–50°C). A good approximation is [60]

$$E = a + bT + cT^2 \qquad\qquad (36)$$

where the a, b, c coefficients for the most popular junction types
are given in Table 5. Equation 36 is readily inverted to give a
quadratic relation for T(E). More accurate formulas and tables are
available [61].

A major problem with thermocouple circuits is that all connections
represent additional junctions and are therefore potential temperature
sensors, which may give rise to error or "parasitic" emf's. These
added junctions will occur in pairs; each pair must separately be isothermal to avoid error emf's. For example, in Fig. 8, each probe
lead connection must be isothermal, and paired meter connections
J_1, J_2 must be at the same temperature [68]. Fortunately, temperature gradients *within* leads contribute no thermoelectric emf's [61].

TABLE 5 Characteristics of Popular Thermocouples[a]

Type	a (μV)	b (μV °C^{-1})	c (μV °C^{-2})
T Copper-constantan	-0.09	38.7	0.041
K Chromel-alumel	1.28	39.5	0.019
E Chromel-constantan	0.61	58.6	0.046
J Iron-constantan	1.02	50.4	0.026

[a]The thermoelectric emf E is given by $E = a + bT + cT^2$, where T is the temperature (°C) and the reference junction is at 0°C.

Type T, formerly the most popular, can help to minimize para- sitic emf's if it has copper lead wires connected to a predominantly copper conductor circuit. Increasingly reliable, and popular with manufacturers, are type K, which is more linear and resilient than type T; and type E, which offers greater uniformity, higher tem- perature coefficient, and potentially greater resolution.

Connection of N junctions in series gives a series thermopile: total emf is then the sum of the N emf's. This is used to amplify output in the soil heat flux plate (Section III.C). For spatial aver- aging of temperature, parallel connection should be used [69]. This maintains the signal level of a single device, requiring only one ERJ, Fig. 9. However, imbalance between thermocouple resistances will lead to error, so each junction should be series-connected to a matched (±1%) "swamping" resistor, with value about 20 times the thermocouple resistance [61].

The main problems in using thermocouples are parasitic emf's, the need for a reference junction, poor uniformity between wire batches for some types, and the need to avoid electrical interfer- ence in the low-voltage (mV) dc output signal. Additional precau- tions required include rigorous insulation of leads and junctions to eliminate connections to ground, especially in wet soil, and use of screening or twisted pairs to minimize interference. With high-input impedance measurement circuitry, which minimizes thermocouple load- ing, very long lead wires may be used, though this increases the risk of interference. Details of construction are given in Refs. 60, 61, and 68.

(2) Resistance thermometers. These circuit elements are formed from metal conductors, usually in wire but also in film form. The most popular is the platinum resistance thermometer (PRT), described here, though other metals (e.g., nickel, tungsten) are used. The resistance R(T) has very good linearity but is more accurately

described by a quadratic relation [60]. Though the temperature coefficient is low (Table 4), resistance can be measured with great accuracy. Most commercial PRT units have R(0°C) = 100 Ω and an interchangeability of 0.1 Ω (= 0.3 K) at 0°C. A disadvantage of resistance thermometers is their low resistance. With two-wire measurement, increase in lead length beyond a few meters will cause increasing error due to cable resistance, which, though varying predictably with length, will change unpredictably with cable temperature. Compensating leads can then be used. Cable error is reduced with commercial three-wire probes in conjunction with a three-wire bridge or eliminated entirely with four-wire compensation [60,62]. Errors from (fixed polarity) thermal emf's in connectors and sensor junctions can be minimized by consecutive measurements with reversed excitation currents or by using an ac bridge [60]. The large size of PRTs makes them unsuitable for "point" measurements (e.g., at the soil surface) or for steep subsurface gradients.

(3) Semiconductor sensors. The two basic types are the thermistor ("thermally sensitive resistor") and the junction device, based on the temperature sensitivity of a diode or transistor junction. The *thermistor* exploits the strong negative temperature coefficient of the resistance of semiconductor (metal oxide) material. The rate of promotion of electrons from the valence to the conduction band, across an effective energy gap E, hence the electrical conductivity, is governed by the Boltzmann probability factor $\exp(-E/T)$. Thus resistance, inversely proportional to conductivity, varies approximately as

$$R(T) = R(T_0) \exp B\left[\frac{1}{T} - \frac{1}{T_0}\right] \tag{37}$$

where T is absolute temperature and B is a constant. The thermistor is frequently specified by its resistance $R(T_0)$ (usually several kilohms) at $T_0 = 298.15$ K = 25°C. Following Eq. 37, a useful representation of calibration data obtained over the expected measurement range is a linear regression of the form $\log R = mT^{-1} + c$, which is easily inverted to give T(R). Small corrections can then be made for deviations from this linearity. Other calibration relationships are possible [e.g., a fifth-order polynomial for T(R)] [70].

Advantages of thermistors are their low cost, stability, robustness, and high temperature coefficient (~ 10 times that of a PRT). They come in a wide variety of sizes suitable for soil use, down to catheter types of less than 1 mm diameter, which are excellent for point and surface contact measurement [27]. Their large resistance minimizes interference and thermal emf errors, and swamps connector and cable resistances, with cable errors typically only about 0.001°C m^{-1}. Self-heating effects are negligible with modern logging control methods, which apply only a short-pulse excitation at sampling times.

Junction devices measure temperature via the temperature sensitivity of the p-n junction [60,61,66]. The voltage across a forward-biased silicon diode carrying constant current decreases linearly by about 2 mV $^\circ C^{-1}$. Since the base-emitter junction of a bipolar transistor behaves essentially as a diode, both diodes and transistors can be used. More recent integrated circuit (IC) sensors use the T sensitivity of transistors fabricated on the chip and have integral amplifier circuitry. The Analog Devices AD590 (Table 4) is a two-terminal IC that will operate with supply voltages between +4 and 30 V and has a nominal "PTAT" (proportional to absolute T) current output of 1 μA K^{-1} (e.g., 298.2 μA at 25°C). With simple external trim circuitry and a two-point calibration, the nonlinearity error can be reduced to about ±0.2°C over the environmental range [71]. Alternative, voltage-output devices produced by National Semiconductor (LX and LM series) behave like a zener diode (a "constant" voltage device) with a temperature coefficient of 10 mV K^{-1} (e.g., output 2.982 V at 25°C) [66]. These IC devices, about the size of a small transistor, remove the need for linearizing or resistance-measurement circuitry, cold-junction compensation, and, in the case of the AD590, for lead wire compensation.

(4) Infrared thermometers (IRT) [9,60,63]. These devices measure the temperature T_0 of a surface by remote sensing of the thermal radiation L_r transmitted to a detector, where

$$L_r = \tau[\epsilon\sigma T_0^4 + (1 - \epsilon)L_d] \qquad (38)$$

Here ϵ is the emissivity of the surface and $(1 - \epsilon)$ is, correspondingly, its reflectivity for long-wave irradiance from the surroundings, equal to the sky radiation L_d for a horizontal surface. σ is the Stefan-Boltzmann constant. Radiation enters the detector via a sharp bandpass IR filter, which has transmissivity τ, usually over a wavelength band within the 8-14 μm range (i.e., the atmospheric "window" from which major water vapor and CO_2 absorption bands are absent). This window conveniently includes the blackbody radiation peak for terrestrial radiation, which varies according to Wien's law, $\lambda_{max} = 2900/T$ μm, between 10.6 μm (at 273 K) and 8.7 μm (at 333 K). Most commercial IRTs are calibrated for "gray-body" emission (i.e,, $L_r = \tau\epsilon\sigma T_0^4$). The ϵ value is preset on a switch or dial. The reflected component in Eq. 38 is thus ignored, though more sophisticated devices detect and correct for sky radiation automatically. Unfortunately ϵ is very difficult to measure accurately, and literature values must usually be used. For a typical bare loam soil, it may vary between 0.90 (dry) and 0.95 (wet) [72]. Extreme values are 0.85 for pure, dry quartz sand and 0.98 for water [73]. Thus errors in the IRT readout will arise from uncertainty in ϵ and in L_d, if retrospective corrections are made for reflected radiation. Errors of both types may be of the order of 1°C [60]. Care must be taken to distinguish between the *resolution* of the IRT, which may be as small as 0.1°C for a field device [9], and its *accuracy* under the prevailing conditions. The latter may be as good as ±0.5°C using a high-quality IRT, plus an accurate ε value and reflection correction, but may be ±1°C or poorer for some commercial devices [9]. However, for comparative measurements under given radiative conditions (e.g., of spatial variability) resolution rather than accuracy is a better guide to discrimination ability. Devices, which may be hand-held, are available with both narrow- and wide-angle fields of view (e.g., 3° and 60°).

Air- or satellite-borne scanning IRTs are used in remote sensing and record spatial averages (e.g., the Skylab scanner with a 72 m resolution). The longer radiation travel paths require either corrections for atmospheric absorption and emission or calibration with simultaneous "ground truth" measurements [74].

(5) Other sensor types. The *quartz-crystal* thermometer produces a temperature-dependent oscillation frequency [60] and is capable of very high accuracy (±0.02°C) and resolution (10^{-4} °C). Despite its great expense, it has been used in soils [75]. Large size (e.g., 1 cm diameter) makes it unsuitable for "point" use.

Several designs of *integrating thermometer* have been developed, giving either average temperature, or the integral above some lower limit (e.g., "thermal time" measured in degree-days [76]). Design principles include the "sugar-tube," based on the T-dependent rate of inversion of sucrose [77], and a device based on the flow of a viscous liquid through a narrow capillary tube [78]. Electronic integration of electrical sensor output is an obvious choice (see, e.g., Ref. 79).

3. Comparison of Sensors: Calibration

For field measurement of soil temperature, where a minimum error of ±0.1°C is acceptable (see Table 3), the author's preference is to use closely matched thermistors. These low-cost devices are robust, reliable, and available in small sizes for good "point" (including surface) measurements. Their high resistance enables simple two-wire connection without the need for the lead compensation associated with resistance thermometers. They also enable measurement in the volt range, rather than the millivolt range obtained from thermocouples, giving greater immunity to interference, including thermal emf's. Also, reference junctions, often requiring an additional accurate temperature measurement, are avoided. However, where smaller error is required (e.g., in measuring localized gradients or small spatial contrasts), thermocouples should be used. Temperature difference between two points can be measured more accurately by direct differential use of a thermocouple junction pair, rather than two separately referenced junctions.

For surface thermometry, contact measurement with small, fast-response thermistors gives very good results. With careful installation [27], the error ($\sim \pm 0.3°C$) is better than for typical use of an IRT, though the latter has clear advantage where rapid scanning or areal averaging are required.

Details of *calibration* may be found elsewhere [63,64,80]. Where accuracy of about $\pm 0.2°C$ is acceptable, laboratory calibration against a secondary-reference mercury-in-glass thermometer in a closely controlled temperature bath should be adequate.

B. Sampling and Smoothing

Here we consider the requirements for adequate space and time sampling of soil temperature, and also methods of data smoothing. Sampling of heat flux with transducers is considered in Section III.C. A discussion of electrical and electronic aspects of sampling (i.e., screening and grounding to reduce noise and interference, and sensor-recorder interfacing) may be found in Refs. 60, 61, and 81.

1. Spatial

Vertical Sampling. Because steeper temperature gradients require denser sampling, the exponential decay $A_1 \exp(-z/D_1)$ of the dominant fundamental in the diurnal wave (Eq. 22) requires decreasing spacing toward the soil surface. The optimum depth sequence will depend on objectives and should ideally be scaled to the damping depth D_1 (e.g., closer spacing in a peat than a loam). For validation of models predicting diurnal temperature variation, the author found the geometrical progression 0, 2, 4, 8, 16, 32, 64 cm in a loam (D_1 = 12 cm) to be adequate [27]. However, closer spacing will be required for accurate heat flux determination by gradiometric or calorimetric methods (Section III.C). One possible procedure is to space sensors at points marking constant decrement in amplitude, by a fraction f (e.g., 1/20th) of the surface value A_1. Defining dimensionless depth z^* as z/D_1 (i.e., with damping depth as a scale factor) this generates the sequence

$$z^*_{n+1} = z^*_n - \ln(1 - f \exp z^*_n) \tag{39}$$

which may begin with $z^*_0 = 0$. This oversamples toward the surface compared to conventional practice, but with, say f = 1/20th, only a partial sampling of the implied 20 depths need be used. Where the effects of seasonal progression on the diurnal wave are significant (e.g., on noncyclic heat flow, Section II.E), measurements to 1 m or deeper will be required.

The WMO-recommended depths are 5, 10, 20, 50, and 100 cm, plus discretionary "additional depths" under bare soil or grass [67].

Smoothing and interpolation of T-profile data can be achieved with the cubic-spline method [82]. Measurements are grouped into successive sets of three, and curve segments (cubic polynomials) are fitted to each triplet. Imposition of continuity of slope at the outer two data points, where segments meet, determines the fourth parameter in each polynomial. This matches physical reality by ensuring continuity of dT/dz, hence of soil heat flux, whose profile can then be calculated more accurately.

Horizontal Sampling. Horizontal replication is desirable to cope with two problems: variability, and "rogue" or faulty data. Both worsen toward the surface. Variability has both a deterministic and a stochastic character [83]. Localized measurements are concerned mainly with the latter, arising (for temperature) from variations in both surface condition (e.g., wetness, compaction) and in bulk physical, especially thermal properties. Variability of T_0 at the surface is amplified by increased radiative forcing. Thus Buchan [27], sampling T_0 with three sensors (spacings 10 cm), found a range averaging 0.4 and 0.2 K under daylight and dark conditions, respectively. Instantaneous values had a range up to about 1 K. Such surface variability should attenuate rapidly with depth because of the isothermalizing effect of horizontal heat flow. Thus for contact thermometry at the surface, where faulty exposure can be an added problem, a minimum three-point replication is recommended to enable averaging and possible "rogue rejection." Two-point minimum is recommended at depths down to about D_1. Note that replication with parallel-wired thermocouples (Section III.A) enables averaging but not rogue rejection.

Sensor Installation. The main sources of installation error are probe size, soil disturbance, "stem" conduction, and, for a surface-placed probe, errors due to poor positioning, unrepresentative radiative characteristics, and "nonevaporation" from its surface. For near-surface measurement, including accurate gradiometry, small probes are required (i.e., thermocouples or bead thermistors). Horizontal installation of rod-shaped probes and probe leads will minimize conduction errors. At the surface, reliable contact thermometry was achieved by the author using a miniature (2 mm) bead thermistor, made to resemble a small surface crumb by gluing onto it a coating of fine soil [27]. For installation, a steel rod was pushed into the face of a small pit, at roughly 30° to the horizontal, to emerge from the undisturbed soil surface about 10-20 cm from the pit edge. The bead, at the tip of a fine (1.4 mm) flexible nylon tube containing the paired lead wires, was then pushed through the needle hole to lie flush with the surface. This measures T_0 as accurately as is possible by a contact method.

2. Temporal

Temperature fluctuations in soil are damped by its thermal mass; thus the highest frequencies to be recorded are likely to be considerably lower than the cutoff frequency (related to the time constant [61]) of a well-chosen sensor. The problem then is to find a suitable sampling rate or frequency f_s, which must be high enough to avoid "information loss" on the time variation, but low enough to avoid "information excess," particularly with limited data storage. The sampling theorem [60,61] states that faithful recording of the detail of a signal containing a maximum frequency component f_{max} requires $f_s \geq 2f_{max}$ (i.e., the so-called Nyquist frequency). (This is suggested by the Fourier series method, Eq. 21.) If f_s is less than this, one risks incurring the "aliasing effect": signal frequencies between $f_s/2$ and f_{max} do not go unmeasured, but by intermittent sampling (at rate f_s), they appear "disguised" as lower frequency variations [61]. However, this is unlikely to be a problem in measuring soil temperature. So in practice f_s should be chosen to match the highest significant frequency component f_M required for adequate reconstruction (i.e., consistent with the sampling theorem, $f_s \geq 2f_M$). For the diurnal variation, specific recommendations are as follows. For the "climatic" (i.e., multiday average) diurnal wave, $M = 2$ or 3 harmonics in the Fourier series has been found to be adequate [11,27]. This implies a minimum of 2M (say 6) measurements per day. For the "weather" variation of a single day, a minimum of 6 harmonics is recommended, particularly on cloudy days, implying 12 or more samples per day [28,84]. In practice, the convenient sampling interval $1/f_s = \Delta t = 0.5$ hour or 1 hour should be used, adequate even at the surface. A useful smoothing can then be achieved by fitting a Fourier series, Eq. 22, to the data and truncating it at a cutoff frequency to suppress higher, noiselike harmonics. Cutoff at the harmonic $M = 10$ ($\equiv 0.42$ cycles h^{-1}) has been recommended and used [85,86].

An incorrect procedure [33], for both air and soil temperature, is to use $(T_{min} + T_{max})/2$ as an indicator of mean temperature $\bar{T}$ (i.e., effectively a two-point characterization of the diurnal variation). While correct for a simple sinusoidal wave, this procedure is invalid for a multiharmonic or more complex variation and in practice overweights T_{max}. For the multiday mean surface temperature wave, it has been found by the author to overestimate $\bar{T}$ by 0.5-1.0 K. For single-day variation the error could be much greater. An improved procedure is to use a weighted average, $\bar{T} = [\eta T_{min} + (1 - \eta)T_{max}]/2$, with a weighting factor $\eta > 0.5$ (e.g, $\eta = 0.59$ at the soil surface [87]).

Use of a daily-averaged or integrated temperature—for example, in the calculation of "thermal time" for biological processes [76]—represents a single-datum characterization of the diurnal wave. However, this will be inadequate where the process response to temperature is nonlinear.

C. Soil Heat Flux

The main practical use of measurements of heat flux G is to determine the flux at the surface ($G_0(t)$ in Eq. 1b) usually for more accurate assessment of assessment of evaporation. Heat flux at depth $G(z,t)$ is important for research into heat and coupled flows. Diurnal and annual cycles of the conductive component $G_T(z,t)$ may be decomposed into a set of harmonics (Eq. 23), the amplitude of each decaying according to $\exp(-z/D_n)$. The amplitude of the dominant fundamental component decays to 10% of its surface value at $z = 2.3D_1$ (i.e., roughly 30 cm for the diurnal wave in a loam), so measurements may usually be confined to shallower depths. While daily average values of G_0 may be small (but see Section II.E on noncyclic heat storage), instantaneous values may be up to about 250 W m^{-2} during peak daytime inflow [11,88]. As a relative measure, the flux ratio G_0/R_n is often used and varies from about 0.5 for bare soils down to less than 0.1 under dense vegetation. Thus, contrary to frequent assumption, G_0 may be a large fraction of the energy balance, and indeed dominant at night for sparsely vegetated surfaces.

As discussed in Sections II.A and II.C, care must be taken with the heat flux concept where moisture gradients are large, as the *total* heat flux G_{tot} may contain a large isothermal latent heat component. This is not accounted for in the "conductive" component $G_T = \lambda dT/dz$, which incorporates only the thermal latent heat flux. Thus in the common use of Eq. 1b to determine total soil evaporation E_{tot}, the frequent mispractice of using G_T instead of G_{tot}, may give rise to substantial "phase-change errors." These errors arise from neglect of subsurface evaporation (or conversely condensation), but could similarly arise from neglect of freeze/thaw effects.

Five methods of measurement or estimation of heat flux can be identified: transducer, the temperature gradient method, calorimetric methods, the Fourier analysis method, and the "flux regression" (i.e., G_0 versus R_n) method. Note that the first two methods determine G_T via the temperature gradient and so automatically exclude the isothermal latent heat component G_{vp} (Eq. 12).

1. Transducers

The principal design of a heat flux meter consists of a thermopile embedded in a thin, flat plate, usually in the form of a disk of glass or resin about 50 mm in diameter (d) and 5 mm in thickness (h). The heat flux plate (HFP) impedes both liquid and vapor water flow, including evaporative supply to the surface, and so should not be placed where moisture gradients are large. Placement at 5-10 cm or deeper is recommended [86,88]. Despite this, installation depths of 1 cm or less have been used [75,89]. To obtain G_0, the "combination method" [88,90] can be used; that is, the rate of change of heat storage above the HFP can be determined by calorimetry. A major problem is the heat flux disturbance caused by difference between the plate conductivity λ_p and that of the soil, λ. Philip [91], using the model of a thin spheroidal plate, derived an expression for the ratio of heat flux in the soil to that in the plate, G_p (the so-called heat flow disturbance factor [92]):

$$\frac{G}{G_p} = 1 - 1.7\left(\frac{h}{d}\right)\left(1 - \frac{\lambda}{\lambda_p}\right) \qquad (40)$$

Plate thermopile output is, in fact, a measure of ΔT, the temperature difference across the plate, and so of $G_p = \lambda_p \Delta T / h$. Hence, use of a single calibration factor for the HFP presupposes a unique ratio G/G_p. However, since λ is variable, G/G_p should be close to 1 to minimize errors. This implies three design requirements: (1) the plate should be thin (h/d << 1), (2) λ/λ_p should be close to 1 (e.g., by ensuring that plate conductivity is close to the average value for the soil), and (3) since in practice heat flux disturbance will still occur (i.e., $G/G_p \neq 1$), the HFP should be calibrated in a medium with conductivity close to the soil average. As an extreme example of the effect of calibration medium, Weaver and Campbell [92] found that calibration of a plate with thickness h = 4 mm and conductivity $\lambda_p = 0.4$ W m^{-1} K^{-1}, first in dry and then wet sand ($\lambda = 0.4$ and 2.0 W m^{-1} K^{-1}, respectively) led to a doubling of the calibration constant—that is, a potential for mismeasurement by a factor of 2. One possible strategy to reduce errors due to λ variations within a given soil would be to use Eq. 40 to adjust continually the calibration factor, using updated estimates of λ from measured water content.

As an alternative to the thermopile HFP, commercial Peltier coolers have been used [92]. These are thermopile-like devices, but with junctions of dissimilar (n- and p-type) semiconductor materials, rather than dissimilar metals. They are designed for "cold-junction" cooling under an applied electric current, that is, exploiting the Peltier effect, which is the inverse of the Seebeck effect (Section III.A). They can, however, be used in Seebeck (thermoelectric) mode. The devices used by Weaver and Campbell had sensitivities about 70 mV/(kW·m^{-2}) (i.e., ~ 8 times that of a thermopile).

However, their temperature coefficient, at about 0.25% K^{-1}, was also greater, by about a factor of 4. While Peltier coolers are cheaper and more sensitive than thermopiles, their conductivity (~ 0.4 W m^{-1} K^{-1}) is lower than typical soil values, so heat flux disturbance may be large.

Close thermal contact between plate and soil is essential. Air gaps, particularly in drying soil close to the surface, can lead to large errors. A minimum threefold replication is recommended for reliable near-surface measurements. To increase sensitivity at deeper levels, where G_T is lower, series connection of plates (in fours) has been used [93].

Details of flux plate calibration [60,61,94,95] and construction [88,95] are given elsewhere.

2. Temperature Gradient Method

Quite simply, this method computes the thermal heat flux $G_T(z)$ = $-\lambda dT/dz$ from the conductivity λ and the temperature gradient. Accurate gradiometry, especially close to the surface, requires accurate thermometry, with small sensors sufficiently close to avoid errors due to T-profile curvature. Local conductivity may be either measured (Section III.D) or calculated (Section II.D).

3. Calorimetric Methods

These are based on the depth integral of the heat conservation equation (Eq. 14), that is,

$$
G_{tot}(z) = G_{tot}(z_r) - \int_{z_r}^{z} C\left(\frac{\partial T}{\partial t}\right) dz
$$

$$
= G_{tot}(z_r) - \sum C_i \Delta z_i \left(\frac{\Delta T_i}{\Delta t}\right) \tag{41}
$$

where z_r is a reference depth at which G_{tot} must be known. In practice, discrete sampling requires the second, finite-difference form of Eq. 41: a soil layer i, thickness Δz_i, is ascribed to each temperature sensor, recording temperature change ΔT_i over sampling interval Δt. Heat capacities C_i can be calculated from Eq. 17, using measured moisture contents for accuracy. Large errors may arise using short time intervals (e.g., Δt < 30 minutes) because the ΔT_i may become too small to be measured accurately [90].

Equation 41 may be used or misused as follows. The "correct use" recognizes G_{tot} as the total heat flux (i.e., $G_T + G_{vp}$, Eq. 12). However, most practitioners of calorimetry have neglected subsurface evaporation and assumed in effect that G_{tot} is the conductive flux, G_T [2]. This is the "incorrect use." While this becomes correct when $G_{vp} = 0$ (i.e., when the heat summation is through soil layers effectively free from phase change), it can give rise to large phase-change errors where summation is through strong sites of net evaporation (e.g., drying near-surface layers). Following Eqs. 41 and 14, the correct equation for $G_T(z)$ is

$$G_T(z) = G_T(z_r) - \int_{z_r}^{z} C\left(\frac{\delta T}{\delta t}\right) dz - H(z, z_r) \tag{42}$$

where

$$H(z, z_r) = L_v[J_{vp}(z) - J_{vp}(z_r)] = L_v \int_{z_r}^{z} E_s(z) \, dz \tag{43}$$

Here H is the latent heat consumed by subsurface evaporation between z_r and z (but excluding that induced by the thermal vapor flux). Thus $H(z_1, 0) = L_v E_{s0}$ (i.e., the total heat consumption by subsurface evaporation), where z_1 is a depth below which evaporation effectively ceases (i.e., "the lowest site of net phase change" [2]). This may be the dominant soil evaporative heat demand in a soil with a dry surface.

Thus the correct calorimetric procedure for the conductive flux G_T requires, in the presence of subsurface evaporation, separate monitoring of the water (matric) potential profile to evaluate divergence in the vapor flux J_{vp}, hence Eq. 43. However, in the main practical use of calorimetry (i.e., to obtain G_0 at the surface for evaluation of *total* soil evaporation E_{tot}), it turns out (Section II.A) that $G_{tot}(z) = 0$ is required. Fortuitously, this is correctly delivered by the "simple" calorimetry of Eq. 41 if z_r is below the lowest site of evaporation (i.e., details of subsurface evaporation and the moisture profile are not required).

The key to calorimetry is knowledge of G at depth z_r. Several choices of z_r are possible. We look at each practice in turn, pointing out where subsoil evaporation can vitiate results. First, a large value may be chosen for z_r (e.g., a depth of 100 cm), where $G_{tot} = G_T \simeq 0$, and may either be neglected or, preferably, estimated from $\lambda dT/dz$.

Second, the "null-point" method, as originally devised, locates z_r at a null point (i.e., where dT/dz = 0). While this implies $G_T = 0$, error will result in Eq. 41 if z_r is within the evaporating layers. Unfortunately, as depicted in Fig. 10, weak curvature in T(z) means that null points are poorly defined, except at times a few hours after sunrise or before sunset. To enable calculation of G(z) throughout the day, Kimball and Jackson [90] introduced the so-called *null-alignment method*. Serious criticisms of this method have been made

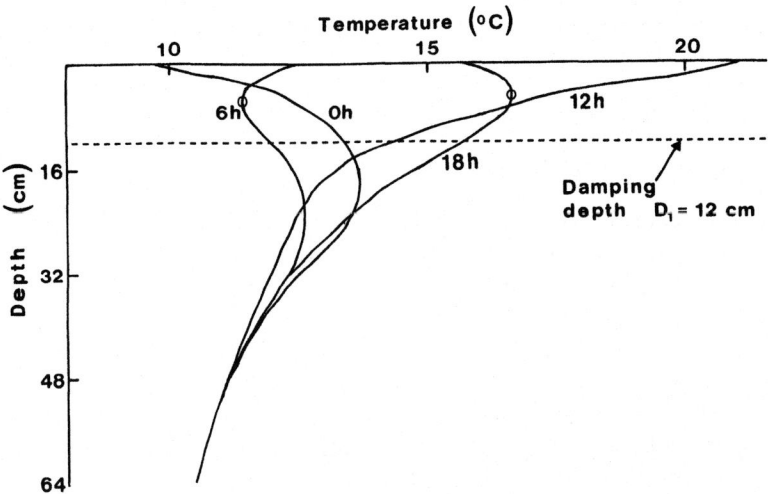

FIG. 10   Tautochrones (depth profiles) of soil temperature, showing "null points" (circled) 2.5 hours after sunrise and before sunset. Data are for a 15-day average diurnal variation, as in Fig. 4.

[2], as its practitioners have usually implemented it in soils undergoing strong subsurface evaporation, while neglecting phase change by incorrect use of Eq. 41. This practice can accrue serious errors, as can be seen in the following outline of the method. Essentially, a well-defined null point (around sunrise or sunset) is used as a primary reference depth z_r to define a deeper secondary reference depth z_r' for subsequent use in Eq. 41. The procedure is:

1.   Obtain $G_T(z_r')$ from Eq. 41 when $G_T(z_r) = 0$.
2.   Compute $\lambda(z_r') = -G_T(z_r')/(dT/dz)$ from the simultaneously measured T gradient at z_r'.
3.   From then on, use z' as the reference depth in Eq. 41, with $G_T(z_r') = -\lambda(z_r')(dT/dz)$ calculated from the current T gradient and λ from step 2.

Thus with z_r close to the surface, and incorrect use of Eq. 41, large errors accrue due to heat summation both down (step 2) and up (step 3) through the evaporating layers. These errors are both in G(z), and in derived λ values, which may be underestimated by a factor of 0.5 or more [2].

The *combination method* determines the surface flux G_0 using calorimetry, with $G_T(z_r)$ measured by a heat flux plate, and z = 0 in Eq. 41. However, if the plate is too shallow (z_r above the lowest site of net phase change), this method will give the required total heat flux at the surface, *less* the isothermal latent heat flux at z_r not measured by the plate [i.e., $G_0 = G_{tot}(0) - L_v J_{vp}(z_r)$].

In summary, calorimetry-based methods are attractive because they appear to require only monitoring of the temperature profile, plus knowledge of $C(z)$ and of G at one reference level. The null-alignment method appears to dispense with even the last requirement. However, calorimetry, particularly the null-alignment method, can fail badly in soils with significant subsurface evaporation. Then correct procedure requires attention to (and some measurement of) vapor diffusion induced by water potential gradients. However, it turns out that for the main practical use of calorimetry [i.e., to determine $G_{tot}(0)$ for calculating evaporation], this is not required: simple calorimetry using a reference level below the lowest site of evaporation should give the correct value.

4. Fourier Analysis Method

This method is based on fitting Fourier series (Eq. 22) to measured temperatures, followed by substitution of the derived amplitude and phase parameters in Eq. 23 to obtain $G_T(z,t)$. Thus for accuracy the three assumptions listed after Eq. 22 should be approximately satisfied. The soil thermal properties (assumed to be uniform) must be known to evaluate Eq. 23. If both λ and C are known, $G(z,t)$ may be calculated using Fourier parameters obtained from $T(z,t)$ at a single measurement depth. Alternatively, if, say, only C is known, Fourier analysis of temperatures at two or more measurement depths can be used to evaluate κ (Section III.D), and then $\lambda = C\kappa$. The method has been used with a 10-harmonic fit to single-day data [86], giving good agreement with calorimetry and with a 2-harmonic fit to multiday averaged diurnal variations [11]. Note that the presence of significant subsurface evaporation will vitiate this method, since Eqs. 22 and 23 are solutions to the uncoupled heat diffusion equation.

5. Flux Regression Method

Several studies have sought approximate statistical relationships between the soil heat flux G and net radiation R_n, often as a simple ratio G/R_n. The main motivations are: (1) the diurnal oscillations of G_0 and R_n are roughly sympathetic, and (2) $R_n - G_0$ is the "available energy" for partitioning between $L_v E$ and H in Eq. 1b, so that relationships involving the *surface* flux G_0 will enable better assessment of evaporation, either in ground-based or remote-sensing applications [75,89,96,97]. However, relationships are complicated by their strong dependence on time of day, vegetative cover and height (e.g., G_0/R_n ranges from about 0.5 for dry bare soil [89] to about 0.1 for grass or cropped soil [96,97]), the depth at which G is measured, and soil water content θ. The θ dependence appears to apply only to bare soils [97]; with G_T measured at 1 cm depth by heat flux plate, Idso et al. [89] found that the ratio G_T/R_n ranged from 0.5 for dry soil to 0.3 for wet soil.

Comparison of studies is further complicated by the different methods used to measure G, giving different components of the heat flux. Thus while Idso et al. [89] measured G_T (z = 1 cm), Clothier et al. [97] used the combination method with plates at 5 cm, which would (see above, Section III.C.3) have given $G_{tot}(0)$ less the isothermal latent heat flux at 5 cm.

D. Thermal Properties

Thermal conductivity and diffusivity are measured by inference, that is, by interpretation of observed temperature variations via the heat conduction equations. As these uncoupled equations do not account for the evaporative heat sink effect accompanying moisture redistribution, the following methods must be used carefully in the presence of moisture gradients, particularly subsurface evaporation.

1. Heat Capacity C

This is invariably calculated from Eq. 17, using volumetric sampling data, although calorimetric measurement is possible [98].

2. Thermal Conductivity λ

Methods can be classified as steady-state, transient-state, the heat flux method, and the diffusivity method.

Steady-State Methods. These laboratory methods establish a steady, uniform temperature gradient dT/dx in a containerized sample by supplying heat at a rate Q (W m^{-2}) at one end. Then λ = -Q/(dT/dx). Cylindrical (Lucite tube) containers have been used, with a planar electrical heat source placed at one end and dT/dx measured close to the source [99]. The "divided bar" method is used for both rocks and soils [100]. The sample is placed between plates of a uniform material of known conductivity, and a steady-state established across the "sandwich," Q being calculated from dT/dx in the outer plates.

A serious problem with steady-state methods is the moisture redistribution caused by migration of water from warmer to colder zones, leading to nonuniformity. Hadas [99] overcame this by measuring both temperature and water gradients close to the heat source to obtain λ(θ). While stationary methods are acceptable for saturated or very dry soils, transient methods should be used for moist soils.

Transient-State Methods. The *cylindrical heat probe* has been used extensively in both laboratory and field [25,31,41,101,102].

Its use of transient-state heat flow avoids major moisture redistribution. It consists of a line heat source, usually a narrow cylindrical probe containing an electrical heating wire or element. The
rate of temperature increase T(t) is then measured with a small,
usually built-in sensor next to the heating wire. From Eq. 32 [103],
a plot of T against log t is a straight line whose slope gives λ.
However, this is valid only for probes of small (strictly, infinitesimal) diameter. Large-diameter probes require a more complex solution to the conduction equation [31,103]. Imperfect probe-soil contact is a potential problem, particularly in drier soil: the extra
thermal resistance involves an additional "contact factor" in the solution [99,101]. At high water contents, air entrapment can cause
inaccuracy [104]. A laboratory variant of the single probe is the
"twin transient-state cylindrical-probe" method [105]. One probe
is placed axially in a cylinder of the soil, and its temperature
change is referenced to that of an identical probe placed in a cylinder of standard material (agar gel).

An alternative, "powerless probe" technique uses a cylindrical
glass or aluminum probe without a heating wire [44]. Instead, a
thicker probe is preheated (or cooled) and thrust into the soil. As
it equilibrates, its temperature is monitored via a built-in thermocouple. Since no closed-form solution to the conduction equation
exists for this situation, a finite-element simulation is also required.
λ is equated to that value giving closest simulation of the measured
temperature trend. Advantages of the method are use of a large,
more durable probe, and absence of heater circuitry.

Heat Flux Method. This is the inverse of the temperature gradient method for measuring the conductive heat flux G_T. Thus
$\lambda = -G/(dT/dz)$ is obtained from simultaneous measurement of G and
the temperature gradient. However, G must be the conductive component G_T of the heat flux (e.g., as measured with a heat flux
plate). In field soils with strong subsurface evaporation, use of
an incorrect G (e.g., from erroneous calorimetry) has been shown
to give serious errors in "measured" λ values [2].

Diffusivity Method. Here, conductivity is obtained as $\lambda = \kappa C$,
where the thermal diffusivity κ can be measured as described below.

3. Thermal Diffusivity κ

Diffusivity controls the dynamic redistribution of heat in soil. Its
value is therefore deduced from the observed propagation of temperature variations, either periodic or transient. Field methods are
based on solutions (Section II.E) to the one-dimensional, uncoupled
heat conduction equation (Eq. 7). Thus they are prone to phase-
change error, particularly as measurements are usually made in
upper soil layers.

Propagation of Periodic Variations. Methods here are based on depth penetration of the diurnal or annual wave, represented in Fourier series form (Eq. 22). Thus, in addition to the assumption of uncoupled heat flow, the three assumptions following Eq. 22 are implicit. κ is determined from temperature measurements at two depths z_1, z_2, by one of three possible methods: amplitude decay, phase lag, or matching of wave penetration. The first two methods yield κ via the damping depth $D_n = \sqrt{(2\kappa/n\omega_1)}$, usually for the fundamental (n = 1), using Eqs. 44 and 45, respectively. From Eq. 22, D_n is given by

$$D_n = \frac{[z_2 - z_1]}{\ln[A_n(z_1)/A_n(z_2)]} \qquad (44)$$

$$D_n = \frac{[z_2 - z_1]}{[\gamma_n(z_1) - \gamma_n(z_2)]} \qquad (45)$$

where $A_n(z)$ is the amplitude and $\gamma_n(z) = \phi_n - z/D_n$ is the phase at depth z for the nth harmonic. In the third method, a Fourier series is fitted to measured T at depth z_1, and κ is selected as that value in Eq. 22 which best reproduces measurements at z_2 (e.g., by minimizing the sum of squared differences) [84].

The first two methods have been found by the author and others [84] to give erratic results for the diurnal wave, chiefly due to the large relative errors in the ratio and difference terms in Eqs. 44 and 45. However, Horton et al. [84] found that the third method gave consistent, reliable results, provided the number of observations per day at both depths was between 8 (on clear days) and 12 (on cloudy days, with more irregular variations). A simplified practice to be avoided uses Eq. 44 with the approximation $A_1(z) = (T_{max} - T_{min})/2$ for the fundamental. While correct for a simple sine wave, this is clearly in error for a typical, complex diurnal variation. Not surprisingly, this gives an unreliable κ value [84].

Propagation of Transients. The principal field method is based on Laplace transform analysis of the shallow penetration of a transient change in the heating of the soil surface, thus giving an effective κ for, typically, the upper 5-10 cm [38,40,106]. In the original technique [106], for field or laboratory use, an artificial "heat pulse" is imposed by irradiating the surface with an incandescent lamp for a short period (5-30 minutes). (Conversely, soil shading can be applied on bright days.) Temperature T(z,t) is measured at two or more depths reached by the pulse. Following Eq. 31 et seq., a plot of L<T'> versus depth should then give a straight line of slope $-\sqrt{(s/\kappa)}$. The method is easier to apply if the upper layers are initially isothermal. If not, the "superposition trick" can be used, although this introduces possible error from interpolation of the trend of "background" temperature, which is to be subtracted from T to give the transient component T'. If the soil heat flux variation at the surface is measured simultaneously (e.g., using a flux plate beneath the heating lamp), λ can also be calculated [106]. Natural transients caused by intermittent radiation under broken cloud can also be used. Because the method is based on the un-coupled conduction equation (Eq. 7), phase-change error will result in soils undergoing strong subsurface evaporation, especially as T is measured in the upper layers. Reported disagreement with the de Vries conductivity model [40] is therefore not surprising.

An alternative mathematical approach analyzes a short-term portion of the diurnal variation of $T(z,t)$ in upper soil layers, starting at a time of isothermal conditions [107]. Two parameterized curves are fitted to observed temperature variations: (1) an analytic function to $T_0(t)$ (e.g., linear rise or fall, or sine wave), and (2) a cubic spline to the near-surface T profile at a fixed time $t_0 > 0$, to give dT/dz at the surface at t_0. Analytic expressions give κ in terms of the parameters of the fitted curves.

The diffusivity of soils packed in long cylinders can be determined using an un-steady-state method [108,109]. The sealed column is equilibrated in a water bath at temperature T_1, then transferred to a bath at temperature T_2, the heating or cooling curve $T(t)$ at the column center being measured with a sensor. A plot of $\log[(T_2 - T)/(T_2 - T_1)]$ against t is a straight line whose slope gives κ.

E. Frost and Thaw Penetration Depth

Five main methods of measuring the depth of the frozen-unfrozen interface in soil are briefly reviewed here. First, the $0°C$ *isotherm* can be located by interpolation on a measured T profile [57,110]. Depression of the freezing point (Section II.F) may make this an inaccurate indicator of the interface [111], particularly when temperature gradients are small and the freezing front diffuse. The *frost tube* [57,112,113] consists of a bottom-sealed length of PVC pipe, installed in the soil and containing a removable inner acrylic or Lexan tube filled with sand saturated with fluorescein solution. Upon freezing, the dye changes from green to pale yellow, indicating the depth of the frozen-unfrozen interface within the tube. The *time domain reflectometry* (TDR) method, originally devised for measuring water content by its effect on soil dielectric constant (Chap. 1), also responds to the different dielectric characteristics of water in the liquid and frozen state [110]. The interface is detected by either of two techniques: the change in apparent *liquid*

water content occurring at the frost or thaw front, or a change in
the reflection coefficient for electromagnetic waves at the interface,
shown as a kink in the TDR output trace. A comparative assessment of the isotherm, frost-tube, and TDR methods showed that
they gave comparable results for frost penetration depth, but that
the TDR method was superior in detecting the presence and depth
of the unfrozen-frozen interface in thawing soil [57]. An electrical
capacitance probe developed by Brach et al. [111] also exploits the
large drop in dielectric constant (from about 80 to 3) as water
freezes. The apparent capacitance of the soil indicates the state
of pore water. *Gypsum blocks* placed in soil undergo an abrupt
change in electrical conductivity as water freezes within them and
can be used with a low-power dc supply [114]. However, they
lose discrimination when dry (i.e., in dry soils).

IV. APPLICATIONS

A knowledge of soil thermal regime finds practical application in
four main areas: (1) its effects on biological processes, (2) its
effects on chemical and physical processes, including soil weathering, (3) its role in determining above-ground climate, and (4) remote sensing applications. Applications may be classified into one
of two categories: "passive" measurement and modeling, or "active"
manipulation.

A. Measurement and Modeling

Many workers have characterized soil temperature regimes under
contrasting types of ground cover (e.g., forest [115,116], grassland and field crops [87,117-119], bare soil [27,120]) and in response to variations in environmental factors (e.g., altitude [121],
latitude and geographical location [122,123], snow cover and freezing [124]). However, the theory of heat and water flow in the soil,
coupled to energy exchanges at the surface (particularly below complex canopies) is insufficiently developed to provide a comprehensive
framework capable of accurate prediction of these contrasting regimes.

B. Manipulation

Manipulation is directed at areas 1 and 3 listed above (Section IV).
Many reports and studies have assessed the effects on soil temperature regime of drainage [125], mulching [13,126-129], tillage [11,47,
69,130], alteration of surface characteristics [131], plant cover and
height [126,132], shading [133], and artificial heating [126]. The
mitigation of both ground and air frost is a prime example of manipulation [126,128]. However, the multiprocess nature of the soil-at-
mosphere system often leads to a lack of clear, unequivocal results.

Again, there is a distinct need for careful development of mechanistic models, reduced by appropriate simplifications, capable of resolving the effects of changes (often subtle) in surface characteristics, soil thermal properties, and the movement of water in the soil toward evaporation sites.

V. CONCLUSION

Practical interest in soil thermal regime focuses on two quantities: first, temperature, which controls T-sensitive processes within soil; second, soil heat flux, which controls energy exchanges within the soil and at its surface. The main utility of the latter is in improving estimates of soil surface evaporation E as a component of total evapotranspiration.

However, it is essential to realize that E does not originate purely at the surface. Subsurface evaporation, often dominant in drying soils, can, via its heat sink effect, profoundly influence heat flux and temperature profiles. Unfortunately, in many methods of measuring heat flux and thermal properties, and of analyzing soil temperature variations, this phase-change effect is overlooked. This chapter has attempted to clarify the concepts essential to proper design and interpretation of experiments, in particular the concept of total soil heat flux, and its role in correct interpretation of the surface energy balance equation. Potential sources of error in the main experimental methods have been indicated.

Soil temperature regime is sometimes misconceived as one of the "solved" disciplines of soil physics. In fact, it has many unsolved problems and deficiencies. Some of these are summarized as follows. The greatest problems are posed by soils undergoing aqueous phase change, requiring description of coupled heat and mass transfer processes. The mechanisms of phase change and transport in freezing soils [52] undoubtedly qualify as the outstanding "unknown." In unfrozen field soil, more detailed analysis and measurement of the extent of subsurface evaporation is clearly required [2].

The coupling of soil temperature models to the surface energy balance offers prospects of explaining hitherto empirical observations (e.g., of relations between soil and air temperatures). As an example, concurrent diurnal variations of temperature in soils under different surface treatments have been found to be closely similar when plotted as "normalized" curves (i.e., using the difference between daily maximum and minimum temperatures at the point of measurement as a normalizing factor) [11]. Also, normalized soil and air temperatures are correlated. Such observations suggest possible mechanistic explanation using models based on energy balance. Relationships between the harmonic composition of soil temperature and that of its

driving variables (solar radiation, air temperature [34]), discussed
briefly above (Section II.E), deserve further study in, for example,
climatological analysis of geographical and long-term variations of
the annual wave.

The effects of *time-varying* temperature on biological processes
are relatively poorly understood. Finally, industrial thermographic
imaging techniques offer methods to study variability in temperature,
hence physical properties, of soil.

ACKNOWLEDGMENTS

The author is grateful for earlier support from colleagues at the
Macaulay Institute for Soil Research, Aberdeen (now the Macaulay
Land Use Research Institute), and the Department of Applied
Physics, University of Strathclyde.

LIST OF SYMBOLS

A	amplitude (d, diurnal; a, annual), °C
C	volumetric heat capacity, $J\ m^{-3}\ K^{-1}$
D	damping depth (D_n for nth harmonic), m
D_v	apparent vapor diffusivity in soil air, $kg\ m^{-1}\ s^{-1}\ Pa^{-1}$
D_{va}	vapor diffusivity in bulk air, $kg\ m^{-1}\ s^{-1}\ Pa^{-1}$
E	generalized rate of evaporation to air, $kg\ m^{-2}\ s^{-1}$
E_{tot}	$= E_0 + E_{s0} =$ total soil evaporation to air, $kg\ m^{-2}\ s^{-1}$
E_0	component sourced at soil "surface," $kg\ m^{-2}\ s^{-1}$
E_{s0}	subsurface-sourced component, $kg\ m^{-2}\ s^{-1}$
$E_s(z)$	rate of net evaporation at depth z, driven by gradients in moisture content, $kg\ m^{-3}\ s^{-1}$
e	vapor pressure, Pa
e_s	saturation vapor pressure, Pa
G	generalized soil heat flux, $W\ m^{-2}$
G_0	surface value of G, $W\ m^{-2}$
G_{tot}	$= G_T + G_{vp} =$ total soil heat flux, $W\ m^{-2}$
G_T	$= -\lambda dT/dz =$ thermal soil heat flux, $W\ m^{-2}$
G_{vp}	$= L_v J_{vp} =$ isothermal latent heat flux induced by a moisture gradient, $W\ m^{-2}$
G_{vT}	$= L_v J_{vT} =$ thermal latent heat flux induced by a temperature gradient, $W\ m^{-2}$

G_c	"pure" conduction component of G, W m^{-2}
g	granule shape factor
H	sensible heat flux from soil surface to air, W m^{-2}
h	relative humidity
k	weighting factor for thermal conductivity
J_v	total vapor flux in soil air phase, kg m^{-2} s^{-1}
J_{vT}	thermal vapor flux, kg m^{-2} s^{-1}
J_{vp}	isothermal vapor flux, kg m^{-2} s^{-1}
L<f>	Laplace transform of function f
L_d	downward long-wave irradiance ("sky radiation"), W m^{-2}
L_u	upward long-wave irradiance, W m^{-2}
L_f	latent heat of fusion of ice, J kg^{-1}
L_v	latent heat of vaporization, J kg^{-1}
M_w	molecular weight of water, kg mol^{-1}
n	harmonic number
N	number of data points
N	number of parameters in Fourier series
p	total air pressure in soil, Pa
r	radius, m
R	gas constant, 8.314, J mol^{-1} K^{-1}
R_n	net radiation, W m^{-2}
R_s	solar radiation, W m^{-2}
s	= de_s/dT = slope of svp curve, Pa K^{-1}
s	parameter in Laplace transform
S	rate of heat absorption or production, W m^{-3}
t	time, s
T	temperature (subscripts: o, soil surface; a, air), °C or K
x	volume fraction of a soil component (subscripts: m, mineral solids; w, water; a, air; o, organic matter; i, ice; c, continuous phase)
z	soil depth, m
z_1	depth of lowest site of net evaporation, m
α	short-wave reflection coefficient
α	pore space tortuosity factor
ε	long-wave emissivity

ΔS	daily change in soil heat storage, $J \ m^{-2} \ d^{-1}$
η	weighting factor
θ	volumetric water content ($= x_w$)
θ_a	air-filled porosity
κ	thermal diffusivity, $m^2 \ s^{-1}$
λ	thermal conductivity, $W \ m^{-1} \ K^{-1}$
ν	mass flow factor
ρ	density (subscripts as for x), $kg \ m^{-3}$
σ	Stefan-Boltzmann constant, $W \ m^{-2} \ K^{-4}$
τ	period, time constant, transmissivity, s
ϕ	phase angle
ψ_m	matric potential, $J \ kg^{-1}$
π	osmotic potential
ω	angular frequency, s^{-1}

REFERENCES

1. Philip, J. R., and D. A. de Vries, Moisture movement in porous materials under temperature gradients, *Trans. Am. Geophys. Union*, 38: 222-232 (1957).
2. de Vries, D. A., and J. R. Philip, Soil heat flux, thermal conductivity, and the null-alignment method, *Soil Sci. Soc. Am. J.*, 50: 12-18 (1986).
3. Farouki, O. T., *Thermal Properties of Soils*, Trans Tech Publications, Clausthal-Zellerfeld, Germany, 1986.
4. Gilman, K., *Movement of Heat in Soils*, Report No. 44, Instit. Hydrol., Wallingford, Oxfordshire, U.K., 1977.
5. Chudnovskii, A. F., *Heat Transfer in the Soil*, Israel Program for Scientific Translations, Jerusalem, 1962.
6. Shul'gin, A. M., *The Temperature Regime of Soils*, Israel Program for Scientific Translations, Jerusalem, 1965.
7. Hillel, D., *Fundamentals of Soil Physics*, Academic Press, New York, 1980, pp. 287-317 (Chap. 12, Soil temperature and heat flow).
8. Ghildyal, B. P., and R. P. Tripathi, *Soil Physics*, Wiley Eastern Ltd., New Delhi, 1987.
9. Huband, N., An infra-red radiometer for measuring surface temperature in the field: I. Design and construction; II. Calibration and performance, *Agric. For. Meteorol.*, 34: 215-234 (1985).

10. Hasfurther, V. R., and R. D. Burman, Soil temperature modeling using air temperature as a driving mechanism, *Trans. ASAE, Struct. Environ. Div.*, 16: 78-81 (1973).

11. Gupta, S. C., W. E. Larson, and R. R. Allmaras, Predicting soil temperature and soil heat flux under different tillage-surface residue conditions, *Soil Sci. Soc. Am. J.*, 48: 223-232 (1984).

12. de Vries, D. A., Heat transfer in soils, in *Heat and Mass Transfer in the Biosphere* (D. A. de Vries and N. H. Afgan, Eds.), Scripta Book Co., New York, 1975, pp. 5-28.

13. Bristow, K. L., G. S. Campbell, R. I. Papendick, and L. F. Elliott, Simulation of heat and moisture transfer through a surface residue-soil system, *Agric. For. Meteorol.*, 36: 193-214 (1986).

14. Cass, A., G. S. Campbell, and T. L. Jones, Enhancement of thermal water vapor diffusion in soil, *Soil Sci. Soc. Am. J.*, 48: 25-31 (1984).

15. Kimball, B. A., R. D. Jackson, R. J. Reginato, F. S. Nakayama, and S. B. Idso, Comparison of field-measured and calculated soil heat fluxes, *Soil Sci. Soc. Am. J.*, 40: 18-25 (1976).

16. Milly, P. C., Moisture and heat transport in hysteretic, inhomogeneous porous media: A matric head-based formulation and a numerical model, *Water Resour. Res.*, 18: 489-498 (1982).

17. de Vries, D. A., Thermal properties of soils, in *Physics of Plant Environment* (W. R. van Wijk, Ed.), North Holland Publishing Co., Amsterdam, 1963, pp. 210-235.

18. Raats, P. A. C., Transformations of fluxes and forces describing the simultaneous transport of water and heat in unsaturated porous media, *Water Resour. Res.*, 11: 938-942 (1975).

19. Sidiropoulos, E., and C. Tzimopoulos, Sensitivity analysis of a coupled heat and mass transfer model in unsaturated porous media, *J. Hydrol.*, 64: 281-298 (1983).

20. Hopmans, J. W., and J. H. Dane, Thermal conductivity of two porous media as a function of water content, temperature and density, *Soil Sci.*, 142: 187-195 (1986).

21. Ross, P. J., and B. J. Bridge, Thermal properties of swelling clay soils, *Aust. J. Soil Res.*, 25: 29-41 (1987).

22. Campbell, G. S., *Soil Physics with BASIC*, Elsevier, Amsterdam, 1985.

23. McInnes, K. J., Thermal conductivities of soils from dryland wheat regions of Eastern Washington, M.S. thesis, Washington State University, Pullman, 1981.

24. Kersten, M. S., *Thermal Properties of Soils*, University of Minnesota, Eng. Exp. Stn. Bull. No. 28, 1949.

25. Al-Nakshabande, G., and H. Kohnke, Thermal conductivity
     and diffusivity of soils as related to moisture tension and other
     physical properties, *Agric. Meteorol.*, 2: 271-279 (1965).

26. Buchan, G. D., Predicting bare soil temperature: I. Theory
     and models for the multi-day mean diurnal variation, *J. Soil
     Sci.*, 33: 185-197 (1982).

27. Buchan, G. D., Predicting bare soil temperature: II. Exper-
     imental testing of multi-day models, *J. Soil Sci.*, 33: 199-209
     (1982).

28. Buchan, G. D., Predicting bare soil temperature: III. Exten-
     sion to single-day variation, *J. Soil Sci.*, 33: 365-373 (1982).

29. Bloomfield, P., *Fourier Analysis of Time Series: An Introduc-
     tion*, Wiley, New York, 1976.

30. Carslaw, H. S., and J. C. Jaeger, *Conduction of Heat in
     Solids*, 2nd ed., Clarendon Press, Oxford, 1967.

31. van Wijk, W. R., and D. A. de Vries, Periodic temperature
     variations in a homogeneous soil, in *Physics of Plant Environ-
     ment* (W. R. van Wijk, Ed.), North Holland Publishing Co.,
     Amsterdam, 1963, pp. 102-143.

32. Carson, J. E., Analysis of soil and air temperatures by Fourier
     techniques, *J. Geophys. Res.*, 68: 2217-2232 (1963).

33. Persaud, N., and A. C. Chang, Computing mean apparent soil
     thermal diffusivity from daily observations of soil temperature
     at two depths, *Soil Sci.*, 139: 297-304 (1985).

34. Tabony, R. C., Non-sinusoidal features of the seasonal varia-
     tion of temperature in mid-latitudes, *Meteorol. Mag.*, 113: 64-
     71 (1984).

35. Miller, D. H., *Energy at the Surface of the Earth*, Academic
     Press, New York, 1981.

36. Menenti, M., *Physical Aspects and Determination of Evaporation
     in Deserts Applying Remote Sensing Techniques*, Report No. 10,
     ICW, Wageningen, The Netherlands, 1984.

37. Novak, M. D., Theoretical values of daily atmospheric and soil
     thermal admittances, *Boundary Layer Meteorol.*, 34: 17-34 (1986).

38. van Wijk, W. R., General temperature variations in a homogene-
     ous soil, in *Physics of Plant Environment* (W. R. van Wijk, Ed.),
     North Holland Publishing Co., Amsterdam, 1963, pp. 144-170.

39. Moench, A. F., and D. D. Evans, Thermal conductivity and
     diffusivity of soil using a cylindrical heat source, *Soil Sci. Soc.
     Am. Proc.*, 34: 377-381 (1970).

40. Asrar, G., and E. T. Kanemasu, Estimating thermal diffusivity
     near the soil surface using Laplace transform: Uniform initial
     conditions, *Soil Sci. Soc. Am. J.*, 47: 397-401 (1983).

41. Sepaskah, A. R., and L. Boersma, Thermal conductivity of
     soils as a function of temperature and water content, *Soil Sci.
     Soc. Am. J.*, 43: 439-444 (1979).

42. van Wijk, W. R., and W. J. Derksen, Sinusoidal temperature variation in a layered soil, in *Physics of Plant Environment* (W. R. van Wijk, Ed.), North Holland Publishing Co., Amsterdam, 1963, pp. 171-209.

43. Lettau, H. H., A theoretical model of thermal diffusion in non-homogeneous conductors, *Gerlands Beitr. Geophys.*, 71: 257-271 (1962).

44. Riha, S. J., K. J. McInnes, S. W. Childs, and G. S. Campbell, A finite element calculation for determining thermal conductivity, *Soil Sci. Soc. Am. J.*, 44: 1323-1325 (1980).

45. de Wit, C. T., and H. van Keulen, *Simulation of Transport Processes in Soils*, Centre for Agricultural Publishing and Documentation, Wageningen, The Netherlands, 1972.

46. Gerald, C. F., and P. O. Wheatley, *Applied Numerical Analysis*, 3rd ed., Addison-Wesley, Reading, MA, 1985.

47. Mahrer, Y., A theoretical study of the effect of soil surface shape upon the soil temperature profile, *Soil Sci.*, 134: 381-387 (1982).

48. Lascano, R. J., and C. van Bavel, Experimental verification of a model to predict soil moisture and temperature profiles, *Soil Sci. Soc. Am. J.*, 47: 441-448 (1983).

49. Wierenga, P. J., and C. T. de Wit, Simulation of heat transfer in soils, *Soil Sci. Soc. Am. Proc.*, 34: 845-848 (1970).

50. Milly, P. C., A simulation analysis of thermal effects on evaporation from soil, *Water Resour. Res.*, 20: 1087-1098 (1984).

51. Miller, R. D., Freezing phenomena in soils, in *Applications of Soil Physics* (D. Hillel, Ed.), Academic Press, New York, 1980, pp. 254-299.

52. Fuchs, M., G. S. Campbell, and R. I. Papendick, An analysis of sensible and latent heat flow in a partially frozen unsaturated soil, *Soil Sci. Soc. Am. J.*, 42: 379-385 (1978).

53. Yong, R. N., and B. P. Warkentin, *Soil Properties and Behaviour*, Elsevier, Amsterdam, 1975.

54. Jumikis, A. R., *Thermal Geotechnics*, Rutgers University Press, New Brunswick, NJ, 1977.

55. Penner, E., Thermal conductivity of frozen soils, *Can. J. Earth Sci.*, 7: 982-987 (1970).

56. Bell, G. E., The prediction of frost penetration, *Inst. J. Numer. Anal. Methods Geomech.*, 8: 287-290 (1982).

57. Hayhoe, H. N., G. C. Topp, and S. N. Edey, Analysis of measurement and numerical schemes to estimate frost and thaw penetration of a soil, *Can. J. Soil Sci.*, 63: 67-77 (1983).

58. Kung, S. K., and T. S. Steenhuis, Heat and moisture transfer in a partly frozen nonheaving soil, *Soil Sci. Soc. Am. J.*, 50: 1114-1122 (1986).

59. Jame, Y. W., and D. Norman, Heat and mass transfer in a freezing, unsaturated brown medium, *Water Resour. Res.*, 16: 811-819 (1980).

60. Woodward, F. I., and J. E. Sheehy, *Principles and Measurements in Environmental Biology*, Butterworths, London, 1983.

61. Fritschen, L. J., and L. W. Gay, *Environmental Instrumentation*, Springer-Verlag, Berlin, 1979.

62. Meteorological Office, *Handbook of Meteorological Instruments*, Vol. 2, *Temperature*, 2nd ed., HMSO, London, 1981.

63. Bell, C. J., and D. A. Rose, The measurement of temperature, in *Instrumentation for Environmental Physiology* (B. Marshall and F. I. Woodward, Eds.), Cambridge University Press, Cambridge, 1985, pp. 79-99.

64. Benedict, R. P., *Fundamentals of Temperature, Pressure and Flow Measurements*, 2nd ed., Wiley, New York, 1984.

65. Sydenham, P. H., History and technique of temperature measurement, in *Transducers in Measurement and Control*, Adam Hilger Ltd., Bristol, U.K., 1980, pp. 36-44.

66. Horowitz, P., and W. Hill, *The Art of Electronics*, Cambridge University Press, Cambridge, 1980.

67. World Meteorological Organization, *Guide to Meteorological Instrument and Observing Practices*, WMO Publication No. 8.TP.3 (1971).

68. Schimmelpfennig, H., *Basic Thermocouple Thermometry*, Application Note No. W-15, Westcor, Inc., Logan, UT, 1976.

69. Radke, J. K., A. R. Dexter, and O. J. Devine, Tillage effects on soil temperature, soil water, and wheat growth in South Australia, *Soil Sci. Sci. Am. J.*, 49: 1542-1547 (1985).

70. Campbell Scientific Instruments, *Model 107 Temperature Probe*, Logan, UT, 1985.

71. R. S. Components, *Semiconductor Temperature Sensor*, Data Sheet No. 3992, R. S. Components, London, 1983.

72. Brooks, F. A., *An Introduction to Physical Microclimatology*, Syllabus No. 397, University of California, Davis, 1960.

73. Becker, F., Thermal infra-red remote sensing principles and applications, in *Remote Sensing Application in Agriculture and Hydrology* (G. Fraysse, Ed.), A. A. Balkema, Rotterdam, 1980, pp. 153-213.

74. Curran, P. J., *Principles of Remote Sensing*, Longman, London, 1985.

75. Oliver, S. A., H. R. Oliver, J. S. Wallace, and A. M. Roberts, Soil heat flux and temperature variation with vegetation, soil type and climate, *Agric. For. Meteorol.*, 39: 257-269 (1987).

76. Johnson, I. R., and J. Thornley, Temperature dependence of plant and crop processes, *Ann. Bot.*, 55: 1-24 (1985).

77. Jones, R., The measurement of mean temperatures by the sucrose inversion method: A review, *Soils Fert.*, 35: 615-619 (1972).

78. Hartley, G. S., and J. MacLauchlan, A simple integrating thermometer for field use, *J. Ecol.*, 57: 151-154 (1969).

79. Green, C. F., P. N. Schaare, and C. N. Bates, A temperature sensor for temperature integration in the field, *J. Exp. Bot.*, 34: 226-229 (1983).

80. Taylor, S. A., and R. D. Jackson, Temperature, in *Methods of Soil Analysis*, Part 1, 2nd ed. (A. Klute, Ed.), Am. Soc. Agron, Madison, WI, 1986, pp. 927-940.

81. Weichert, L., The avoidance of electrical interference in instruments, *J. Phys. E: Sci. Instrum.*, 16: 1003-1012 (1983).

82. Kimball, B. A., Smoothing data with cubic splines, *Agron. J.*, 68: 126-129 (1976).

83. Philip, J. R., Field heterogeneity: Some basic issues, *Water Resour. Res.*, 16: 443-448 (1980).

84. Horton, R., P. J. Wierenga, and D. R. Nielsen, Evaluation of methods for determining the apparent thermal diffusivity of soil near the surface, *Soil Sci. Soc. Am. J.*, 47: 25-32 (1983).

85. Kimball, B. A., Smoothing data with Fourier transformations, *Agron. J.*, 66: 259-262 (1974).

86. Horton, R., and P. J. Wierenga, Estimating the soil heat flux from observations of soil temperature near the surface, *Soil Sci. Soc. Am. J.*, 47: 14-20 (1983).

87. Parton, W. J., Predicting soil temperatures in a shortgrass steppe, *Soil Sci.*, 138: 93-101 (1984).

88. Tanner, C. B., *Basic Instrumentation and Measurements for Plant Environment and Micrometeorology*, Soils Bull. No. 6, University of Wisconsin, Madison, 1963.

89. Idso, S. B., J. K. Aase, and R. D. Jackson, Net radiation-soil heat flux relations as influenced by soil water content variations, *Boundary Layer Meteorol.*, 9: 113-122 (1975).

90. Kimball, B. A., and R. D. Jackson, Soil heat flux determination: A null-alignment method, *Agric. Meteorol.*, 15: 1-9 (1975).

91. Philip, J. R., The theory of heat flux meters, *J. Geophys. Res.*, 66: 571-579 (1961).

92. Weaver, H. L., and G. S. Campbell, Use of Peltier coolers as soil heat flux transducers, *Soil Sci. Soc. Am. J.*, 49: 1065-1067 (1985).

93. Fuchs, M., and A. Hadas, The heat flux density in a nonhomogeneous bare loessial soil, *Boundary Layer Meteorol.*, 3: 191-200 (1972).

94. Biscoe, P. V., R. A. Safell, and P. D. Smith, An apparatus for calibrating soil heat flux plates, *Agric. Meteorol.*, 18: 49-54 (1977).

95.   Fuchs, M., and C. B. Tanner, Calibration and field test of soil heat flux plates, *Soil Sci. Soc. Am. Proc.*, 32: 326-328 (1968).

96.   de Bruin, H. A. R., and A. Holtslag, A simple parameterization of the surface fluxes of sensible and latent heat during daytime compared with the Penman-Monteith concept, *J. Appl. Meteorol.*, 21: 1610-1621 (1982).

97.   Clothier, B. E., K. L. Clawson, P. J. Pinter, M. S. Moran, R. J. Reginato, and R. D. Jackson, Estimation of soil heat flux from net radiation during the growth of alfalfa, *Agric. For. Meteorol.*, 37: 319-329 (1986).

98.   Taylor, S. A., and R. D. Jackson, Heat capacity and specific heat, in *Methods of Soil Analysis*, Part 1, 2nd ed. (A. Klute, Ed.), Am. Soc. Agron., Madison, WI, 1986, pp. 941-944.

99.   Hadas, A., Problems involved in measuring the soil thermal conductivity and diffusivity in a moist soil, *Agric. Meteorol.*, 13: 105-113 (1974).

100. Williams, P. J., *The Surface of the Earth: An Introduction to Geotechnical Science*, Longman, London, 1982.

101. de Vries, D. A., and A. J. Peck, On the cylindrical probe method of measuring thermal conductivity with special reference to soil: I. Extension of theory and discussion of probe characteristics; II. Analysis of moisture effects, *Aust. J. Phys.*, 11: 255-271, 409-423 (1958).

102. Fritton, D. D., W. J. Busscher, and J. E. Alpert, An inexpensive but durable thermal conductivity probe for field use, *Soil Sci. Soc. Am. Proc.*, 38: 854-855 (1974).

103. Jackson, R. D., and S. A. Taylor, Thermal conductivity and diffusivity, in *Methods of Soil Analysis*, Part 1, 2nd ed. (A. Klute, Ed.), Am. Soc. Agron., Madison, WI, 1986, pp. 945-956.

104. Nagpal, N. K., and L. Boersma, Air entrapment as a possible source of error in the use of a cylindrical heat probe, *Soil Sci. Soc. Am. Proc.*, 37: 828-832 (1973).

105. Kasabuchi, T., Heat conduction model of saturated soil and estimation of thermal conductivity of soil solid phase, *Soil Sci.*, 138: 240-247 (1984).

106. van Wijk, W. R., and W. J. Derksen, Thermal properties of a soil near the surface, *Agric. Meteorol.*, 3: 333-342 (1966).

107. Singh, S. R., and B. K. Sinha, Soil thermal diffusivity determination from overspecification of boundary data, *Soil Sci. Soc. Am. J.*, 41: 831-834 (1977).

108. Hopmans, J. W., and J. H. Dane, Thermal conductivity of two porous media as a function of water content, temperature and density, *Soil Sci.*, 142: 187-195 (1986).

109. Parikh, R. J., J. A. Havens, and H. D. Scott, Thermal diffusivity and conductivity of moist porous media, *Soil Sci. Soc. Am. J.*, 43: 1050-1052 (1979).

110. Hayhoe, H. N., G. C. Topp, and W. G. Bailey, Measurement of soil water contents and frozen soil depth during a thaw using time-domain reflectrometry, *Atmosph.-Ocean.*, 21: 299-311 (1983).

111. Brach, E. J., A. R. Mack, H. Hayhoe, and B. Scobie, Electrical determination for frost depth in soil, *Agric. For. Meteorol.*, 34: 173-181 (1985).

112. Caprio, J., *Soil Freeze Tube Depths*, Abstracts, Natl. Conf. Agric. For. Meteorol., Am. Meteorol. Soc., Tucson, AZ, October 1977.

113. Rickard, W., and J. Brown, The performance of a frost tube for the determination of soil freezing and thawing depth, *Soil Sci.*, 113: 149-154 (1972).

114. Burgess, M. D., and C. L. Hanson, Automatic soil-frost measuring systems, *Agric. Meteorol.*, 20: 313-318 (1979).

115. Bocock, K. L., J. Jeffers, D. K. Lindley, J. Adamson, and C. Gill, Estimating woodland soil temperature from air temperature and other climatic variables, *Agric. Meteorol.*, 18: 351-372 (1977).

116. Halldin, S. (Ed.), *Comparison of Forest Water and Energy Exchange Models*, Elsevier, Amsterdam, 1979.

117. Deardorff, J. W., Efficient prediction of ground surface temperature and moisture with inclusion of a layer of vegetation, *J. Geophys. Res.*, 83: 1889-1903 (1978).

118. Roodenburg, J., Estimating 10-cm soil temperatures under grass, *Agric. For. Meteorol.*, 34: 41-52 (1985).

119. Horton, R., O. Aguirre-Luna, and P. J. Wierenga, Observed and predicted two-dimensional soil temperature distributions under a row crop, *Soil Sci. Soc. Am. J.*, 48: 1147-1151 (1984).

120. Schieldge, J. P., A. B. Kahle, and R. E. Alley, A numerical simulation of soil temperature and moisture variations for a bare field, *Soil Sci.*, 133: 197-207 (1982).

121. Green, F. H. W., and R. J. Harding, The effects of altitude on soil temperature, *Meteorol. Mag.*, 108: 81-91 (1979).

122. Toy, T. J., A. J. Kuhaida, and B. E. Munson, The prediction of mean monthly soil temperature from mean monthly air temperature, *Soil Sci.*, 126: 181-189 (1978).

123. Meikle, R. W., and A. J. Gilchrist, A mathematical method for estimation of soil temperatures in England and Scotland, *Agric. Meteorol.*, 30: 221-226 (1983).

124. Reiger, S., *The Genesis and Classification of Cold Soils*, Academic Press, New York, 1983.

125. Scotter, D. R., and D. J. Horne, The effect of mole drainage on soil temperatures under pasture, *J. Soil Sci.*, 36: 319-327 (1985).

126. Rosenberg, N. J., B. L. Blad, and S. B. Verma, *Microclimate: The Biological Environment*, 2nd ed., Wiley, New York, 1983.

127. Davies, J. W., *Mulching Effects on Plant Climate and Yield*, Tech. Note No. 136, WMO, Geneva, 1975.

128. Tanner, C. B., Microclimatic modification: Basic concepts, *Hortic. Sci.*, 9: 555-560 (1974).

129. Liakatis, A., J. A. Clark, and J. L. Monteith, Measurements of the heat balance under plastic mulches: I. Radiation balance and soil heat flux, *Agric. For. Meteorol.*, 36: 227-239 (1986).

130. Allmaras, R. R., E. A. Hallauer, W. W. Nelson, and S. D. Evans, *Surface Energy Balance and Soil Thermal Property Modifications by Tillage-Induced Soil Structure*, University of Minnesota, Agric. Exp. Stn., Tech. Bull. No. 306, 1977.

131. Potter, K. N., R. Horton, and R. M. Cruse, Soil surface roughness effects on radiation reflectance and soil heat flux, *Soil Sci. Soc. Am. J.*, 51: 855-860 (1987).

132. Green, F., R. J. Harding, and H. R. Oliver, The relationship of soil temperature to vegetation height, *J. Climatol.*, 4: 229-240 (1984).

133. Stigter, C. J., Shading: A traditional method of microclimate manipulation, *Neth. J. Agric. Sci.*, 32: 81-86 (1984).

Index

Air capacity, 150
Air-filled porosity, 511-512,
    528-530 (*see also* Gas
    movement, by mass flow)
Air permeability, comparison
    with porosity and diffusivity
    measurements, 541-542 (*see
    also* Gas movement, by mass
    flow)
Available water capacity,
    definition, 148

Bulk density, 329-366, 449-450
  applications, 354-361
    compaction of wheels, 355-
      359
    effects of tillage, 359-361
  direct methods of measurement,
    330, 341-346
    balloon method, 342
    clod method, 346
    core sampling, 341-342
    sand replacement, 342-346
    theory, 330
  dry, definition, 329
  effect on root system develop-
    ment, 449-450
  gamma-ray methods, 330-341
    access holes for, 347
    backscatter gauges, 331-332
    calibration, 336-337, 347-350

[Bulk density]
    commercial equipment, 344-
      345
    comparison of methods, 352-
      354
    effect of soil composition,
      334-335
    effect of stones, 351
    energy discriminating vs GM
      detectors, 334
    gauge design, 337-340
    precision of, 350-351
    radioactive sources, 337-338
    reference measurement, 335-
      336
    safety, 351
    sample preparation, 347
    water content determination
      for, 340-341
  wet, definition, 329

California bearing ratio (CBR),
    405, 407, 422
Capacitance method for water
    content, 39-45
  access tubes, installation, 41
  calibration, 41-43
  instrument design, 39-41
  principles, 37-39
  resolution, 43-44
  speed, 44

613